The Analysis of Algorithms

The Analysis of Algorithms

Paul Walton Purdom, Jr.
Indiana University

Cynthia A. Brown
GTE Laboratories, Inc.
and Northeastern University

HOLT, RINEHART AND WINSTON
New York Chicago San Francisco Philadelphia
Montreal Toronto London Sydney Tokyo
Mexico City Rio de Janeiro Madrid

Address correspondence to:
383 Madison Avenue, New York, NY 10017

Library of Congress Cataloging in Publication Data

Purdom, Paul Walton, Jr.
The analysis of algorithms.

Bibliograph: p.
Includes indexes.
1. Electronic digital computers—Programming.
2. Algorithms. I. Brown, Cynthia A. II. Title.
QA76.6.P86 1984 519.4 84-19153

ISBN 0-03-072044-3

Printed in the United States of America
Published simultaneously in Canada

5 6 7 8 039 9 8 7 6 5 4 3 2 1

CBS COLLEGE PUBLISHING
Holt, Rinehart and Winston
The Dryden Press
Saunders College Publishing

To Don Knuth:
teacher, algorithm analyzer, and inventor of TEX.

Contents

Preface

Computers have undergone phenomenal improvement in speed, availability, and price. This has resulted in a great deal of emphasis being placed on rapid development of correct programs, with performance playing a secondary role. Nevertheless, there are many problems for which a naïve approach is so inefficient that it is unsuitable for even the fastest and cheapest computers. Moreover, there are some problems so common that even a modest improvement in the algorithms used to solve them can lead to important savings. An expert programmer needs to be able to estimate the resources a program will consume. As the field of computer science matures, such considerations will assume increasing importance.

Analysis of algorithms brings mathematical techniques to bear on the problem of deciding how much time and space an algorithm requires. It addresses such questions as the maximum time and space needed by an algorithm to solve a problem of a given size, the average amount of resources consumed by an algorithm, and the amount of resources needed by the best possible algorithm to solve a given problem.

The purpose of this book is to teach the techniques needed to analyze algorithms. Students are expected to have had courses in computer science up through data structures and in mathematics through calculus. (Students who are very well prepared mathematically can handle the material without as much programming experience.) The book is organized by analysis techniques and includes a systematic and largely self-contained treatment of the mathematics needed for elementary and intermediate analyses, as well as brief guides to the sources for more advanced techniques.

The focus of our presentation is on practical application of the techniques. Each technique, as far as possible, is illustrated by being applied to the analysis of a realistic algorithm. We also provide explicit guidance on the use of various methods. For example, the discussion of mathematical induction emphasizes the process of finding the induction hypothesis; the chapter on formulas for simplifying summations includes a section on deciding which formulas to apply. Many of the exercises give the student an opportunity to apply the techniques in developing original algorithm analyses.

Although the organization of the material emphasizes analysis techniques, most of the algorithms covered in a standard algorithms course appear in the book. (The major exception is graph algorithms.) Principles of algorithm design are brought out at points where the analyses support the ideas. Sections such as 3.1.2 and 7.4.3 give additional emphasis to good design and its relation to analysis.

The style of presentation of algorithms is a straightforward step format reminiscent of that used in Knuth's series of books, but with the addition of modern control structures. The style was chosen as the best compromise between clarity of presentation and ease of analysis. The student should be able to implement the algorithms in any modern language without difficulty.

The book is organized as follows. Chapter 1 introduces the basic concepts of algorithm analysis and also provides a review of fundamental mathematical material. Chapter 2 covers basic techniques for doing finite summations. Chapter 3 extends the techniques of Chapter 2 and also covers elementary combinatorics. The material in Chapter 3 encourages the student to develop skill in manipulating algebraic expressions and in discovering applications for formulas. The student who masters the first three chapters will be expert at analyzing simple algorithms.

Chapter 4 discusses the process of finding bounds and asymptotic approximations, both for closed expressions and for sums. The material in this chapter is the most mathematically sophisticated in the book; many instructors will wish to concentrate on the easier parts of the chapter.

Chapters 5 through 8 provide a thorough coverage of recurrence equations (also called difference equations). In some ways, these chapters are the heart of the book. We have assembled a catalog of methods for solving these problems and presented them in a unified treatment. The material ranges from elementary to difficult.

Chapter 9 covers a number of cases where the overall structure of the problem is important. It contains material on amortized cost analysis and the Fast Fourier Transform, among other topics. Chapter 10 is on lower bounds, NP-completeness, and undecidability. The emphasis is on techniques needed to obtain NP-completeness proofs. Chapter 11 has a brief introduction to statistics and includes a case study of the statistical analysis of an algorithm.

The symbols in the table of contents divide the sections into three types. Unmarked sections are fundamental (or are examples of fundamental techniques) and use relatively easy math. The sections marked with an asterisk are less fundamental or more difficult (usually both). Those marked with a dagger are fairly difficult but also fundamental.

Important equations have boldface numbers; most of them are collected in the Math Index, along with definitions and conventions. Sections end with a reference to sources that contain further reading or another presentation of the material covered in the section, including the sources we drew upon.

A one-semester course cannot cover all the material in the book. Starting at the beginning and proceeding straight through, it is possible to cover almost all of the first five chapters, including the starred sections. (The remainder can then be used for a second semester course.) An alternative is to cover the unstarred parts of the first four chapters, Chapter 5 through Section 5.3.3, Section 7.4.3, and one of Chapters 9 to 11. This more closely approximates the material covered in the typical algorithms course.

The exercises are an integral part of the book. The student should read all of them, since many of them contain techniques or results that will be helpful in later chapters. Thinking briefly about how to solve a problem can be quite helpful, even if one does not work out a complete solution. The student should also work some of the exercises. We usually assign about one exercise per section. Often the purpose of an exercise is to provide the student an opportunity to develop approaches to problem solving independently. When appropriate we provide rather broad hints to ensure that the student will succeed in this effort.

This book has benefited from the direct and indirect help provided by many people. Many of the researchers who have helped create the field of analysis of algorithms are mentioned in the references; many others are not. We owe particular thanks to Don Knuth, Al Aho, John Hopcroft, Jeff Ullman, and Sara Baase; their earlier efforts to organize the field contributed much to the structure of our book. In particular, the first volume of Knuth's series of books was the initial impetus for this one.

The writing of our book was made at least six times easier than it would have been in 1970 by the development of computer text preparation tools. The book has been typeset using the TeX system developed by Don Knuth. The drafts were prepared with the EMACS editor on a VAX 11/780 running Berkeley Unix. The initial drafts were typeset on an Imagen laser printer. The equipment was purchased with funds provided by the National Science Foundation and by Indiana University.

Our book would have many more errors than it does if it were not for the careful reading of parts of the manuscript by many people, including the 1983–1984 classes in Analysis of Algorithms at Indiana University, the 1984 class at Brandeis University, Jim Burns, Khaled Bugrara, Anne Cross, George Epstein, Moon Sung Han, Rao Kosaraju, Ian Munro, John Rief, Robert Tarjan, and Chun-Hung Tzeng.

The authors were at Indiana University (PWP) and GTE Laboratories (CAB) during most of the writing. We received partial support from the National Science Foundation. We wish to thank these organizations for helping make this book possible. Finally, we wish to thank our spouses, Donna Purdom and Arlin Brown, and also Swami Chetanananda for much-needed moral support.

The Analysis of Algorithms

CHAPTER 1

Introduction

The purpose of analyzing algorithms is to predict how much time or space is required by a computer to run a program or to solve a problem. On small problems, like the ones usually encountered in introductory computer science courses, this is ordinarily not too important. If a student is asked to write a program that sorts a list, for example, then any correct program will probably do. In principle, most programs that will sort a list of ten elements correctly will sort correctly a list of any length that will fit inside the computer.

While the previous sentence is true in theory, it is misleading in practice. The running time of an algorithm is usually a function of the length of the input. If the length of the input is n, a number of good sorting algorithms run in time proportional to $n \lg n$ (we use *lg* for the logarithm with base 2). Some sorting programs run in time proportional to n^2, and students occasionally write algorithms that run in time proportional to n^3. If $n = 10$, then $n \lg n \approx 33.2$, $n^2 = 100$, and $n^3 = 1000$. If $n = 100$, then $n \lg n \approx 664$, $n^2 = 10,000$, and $n^3 = 1,000,000$. If $n = 1000$, then $n \lg n \approx 9966$, $n^2 = 1,000,000$, and $n^3 = 1,000,000,000$. For the small problem with input length 10, the difference in running time will hardly be noticeable on a fast computer. For the input of length 1000 the fast algorithm still runs in a very reasonable time, while the n^3 algorithm is at best eating up about 1000 CPU seconds (assuming 1 microsecond per step). One may want to sort 100,000 items. In that case, $n \lg n \approx 1,660,964$, $n^2 = 10,000,000,000$, and $n^3 = 1,000,000,000,000,000$. The first algorithm still is reasonably fast (under 1.7 CPU seconds with the above assumptions). The second algorithm takes about 2.8 CPU hours, which is much too long. The third algorithm takes more than 31.7 CPU *years*, which is ridiculous. There is no way the n^3 algorithm will be used when you need to sort 100,000 items.

The running times of most algorithms discussed in this book will be given in terms of a few common functions: polynomials, logarithms, exponentials, and products of these functions. Table 1.1 illustrates the relative growth rates of functions of each type over a range of values. A feeling for these growth rates

n	1	10	100	1000	10,000
1	1	1	1	1	1
$\lg n$	0	3.32	6.64	9.97	13.3
$n^{0.1}$	1	1.26	1.58	2.00	2.51
n	1	10	100	1.00×10^3	1.00×10^4
$n \lg n$	0	33.2	664.	9.97×10^3	1.33×10^5
n^2	1	100	1.00×10^4	1.00×10^6	1.00×10^8
n^3	1	1.00×10^3	1.00×10^6	1.00×10^9	1.00×10^{12}
n^{10}	1	1.00×10^{10}	1.00×10^{20}	1.00×10^{30}	1.00×10^{40}
$e^{0.01n}$	1.01	1.11	2.72	2.20×10^4	2.69×10^{43}
$e^{0.1n}$	1.11	2.72	2.20×10^4	2.69×10^{43}	1.97×10^{434}
2^n	2	1.02×10^3	1.27×10^{30}	1.07×10^{301}	1.99×10^{3010}
e^n	2.72	2.20×10^4	2.69×10^{43}	1.97×10^{434}	8.81×10^{4342}

TABLE 1.1. Functions and values. The relative growth rates of some representative functions are shown.

is important in understanding the significance of the mathematical results. The last column in Table 1.1 gives a good indication of the relative rates of growth for large n in all but one case. (For $n > 5 \times 10^{17}$, $\lg n < n^{0.1}$.)

We are mainly interested in how fast algorithms run for moderately large problems. Most small problems can be solved so rapidly by any reasonable method that it is not worth worrying about which method is best. Extremely large problems take so long to solve with even the best methods that no one can solve them. As Table 1.1 suggests, the time taken on extremely large problems is often a good guide to how fast an algorithm will be on moderately large problems.

An algorithm that runs in *constant* time is very fast. Some hash table algorithms can look up one item from a table of n items in an average time which is constant (independent of the table size). A single processor computer cannot produce in constant time any results that depend on n inputs because constant time is not adequate to even read the input. There are, however, interesting problems that can be solved in constant time on a parallel computer (assuming that enough processors are available).

Logarithmic time is also quite fast. Logarithmic time is typical of many algorithms that use binary trees. For example, any single item in an n-node balanced binary tree can be found in time proportional to $\log n$.

Linear time is typical of fast algorithms on single-processor computers. Any algorithm which processes n items on a single-processor computer will use an amount of time that increases at least as fast as linearly with n, assuming that it is necessary to read all the input to do the calculation. Modern computers can do several million operations per second, so they can solve quite large problems when linear time algorithms are available.

Time $n \log n$ is typical of the best sorting algorithms. Since the logarithm base 2 of one million is less than 20, computers can also solve large problems when $n \log n$ algorithms are available.

An algorithm takes *polynomial* time when a problem of size n can be solved in time n^k for some k and all large positive n. When the exponent is small ($0 \le k \le 3$), then moderate size problems can be solved. When the exponent is large ($k = 10$, for example), only small problems can be solved.

Exponential time algorithms (such as those that use time 2^n or e^n) are suitable only for small problems, even when the fastest computers are used. Unfortunately, the best algorithms that are known for many problems use exponential time. Since there is more incentive to develop better algorithms for problems that presently require a large amount of time to solve, much work goes into developing better algorithms for the problems that require exponential time with present algorithms. There is a large variation in the size of various exponential functions (compare $e^{0.01n}$ with e^n), but in all cases exponential functions become huge when their parameter is large enough.

Knowing (even roughly) how long a program is likely to run is useful in several ways. First, it can help in deciding which of several correct algorithms to use in solving a problem. As the example above shows, the amount of time required by different algorithms can vary tremendously. The most obvious approach may work perfectly well on small problems but require years or even centuries on large ones. Analysis can determine whether an algorithm is practical for the size problem it will be run on. If it is not practical, one can avoid the expense and frustration of starting a computer run that can never finish and can concentrate instead on looking for a better algorithm.

Sometimes the best possible algorithm for a problem is slow. Discovering a meaningful lower bound for the running time of every possible algorithm for a given problem is usually harder than analyzing an individual algorithm, but if such results are available, they can save much fruitless searching for a better algorithm. If the best possible algorithm is too slow, one can sometimes reformulate the original problem to use the solution of a slightly different, easier problem. This is especially likely to be possible if the complex problem arose as a subpart of a method for solving another problem.

In this book we will concentrate mostly on analyzing the running time of algorithms, although occasionally we will also discuss the amount of memory space an algorithm uses. For many problems it is possible to find an incredibly fast algorithm that uses an enormous amount of space (by looking up the answer in a table, for example) and also an algorithm that uses comparatively little space but requires a huge amount of time. The best algorithms make good use of both time and space.

The basic approach to analyzing the running time of an algorithm is to figure out how long each step of the algorithm takes and how often each step is performed. The total time then is given by multiplying the time for each step by the number of times the step is done and summing the products. Thus

$$T = n_1 t_1 + n_2 t_2 + \cdots + n_k t_k = \sum_{1 \le j \le k} n_j t_j, \qquad \textbf{(1)}$$

where T is the total time used by the algorithm, n_j is the number of times the

j^{th} step of the algorithm is performed, t_j is the time required to perform the jth step one time, and k is the number of steps in the algorithm.

Usually it is straightforward to determine the t_j. For example, if the individual steps of the algorithm correspond to simple machine language instructions for some computer, it is usually possible to look up the time required for each instruction in a manual for the computer. The techniques for determining how often each step is done are sometimes quite simple and sometimes complex. Determining the n_j and simplifying eq. (1) are the main subjects of this book.

Sources for this section include Aho et al. [1, Section 1.1] and Sedgewick [62, Chapter 1]. Also see Horowitz and Sahni [7, Section 1.4].

EXERCISE

1. When you have a choice, should you use a polynomial time algorithm or an exponential time algorithm to solve a large problem?

1.1 STRAIGHT-LINE PROGRAMS

The simplest programs to analyze are straight-line programs—those where each step is done once. Since the n_j for eq. (1) are 1 in this case, only the t_j have to be computed.

Consider the problem of generating random numbers on a computer. Most computers are not suited to generating truly random numbers. They can, however, generate pseudo-random number sequences—sequences of numbers that are calculated by some carefully chosen rule, so that there does not *appear* to be any relation between successive numbers in the sequence.

One common method of generating pseudo-random sequences is to select some starting number x_0 in whatever way is convenient (for example, the time of day might be a good choice) and then generate successive numbers in the sequence using the rule

$$x_{i+1} = ax_i \bmod w \tag{2}$$

where x_i is the ith pseudo-random number, w is a large number selected to make the method easy to program, and a is carefully selected to make the method generate numbers that appear to be random. One usually chooses $w = 2^n$, where n is near a multiple of the word size of the computer and is fairly large (say 45 or larger). In this case $ax_i \bmod w$ is the last n bits of ax_i.

Suppose for example you want to use the method of eq. (2) on an old CDC 3600 computer, which has a 48-bit word size. The algorithm for $a = 273673163155$ and $w = 2^{47}$ can be written in assembly language as follows:

Label	Op Code	Address	Time	Comments
	LDA	X	2.00	
	MUI	=273673163155	6.40	ax
	SCL	=03777777777777777	2.13	$ax \bmod 2^{47}$ (set sign bit)
	STA	X	1.88	$x \leftarrow ax \bmod 2^{47}$

The time (in microseconds) required for each instruction is obtained from the CDC 3600 reference manual. From this code it is clear that the algorithm requires 12.41 microseconds (provided the instruction timings from the manual are correct).

Whenever you need to know exactly how long a computer program will take, you can apply a similar method. First obtain an assembly language listing of the program (easy to do if you wrote the program yourself in assembly language, harder in most other cases), look up the time required for each instruction (usually not too hard, but we still cannot find a book that gives the instruction timings for the VAX computer that we are using), figure out how often each instruction is executed (which is why you have this book) and plug the numbers into eq. (1). This is a good approach when it is very important to you to predict exactly how much time your program takes. Often, however, there are practical difficulties in applying the method.

Let us now briefly consider some of the difficulties and what to do about them. First a minor difficulty: your computer manual may give formulas rather than numbers for some of the instruction timings. Evaluating these formulas requires a simple application of the techniques explained in this book. Determining instruction timings is more complex for pipelined computers, because for such computers the time required for an instruction depends on which instructions immediately precede it. Even with pipelined computers, however, there is in principle no reason why you cannot compute timings from the assembly code and a suitable manual.

A more serious problem arises when you cannot get the assembly code or you do not have a manual that gives correct instruction timings. It is usually still possible to determine how much time a program requires by using a combination of analysis and measurement. The algorithm for generating pseudo-random numbers, for example, requires the same amount of time (or close to the same amount of time) each time it is run. Thus one can run the algorithm, time how long it takes, and use the measured time to predict how long it will take any time it is run again. For more complex programs it is necessary to analyze the measurements in a more complex way, but the basic idea is the same.

There are several difficulties with using measurement that should be mentioned. First, the clock used for timing the program may not tick very often. For example, the clock on the CDC 3600 ticks once every 1000 microseconds. This can make it necessary to run short programs several thousand times to obtain any useful information about their running time. Second, the timing method itself may use up some time. To compensate for this, you can compare the time required by the program being measured with the time required by a program that does not do anything except measure its own time. This is particularly important when timing short programs. Third, many computers run several jobs at once, while the clock measures the elapsed time. Thus your measured time will depend on both the program being measured and on the other programs being run. In such cases it is best, when possible, to make timing runs when no other jobs are on the system. A knowledge of statistics is very helpful in dealing with these practical problems.

Often you may be satisfied with less precise information than the exact running time for a program. For example, it is obvious that method (2) runs in (close to) constant time on any computer and that this time is not very large. This result is much less precise than the statement that it requires 12.41 microseconds on a CDC 3600, but it is much more general. Often this is the type of answer that people are interested in.

From now on we will be mainly concerned with obtaining the more general but less precise answers, i.e., computing the n_j for eq. (1). When necessary the more general answers can be converted into precise answers by using the appropriate instruction timings in eq. (1).

It is convenient to explain algorithms using a higher-level language. It is best if the language is such that the elementary steps in the language take constant time (or close to constant time). This is desirable so that the algorithm analyzer can concentrate on determining how often the various steps of the algorithm are done rather than on how much time each step requires. The high-level language will permit us to get on with the task of analyzing algorithms without being too concerned with the details of machine language coding.

The random number generator used in this section was designed using techniques discussed by Knuth [10, Sections 3.2.1.1 and 3.3.4].

EXERCISES

1. Take a simple straight-line program of your choice [no more complex than the method of eq. (2)] and calculate how long it takes to run on some computer that you already know how to program. Use the instruction timings from a manual.
2. Take a simple straight-line program and measure how long it takes. Explain what difficulties you had to overcome to perform your measurements.
3. Do both of the previous problems using the same computer and the same program. Do your answers from the two methods agree? If not, why not? Which method was easiest to use? Which method gave the most reliable answer?

1.2 SIMPLE LOOPS

If all algorithms consisted just of straight-line code, there would be little need for people to analyze algorithms. Just about any modern computer can execute in under one minute any straight-line program that fits in its memory. Most programs have loops. As soon as a program has loops, its analysis becomes somewhat more complex, because some statements are executed more than once.

1.2.1 Horner's Method

Consider the following algorithm, which evaluates a polynomial:

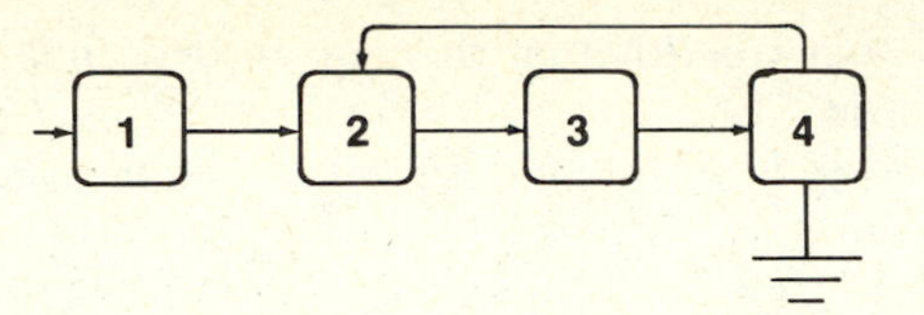

FIGURE 1.1 The flow graph for Algorithm 1.1.

Algorithm 1.1 Horner's Method: Input: The value of a variable x and a polynomial

$$p(x) = a_n x^n + a_{n-1} x^{n-1} + \cdots + a_1 x + a_0 = \sum_{0 \le i \le n} a_i x^i \qquad \textbf{(3)}$$

given by its degree $n \geq 0$ and its coefficients $a_0, \ldots, a_n$. Output: The value, v, of the polynomial evaluated at x.

Step **1.** Set $i \leftarrow n$ and $v \leftarrow 0$.

Step **2.** Set $v \leftarrow vx + a_i$ and $i \leftarrow i - 1$.

Step **3.** If $i \geq 0$, then go to Step 2.

Step **4.** Output v.

Figure 1.1 shows a flow graph for Algorithm 1.1. Each box (node) represents a step in the algorithm, and each arc shows a way that the flow of control can proceed. The incoming arrow on node one shows that execution starts there. The ground sign on node four shows that execution can terminate there. The flow graph suppresses many details about the program, but it makes clear which ways the flow of control can proceed.

Now consider how often each step of the algorithm is done. Step 1 is executed at the beginning, and once we have left it and gone to Step 2 there is no way to return. This shows that Step 1 is executed exactly once. Step 4 can only be entered from Step 3; once it is finished, the program terminates. Thus it is executed at most once. We enter Step 4 when the test in Step 3 shows that $i < 0$. Since i is decremented each time through the loop and is never incremented, it must eventually become negative. When that happens, Step 4 must be executed. So Step 4 is executed exactly once.

Steps 2 and 3 make up the loop. Variable i is set to n before the loop is entered and is decreased by one each time Step 2 is executed. Its value is not changed elsewhere in the loop. Step 2 must be executed $n + 1$ times before i is -1 on exit from Step 2. (The program assumes that $n \geq 0$ and so does this analysis.) Step 3 is done the same number of times as Step 2.

When analyzing the number of times a program executes a loop, it is useful to determine an invariant condition for a statement in the loop. An *invariant condition* for a statement is a condition that is true each time the statement is executed. When the invariant condition is an equation that contains the number of times a loop has been executed, then the resulting equation can be combined with the exit condition on the loop to determine the number of times the loop is executed.

For Step 3 of Horner's Algorithm, an invariant condition is

$$i + k = n, \tag{4}$$

where k is the number of times the loop has been executed so far, counting the current time. When we start the loop, $k = 1$ and $i = n - 1$ at Step 3. The exit condition on the loop is $i = -1$. Thus at the end of the loop k is equal to $n+1$.

If we call the times required by the steps in the above algorithm t_1, t_2, t_3, and t_4, then the total time required by the algorithm is

$$T = t_1 + t_4 + (t_2 + t_3)(n + 1) \tag{5}$$

$$= (t_1 + t_2 + t_3 + t_4) + (t_2 + t_3)n. \tag{6}$$

It is important to express the results of an algorithm analysis in the simplest possible form. Most people find eq. (6) simpler than eq. (5). If you want to give a more detailed answer, then you can apply the techniques of the previous section to obtain the values of the t_i. If you want to give a less detailed answer, you can say that the algorithm uses linear time. This means that the time used by the algorithm is of the form $c_1 + c_2 m$ where c_1 and c_2 are unspecified constants and m is a parameter associated with the size of the problem. It is important that the context make clear what parameter (n in the case of Horner's rule) is implied by the phrase "linear time".

You may have noticed that Step 3 does not take exactly the same time in all cases. The time when $i < 0$ is unlikely to be exactly the same as when $i \geq 0$. A more precise analysis would give a time of $(t_1 + t_2 + t_{3a} + t_4) + (t_2 + t_{3b})n$, where t_{3a} is the time Step 3 uses when $i < 0$ and t_{3b} is the time it uses when $i \geq 0$. Usually the difference between t_{3a} and t_{3b} is not large enough for such precision to be important. The result is still linear in n.

Many programs with loops are no more difficult than this to analyze. It may sometimes be a little tricky to decide exactly how many times the loop is executed: is it $n + 1$, n, or $n - 1$? The easiest way to check your answer is to hand execute the algorithm using a small value of n, keeping track of how often a step is actually done. With more complex algorithms it is a good idea to hand execute them using representative data to be sure you understand exactly how the algorithm works. Also, with complex algorithms it is usually helpful to program and run a version of the algorithm that counts how often it executes each step. Comparing the counts with the analysis will often uncover errors in the analysis (and in the programming).

EXERCISE

1. If each assignment, arithmetic operation, comparison, and go to requires one unit of time, how many time units are required by Algorithm 1.1? (Your answer should be a function of n.)

1.3 NUMBERS

A positive integer (whole number) is usually represented by a string of digits: $d_n\ d_{n-1}\ \ldots\ d_1 d_0$. For example, the number one hundred and sixty five is represented as 165. The value of a number is given by

$$d_n b^n + d_{n-1} b^{n-1} + \cdots + d_1 b + d_0 = \sum_{0 \le i \le n} d_i b^i \qquad (7)$$

where b is the base of the number system. People usually use base 10, although those who work with computers a lot often use bases 2, 8, and 16. Applying eq. (7) gives 165(base 10) $= 1 \times 10^2 + 6 \times 10^1 + 5 \times 10^0$ while 165(base 8) $= 1 \times 8^2 + 6 \times 8^1 + 5 \times 8^0 = 117$(base 10).

The digits of a number given in base b are usually required to be in the range $0 \le d < b$, and the base is usually required to be a positive integer. These requirements result in each integer having a unique representation in a given base.

A *rational number* is the ratio of two integers; it is represented by p/q where p is an integer and q is a positive integer. Usually rational numbers are represented in lowest terms, i.e., there is no integer greater than 1 that evenly divides both p and q.

A positive *real number* is usually represented by its decimal expansion $d_n d_{n-1} \ldots d_1 d_0 . d_{-1} d_{-2} \ldots$. The number of digits to the right of the decimal point is infinite in some cases. Thus the fraction 1/3 is equal to 0.333... (base 10). A real number written in base b is not permitted to contain an infinite sequence of successive digits equal to $b - 1$. Thus in base 10 it is necessary to write 2.0 rather than 1.999.... This rule is imposed so that there will be a unique way to represent each real number.

Much time is spent in elementary school teaching algorithms to do arithmetic in base 10. These algorithms need but slight change to work in any base.

1.3.1 Integer Value

The following algorithm computes the value of a number from its digits and the base.

Algorithm 1.2 Integer Value: Input: An integer which is represented by its base, b, its length minus one, n, and its digits d_n, d_{n-1}, ..., d_1, d_0, where $n \ge 0$. Output: v, the value of the integer. It is assumed that the value of the integer will fit in one computer word.

Step **1.** Set $i \leftarrow n$ and $v \leftarrow 0$.

Step **2.** Set $v \leftarrow vb + d_i$ and $i \leftarrow i - 1$.

Step **3.** If $i \ge 0$, then go to Step 2.

Step **4.** Output v.

Notice that the Integer Value Algorithm is really the same algorithm as Horner's Method, given in the last section. The program structure is the same; only the names and meanings of the variables have been changed. Thus this algorithm also takes linear time.

Algorithms that loop can usually be expressed more clearly using structured control statements such as for and while loops rather than tests and branches. Such control structures often make algorithms easier to understand, but it is necessary to establish their exact meaning in order to analyze algorithms that use them. In this book, we assume that the heads of for and while loops execute one time for each time through the body of the loop plus one extra time; the extra time takes care of the final test that terminates the loop. Here is the Integer Value Algorithm reexpressed with a for loop.

Algorithm 1.3 Integer Value: Input: An integer which is represented by its base, b, its length minus one, n, and its digits d_n, d_{n-1}, ..., d_1, d_0, where $n \geq 0$. Output: v, the value of the integer. It is assumed that the value of the integer will fit in one computer word.

Step **1.** Set $v \leftarrow 0$.

Step **2.** For $i \leftarrow n$ down to 0 do

Step **3.** Set $v \leftarrow vb + d_i$.

Step **4.** End for.

Step **5.** Output v.

The analysis of Algorithm 1.3 is slightly different from that of Algorithm 1.2 because the for statement tests the value of i before doing the loop the first time. Algorithm 1.3 works in a reasonable way even for $n < 0$, whereas Algorithm 1.2 requires $n \geq 0$. In Algorithm 1.3, Step 2 is done $n + 2$ times, once for each value of i in the range 0 to n plus one time to test that i is below 0. [More precisely, the time for Step 2 is $(n+1)t_{2a} + t_{2b}$, where t_{2a} is the time used for loop control when the interior of the loop is to be executed, and t_{2b} is the time for loop initialization plus the time used for loop control when exiting the loop. When we do not need such precision, we will just say that the time is $(n+2)t_2$, where $t_2 = t_{2a}$.] Step 4 is done $n+1$ times, once for each time around the loop. The remaining steps in Algorithm 1.3 are done the same number of times as the corresponding steps in Algorithm 1.2.

Knuth [10, Section 4.4] has a short but complete discussion of the problem of computing the value of a base b number from its digits.

EXERCISES

1. If each assignment, arithmetic operation, comparison, and go to requires one unit of time, how many time units are required by Algorithm 1.2? (Your answer should be a function of n.)
2. If each assignment, arithmetic operation, comparison, and go to requires one unit of time, how many time units are required by Algorithm 1.3? (Your answer should be a function of n.)
3. Program a version of the Integer Value Algorithm that counts how often each step is done. Also analyze the algorithm. Run the program for several values of n. Compare the counts from the runs with the results from the analysis.

1.4 DIVISION: QUOTIENTS AND REMAINDERS

The division of positive integer x by positive integer y produces a quotient q and a remainder r such that

$$x = qy + r, \quad \text{where } 0 \leq r < y. \tag{8}$$

Most computers have an integer divide instruction that computes both the quotient and the remainder for integers. Division can also be done when one or both of the integers is negative. There are, however, several different conventions used to define the quotient and remainder in such cases.

We use x/y to represent the real number that results from dividing x by y. The quotient that results from integer division of positive numbers is represented by $\lfloor x/y \rfloor$. The square brackets $\lfloor\ \rfloor$ that are missing their tops stand for the *floor* function. The value of the floor function is obtained by rounding its argument downward. Thus

$$\begin{array}{lllll} \lfloor 1 \rfloor = 1 & \lfloor 1.2 \rfloor = 1 & \lfloor 1.5 \rfloor = 1 & \lfloor 1.9 \rfloor = 1 & \lfloor 2 \rfloor = 2 \\ \lfloor -1 \rfloor = -1 & \lfloor -1.2 \rfloor = -2 & \lfloor -1.5 \rfloor = -2 & \lfloor -1.9 \rfloor = -2 & \lfloor -2 \rfloor = -2\,. \end{array}$$

The *ceiling* function, denoted by $\lceil\ \rceil$, is used to represent the results of rounding upward. Thus

$$\begin{array}{lllll} \lceil 1 \rceil = 1 & \lceil 1.2 \rceil = 2 & \lceil 1.5 \rceil = 2 & \lceil 1.9 \rceil = 2 & \lceil 2 \rceil = 2 \\ \lceil -1 \rceil = -1 & \lceil -1.2 \rceil = -1 & \lceil -1.5 \rceil = -1 & \lceil -1.9 \rceil = -1 & \lceil -2 \rceil = -2\,. \end{array}$$

The remainder resulting from x divided by y is written as $x \bmod y$. The *mod operation* is related to the floor function by the equation

$$x \bmod y = x - y\left\lfloor \frac{x}{y} \right\rfloor, \quad \text{if } y \neq 0; \qquad x \bmod 0 = x. \tag{9}$$

Eq. (9) provides a definition of the mod function for all real numbers x and y. Notice that for any number x,

$$x = \lfloor x \rfloor + (x \bmod 1). \tag{10}$$

The quantity $x \bmod 1$ is called the *fractional part* of x.

For example,

$$\begin{array}{rlrlrl} 5 \bmod 2 &= 1 & 5 \bmod 3 &= 2 & 5 \bmod 4 &= 1 \\ 8 \bmod 2 &= 0 & 8 \bmod 3 &= 2 & 8 \bmod 5 &= 3 \\ 8 \bmod 9 &= 8 & 8 \bmod 1 &= 0 & 8 \bmod 0 &= 8 \\ 7 \bmod 3 &= 1 & 7 \bmod -3 &= -2 & -7 \bmod 3 &= 2 \\ -7 \bmod -3 &= -1 & 1.3 \bmod 1 &= 0.3 & 1.35 \bmod 0.1 &= 0.05 \end{array}$$

and

$$8 \bmod \pi = 8 - 2\pi \approx 1.7168147\,.$$

The notation $x \equiv a \pmod{n}$ (read x is congruent to a modulo n) stands for $x \bmod n = a \bmod n$.

Knuth [9, Section 1.2.4] was a source for this section.

EXERCISES

1. Show that for $W > 0$, $W \bmod 2^k$ is the bottom k bits of W when W is written in binary. That is, if W has the binary representation $d_n d_{n-1} \ldots d_0$, then the number $W \bmod 2^k$ has the binary representation $d_{k-1} d_{k-2} \ldots d_0$.
2. Show that for $W > 0$, $\lfloor W/2^k \rfloor$ is W with its bottom k bits removed. That is, if W has the binary representation $d_n d_{n-1} \ldots d_0$, then $\lfloor W/2^k \rfloor = 0$ if $k > n$ and $\lfloor W/2^k \rfloor$ has the binary representation $d_n d_{n-1} \ldots d_k$ if $k \le n$.
3. Prove that $\lfloor -x \rfloor = -\lceil x \rceil$.

1.4.1 Addition

The following algorithm uses arithmetic with small numbers (one or two digits) to add big numbers.

Algorithm 1.4 Addition: Input: A base b, a length n, and two positive n digit integers X and Y represented by their digits:

$$X = x_{n-1}b^{n-1} + x_{n-2}b^{n-2} + \cdots + x_1 b + x_0 = \sum_{0 \le i < n} x_i b^i, \tag{11}$$

$$Y = y_{n-1}b^{n-1} + y_{n-2}b^{n-2} + \cdots + y_1 b + y_0 = \sum_{0 \le i < n} y_i b^i. \tag{12}$$

Intermediate variable: c, the carry. Output: An integer Z represented by its digits such that $Z = X + Y$, or an error if $X + Y$ cannot be represented as an n digit, base b integer.

Step **1.** [Initialize] Set $c \leftarrow 0$.

Step **2.** [Loop] For $i \leftarrow 0$ to $n - 1$ do:

Step **3.** [Add digits] Set $s \leftarrow x_i + y_i + c$.

Step **4.** [Compute digit and carry] Set $z_i \leftarrow s \bmod b$ and $c \leftarrow \lfloor s/b \rfloor$.

Step **5.** End for.

Step **6.** [Check answer] If $c = 0$, then $z_{n-1}, \ldots, z_0$ contains the correct answer. Otherwise it is too low by b^n and produces an error message.

Algorithms like this are often used when it is necessary to do arithmetic on numbers that are too large to fit in a computer word. For such applications one wants the base to be as large as possible. It is straightforward to program the algorithm correctly when the number $2b - 1$ will fit in one computer word because then the sum of a pair of the largest digits $(b - 1)$ and a carry will not cause an overflow. With clever assembly language coding the above algorithm can be adapted to work when b is one larger than the biggest number that will fit in one word by making the appropriate correction each time an overflow occurs.

In the Addition Algorithm Steps 3, 4, and 5 are done n times and Step 2 is done $n+1$ times, so the total time required is $(t_1 + t_2 + t_6) + (t_2 + t_3 + t_4 + t_5)n$. To do a more precise analysis, notice that the time for Step 6 depends on whether the addition results in an error condition. It is common to do analyses under the assumption that the error conditions do not occur.

Knuth [10, Section 4.3.1] gives the algorithm for adding multidigit numbers.

EXERCISES

1. If each assignment, arithmetic operation, comparison, and go to requires one unit of time, how many time units are required by Algorithm 1.4? (Your answer should be a function of n.)
2. Program a version of the Addition Algorithm that counts how often each step is done. Also analyze the algorithm. Run the program for several values of n. Compare the counts from the runs with the results from the analysis.

1.4.2* Arithmetic Modulo n

The integers modulo n have many properties in common with the integers. One of the advantages of the numbers modulo n is that there are only n different ones. This fits in very nicely with the fact that a word on a computer can hold only a fixed number of different values. Much of arithmetic is the same for the numbers modulo n as it is for integers.

These similarities are summarized by

Theorem 1.1 If $a \equiv b \pmod{m}$ and $x \equiv y \pmod{m}$, then

$$a + x \equiv b + y \pmod{m}, \tag{13}$$

$$a - x \equiv b - y \pmod{m}, \tag{14}$$

$$ax \equiv by \pmod{m}. \tag{15}$$

Addition of a and b modulo m means to compute $(a + b) \pmod{m}$, and similarly for other operations. Theorem 1.1 implies that for any series of additions, subtractions, and multiplications, you can replace each operation by the operation modulo m, provided you know that the final answer is in the range 0 to $m - 1$. (Other ranges for the answer can also be handled provided the length of the range is no more than m.) This is true no matter how big the intermediate value becomes.

For example, suppose you want to compute $25 \times 25 - 24 \times 24 - 40$ and you know the answer is in the range 0 to 9. (How you might know that is another problem.) Then, based on Theorem 1.1 with $m = 10$, we can do the calculations as follows:

$$\begin{array}{lll} 25 \equiv 5 \pmod{10} & \text{so} & 25 \times 25 \equiv 5 \times 5 \pmod{10} = 25 \equiv 5 \pmod{10}, \\ 24 \equiv 4 \pmod{10} & \text{so} & 24 \times 24 \equiv 4 \times 4 \pmod{10} = 16 \equiv 6 \pmod{10}, \\ 40 \equiv 0 \pmod{10} & \text{so} & 5 - 6 - 0 \equiv 9 \pmod{10}. \end{array}$$

Thus $25 \times 25 - 24 \times 24 - 40 = 9$. Even if you didn't know that the final answer was in the range 0 to 9, you would know that $25 \times 25 - 24 \times 24 - 40 \equiv 9 \pmod{10}$.

People find arithmetic modulo 10 (and 100, 1000, etc.) particularly convenient because they use numbers in base 10. Likewise, computers can usually do arithmetic modulo a power of 2 particularly quickly. Many programs do arithmetic modulo the word size of their computer.

Division sometimes works in arithmetic modulo m. The number a is *relatively prime* to m if their only common factor is 1.

Theorem 1.2 If $ax \equiv by \pmod{m}$, $a \equiv b \pmod{m}$, and a is relatively prime to m, then $x \equiv y \pmod{m}$.

This is like the usual cancellation law for division with real numbers, except the requirement that the number being cancelled must not be zero is replaced by the requirement that it must be relatively prime to the modulus.

To see that the requirement that a must be relatively prime to m is necessary, consider $a = 5$, $b = 15$, $x = 3$, $y = 1$, and $m = 10$. These numbers satisfy all the conditions of the theorem except that 5 is not relatively prime to 10. So, although $5 \times 3 \equiv 15 \times 1 \pmod{10}$ and $5 \equiv 15 \pmod{10}$ we have $3 \not\equiv 1 \pmod{10}$.

One way to avoid trouble with division modulo m is to choose a prime number p as the modulus. Any $a \not\equiv 0 \pmod{p}$ is relatively prime to p. So with a prime modulus, you can do problems with addition, subtraction, multiplication, and division by replacing each operation by the operation modulo p, *provided* you know that the range of the final answer is small enough (0 to $p - 1$), and *provided* you know that the final answer is an integer.

Two more theorems are useful when converting numbers from one modulus to another.

Theorem 1.3 For $n \neq 0$,

$$a \equiv b \pmod{m} \quad \text{if and only if} \quad an \equiv bn \pmod{mn}. \qquad \textbf{(16)}$$

Theorem 1.4 If r is relatively prime to s, then $a \equiv b \pmod{rs}$ if and only if both $a \equiv b \pmod{r}$ and $a \equiv b \pmod{s}$.

Properties of the integers modulo n are discussed by Knuth [9, Section 1.2.4] and by Birkhoff and Mac Lane [22, Sections 1.9 and 1.10].

EXERCISES

1. Prove Theorem 1.1. Hint: Start by writing $a = b + k_1 m$ and $x = y + k_2 m$ for some integers k_1 and k_2.

2. Prove Theorem 1.2.

3. Show that for a prime number p, $x \equiv a \pmod{p}$, $y \equiv b \pmod{p}$, and $y \not\equiv 0 \pmod{p}$ implies $x/y \equiv a/b \pmod{p}$.

4. Let n have the base b representation $n_k n_{k-1} \ldots n_0$. Prove that

$$n \equiv n_0 + n_1 + \cdots + n_k \pmod{b-1}.$$

This gives a quick way to compute the mod function when the modulus is one less than the base. For example,

$$931 \equiv 13 \equiv 4 \pmod{9}$$

and

$$931 \equiv 40 \pmod{99}.$$

Hint: Consider using eq. (7) and polynomial division.

5. Let n have the base b representation $n_k n_{k-1} \ldots n_0$. Prove that

$$n \equiv n_0 - n_1 + \cdots + (-1)^k n_k \pmod{b+1}.$$

This gives a quick way to compute the mod function when the modulus is one more than the base. For example,

$$931 \equiv 7 \pmod{11}$$

and

$$931 \equiv 22 \pmod{101}.$$

1.4.3* The Euclidean Algorithm

One important calculation in arithmetic is the computation of the greatest common divisor of two numbers. For example, you need the greatest common divisor when you want to reduce a fraction to lowest terms. Schools often teach a method of finding the greatest common divisor based on factoring. Such methods are easy to understand, but they are very slow for large numbers. (No one knows a fast method of factoring large numbers.) However, a fast algorithm for computing the greatest common divisor was already known by the time of Euclid.

We present the extended version of the algorithm, which not only computes d, the greatest common divisor of m and n, but also computes numbers a and b such that $am + bn = d$. When you don't need a and b, you can drop those parts of the algorithm that compute them, thereby obtaining the ordinary Euclidean algorithm.

Algorithm 1.5 Euclidean: Input: Positive integers m and n. Output: Integers d, a, and b such that d is the greatest common divisor of m and n, and such that $am + bn = d$.

Step **1.** [Initialize] Set $a' \leftarrow b \leftarrow 0$, $a \leftarrow b' \leftarrow 1$, $d \leftarrow m$, $r \leftarrow n$, and $q \leftarrow 0$.

Step **2.** [Loop] While $r > 0$ do:

Step **3.** [Recycle] Set $c \leftarrow d$, $d \leftarrow r$, $t \leftarrow a' - qa$,
$a' \leftarrow a$, $a \leftarrow t$, $t \leftarrow b' - qb$, $b' \leftarrow b$, and $b \leftarrow t$.

Step **4.** [Divide] Set $q \leftarrow \lfloor c/d \rfloor$ and $r \leftarrow c - qd$. (Thus q is the quotient and r is the remainder of c divided by d.)

Step **5.** [Loop] End while.

To see that the algorithm works, first notice that at the end of each step we have $a'm + b'n = c$ and $am + bn = d$. Next notice that d always decreases. Finally notice that after Step 4, d and r have the same greatest common divisor as c and d. This condition is preserved by all the other steps. Since the values of c and d are replaced by d and r, the value of the greatest common divisor does not change. The initial values of c and d are m and n. When the algorithm stops (which it does since d is a decreasing integer), r is zero, so it is evenly divisible by any number. Therefore d must be the greatest common divisor of m and n.

The number d, the greatest common divisor of m and n, is the smallest positive number such that the equation $am + bn = d$ has a solution where a and b are integers. The left side of this equation is divisible by the greatest common divisor of m and n. Any integer equal to $am + bn$ for integer a and b must be divisible by d.

One use of the extended Euclidean Algorithm is to find the inverse of a number modulo a second number m'. Suppose you have the number x, and you want to find a number y such that

$$xy \equiv 1 \pmod{m'}. \qquad \textbf{(17)}$$

Such a number y is an *inverse* (modulo m') of x. To compute it, use the extended Euclidean Algorithm with the inputs: $m = x$ and $n = m'$. The value for y is the value a returned by the algorithm, provided $d = 1$. If $d \neq 1$, then your problem has no solution. In particular if m' is a prime, then every integer less than m' has an integer inverse modulo m'.

Table 1.2 gives the inverses of the integers modulo 10. There are several interesting things to notice. First, 1 is its own inverse, which is not surprising. Likewise 9, which is equivalent to -1 modulo 10, is its own inverse. The inverse of 3 is 7, and, of course, the inverse of 7 is 3. Every number that is relatively prime to 10 has an integer inverse modulo 10.

The Euclidean Algorithm is discussed by Knuth [9, Sections 1.1 and 1.2.1; 10, Section 4.5.3], Aho et al. [1, Section 8.8], and Horowitz and Sahni [7, Section 9.4]. The Euclidean Algorithm is adapted from Knuth [9, Section 1.2.1].

Integer	*Inverse*	*Integer*	*Inverse*
0	none	5	none
1	1	6	none
2	none	7	3
3	7	8	none
4	none	9	9

TABLE 1.2. Inverses of the integers modulo 10.

EXERCISES

1. Find the greatest common divisor of 5,396,834,063 and 3,268,703,989.

2. Which integers have an inverse modulo 9? What is the inverse of those integers?

1.5 EXPONENTS AND LOGARITHMS

For a positive integer n, the notation x^n means the product of n x's. Thus

$$x^1 = x, \quad x^2 = xx, \quad x^3 = xxx, \quad \text{etc.} \tag{18}$$

We define $x^0 = 1$ (for $x \neq 0$) and $x^{-n} = 1/x^n$. Usually it is also appropriate to use $0^0 = 1$, but each case where this expression arises will need individual consideration.

The value y of $x^{p/q}$ is defined to be the solution of the equation

$$y^q = x^p. \tag{19}$$

The quantity $x^{1/n}$ is called the nth *root* of x. The value of x^z for a real number z is defined by requiring that x^z be a continuous function of z (when $x \neq 0$). The *laws of exponents* say that:

$$x^{y+z} = x^y x^z \quad \text{and} \quad (x^y)^z = x^{yz} \tag{20}$$

The notation $y = \log_b x$ ($b > 1$) means that y is the solution of the equation

$$x = b^y. \tag{21}$$

We call y the *logarithm* to base b of x. Notice that $\log_b 1 = 0$ and $\log_b b = 1$. The *laws of logarithms* state that

$$\log_b(xy) = \log_b x + \log_b y, \quad \text{if } x > 0, \quad y > 0 \tag{22}$$

and

$$\log_b(x^y) = y \log_b x, \quad \text{if } x > 0. \tag{23}$$

These laws of logarithms follow directly from the laws of exponents and the definition of logarithm.

Logarithms occur often in the analysis of algorithms. One reason has to do with their relation to trees. A complete binary tree of height h has $2^{h+1} - 1$ nodes and 2^h leaves. Putting it another way, the height of a complete binary tree with n nodes is $\log_2(n+1)-1$. We will use the notation "lg" for the logarithm base 2; the height of a complete binary tree with n nodes is

$$\lg(n+1)-1. \tag{24}$$

Binary trees also can be used to describe the operation of algorithms that repeatedly choose between two alternatives. Since a complete binary tree contains the maximum number of leaves for a given height, the time of an efficient algorithm is frequently related to the logarithm of the number of alternatives that the algorithm must choose between. (See the following section.)

Base e, where e is the irrational number 2.71828..., also occurs commonly. It arises because e is the correct base to use in the equation

$$\int_1^x \frac{du}{u} = \log_e x. \tag{25}$$

The notation "ln" is used for $\log_e$. The logarithm to the base e is called the *natural logarithm.*

There is a simple relation between the values of logarithms to different bases. Let $y = \log_a x$. Then $x = a^y$. To obtain the value of $\log_b x$, just take $\log_b$ of this equation to get $\log_b x = \log_b(a^y) = y \log_b a = \log_a x \log_b a$. By considering this equation with $x = b$, you can see that $\log_a b = 1/\log_b a$. In summary,

$$\log_b x = \frac{\log_a x}{\log_a b}. \tag{26}$$

Knuth [9, Section 1.2.2] was a source for this section.

EXERCISES

1. Show that $1 + \lg(3/2) = \lg 3$.

2. Show that $(n+1)/2 \leq 2^{\lfloor \lg n \rfloor} \leq n$ for integer $n \geq 1$.

3. Express $2^{2 \lg x}$ in the form x^n and give the value of n.

4. Show that

$$2^{\frac{\ln\left(1+\frac{t(-\ln(1-p))}{\ln 2}\right)}{-\ln(1-p)}} = \left(\frac{t(-\ln(1-p))}{\ln 2 + t(-\ln(1-p))}\right)^{\ln 2/(-\ln(1-p))}$$

5. Show that $x^{\ln y} = y^{\ln x}$.

6. Show that $n(3/2)^{\lg n} = n^{\lg 3}$.

1.5.1 Binary Search

For each of the last three algorithms, determining how often the steps inside the loop are done can be reduced to solving an equation. The amount of change in the loop index has to equal the difference between its initial value and its final value. For example, for the first two algorithms the amount of change is $n+1$, and the index changes by one each time through the loop. Therefore the number of times through the loop, i, is the solution of the equation

$$i = n + 1, \tag{27}$$

an equation so simple that you do not need to write it down to solve it. For more complex programs with a single loop, it is usually also possible to write an equation whose solution says how often the program goes through the loop. If the equation is complicated, you should write it down before trying to solve it. Consider the following algorithm, which is slightly more complex than the previous three:

Algorithm 1.6 Binary Search: Input: A number q, called a *query*, a length $n > 1$, and a sorted array X of n numbers, x_0, x_1, ..., x_{n-1}, where $x_i \leq x_{i+1}$ for $0 \leq i \leq n-2$. Output: The result "success" and an index b if there is a b such that $q = x_b$, and the result "failure" otherwise.

Step **1.** [Initialize] Set the bottom pointer $b \leftarrow 0$ and the top pointer $t \leftarrow n$.

Step **2.** [Loop] While $t - b > 1$ do:

Step **3.** [Halve region] Compute the midpoint $m \leftarrow \left\lfloor \frac{b+t}{2} \right\rfloor$. If $x_m \leq q$, then set $b \leftarrow m$. Otherwise set $t \leftarrow m$.

Step **4.** End while.

Step **5.** [Test if found] If $q = x_b$, then report "success"; otherwise report "failure".

This algorithm is somewhat more difficult to analyze than the previous two. There is no linear equation that leads directly to the solution. The key in this case is to notice the behavior of the quantity $t - b$. When n is a power of 2, $t - b$ always decreases by a factor of 2 each time through the loop. To avoid the more difficult cases, let us assume that n is a power of 2. The initial value of $t - b$ is n. Letting k be the number times the algorithm has started around the loop, we have the invariant

$$(t-b)2^k = n \tag{28}$$

at the end of Step 3. The exit condition is $t - b = 1$. So, letting i be the total number of times around the loop, i obeys the equation

$$n/2^i = 1. \tag{29}$$

Taking the logarithm base 2 of both sides of eq. (29) and solving for i gives

$$i = \lg n. \tag{30}$$

So the time required by the Binary Search algorithm is

$$(t_1 + t_2 + t_5) + (t_2 + t_3 + t_4) \lg n. \tag{31}$$

When n is not a power of 2, the analysis is a little more complex because the time required depends on which item in the table is being looked up. Techniques that are developed in Section 1.9 are appropriate for that case. Since the time used by Binary Search increases with n, even when n is not a power of 2, the above analysis gives a result accurate enough for most applications. For n between two powers of 2 (say, 2^i and 2^{i+1}), the time required is between the results obtained when the two limits (2^i and 2^{i+1}) are substituted for n in eq. (31).

Binary search is discussed by Knuth [11, Section 6.2.1], Sedgewick [62, Chapter 14], Aho et al. [1, Section 4.3], Baase [2, Section 1.4], Horowitz and Sahni [7, Section 3.2], Reingold and Hansen [61, Section 1.3], and Standish [63, Section 4.2.2].

EXERCISES

1. When the Binary Search Algorithm is given an item q which is not in the array, such that $x_0 \leq q$, it sets b to the place to insert the item into the array. That is, when q is not in the array, b is set so that either $x_b < q$ or $b = 0$, and so that either $q < x_{b+1}$ or $b = n - 1$. Modify the algorithm so that for any item q, the algorithm will set b so that $x_b = q$ if q is already in the array, will insert q into the array X if it is not already in the array, and will set b to the index at which q is inserted. Your algorithm should keep the array sorted and it should keep the first element in x_0. It should set n to the number of items in the array. Compare the running time of your algorithm with the one in the text. How long does it take to insert an item into the table? Does your algorithm take more or less time than the algorithm in the book to look up items that are already in the table?
2. Show that the number of times you go around the outer loop in the Euclidean Algorithm (Algorithm 1.5) is no more than $2 \lg n$ when $m < n$.

1.6 SIMPLE NESTED LOOPS

Many algorithms have nested loops. This does not increase the difficulty of analyzing the algorithm provided the variables set in the outer loop do not affect how often the steps in the inner loop are done. You can just apply the technique of the previous section repeatedly.

1.6.1 Matrix Addition

A *matrix* is a rectangular array of numbers. A single capital letter can be used to stand for the entire matrix, or the elements can be written out in an array to emphasize each element. The element of matrix A in row i and column j is denoted by a_{ij}. Thus the equation

$$A = \begin{pmatrix} a_{11} & a_{12} & \cdots & a_{1m} \\ a_{21} & a_{22} & \cdots & a_{2m} \\ \vdots & \vdots & & \vdots \\ a_{n1} & a_{n2} & \cdots & a_{nm} \end{pmatrix} \tag{32}$$

shows the matrix A represented each way.

An example of an algorithm with a pair of simple nested loops is the following algorithm to add matrices. If A, B, and C are n by m matrices (hereafter denoted $n \times m$) and $C = A + B$, then $c_{ij} = a_{ij} + b_{ij}$ for $1 \le i \le n$, $1 \le j \le m$.

Algorithm 1.7 Matrix Addition: Input: Two $n \times m$ matrices A , B with elements a_{ij} and b_{ij} for $1 \le i \le n$, $1 \le j \le m$. Output: The $n \times m$ matrix C with elements c_{ij} for $1 \le i \le n$, $1 \le j \le m$ where $C = A + B$.

Step **1.** [Start outer loop] For $i \leftarrow 1$ to n do:

Step **2.** [Start inner loop] For $j \leftarrow 1$ to m do:

Step **3.** [Compute element] Set $c_{ij} \leftarrow a_{ij} + b_{ij}$.

Step **4.** End for j.

Step **5.** End for i.

This algorithm is easy to analyze. Step 1 is done $n + 1$ times. Step 5 is done n times. Step 2 is done $m + 1$ times *each* time the outer loop is done, so Step 2 is done a total of $n(m + 1)$ times. Steps 3 and 4 are done nm times.

This illustrates a general principle. When an algorithm has nested loops, and the number of times the inner loop is executed does not depend on variables that are set in next outer loop, a statement in the inner loop is executed a fixed number of times (n in our example) every time the next outer loop is executed. The total number of times the statement in the inner loop is executed is therefore the product of the number of repetitions of the inner loop and the number of repetitions of the next outer loop.

If a program has deeply nested loops (and the number of times each loop is done is independent of the variables set in the outer loops), then you can just about be sure that it is a slow-running program. For example, if each loop is done n times each time the next outer loop is done, and if you have k levels of nesting, then the innermost steps are done n^k times. If you remember that 2^{20}, 4^{10}, and 16^5 are all just over 1 million, you can see that it is easy for a deeply nested program to use up a huge amount of time even when each step in the program can be done very rapidly.

When a program has loops, it can often be speeded up significantly by moving all calculations to the outermost possible loop. Suppose the statement "Set $i \leftarrow i + x \times y$" occurs in an inner loop, and x and y do not change in the loop. Then the statement "Set $t \leftarrow x \times y$" should be placed in the next outer loop (or even further out if the values of x and y do not change in the next outer loop either) and the original statement should be replaced by "Set $i \leftarrow i + t$". Some compilers perform this sort of program optimization automatically.

1:6.2 Multiplication

Another example of an algorithm with simple nested loops is the following algorithm for multiplying two multidigit integers. This algorithm works in a way similar to the algorithm taught in grade school, with one exception. Rather than saving up a column of partial products and then adding them later, this algorithm adds up the partial products as it goes. This saves a good bit of computer memory. People usually do not use this algorithm for hand calculations because they are less likely to get confused when they have some intermediate output to write down, a consideration that does not apply to computers.

Algorithm 1.8 Multiplication: Input: A base b, a length n, and two n-digit integers X and Y represented by their digits:

$$X = x_{n-1}b^{n-1} + x_{n-2}b^{n-2} + \cdots + x_1 b + x_0 = \sum_{0 \le i \le n-1} x_i b^i, \tag{33}$$

$$Y = y_{n-1}b^{n-1} + y_{n-2}b^{n-2} + \cdots + y_1 b + y_0 = \sum_{0 \le i \le n-1} y_i b^i. \tag{34}$$

Output: A $2n$-digit integer Z represented by its digits such that $Z = XY$.

Step **1.** [Initialize] For $i \leftarrow 0$ to $n-1$, set $z_i \leftarrow 0$.

Step **2.** [Top loop] For $i \leftarrow 0$ to $n-1$ do:

Step **3.** [Initialize carry] Set $c \leftarrow 0$.

Step **4.** [Bottom loop] For $j \leftarrow 0$ to $n-1$ do:

Step **5.** [Multiply digits] Set $p \leftarrow x_i y_j + z_{i+j} + c$. This step multiplies digit i of the first number by digit j of the second number and combines the result with the part of the answer that has been accumulated.

Step **6.** [Compute digit and carry] Set $z_{i+j} \leftarrow p \bmod b$ and $c \leftarrow \lfloor p/b \rfloor$.

Step **7.** End for j.

Step **8.** [Next digit] Set $z_{i+n} \leftarrow c$.

Step **9.** End for i.

Step **10.** [Done] Stop. The digits $z_{2n-1}, \ldots, z_0$ contain the correct answer.

This algorithm is easy to analyze. Steps 1 and 10 are each done once. Notice, however, that Step 1 is not a simple step. The typical computer will require at least $n+1$ units of time to carry it out. Step 2 is done $n+1$ times. Steps 3, 8, and 9 are each done n times, once for each value of i between 0 and $n-1$. Step 4 is done $n+1$ times *every* time Step 2 is done, so Step 4 is done n^2+n times. Each of Steps 5 and 6 is done n times *every* time Step 2 is done, once for each value of j between 0 and $n-1$. So each of Steps 5 and 6 is done a total of n^2 times. Since n^2 grows so much faster than n, the total time for the algorithm is dominated by the n^2 time used in the inner loop.

The classical method for multiplication of positive integers is discussed in Knuth [10, Section 4.3.1].

EXERCISES

1. In Step 1 of the Multiplication Algorithm, why do we need to initialize the first n digits of Z, but not the rest of Z?

2. The Multiplication Algorithm is similar to the Addition Algorithm, but Addition checks to be sure that the answer is not too big. Does Multiplication need such a check? If it does not, give a proof that the answer is never bigger than the largest $2n$-digit, base b integer.

3. Modify the Multiplication Algorithm so that it multiplies an n-digit number by an m-digit number. How long does your algorithm take to run?

4. How much memory does the Multiplication Algorithm use? Assume that one computer word is needed for each digit that it stores. Ignore the memory used by the variables i and j and by the code for the algorithm.

5. Write down as an algorithm the method of multiplying large numbers that you learned in grade school. How long does the method take? How much memory does it use?

1.6.3 Matrix Multiplication

For another example consider the following algorithm, which has three simple nested loops. It computes matrix product. If A is an $n \times p$ matrix, B a $p \times m$ matrix, and C an $n \times m$ matrix with $C = AB$, then

$$\begin{aligned} c_{ij} &= a_{i1}b_{1j} + a_{i2}b_{2j} + \cdots + a_{ip}b_{pj} \quad \text{for } 1 \le i \le n,\ 1 \le j \le m \\ &= \sum_{1 \le k \le p} a_{ik}b_{kj} \quad \text{for } 1 \le i \le n,\ 1 \le j \le m. \end{aligned} \tag{35}$$

Algorithm 1.9 Matrix Multiplication: Input: An $n \times p$ matrix A with elements a_{ij} and a $p \times m$ matrix B with elements b_{ij}. Output: The $n \times m$ matrix C with elements c_{ij}, where $C = AB$.

Step **1.** [Outer loop] For $i \leftarrow 1$ to n do:

Step **2.** [Middle loop] For $j \leftarrow 1$ to m do:

Step **3.** [Initialize] Set $c_{ij} \leftarrow 0$.

Step **4.** [Inner loop] For $k \leftarrow 1$ to p do:

Step **5.** [Compute element] Set $c_{ij} \leftarrow c_{ij} + a_{ik}b_{kj}$.

Step **6.** End for k.

Step **7.** End for j.

Step **8.** End for i.

Step 1 is done $n+1$ times. Step 8 is done n times. Step 2 is done $m+1$ times each time Step 1 is done. So Step 2 is done a total of $(m+1)n$ times. Steps 3 and 7 are done mn times. Step 4 is done $p+1$ times each time the middle loop is executed, so it is done a total of $mn(p+1)$ times. Steps 5 and 6 are done mnp times.

Matrix multiplication is discussed in Baase [2, Section 1.3].

1.6.4 Shortest Path

Assume that you have n cities connected by one-way roads and that you have a matrix D with entries d_{ij}, where d_{ij} is the distance from city i to city j by the direct road whenever there is a direct road, and $d_{ij} = \infty$ when there is no direct road. It is required that $d_{ij} \geq 0$. If you have a two-way road between i and j, that is also okay; in that case $d_{ij} = d_{ji}$. The following algorithm modifies d_{ij} so that it gives the distance from city i to city j by the shortest path from i to j, with $d_{ij} = \infty$ when there is no path from i to j. Notice that the algorithm will never increase an entry in D.

Algorithm 1.10 Shortest Path: Input: A distance matrix D with entries d_{ij} for $1 \leq i, j \leq n$ where d_{ij} is the length of the direct path from i to j. Output: A matrix D where entry d_{ij} is the length of the shortest path from i to j.

Step **1.** [Outer loop] For $k \leftarrow 1$ to n do:

Step **2.** [Middle loop] For $i \leftarrow 1$ to n do:

Step **3.** [Inner loop] For $j \leftarrow 1$ to n do:

Step **4.** [Compute element] Set $d_{ij} \leftarrow \min(d_{ij}, d_{ik} + d_{kj})$.

Step **5.** End for j.

Step **6.** End for i.

Step **7.** End for k.

The order of the three loops is quite important to understanding why the shortest path algorithm works. Initially the matrix D contains the length of the shortest path which follows a direct route between each city. After the outer loop is done for $k = 1$, D contains the length of the shortest path following either a direct route or going by way of city 1. After the outer loop is done for $k = 2$, D contains the length of the shortest path following either a direct route or a route going by way of any combination of cities 1 and 2. After p steps, D contains the length of the shortest path following either a direct route or a route going by way of any of the first p cities. The analysis of this algorithm is left to the exercises.

The shortest path problem is discussed by Aho et al. [1, Section 5.8], Baase [2, Section 3.3], Horowitz and Sahni [7, Section 5.3], and Sedgewick [62, Chapter 31]. The algorithm for finding the shortest path between all pairs of points was developed by Floyd [87].

EXERCISES

1. In what way is finding the shortest path similar to multiplying matrices?

2. How often is each step in the Shortest Path Algorithm done?

1.7 LINEAR SUMS

Often the number of times that an algorithm goes around an inner loop is given by an expression that depends on the value of a parameter set in an outer loop. In such cases, to compute the number of times the inner loop is executed, you must sum the expression. As a result, much of the analysis of algorithms is closely connected to summation problems.

We use the notation

$$S = \sum_{1 \le i \le n} x_i \tag{36}$$

for S, the sum of the sequence $x_1, x_2, \ldots, x_n$.

It is important to present your answers in the simplest possible form. One important part of simplifying answers is to replace summations by their closed form equivalent whenever possible. A closed form formula is one where the result can be calculated with a fixed number of elementary operations. The nature of an elementary operation varies somewhat with context, but usually addition, subtraction, multiplication, and division are regarded as elementary operations. Eq. (36) is not in closed form. The number of additions required to evaluate eq. (36) is $n - 1$, which is not a fixed number.

Unless we have more information about the x_i, of course, there is no way to simplify eq. (36). If the x_i are known, then there are a number of techniques that can be used. One of the simplest cases occurs when the x_i are all the same. Then we have

$$\sum_{1 \le i \le n} m = nm \tag{37}$$

from the definition of multiplication.

Often you can simplify a summation by reexpressing it in terms of itself. Consider the summation $\sum_{1\le i\le n} i$. It can be simplified as follows. First,

$$\sum_{1\le i\le n} i = \sum_{1\le i\le n} (n+1-i), \tag{38}$$

because the same numbers are being summed on the right and left sides; only the order in which they are being summed is different. Now

$$\sum_{1\le i\le n} (n+1-i) = \sum_{1\le i\le n} (n+1) - \sum_{1\le i\le n} i \tag{39}$$

since $n+1-i = (n+1)-i$ and since we can change the order in which the summations are performed without changing the answer. Since $\sum_{1\le i\le n}(n+1) = n(n+1)$, eq. (38) and eq. (39) give (using $S = \sum_{1\le i\le n} i$)

$$S = n(n+1) - S. \tag{40}$$

Solving eq. (40) for S gives

$$S = \sum_{1\le i\le n} i = \frac{n(n+1)}{2}. \tag{41}$$

Notice that the general idea that led to eq. (41) was to find a transformation of the sum that reexpresses the sum in terms of itself. When one first tries this method, one often obtains an equation of the form $S = a+S$, which is only good for checking one's algebra (a will be zero if one has made no error). The difficult part of using the method is to obtain a nontrivial result of the form $S = a+bS$, where $b \neq 1$; then you can solve the equation to obtain $S = a/(1-b)$. We will have much more to say about solving sums in Chapters 2 and 3. The sum in eq. (41) will be used in the analysis of several algorithms in the remainder of this chapter.

EXERCISE

1. It is said that when Gauss was in grade school, he was kept after school as punishment. He had to sum the numbers from 1 to 100 before he could go home. He surprised the teacher by quickly producing the correct answer. The method he used was equivalent to eq. (41), and depends on the observation that $1+100 = 2+99 = 3+98 = \cdots$. What is the sum of the numbers from 1 to 100?

1.7.1 Triangular Matrix Addition

A *triangular matrix* is one in which all the elements below the diagonal are zero. When adding such matrices, it is not necessary to add the zeros; the result is always another triangular matrix. Thus, we have the following algorithm for triangular matrix addition.

Algorithm 1.11 Triangular Matrix Addition: Input: A size n and two triangular matrices A and B with elements a_{ij} and b_{ij} for $1 \le j \le i \le n$. Output: Matrix $C = A + B$.

Step **1.** [Outer loop] For $i \leftarrow 1$ to n do:

Step **2.** [Inner loop] For $j \leftarrow 1$ to i do:

Step **3.** [Compute element] Set $c_{ij} \leftarrow a_{ij} + b_{ij}$.

Step **4.** End for j.

Step **5.** End for i.

Step 1 is done $n+1$ times. Step 5 is done n times. Step 2 is done $i+1$ times each time the outer loop is done. So Step 2 is done a total of $\sum_{1\le i\le n}(i+1) = n(n+1)/2 + n = n(n+3)/2$ times [see eqs. (37, 41)]. Steps 3 and 4 are done i times each time the outer loop is done, giving a total of $\sum_{1\le i\le n} i = n(n+1)/2$ times for each step.

EXERCISES

1. The product of the polynomial $a_n x^n + a_{n-1}x^{n-1} + \cdots + a_1 x + a_0$ with the polynomial $b_m x^m + b_{m-1}x^{m-1} + \cdots + b_1 x + b_0$ is the polynomial $c_{m+n}x^{m+n} + c_{m+n-1}x^{m+n-1} + \cdots + c_1 x + c_0$, where

$$c_i = \sum_{\max\{0,i-m\}\le j\le \min\{i,n\}} a_j b_{i-j}.$$

If you compute c_i for $0 \le i \le m+n$ directly from the definition, how many multiplications are done?

2. Show that if you use the convention (1) $a_i = 0$ for $i < 0$ and for $i > n$, (2) $b_i = 0$ for $i < 0$ and for $i > m$, and (3) $\sum_i x_i$ means sum the nonzero x_i for $-\infty < i < \infty$, then the formula for the product of the polynomials in the last exercise can be written as

$$c_i = \sum_j a_j b_{i-j}.$$

3. Give a simple algorithm for computing the product of two polynomials. How often is each step in your algorithm executed?

1.8 PROBABILITY

The analysis of algorithms often involves the study of random events. For example, the time required by a sorting program depends on the initial order of the numbers to be sorted. Usually it is appropriate to assume that the numbers are in random order. In order to study the behavior of an algorithm whose time depends on the nature as well as the size of its input, we need to use the notion of probability.

The *probability* of an event is the fraction of times that the event occurs if the same process is repeated a large number of times. It is important when repeating the process that all relevant conditions be the same each time. Thus when determining how frequently a particular coin comes up heads, you should flip the same coin each time. On the other hand, if you are determining how frequently the sticks from a lot of dynamite explode when the fuse is lit, it is important to use a different stick from the lot each time.

Often you can determine the probability of random events by finding an appropriate set of basic events and noticing that the process that generates the basic events will cause each basic event to occur equally often. Thus when you flip a coin, you expect it to land heads 1/2 of the time and tails 1/2 of the time, because there is no significant difference between the head and tail side of a coin. Likewise, when you roll a die, you expect each of the numbers 1 through 6 to appear equally often. When determining probabilities by such an analysis, it is, of course, important to correctly identify a set of equiprobable basic events. Thus a die can land showing a number that is divisible by 3 or one that is not. These two events are not, however, equally likely basic events. The appropriate basic event for a symmetric die is the particular face showing. Identifying equally likely basic events is the most elementary way to solve those problems to which the method applies.

Some probabilities cannot be determined by considering equally likely basic events. For example, if you have an irregularly shaped die, you will not have a set of equiprobable basic events. Likewise, basic events are of no use if you want to determine the probability that it will snow next Christmas or if you want to know how often the identifier "ERROR1" occurs in typical programs. Usually such probabilities are determined by measurements.

When analyzing algorithms it is often necessary to make assumptions about the input. Sometimes the assumptions are minor and even notational in nature. Thus you may assume that the identifier "ERROR1" occurs with a definite probability p_{ERROR1}. Whoever uses your results will then need to first measure p_{ERROR1}. Often, however, it is necessary to make more drastic assumptions. For example, when analyzing sorting algorithms, it is common to assume that the numbers being sorted are all distinct and that each arrangement of the input numbers is equally likely. This may or may not be the case. One needs to measure a large sample of typical events to know whether or not it is the case. It is, however, a reasonable assumption in that there is no immediate argument that it is wrong, and it does lead to useful analyses of sorting algorithms.

The following are useful principles to remember when making assumptions about probabilities. First, make no more assumptions than are necessary to establish your result. The fewer assumptions you make, the more useful your result will be. Second, only make reasonable assumptions. Deciding what assumptions are reasonable is something of an art. You will see a lot of examples in this book. Third, clearly state whatever assumptions you do make. Statistical analyses depend directly on the statistical assumptions that are made during the analysis. A clear statement of the assumptions will greatly increase the usefulness of your analyses and decrease the chance that they will mislead anyone.

EXERCISE

1. If you feed a program a sequence of n distinct numbers which are in random order, what is the probability that the first number is the largest?

1.8.1 Combining Probabilities

It is often necessary to combine probabilities. For example, the basic events that are appropriate for establishing the values of the probabilities may not be the events that are of final concern.

The basic formula for combining probabilities is

$$\text{Prob}(A \cup B) = \text{Prob}(A) + \text{Prob}(B) - \text{Prob}(A \cap B), \tag{42}$$

where $\text{Prob}(A \cup B)$ is the probability that event A *or* event B (or both) occur, $\text{Prob}(A)$ is the probability that event A occurs, $\text{Prob}(B)$ is the probability that event B occurs, and $\text{Prob}(A \cap B)$ is the probability that both events A *and* B occur. The cases where both event A and event B occur are counted in both $\text{Prob}(A)$ and in $\text{Prob}(B)$, so $\text{Prob}(A \cap B)$ must be subtracted so that such events will not be counted twice.

There are several particular cases that are worth special consideration. If events A and B cannot both occur, then $\text{Prob}(A \cap B) = 0$. Such events are called *mutually exclusive*. For example, when you roll a die, you get a 1 with probability $1/6$, a 2 with probability $1/6$, etc. You never get both a 1 and a 2. The chance of getting either a 1 or a 2 with a single roll is thus $1/6+1/6-0 = 1/3$.

If event B is a subcase of event A, then $\text{Prob}(A \cap B) = \text{Prob}(B)$. For example, the probability of rolling an even number with one roll of a die is $1/2$, and the chance of rolling a 2 is $1/6$. The chance of rolling both a 2 and an even number in one roll is $1/6$, because 2 is an even number.

If the occurrence of event A has no effect on whether or not event B will occur, then events A and B are *independent*. In this case $\text{Prob}(A \cap B) = \text{Prob}(A) \cdot \text{Prob}(B)$. For example, the probability of getting a 1 with one roll of a die is $1/6$. If you roll the die twice, the second roll is unaffected by the first roll, so the probability of rolling a 1 both times on two rolls of a die is $(1/6) \times (1/6) = 1/36$. Likewise the probability that two rolls of a die will result in a 1 the first time and a 2 the second time is $1/36$. On the other hand, the probability of rolling a 1 and a 2 on two rolls of a die is $1/36 + 1/36 = 1/18$, because there are two mutually exclusive events that lead to this outcome (rolling a 1 on the first roll and a 2 on the second, and rolling a 2 on the first roll and a 1 on the second), each of which occurs with probability 1/36.

EXERCISES

1. A *deck of cards* has 52 cards. Each card has one of 13 values and one of 4 suits. Each combination of suit and value occurs once. The suits are clubs, diamonds, hearts, and spades. Clubs and spades are black suits; diamonds and hearts are red suits. Classify the following pairs of events into the categories: (1) mutually exclusive, (2) subcase, (3) independent, or (4) none of the above. In each case a *single* card is selected at random. (A) A 3 is selected, a 4 is selected. (B) A club is selected, a red suit is selected. (C) A club is selected, a black suit is selected. (D) A club is selected, a 3 is selected.

2. A single card is selected at random from a deck of cards. Compute the probability of each of the following events. (A) A 3 is selected and a 4 is selected. (B) A club is selected and a red suit is selected. (C) A club is selected and a black suit is selected. (D) A club is selected and a 3 is selected. (E) A 3 is selected or a 4 is selected. (The convention is that "or" is used in the inclusive sense.) (F) A club is selected or a red suit is selected. (G) A club is selected or a black suit is selected. (H) A club is selected or a 3 is selected.

1.8.2 Conditional Probabilities

The *conditional probability* that event A occurs provided that event B has occurred is given by the formula

$$\text{Prob}(A|B) = \frac{\text{Prob}(A \cap B)}{\text{Prob}(B)}, \quad \text{where } \text{Prob}(B) > 0. \tag{43}$$

This can also be written as

$$\text{Prob}(A \cap B) = \text{Prob}(B) \cdot \text{Prob}(A|B) = \text{Prob}(A) \cdot \text{Prob}(B|A). \tag{44}$$

Eq. (44) can be used to express $\text{Prob}(A|B)$ in terms of $\text{Prob}(B|A)$.

Consider an urn with ten red balls and ten black balls in it. Suppose you stir the balls, draw out a random ball, and look at its color; you put the ball back and stir again; and then you draw out a second ball and look at its color. (This drawing method is called *selection with replacement.*) What is the probability that you draw a red ball both times? Since you put the first ball back and stirred, you have two independent events. The probability is $(10/20) \times (10/20) = 1/4$.

Now suppose you draw two random balls out of the same urn without putting the first one back. (This drawing method is called *selection without replacement.*) What is the probability that you draw two red balls? For the first ball there is no difference. The probability that the first ball is red is $10/20 = 1/2$. To obtain two red balls, the first ball must be red. If you draw a red ball out of the urn, you are left with an urn that has nine red balls and ten black balls. Thus the conditional probability that you draw a red ball the second time, given that you drew a red ball the first time, is $9/19$. Thus by eq. (44) the probability of drawing two red balls is $(1/2) \times (9/19) = 9/38$, which is a little less than the probability in the previous case.

Complete introductions to probability theory are given in Fisz [28, Chapters 1 and 2] and Feller [27, Chapter 1].

EXERCISES

1. If you toss a coin three times, what is the probability that you will get heads all three times?
2. If you toss a coin three times, what is the probability that you will get heads two times and tails once (in any order)?
3. If you roll two dice, what is the probability that the numbers on the two dice will total 7?
4. If you roll a die n times, what is the probability that you will get a 1 every time?
5. If you have an urn with ten red balls and ten black balls, what is the probability that you will get all red balls if you draw three times with replacement?
6. If you have an urn with ten red balls and ten black balls, what is the probability that you will get balls of both colors if you draw three times with replacement?
7. If you have an urn with ten red balls and ten black balls, what is the probability that you will get all red balls if you draw three times without replacement?
8. If you have an urn with ten red balls and ten black balls, what is the probability that you will get balls of both colors if you draw three times without replacement?

1.9 AVERAGE CASE AND WORST CASE

The algorithms considered so far have been very simple, and their running times have depended on the problems being solved in a very simple way. In each case we have been able to give a formula that relates the running time to the problem size. All problems of the same size have taken the same amount of time (except for some small effects that we have neglected). Most problems are not this simple. Various problems of a given size may take quite different amounts of time. In complex situations we are not interested in giving an exact formula for the running time in each case even when it is possible to do so; the resulting formulas would be too long to be of much use.

What is wanted are a few simple formulas that summarize the performance of the algorithm. Usually you want to know how long an algorithm runs in the worst case and how long the algorithm runs in the average case. Occasionally you want to know how much time is required in the best case and how much variation there is in the average time.

To compute the *worst-case time* for running problems of size n, consider all the problems of size n, say $P_1, P_2, \ldots, P_k$, and take the maximum of the times required for the problems. Thus if T_i is the time required to run problem P_i,

$$T_{\text{worst case}} = \max\{T_1, T_2, \ldots, T_k\} = \max_{1 \le i \le k} T_i. \qquad \textbf{(45)}$$

As you will see when we analyze the Insertion Sort Algorithm in Section 1.9.1, we do not usually use eq. (45) naïvely by first computing the time for each case and then taking the maximum over all the times. Instead we first figure out which case requires the most time, prove that that case does indeed require the most time, and then compute how much time the worst case takes. This gives the same answer as using eq. (45) directly, but it often requires less work. Some thought can often save much drudgery.

To compute the *average time* used to solve a set of problems, you must first assign a probability to each problem in the set. The probabilities are a set of numbers that say what fraction of the time each problem occurs. Let p_i be the probability that problem P_i occurs. Since the p_i are probabilities, it is necessary that

$$p_i \geq 0 \quad \text{for } 1 \leq i \leq k, \tag{46}$$

and

$$p_1 + p_2 + \cdots + p_k = \sum_{1 \leq i \leq k} p_i = 1. \tag{47}$$

Eq. (46) says that the chance that any one thing happens cannot be negative. Eq. (47) says that the chance that *something* happens is one. The average, T_{average}, of a sequence T_i is defined by

$$T_{\text{average}} = \sum_i p_i T_i, \tag{48}$$

where the sum is over all values of i such that $p_i > 0$. The average of a sequence is often called the *expected value* or *mean value* of the sequence.

Notice that an average time analysis depends not only on the algorithm and the problem size but also on the probabilities. People are usually more interested in the average time than in the worst-case time for their algorithms, but often they do not know what probabilities to use for an average-case analysis. When the worst-case time for an algorithm is about the same as the average time for some reasonable set of probabilities, the question of the exact values of the probabilities to use in an average time analysis is not of great importance. When the average time can be much less than the worst-case time, then the question of whether appropriate probabilities are used to compute the average is of paramount importance. The probabilities that are appropriate for one application of the algorithm may not be appropriate for a second application. For that reason many important algorithms have had their average time computed for several sets of probabilities. (For example, see Gonnet [5, Section 3.12.2].)

Suppose $T_i = T_{i1} + T_{i2}$. Then the average of T_i is just the sum of the averages of T_{i1} and T_{i2}. This is easy to prove from eq. (48). Using A for the average of the T_i, A_1 for the average of the T_{i1}, and A_2 for the average of the T_{i2}, we have

$$A = \sum_i p_i T_i = \sum_i p_i (T_{i1} + T_{i2}) = \sum_i p_i T_{i1} + \sum_i p_i T_{i2} \tag{49}$$

$$= A_1 + A_2, \tag{50}$$

where A_1 is the average of the T_{i1} and A_2 is the average of the T_{i2}. In general, if T_i is the sum of several parts, then the average of T_i is the sum of the averages of the parts.

To compute the best-case time of an algorithm, use eq. (45) with max replaced by min. Normally one is only interested in the best-case time when it is close to the worst-case time. If the two times are close together, the average time must be close to both of them, no matter what the values of the probabilities are (see Exercise 1).

The variation of the time is a measure of the deviation of the time for each individual problem from the average time. The variation is usually measured by the *standard deviation*, which is the square root of the variance. The *variance* is the average value of the sequence $(T_{\text{average}} - T_i)^2$. Using V for the variance and σ for the standard deviation, we have

$$V = \sum_i p_i(T_{\text{average}} - T_i)^2 \tag{51}$$

$$\sigma = \sqrt{V}. \tag{52}$$

Eq. (51) is usually fine for calculating the variance numerically. For analytical calculations it is convenient to modify eq. (51) in the following way:

$$V = \sum_i p_i(T^2_{\text{average}} - 2T_iT_{\text{average}} + T_i^2) \tag{53}$$

$$= \sum_i p_iT^2_{\text{average}} - \sum_i 2p_iT_iT_{\text{average}} + \sum_i p_iT_i^2 \tag{54}$$

$$= T^2_{\text{average}} \sum_i p_i - 2T_{\text{average}} \sum_i p_iT_i + \sum_i p_iT_i^2. \tag{55}$$

Eq. (55) can be further simplified by using $\sum_i p_i = 1$ (by the definition of probability) and $\sum_i p_iT_i = T_{\text{average}}$ (by the definition of average). This gives

$$V = T^2_{\text{average}} - 2T^2_{\text{average}} + \sum_i p_iT_i^2 \tag{56}$$

$$= \sum_i p_iT_i^2 - T^2_{\text{average}}. \tag{57}$$

Usually the sum in eq. (57) is simpler to work out than the sum in eq. (51).

When T_i is the sum of several parts, you *cannot* in general compute the variance by summing the variance of the parts. To see this, let's try the same approach that we used for the average [eq. (50)]:

$$V = \sum_i p_i(T_{\text{average1}} + T_{\text{average2}} - T_{i1} - T_{i2})^2 \tag{58}$$

$$= \sum_i p_i((T_{\text{average1}} - T_{i1}) + (T_{\text{average2}} - T_{i2}))^2 \tag{59}$$

$$= \sum_i p_i\Big((T_{\text{average1}} - T_{i1})^2 + (T_{\text{average2}} - T_{i2})^2 + 2(T_{\text{average1}} - T_{i1})(T_{\text{average2}} - T_{i2})\Big) \tag{60}$$

$$= \sum_i p_i(T_{\text{average1}} - T_{i1})^2 + \sum_i p_i(T_{\text{average2}} - T_{i2})^2 + 2\sum_i p_i(T_{\text{average1}} - T_{i1})(T_{\text{average2}} - T_{i2}). \tag{61}$$

Summing the variances of the parts to obtain the variance of the whole is okay when the rightmost sum in eq. (61) is zero, *but in general this sum is not zero.* The *covariance* of T_{i1} and T_{i2} is

$$C_{12} = \sum_i p_i(T_{\text{average1}} - T_{i1})(T_{\text{average2}} - T_{i2}). \tag{62}$$

Therefore

$$V = V_1 + V_2 + 2C_{12}, \tag{63}$$

where V_1 is the variance of T_{i1}, V_2 is the variance of T_{i2}, and C_{12} is the covariance of T_{i1} and T_{i2}. The *correlation coefficient* is defined to be

$$\rho = \frac{C_{12}}{\sigma_1\sigma_2}, \tag{64}$$

where σ_1 is the standard deviation of the T_1 and σ_2 is the standard deviation of the T_2. The correlation coefficient is always between $+1$ and -1, inclusive. If the covariance is zero, then T_1 and T_2 are *uncorrelated.* One important case that gives a covariance of zero is to have one sequence be constant. For example, if T_{i2} is a constant sequence, then $T_{\text{average2}} = T_{i2}$ for all i, and the covariance is zero.

Knuth [9, Section 1.2.10] covers average time and variance. Baase [2, Section 1.3] covers worst-case and average-time complexity.

EXERCISES

1. Prove that $T_{\text{best case}} \leq T_{\text{average}} \leq T_{\text{worst case}}$ for any algorithm.

2. When n is not restricted to be a power of 2, what is the average time for Binary Search to look up an item that is in the table? Assume that each item in the table is equally likely to be looked up. Also assume that each step in the algorithm takes constant time. Notice that when n is not a power of 2, some items require going around the loop $\lfloor \lg n \rfloor$ times, while others require $\lceil \lg n \rceil$ times.

3. Under the same assumptions as the previous problem, what is the variance of the time for Binary Search?

1.9.1 Insertion Sort

The worst-case time of the following algorithm is quite different from its best-case time. The algorithm sorts numbers. It is simple, and it is quite fast when only a few items (say five or fewer) need sorting, but, as the analysis shows, it is not a good algorithm to use when you need to sort a large number of items.

The first time through the loop, the algorithm makes sure that the first two items in the list are in correct relative order. The second time through, it inserts the third item into its correct place relative to the first two items. On the kth time, it places item $k+1$ in its correct relative position. It does this by first assigning the value of item $k+1$ to a temporary variable t and then looking at the already sorted part of the list beginning with item k. It moves each item down to the next position until it finds an item smaller than the item in t. At that point it inserts t into the list. You can think of this process as creating a hole in the list at position $k+1$ and then moving the hole up until the position for the item in t is found. The item in t is then used to fill the hole.

In Step 3 of the following algorithm we use a for loop with an added while condition. This means that the test in the while condition must be done each time through the loop; the loop is exited if either the test fails or all the values of j are used.

Algorithm 1.12 Insertion Sort: Input: A size n, $n \geq 2$, and an array of numbers X with elements x_i for $1 \leq i \leq n$. Output: A sorted permutation of X such that $x_i \leq x_j$ for $1 \leq i < j \leq n$.

Step **1.** [Each element] For $j \leftarrow 2$ to n do:

Step **2.** [Next element] Set $t \leftarrow x_j$.

Step **3.** [Find place] For $i \leftarrow j-1$ down to 1 do:

Step **4.** If $t \geq x_i$, then go to Step 8 (exiting the inner loop).

Step **5.** [Move hole up] Set $x_{i+1} \leftarrow x_i$.

Step **6.** End for i. (If we exit the loop because we have tried all values of i, x_j is the smallest element found so far.)

Step **7.** [Put first] Set $i \leftarrow 0$.

Step **8.** [Place found] Set $x_{i+1} \leftarrow t$.

Step **9.** End for j.

Step 1 is done n times. Steps 2, 8, and 9 are done $n-1$ times. The number of times Steps 3, 4, 5, 6, and 7 are done depends on the x_i. There are, however, some things that can be determined without knowing the values of the x_i. Each time Step 2 is done, Step 3 is done the same number or one more time than Step 4. Step 3 is done at least one time and at most j times each time Step 2 is done. Step 4 is done at least one time and at most $j-1$ times each time Step 2 is done. Step 5 is done the same number of or one less time than Step 4. Step 7 is done between 0 and $n-1$ times.

We will now study how often Step 3 is done. (Since Steps 4 and 5 are each done about the same number of times as Step 3, we will not investigate them separately.) To say more about how often Step 3 is done, we must know where in the array the largest x is, where the second largest is, etc. Other than that the number of times Step 3 is done does not depend on the values of the x's.

If we assume that none of the x's are equal, then we can assume without loss of generality that the x's are the integers $1, 2, \ldots, n$. Consider how many different arrangements of these numbers there are. There are n numbers that can go in the first position of an arrangement, but after the number for the first position is selected, there are only $n-1$ possibilities for the second position, and so on; thus there are $n! = n(n-1)(n-2)\cdots 2 \cdot 1$ arrangements of the data. It is not very useful to produce a formula that says how long the Insertion Sort Algorithm takes for each of the $n!$ cases individually. Such a formula would be too large to be of any use.

To proceed with the analysis, we will first determine what the worst-case behavior is. It is conceivable that no input would cause Step 3 to be executed the maximum number of times for every iteration of Step 2. You should try to determine the nature of the input that causes the worst-case behavior before reading further.

The worst case occurs when the input is sorted in backward order, so that $x_i > x_{i+1}$ for $1 \le i < n$. For this input Step 3 is done j times for each time Step 2 is done. Since we know from above that this step cannot be done any more times, this proves that this input is a worst-case input for the algorithm. The total number of times that Step 3 is done in this case is

$$n_3 = \sum_{2 \le j \le n} j = \frac{n(n+1)}{2} - 1. \tag{65}$$

In saying that the above analysis gives the worst case, one simplifying assumption has been made. Steps 3 and 4 take different amounts of time depending on whether the loop is exited by exhausting the values of i, the loop is exited by doing the go to in Step 4, or the loop goes around again. Call the time for Step 3 exiting t_{3a}, the time for Step 4 when no go to is done t_{4a}, and the time when the go to is done t_{4b}. The above analysis assumes that $t_{3a} + t_{4a} + t_5 + t_6 + t_7 \ge t_{4b}$. This is surely the case on most computers, but one would have to refer to the assembly language code for the algorithm to be entirely sure. If $t_{3a} + t_{4a} + t_5 + t_6 + t_7 < t_{4b}$, then the worst case is slightly different.

To compute the average number of times Step 3 is done, we must first decide what statistical properties are obeyed by the x_i. Here we will make one of the simplest possible assumptions: each of the $n!$ orderings of the x_i is equally likely. When the algorithm goes from Step 2 to Step 3, $j-1$ numbers have been sorted. The new number that is about to be inserted, x_j, has probability $1/j$ of being the biggest number so far, probability $1/j$ of being the second biggest, etc., with probability $1/j$ of being the smallest so far (there are j cases, all equally likely). Step 3 is done one time in the first case, two times in the second case, etc. Thus the average number of times that Step 3 is done each time it is entered from Step 2 is

$$\sum_{1\le i\le j} \frac{i}{j} = \frac{j+1}{2}. \tag{66}$$

The average of the total number of times that Step 3 is done is obtained by summing eq. (66) over j.

$$\sum_{2\le j\le n} \frac{j+1}{2} = \frac{1}{2}\sum_{2\le j\le n} j + \frac{1}{2}\sum_{2\le j\le n} 1 = \frac{n^2+3n-4}{4}. \tag{67}$$

Comparing eq. (67) with eq. (65) shows that for large n on the average Step 3 is done about half as often as in the worst case.

Insertion Sort is covered by Knuth [11, Section 5.2.1], Baase [2, Section 2.6], Horowitz and Sahni [7, Section 3.4], Sedgewick [62, Chapter 8], Reingold and Hansen [61, Section 8.1.1], and Aho et al. [59, Section 8.2].

EXERCISES

1. For n elements, what is the best-case number of times that Step 4 of Insertion Sort is done? What is an input that causes this best case time?

1.10 BIG O NOTATION AND LIMITS

When studying the amount of time required by an algorithm, we are usually most concerned about how rapidly the time increases as the amount of input becomes large. Most correct algorithms can solve small problems in a negligible amount of time, whereas the differences between algorithms on large problems can be immense.

A complex formula for the running time of an algorithm can be difficult to apply. Often a simple approximation is more useful than a complex exact answer. For example, in Section 1.2.1 we found that the time for Horner's method is

$$T = an + b \tag{68}$$

where $a = t_2 + t_3$ and $b = t_1 + t_2 + t_3 + t_4$. For large n the b term is not very important; a simpler way to express this result is to say that T is approximately equal to an, i.e.,

$$T \approx an \tag{69}$$

for large n. The difficulty with this answer is that it is a little too imprecise: we have no exact meaning for "approximately equal to".

The big O notation gives us a more precise way to express approximate answers. Suppose we have a function $g(n)$, where $g(n)$ might be the running time for an algorithm on inputs of size n. The statement

$$g(n) = O(f(n)) \tag{70}$$

means that there exist constants C and n_0 such that

$$|g(n)| \leq Cf(n) \tag{71}$$

for all $n \geq n_0$. In other words, $f(n)$ gives the functional form of an upper bound on the values of $g(n)$ for large n.

It is the convention when writing equations with big O to always require that the left side of the equation be as precise or more precise than the right side. Thus one can say $O(n) = O(n^2)$, but it would be wrong to interchange the left and right sides. These *one-way equalities* are actually statements that say it is correct to replace the left side with the right side. They are not ordinary equations, and it is not correct to replace the right side with the left.

For example, from eq. (68) we obtain

$$T = an + O(1) \tag{72}$$

by observing that the constant b can be represented by $O(1)$. To see this, notice that

$$|b| \leq b \cdot 1 \tag{73}$$

with b playing the role of C in eq. (71). (We have $b > 0$ since it is the sum of some execution times.)

In big O notation there is a tradeoff between the accuracy of a formula and its complexity. For example, we can obtain a simpler and less accurate version of the result in eq. (68):

$$T = O(n) \tag{74}$$

because, for all $n \geq 1$,

$$T = an + b \leq (a + b)n. \tag{75}$$

This matches eq. (71) with $n_0 = 1$ and $C = a + b$.

Remember that the big O notation only gives an upper limit. Using eq. (68), we could write $T = O(n^2)$. This upper limit is not as useful as the one given in eq. (74), since it is much less precise. When using big O notation, you should always give the most precise (slowest growing) upper bound you can, consistent with the goal of simplicity. Keep in mind, when you read results expressed in big O notation, that you cannot count on the bound being the most precise one possible.

Although the big O notation is precise enough for many applications, it is not adequate for comparing the performance of two algorithms. If one algorithm runs in time $O(n)$ and a second algorithm runs in time $O(n^2)$ you cannot be sure which algorithm runs faster for large n. Presumably the first does, but perhaps the second algorithm has not been analyzed very carefully. The O notation gives no information about how good an algorithm is. It just gives an upper bound on how bad it can be.

Big O notation does not say anything about the size of the implied constant C; it just says that a finite constant exists. In fact, $O(3n) = O(1{,}000{,}000n) =$

$O(n)$, which is why a constant multiplier is never used on the argument of a big O expression. It is good practice to warn others whenever the implied constant is unexpectedly large.

In Chapter 4 we discuss big O and related notations in much more detail. The material in this section is sufficient for understanding the uses of big O in the first three chapters.

We can often determine when eq. (71) is true by studying the behavior of the ratio $f(n)/g(n)$ as n becomes large. To do this we need to use limits. The *limit* of $f(n)$ as n goes to infinity,

$$\lim_{n \to \infty} f(n) = a, \tag{76}$$

means that for every $\epsilon > 0$ there exists an n_0 such that for $n \geq n_0$, $|f(n) - a| < \epsilon$. That is, by taking n large enough, we can make the value of $f(n)$ arbitrarily close to a.

If, for $f(n) > 0$,

$$\lim_{n \to \infty} \frac{|g(n)|}{f(n)} = a, \tag{77}$$

then $g(n) = O(f(n))$. {The converse is not true; if $g(n) = 1$ for even n and 0 for odd n, and $f(n) = 1$, then $\lim_{n\to\infty} |g(n)|/f(n)$ does not exist [i.e., there is no a satisfying eq. (76)], but $g(n) = O(f(n))$.}We now show that eq. (77) implies eq. (71). Eq. (77) means that, for any $\epsilon > 0$, there exists an n_0 such that for all $n \geq n_0$

$$\left| \frac{|g(n)|}{f(n)} - a \right| < \epsilon. \tag{78}$$

Choose $\epsilon = 1$, and let n_0 be large enough to make eq. (78) true for this value of ϵ. Then, for $n \geq n_0$,

$$\left| \frac{|g(n)|}{f(n)} - a \right| < 1, \tag{79}$$

$$\frac{|g(n)|}{f(n)} - a < 1, \tag{80}$$

$$|g(n)| < (1 + a) f(n). \tag{81}$$

So eq. (71) is satisfied with $C = 1 + a$.

When the limit a in eq. (77) is positive and finite, and $g(n) \geq 0$ for large x, we have that $g(n) = O(f(n))$ and $f(n) = O(g(n))$. When a is zero, only the first formula is true.

The previous definition of limit applies when the variable approaches infinity. The limit when the variable approaches a finite value,

$$\lim_{x \to b} f(x) = a \tag{82}$$

means that for any $\epsilon > 0$ there exists a $\delta > 0$ such that $|x - b| < \delta$ implies $|f(x) - a| < \epsilon$. That is, by making x close enough to b we can make $f(x)$ arbitrarily close to a.

A function is *continuous* at b if

$$\lim_{x \to b} f(x) = f(b). \tag{83}$$

Most common functions are continuous. Sums, differences, and products of continuous functions are continuous. Quotients of continuous functions are continuous except where the denominator is zero. Raising a positive constant to a continuous function gives a continuous function, and the logarithm of a continuous positive function is continuous.

Limits of an expression involving a continuous function, $f(x)$, can often be worked out by plugging in the limit value [i.e., $f(b)$] when $x \to b$ and knowing that $1/x \to 0$ when $x \to \infty$. More complicated situations arise when indeterminate expressions (such as ∞/∞) result from substituting the limiting value. [This will frequently be the case when using eq. (77).] In such cases L'Hôpital's rule can often be applied. This rule states that if $\lim_{x \to b} f(x) = \lim_{x \to b} g(x) = \infty$, and $f(x)$ and $g(x)$ are differentiable, then

$$\lim_{x \to b} \frac{f(x)}{g(x)} = \lim_{x \to b} \frac{f'(x)}{g'(x)}. \tag{84}$$

L'Hôpital's rule also works when $x \to \infty$. It is often helpful because the derivative of a function is usually simpler than the function itself.

For example, suppose we wish to determine which grows faster, x^2 or e^x. Using L'Hôpital's rule gives

$$\lim_{x \to \infty} \frac{x^2}{e^x} = \lim_{x \to \infty} \frac{2x}{e^x} = \lim_{x \to \infty} \frac{2}{e^x} = 0, \tag{85}$$

so $x^2 = O(e^x)$ [and $e^x \neq O(x^2)$].

Introductions to big O notation and asymptotics are contained in Knuth [9, Section 1.11.1], de Bruijn [25, Chapter 1], Aho et al. [59, Section 1.4], and Horowitz and Sahni [7, Section 1.4]. Also, Chapter 4 of this book has a much more complete discussion of asymptotics.

EXERCISES

1. Using big O notation, say how long the Matrix Multiplication Algorithm takes. Give your answer to two different accuracies.

2. Using big O notation, say how long Insertion Sort takes in the worst case and in the average case. Give answers both to accuracy $O(n^2)$ and $O(n)$. Explain how you were able to get a $O(n)$ answer even though we have not provided an analysis of the average time required for Step 4 of Insertion Sort.

3. The big O notation expresses a relationship between the growth rates of functions. Thus, $n^{1.5} = O(n^2)$ means that $n^{1.5}$ grows no faster than n^2 for large values of n. The leftmost column of Table 1.1 is a list of functions. For each pair of functions f and g in that list, determine whether $f(n) = O(g(n))$ or $g(n) = O(f(n))$. (For some pairs both relations hold.) Hint: Use L'Hôpital's rule in most cases.

4. Prove that, when the limit in eq. (77) is positive and finite, and $g(n) \geq 0$ for large n, $f(n) = O(g(n))$ and $g(n) = O(f(n))$.

1.11* RELATIONS AND ORDERINGS

The *relation* R, $R \subseteq A \times B$, is a subset of pairs of elements where the first element is in set A and the second element is in set B. Sets A and B may be the same set, or they may be different sets. The *characteristic function* of the relation R is the function $f_R(a,b)$ which is *true* if the pair $[a,b]$ is in R and which is *false* otherwise. We will not be distinguishing between a relation and its characteristic function, and so we often use the word "relation" when referring to the characteristic function of the relation.

Relations can be written using several notations. If R is a relation on a set S, and x and y are elements of S, then $R(x,y) = true$ may be written as $R(x,y)$, as xRy, as $x \underset{R}{\rightarrow} y$, or (when R is known from context) as $x \rightarrow y$. Some well known relations are *equals*, *less than*, and *greater than or equal to*, but there are many other interesting relations. For example, "child-of", where child-of(x,y) is *true* if and only if x is a child of y, is a relation.

A relation is *transitive* if, whenever $x \rightarrow y$ and $y \rightarrow z$, then also $x \rightarrow z$ for all x, y, z. That is, $(x \rightarrow y) \wedge (y \rightarrow z)$ implies $(x \rightarrow z)$. The relations *equals*, *less than*, and *greater than or equal to* are all transitive. The relation "child-of" is not.

A relation is *symmetric* if, whenever $x \rightarrow y$, then also $y \rightarrow x$ for all x, y. That is, $x \rightarrow y$ implies $y \rightarrow x$. The relation *equals* is symmetric; *less than*, *greater than or equal to*, and "child-of" are not.

A relation is *reflexive* if $x \rightarrow x$ for all x. The relations *equals* and *greater than or equal to* are reflexive; *less than* and "child-of" are not.

Often we are interested in a relation that is a *minimum extension* of a given relation: the new relation is *true* for all the pairs where the original relation is *true* and it is *true* for just enough additional pairs to make some desired property hold. We will now consider several relations that are obtained from other relations.

The relation $x \overset{n}{\rightarrow} y$ is the relation that is *true* if and only if there is a chain of elements x_0, x_1, ..., x_n in S with $x = x_0$, $y = x_n$, where $x_0 \rightarrow x_1 \rightarrow \cdots \rightarrow x_n$. Informally $x \overset{n}{\rightarrow} y$ is *true* if and only if you can go from x to y using exactly n steps with the relation $\rightarrow$.

The *transitive closure* of a relation is the relation that is *true* if and only if there exists an $n \geq 1$ such that $x \overset{n}{\rightarrow} y$. The transitive closure of the relation $\rightarrow$ is written as $\overset{+}{\rightarrow}$. Informally, $x \overset{+}{\rightarrow} y$ is *true* if and only if you can go from x to y in one or more steps with the relation $\rightarrow$. The transitive closure of a relation is transitive, whether or not the original relation is. It is the minimum extension of the original relation that is transitive.

The *reflexive transitive closure* of $\rightarrow$, written $\overset{*}{\rightarrow}$, is the relation that is *true* if and only if there exists an $n \geq 0$ such that $x \overset{n}{\rightarrow} y$. The relation $x \overset{*}{\rightarrow} y$ is *true* if and only if you can go from x to y in zero or more steps using $\rightarrow$. The reflexive transitive closure of any relation is both reflexive and transitive. It is the minimum extension of the original relation that is both reflexive and transitive. (Some authors use transitive closure to mean reflexive transitive closure.)

The *symmetric reflexive transitive closure* (*symmetric closure* for short) of $\rightarrow$, written $\overset{*}{\leftrightarrow}$, is defined so that $x \overset{*}{\leftrightarrow} y$ is *true* if and only if there is a chain of elements $x_0, x_1, \ldots, x_n$ in S with $x = x_0$, $y = x_n$, and $n \geq 0$ where either $x_i \rightarrow x_{i+1}$ or $x_{i+1} \rightarrow x_i$ for $0 \leq i < n$. In other words $x \overset{*}{\leftrightarrow} y$ is *true* when there is a path from x to y disregarding the direction of the arrow. The symmetric closure of any relation is symmetric, transitive, and reflexive. It is the minimum extension of the original relation that has these properties.

The transitive closure of the "child-of" relation is the "descendent-of" relation. The relations *less than*, *greater than or equal to*, and *equals* are all transitive, so each one is the same as its transitive closure. The reflexive transitive closure of *less than* is *less than or equal to.* The relations *greater than or equal to* and *equals* are their own reflexive transitive closures. The relation *equals* is also its own symmetric closure. The symmetric closure of *less than* is a relation that is always *true.*

A relation $R(x,y)$ is an *equivalence* relation if it is reflexive, symmetric, and transitive. *Equals* is an equivalence relation. Another equivalence relation is the relation $R_n(x,y)$ where $R_n(x,y) = true$ if and only if $x \bmod n = y \bmod n$. [This last relation is called *congruence* modulo n and written $x \equiv y \pmod{n}$. See Section 1.4.2.]

The symmetric closure of any relation is an equivalence relation. Sometimes this relation is not very interesting; for example, the symmetric closure of the *less than* relation yields a relation that is always *true.* Presumably the same thing happens with the "child-of" relation. An interesting relation to take the symmetric closure of is the relation $\rightarrow$ where $x \rightarrow y$ means that the formula x can be simplified to the formula y in one step (using some particular set of rules for simplifying formulas). The symmetric closure of $\rightarrow$ relates those formulas that can be shown to be equivalent under the set of simplification rules.

An equivalence relation can be used to partition a set S into *equivalence classes.* Two elements x and y are in the same equivalence class with respect to the equivalence relation R if and only if xRy. All elements in an equivalence class are related to each other, and none are related to an element in any other class.

Orderings are a particularly important kind of relation. A *partial order* on a set S is a relation $\prec$ that is transitive and irreflexive (no element is related to itself). (Some authors define partial orders that are reflexive rather than irreflexive. They usually use $\preceq$ for their relation.)

An *order* (or *total order*) is a partial order where, for every two elements x and y in S, either $x = y$, $x \prec y$ or $y \prec x$. That is, either x and y are the same element, x precedes y, or y precedes x. Although some sets, such as the integers, have a standard ordering defined on them, any relation that satisfies the above definition is an ordering. A set can be ordered many different ways.

A set S is *well-ordered* if it is ordered and, for any nonempty subset A of S, there is an element $x \in A$ with $x \preceq y$ for all $y \in A$, where $x \preceq y$ if $x \prec y$ or $x = y$. Well-ordering is important because it ensures that there is no infinite sequence of elements such that $x_1 \succ x_2 \succ x_3 \succ \cdots$ (such a sequence is called an *infinite descending chain*). A well-ordered set can contain no infinite descending chain because such a chain has no least element.

It is important to distinguish between a set with an *infinite chain*, where there is a single chain of infinite length, and sets with *unbounded* chains, where the length of each chain is finite, although there is no upper bound on the lengths of the chains. A good example of a set of unbounded chains is the set of descending chains that can be formed with the positive integers ordered with the usual *less than* relation. For any particular integer n there is a longest descending chain starting with n $(n > n-1 > n-2 > \cdots > 1)$, which has length n. There are no infinite descending chains, even though there is no fixed upper bound on the length of chains.

Relations that have no infinite descending chains are interesting in their own right, even if they are not orderings. Such relations are called *Noetherian*, after the mathematician Emmy Noether.

A typical example of a set with infinite descending chains arises when you order the positive binary fractions with the usual *less than* relation. Here there is no longest descending chain from any fraction. Consider the fraction 0.1 and the chain $0.1 > 0.01 > 0.001 > 0.0001 > \cdots$. The chain is of infinite length. This, of course, is closely related to the fact that the set of numbers of the form $\{0.0^k1 | k \geq 0\}$ has no least element.

Although the positive integers under their usual order form a well-ordered set, the set of all integers does not form a well-ordered set under its usual ordering. There is no least integer. For example, from the integer 2 there is the infinite descending chain $2 > 1 > 0 > -1 > -2 > \cdots$. It is, however, easy to construct a well-ordering for the integers (see Exercise 1). Pairs of positive integers are more interesting. One way to well-order pairs of integers is to use *lexicographic order*, where

$$(m_1, n_1) \prec (m_2, n_2) \quad \begin{cases} \text{if } m_1 < m_2, \\ \text{if } m_1 = m_2 \text{ and } n_1 < n_2. \end{cases} \tag{86}$$

Another well-ordering is to use $(m_1, n_1) \prec (m_2, n_2)$ if $m_1 + n_1 < m_2 + n_2$ or if $m_1 + n_1 = m_2 + n_2$ and $m_1 < m_2$. There are many other well-orderings of pairs of positive integers.

If you lexicographically order the pairs of integers (m, n), where $1 \leq m$, $1 \leq n$, the ordering is $(1,1)$, $(1,2)$, $(1,3)$, ..., $(2,1)$, $(2,2)$, $(2,3)$, ..., $(3,1)$, $(3,2)$, $(3,3)$, ... An infinite number of points of the form $(1, n)$ are in the order before $(2,1)$. There is, however, no infinite descending chain. Consider any descending chain starting from $(2,1)$. The second element on the chain has the form $(1, n)$ for some n, and the rest of the chain has length at most $n+1$. Thus any particular descending chain from $(2,1)$ has a bounded length, even though there is no single bound on the length of every descending chain from $(2,1)$.

An ordering on pairs of integers that is not a well-ordering is $(m_1, n_1) \prec (m_2, n_2)$ if $m_1/n_1 < m_2/n_2$. This is the ordering used to order fractions. To see that it is not a well-ordering, consider the sequence $(2,1)$, $(12,11)$, $(112,111)$,

$(1112, 1111)$, ... This descending chain from $(2, 1)$ has no bound on its length and the sequence has no least element.

The ordering $(m_1, n_1) \prec (m_2, n_2)$ if $m_1 + n_1 < m_2 + n_2$ or if $m_1 + n_1 = m_2 + n_2$ and $m_1 < m_2$ is much simpler than the two previous ones. It orders the pairs of integers (m, n), where $1 \leq m$, $1 \leq n$, in the order $(1, 1)$, $(1, 2)$, $(2, 1)$, $(1, 3)$, $(2, 2)$, $(3, 1)$, ...

Another interesting set to order is the set of finite strings over an alphabet. The ordinary *dictionary order* is given by

$$\begin{aligned} a_1a_2a_3\ldots a_i \prec b_1b_2b_3\ldots b_j \quad &\text{if, for some } k \geq 0, \\ &a_1 = b_1,\ a_2 = b_2,\ \ldots,\ a_{k-1} = b_{k-1},\ a_k < b_k \text{ or} \\ &a_1 = b_1,\ a_2 = b_2,\ \ldots,\ a_i = b_i, \text{ and } i < j. \end{aligned} \qquad \textbf{(87)}$$

This is *not* a well-ordering. Suppose $a < b$, and consider the sequence of strings b, ab, aab, $aaab$, ..., which is an infinite descending chain.

The mathematical properties of equivalence relations are discussed in Birkhoff and Mac Lane [22, Section 1.11]. Well-ordering is covered by Birkhoff and Mac Lane [22, Section 1.4] and Knuth [9, Exercise 1.2.1-15].

EXERCISES

1. Show that congruence modulo n is an equivalence relation. Show that the integers congruent to i modulo n form an equivalence class under this relation.

2. Suppose we let [0], [1], [2], [3], and [4] denote the equivalence classes of integers congruent (modulo 5) to 0, 1, 2, 3, and 4, respectively. The properties discussed in Exercises 1, 2, and 3 of Section 1.4 show that the operations of addition, subtraction, multiplication, and division are well-defined on these equivalence classes. That is, since $x \equiv 1 \pmod 5$ and $y \equiv 2 \pmod 5$ implies $x + y \equiv 3 \pmod 5$, we may write $[1] + [2] = [3]$. A complete addition table for these equivalence classes is given by

+	[0]	[1]	[2]	[3]	[4]
[0]	[0]	[1]	[2]	[3]	[4]
[1]	[1]	[2]	[3]	[4]	[0]
[2]	[2]	[3]	[4]	[0]	[1]
[3]	[3]	[4]	[0]	[1]	[2]
[4]	[4]	[0]	[1]	[2]	[3]

Construct the multiplication table for the equivalence classes.

3. Construct the addition and multiplication tables for integers modulo 6 (see the previous exercise for the definitions). Notice that is possible for two nonzero integers to have a product that is equal to zero. The effect of this anomaly is that division is not well defined over the integers modulo 6.

4. Give a well-ordering for the integers. Hint: Consider absolute value.

5. Give a well-ordering for the set of finite strings over a finite alphabet. Hint: Consider the length of the string.

6. Give a well-ordering for three-tuples of positive integers.

1.12 PROOF BY INDUCTION

Induction is used to prove a whole sequence (usually infinite) of statements. The basic idea is to first assign an order to the statements, so that the statements are called P_1, P_2, ... Then to prove P_i, you assume that your series of statements is true for all smaller indices and prove the statement for the current index. If you can do this for all values of the indices, then the entire series of statements is true. Notice that in proving the first statement in the series, you may make no assumptions at all. To prove the second statement, you are permitted to assume the first statement is true. To prove the third, you may assume that the first *two* statements are true, etc.

Of course, you don't prove your statements one at a time; that would take forever if you had an infinite sequence of statements. Rather you give a proof that works for all indices. Often the only assumption you need for case i is that P_{i-1} is true. But be careful. You *cannot* assume that P_{i-1} is true for all i; you can only assume it is true when $i > 1$. You must prove P_1 without making any special assumptions. In a proof by induction, a case that is proved without any assumptions about the preceding cases is called a *base case*.

Proof by induction is a very powerful way of constructing proofs. It does have a major limitation: to apply it, you must first know the answer. Once you know the answer, you can prove the answer is correct by using proof by induction. Many of the methods you might use to come up with answers have the characteristic that they always provide correct answers. We all prefer to use such methods, and when you get your answer from such a method, it would be redundant to use proof by induction on it. There are, however, several powerful methods for obtaining answers that do not provide proofs. One of these is guessing. Some people are very good at guessing. If you have a good guess, then you can use proof by induction to show that your guess is correct. A refinement of guessing is to compute the answer to your problem for several small cases and to look at the answers to see if any pattern appears. If you find a pattern, then you need a proof method to show that your pattern does not break down for some case you did not calculate. Proof by induction should be considered in such cases. This is the way many important results are obtained. (See Exercise 3.2.2–1.)

To practice proof by induction, let's re-prove eq. (41). Our statement P_n is

$$\sum_{1 \le i \le n} i = \frac{n(n+1)}{2}. \tag{88}$$

Assume that it is true for all $n < m$. In particular we will need to assume that it is true for $n = m - 1$. We want to show that

$$\sum_{1 \le i \le m} i = \frac{m(m+1)}{2}. \tag{89}$$

Now we know that $\sum_{1\le i\le m} a_i = a_m + \sum_{1\le i\le m-1} a_i$ for any sequence of a_i, so

$$\sum_{1\le i\le m} i = m + \sum_{1\le i\le m-1} i. \tag{90}$$

Now the right side of eq. (90) has a sum that just goes up to $m-1$, so we can use our induction hypothesis that $\sum_{1\le i\le n} i = \frac{n(n+1)}{2}$ for $n < m$ to obtain

$$\sum_{1\le i\le m} i = m + \frac{(m-1)m}{2}. \tag{91}$$

Adding together the terms on the right side gives

$$\sum_{1\le i\le m} i = \frac{m(m+1)}{2}, \tag{92}$$

which is what we needed to prove. But wait; we are not done. Our assumption that P_{m-1} is true cannot be used for $m = 1$, so we still need to show that

$$\sum_{1\le i\le 1} i = 1. \tag{93}$$

But this is obvious: the only term on the left side is the $i = 1$ term, which is 1.

Sometimes you can use the basic approach of proof by induction even when you have just a rough idea of the nature of the answer. Write your guess for the answer as a function that contains some unknown parameters. As you go through your proof by induction, you will probably obtain some restrictions on the values of the parameters. If you are fortunate, you will end up with a precise answer and a proof. See Section 3.7.1 for an example of this technique.

Proof by induction is covered by Birkhoff and Mac Lane [22, Section 1.5], Knuth [9, Section 1.2.1], Baase [2, Section 2.8], Cohen [23, Appendix], Wand [58, Chapter 2], and Steward [55, Chapter 3]

EXERCISES

1. Prove that $\sum_{1\le i\le n} i^2 = (2n^3 + 3n^2 + n)/6$.

2. Prove that $\sum_{0\le i\le n} x^i = (x^{n+1} - 1)/(x-1)$ for $x \ne 1$.

3. Suppose we have an alphabet f_1, f_2, ..., f_M of operators with degrees d_1, d_2, ..., d_M. Each degree is a nonnegative integer. Let N be the largest degree. The degree zero operators play the role of constants. A *Polish prefix* word α over this alphabet either is a degree zero operator or is of the form $f_j\alpha_1\alpha_2\ldots\alpha_{d_j}$, where α_1, α_2, ..., α_{d_j} are words. For example, consider the alphabet T, F, $\neg$, $\vee$, $\wedge$ with corresponding degrees 0, 0, 1, 2, 2. Some sample words are T, $\wedge TF$, $\vee\wedge\neg TFT$, and $\neg\wedge T\vee TF$. Some sample nonwords are TF, $\wedge\vee TF$, and $\vee\wedge\neg TFTF$. In this example we can interpret T and F as the logical constants *true* and *false*; we can interpret $\neg$, $\vee$, and $\wedge$ as the logical operators *not*, *or*, and *and*; and we can interpret words as logical expressions in prefix form. Let $n_j(\alpha)$

be the number of symbols of degree j in word α. [We will drop the (α) when α is understood from context.] Prove that for any α,

$$n_0 + n_1 + n_2 + \cdots + n_N = 1 + 0n_0 + 1n_1 + 2n_2 + \cdots + Nn_N,$$

i.e.,

$$n_0 = 1 + n_2 + 2n_3 + \cdots + (N-1)n_N.$$

Hint: This can be done by induction where the induction variable is the length of the word.

1.12.1* Induction with Well-Orderings

So far we have seen examples of induction proofs where the order of the statements to be proved was obvious, corresponding to the positive integers. Induction can be used in a more general setting, as long as a suitable ordering can be placed on the statments. To use induction, the ordering must be a well-ordering (or at least a Noetherian relation). It must be possible to prove certain cases directly (the *base* cases) and then show that each other case can be derived from the base cases by a finite descending chain of intermediate steps. Two examples of such proofs are given in the following sections.

In this book, most proofs by induction are concerned with proving that the running time for an algorithm is given by some formula. For complex problems of this type, it is often helpful to divide your efforts into five stages.

1. Develop recurrence equations for your problem. Find a way to relate the time for some values of the parameters to the time for other values of the parameters. In the remaining cases, find a way to obtain the time directly. These direct cases will be the base cases for your proof by induction. For example, eq. (90) relates the value of a sum from one to m to the value of a sum from one to $m-1$. Eq. (93) is an example of a case that is proved directly.

2. Use the equations developed at Step 1 to compute solutions for small cases of your problem. If your equations are complete enough for use in proof by induction, they will be suitable for calculating small cases of your problem. If you cannot use your equations and base cases to calculate your result in all small cases, go back to Step 1 and develop a more complete set of equations. If you are able to calculate values for all the small cases you try, then check that the values you calculate are correct. (Go back and correct errors if necessary.)

3. Guess at the general solution of your problem. Eq. (88) is an example of a proposed general solution. Be sure that your proposed solution agrees with the results that you calculated in Step 2.

4. Plug your proposed solution into the equations you found at Step 1. If any contradictions arise, your proposed solution is false. Otherwise we will say that your proposed solution is *compatible* with your problem. We do not say it is consistent, because until we are successful at Step 5 of this procedure, we do not know whether some other way of plugging in might lead to a contradiction. Eqs. (91, 92) are examples of plugging in the proposed solution.

5. Check that the order in which you did things is consistent with proof by induction. In other words, the way in which you plugged in at Step 4 must correspond to an ordering with no infinite descending chains. If not, redo the earlier steps in a way that will correspond to a well-ordering.

Sometimes your problem is stated in such a way that Step 1 is already done for you. The next section has such a problem. Sometimes you are also given a proposed solution, in which case you can skip Steps 2 and 3.

When actually carrying out an induction proof, people usually mix the steps somewhat. The amount of mixing depends on the person and the problem. Keeping the steps separate reduces the number of things you need to think about at one time, but it increases the chance that you will need to repeat some of your work. Often people do Step 5 before Step 4. Any correct approach that works well for you is fine.

The above approach is designed for discovering proofs by induction. Once you have found your proof, it is usually easier for someone else to follow if you first explain your inductive hypothesis, then give the ordering you use, and finally give the details of the proof of the inductive hypothesis (i.e., the proof that for any case the inductive hypothesis is true under the assumption that it is true for all previous cases). The purpose of a final write-up of a proof is to convince someone else that the proof is correct. Such write-ups are sometimes far removed from the considerations that originally led to the proof. The main purpose of the proofs in this book is to show you how to discover your own proofs, so many of them are given in the order that is most useful to someone trying to discover the proof in the first place.

1.12.1.1* *Binary Merge—Part 1*

The algorithm that we use for our example is the Binary Merge Algorithm, which merges two files. It uses a combination of the ideas of ordinary merging, which works well when the two files have the same size (number of records), and of binary search, which works well when one of the files is of size 1. The Binary Merge Algorithm is particularly useful when both files are large, but one is much larger than the other.

The basic idea of the algorithm is as follows. Suppose the files we are merging are called A and B, where A is currently the larger file. (Which file is larger may change back and forth during execution of the algorithm.) If there are m elements in A and n elements in B, then a position about m/n from the beginning of A is a good place to consider putting the first element of B. If the first element of B goes before the $\lfloor m/n \rfloor^{\text{th}}$ element of A, then binary search can quickly determine where it goes. The algorithm can output the part of A prior to the place where the element goes and the first element of B. If the first element of B goes after the $\lfloor m/n \rfloor^{\text{th}}$ element of A, then the first $\lfloor m/n \rfloor$

elements of A can be output. In either case the algorithm then continues in the same way, switching the roles of A and B if B is now larger.

We are interested in Binary Merge because it leads to several illustrations of proof by induction.

Algorithm 1.13 Binary Merge: Input: File A with elements a_1 through a_m and file B with elements b_1 through b_n. Each file is sorted in increasing order. The first element in each file is element 1. When elements are output from a file, you should think of the remaining elements as being renumbered so that once again the first remaining element is element 1. Output: File C in sorted order with all the elements from A and B. Functions: Binarysearch(F,r,q) where F is the name of a sorted file, r is the length of the part of the file being searched, and q is a query. The result i of Binarysearch is the place to put the query into the file so as to maintain the sorted order $(F_i < q \leq F_{i+1})$ with $1 \leq i \leq r$. The function finds this place with $\lceil \lg r \rceil$ comparisons. (We do not give Binarysearch here because it is almost the same as Algorithm 1.6, and even more like the algorithm for Exercise 1.5.1–1.)

Step **1.** While both $m > 0$ and $n > 0$ do:

Step **2.** If the relation between m and n is:

Step **3.** $m > n$, then set $t \leftarrow \lfloor \lg(m/n) \rfloor$. (Steps 3–5 and 6–7 are the same except that the roles of A and B and of m and n are interchanged.) If the relation between b_1 and a_{2^t} is:

Step **4.** $b_1 \geq a_{2^t}$, then set $m \leftarrow m - 2^t$, and output to file C: $a_1, a_2, \ldots, a_{2^t}$. Go to Step 9.

Step **5.** $b_1 < a_{2^t}$, then set $i \leftarrow$ Binarysearch$(A, 2^t, b_1)$. Output to file C: $a_1, a_2, \ldots, a_i, b_1$ (i may be zero). Set $n \leftarrow n - 1$, $m \leftarrow m - i$. Go to Step 9.

Step **6.** $m \leq n$, then set $t \leftarrow \lfloor \lg(n/m) \rfloor$. If the relation between a_1 and b_{2^t} is:

Step **7.** $a_1 \geq b_{2^t}$, then set $n \leftarrow n - 2^t$, and output to file C: $b_1, b_2, \ldots, b_{2^t}$. Go to Step 9.

Step **8.** $a_1 < b_{2^t}$, then set $i \leftarrow$ Binarysearch$(B, 2^t, a_1)$. Output to file C: $b_1, b_2, \ldots, b_i, a_1$ (i may be zero). Set $m \leftarrow m - 1$, $n \leftarrow n - i$.

Step **9.** End while.

Step **10.** If $m > 0$, then output to file C: $a_1, a_2, \ldots, a_m$. If $n > 0$, then output to file C: $b_1, b_2, \ldots, b_n$.

The "go to Step 9" in Steps 4, 5, and 7 results in jumping to Step 9, which loops back to Step 1 to perform the next iteration of the while loop.

We will now give a proof by induction that in the worst case the number of comparisons done by Binary Merge is given by

$$C_{m,n} = (1+t)m + \left\lfloor \frac{n}{2^t} \right\rfloor - 1, \tag{94}$$

where $t = \lfloor \lg(n/m) \rfloor$ and $1 \leq m \leq n$. The proof has two parts, each of which uses proof by induction. Since the second part of the proof provides a more elementary example of proof by induction than the first part, we will give the second part in this section and the first part in the following section. The proofs are presented in a way that corresponds to the steps in Section 1.12.1. This illustrates the procedure you might use to discover a proof. (A formal write-up would be organized differently, as we explained earlier.) For now let's assume that the number of comparisons obeys the equations

$$C_{m,2n} = C_{m,n} + m \qquad \text{for } m \leq n \tag{95}$$

$$C_{m,2n+1} = C_{m,n} + m \qquad \text{for } m \leq n \tag{96}$$

$$C_{m,n} = m + n - 1 \qquad \text{for } m \leq n < 2m. \tag{97}$$

We will prove later that these equations give the number of comparisons used by Binary Merge; for now we will just prove that eq. (94) is the solution of these equations.

Let's see whether eq. (94) is compatible with each equation [eqs. (95–97)]. If we find any of eqs. (95–97) that are not compatible with eq. (94), then eq. (94) is definitely not a solution of eqs. (95–97). Consider eq. (95) first: if we use eq. (94) on the right side of eq. (95), we get

$$(1+t)m + \left\lfloor \frac{n}{2^t} \right\rfloor - 1 + m = (2+t)m + \left\lfloor \frac{n}{2^t} \right\rfloor - 1. \tag{98}$$

Let $n' = 2n$ and $t' = \lfloor \lg(n'/m) \rfloor = 1 + \lfloor \lg(n/m) \rfloor$. Then eq. (98) can be written as

$$(1+t)m + \left\lfloor \frac{n}{2^t} \right\rfloor - 1 + m = (1+t')m + \left\lfloor \frac{n'}{2^{t'}} \right\rfloor - 1. \tag{99}$$

Since this is what eq. (94) predicts for $C_{m,2n}$, eq. (94) is compatible with eq. (95).

Now let's see if eq. (94) is compatible with eq. (96). Applying eq. (94) to the right side, we get

$$(1+t)m + \left\lfloor \frac{n}{2^t} \right\rfloor - 1 + m = (2+t)m + \left\lfloor \frac{n}{2^t} \right\rfloor - 1. \tag{100}$$

Let $n' = 2n+1$ and $t' = \lfloor \lg(n'/m) \rfloor = 1 + \lfloor \lg(n/m) \rfloor$. (Notice that $\lfloor \lg((2n+1)/m) \rfloor = \lfloor \lg(2n/m) \rfloor$.) Thus eq. (100) can be written as

$$(1+t)m + \left\lfloor \frac{n}{2^t} \right\rfloor - 1 + m = (1+t')m + \left\lfloor \frac{n'}{2^{t'}} \right\rfloor - 1. \tag{101}$$

Since this is what eq. (94) predicts for $C_{m,2n+1}$, eq. (94) is compatible with eq. (96).

Now let's consider eq. (97). Replacing the left side of eq. (97) with the right side of eq. (94), we get

$$m + n - 1 = m + n - 1 \tag{102}$$

because $t = 0$ for $m \leq n < 2m$. Thus eq. (94) agrees with all three equations. So far we have proved that eq. (94) generates a set of numbers that are compatible with eqs. (95–97).

To finish the proof, we must show that every pair of integers (m, n) with $1 \leq m \leq n$ is related by a finite chain of steps to pairs for which the theorem can be proved directly. In our proof, the base cases (those for which the theorem is proved directly) are those for which $m \leq n < 2m$: the cases from eq. (97). Eq. (95) and eq. (96) relate the cases for larger n (for which $n > 2m$) to smaller cases: if n is even, then C_{mn} is obtained from $C_{m,n/2}$, while if n is odd, it is obtained from $C_{m,(n-1)/2}$. If $n > 2m$ [so that eq. (95) or eq. (96) must be applied], then $n/2 > m$ [and $(n-1)/2 > m$ if n is odd]. We can therefore use the partial well-order defined by $(m, n_2) \prec (m, n_1)$ if and only if $n_2 < n_1$.

When $m \leq n \leq 2m$, the theorem is proved directly. When $n > 2m$, assume that the theorem is true for all n' with $m \leq n' < n$. Then eq. (95) holds for $(m, n/2)$ (if n is even), and eq. (96) holds for $(m, (n-1)/2)$ (if n is odd). The partial order we have defined is a partial well-order [the number of elements less than (m, n) is no more than $n + 1$, which is finite], so eq. (94) is true for all pairs (m, n) with $m \leq n$. Notice that, although we could have extended the partial order to a total well-order, there is no need to; the values of C_{m_1,n_1} and C_{m_2,n_2} do not depend on each other when $m_1 \neq m_2$. This finishes the proof that eqs. (95–97) imply eq. (94).

1.12.1.2* *Binary Merge—Part 2*

We still must show that eqs. (95–97) give the number of comparisons used by Binary Merge. Again we use proof by induction. This proof is much longer than the one in the previous section because of the large number of cases that arise in the proof. If you don't want to see all the details, you can now skip the rest of this section. On the other hand, if you want to know why eqs. (95–97) give the running time for the Binary Merge Algorithm, continue reading. Just be prepared for a lot of detailed calculations.

From the algorithm it is clear that in the worst case

$$C_{m,n} = 1 + \max\{C_{m,n-2^t}, t + C_{m-1,n}\} \qquad \text{for } m > 0 \quad \text{and } n \geq m, \tag{103}$$

where $t = \lfloor \lg(n/m) \rfloor$. Eq. (103) has a max operation in it because the worst-case data can force the algorithm into whichever case is going to take more time. The first term in the maximum is the number of additional comparisons that the algorithm will use if at Step 6 it takes the greater than or equal branch (Step 7). The second term is the number of additional comparisons the algorithm will use if it takes the less than branch (Step 8), and only one element gets written out as a result of that step. Notice that $C_{0,n}$ and $C_{m,0}$ are both zero. It is clear from the algorithm that $C_{m,n} = C_{n,m}$. (Compare Steps 3–5 with 6–8.)

We need to show that eq. (103) and eqs. (95–97) have the same solution. We will do this by induction. First we will show that eqs. (95–97) are compatible with eq. (103). This takes care of the first part of the proof. While checking the most general case, we will find that we need to assume that the formula holds for some special cases. After checking the general case, we will do the special cases, which in turn will lead to more special cases. When all the special cases have been checked for compatibility, we will then check that we can order the cases to form an induction proof.

Notice as we go through the first part of the proof that when we are showing the compatibility of the formulas for indices (m_0, n_0), we only assume the proposed solution for points of the form (m, n) where $m \leq m_0$ and $n \leq n_0$ and either $m \neq m_0$ or $n \neq n_0$. If we think of the rectangle of integer points in the plane with one corner at (m_0, n_0) and the other at the origin, we assume the proposed solution only at points in the rectangle other than (m_0, n_0). Thus, although we save the detailed induction argument until after we establish compatibility, we have this strategy for well-ordering the cases in mind as we go through the first part.

We have two general formulas, eq. (95) and eq. (96). One applies when n is even and the other when n is odd. Assume n is even, and rewrite the right side of eq. (103) using eq. (95). This gives

$$C_{m,n} = 1 + \max\{C_{m,n-2^t}, t + C_{m-1,n}\} \tag{104}$$

$$= 1 + \max\{C_{m,n/2-2^{t-1}} + m, t + C_{m-1,n/2} + m\}. \tag{105}$$

Under what conditions are the transformations that lead to eq. (105) valid?

In order to use eq. (95) to evaluate $C_{m,n-2^t}$, it is necessary for $n - 2^t$ to be even. It is also required that $m \leq n/2 - 2^{t-1}$ [the condition on eq. (95) says that one half of the second index must be greater than or equal to the first index for the formula to apply]. This requirement can be written as $2m \leq n - 2^t$. Clearly this means that $2m \leq n$, whatever the value of t is. But then, since $t = \lfloor \lg(n/m) \rfloor$, we have $t \geq 1$.

Now consider the possible values of t, and what conditions we need on m and n to make $2m \leq n - 2^t$. For $t = 1$, we have $2m \leq n - 2$, or $2m + 2 \leq n$. For $t \geq 2$ the fact that $t = \lfloor \lg(n/m) \rfloor \leq \lg(n/m)$ implies $2^t \leq 2^{\lg(n/m)} = n/m$, so $n - 2^t \geq n(1 - 1/m)$. Also $n \geq 4m$ (since $t = \lfloor \lg(n/m) \rfloor \geq 2$), so $n - 2^t \geq 4m(1 - 1/m)$. Thus for $t \geq 2$ the condition $2m \leq n - 2^t$ is satisfied when $2m \leq 4m(1 - 1/m)$.

Factoring out $2m$, we have $1 \leq 2 - 2/m$, or $m \geq 2$. Thus the use of eq. (95) on $C_{m,n-2^t}$ is okay if $m \geq 2$ and $n \geq 2m+2$, provided $n - 2^t$ is even. But for $t \geq 1$, $n - 2^t$ is even when n is even, and $n \geq 2m + 2$ implies that $t \geq 1$. Thus eq. (95) can be used on $C_{m,n-2^t}$ whenever $m \geq 2$, $n \geq 2m + 2$, and n is even.

It is okay to use eq. (95) on the other term in eq. (104) ($C_{m-1,n}$) provided n is even and $2(m - 1) \leq n$, which is less restrictive than the condition in the last paragraph.

So we have that

$$C_{m,n} = 1 + \max\{C_{m,n/2-2^{t-1}} + m, t + C_{m-1,n/2} + m\} \quad (106)$$
$$= 1 + \max\{C_{m,n/2-2^{t-1}}, t + C_{m-1,n/2}\} + m \quad (107)$$

provided n is even, $m \geq 2$, and $n \geq 2m+2$. Now comes a clever portion of the proof, a crucial observation: by eq. (103) $1+\max\{C_{m,n/2-2^{t-1}}, t+C_{m-1,n/2}\} = C_{m,n/2}$, so we have

$$C_{m,n} = C_{m,n/2} + m \quad (108)$$

when n is even, $m \geq 2$, and $n \geq 2m + 2$, provided our induction hypothesis is true.

Next we need to consider the corresponding case for odd n, but that is so similar to the last case that we will leave it as an exercise. After you fill in the details, you will get that

$$C_{m,n} = C_{m,(n-1)/2} + m \quad (109)$$

when n is odd, $m \geq 2$, and $n \geq 2m + 3$.

Before continuing the proof, let's notice how much progress we have made. We now know that eqs. (95, 96) are compatible with eq. (103) for most values of m and n ($m \geq 2$, $n \geq 2m+2$). This is quite promising, but we don't have a result until we complete the proof.

Now let's consider the case of $m = 1$. In this case the previous part of the proof still works unless $2m = 2 > n - 2^t$ (i.e., unless n is a power of 2 or one more than a power of 2). Suppose $m = 1$ and $n = 2^t$ for some integer t. Then using eq. (103) gives

$$C_{1,2^t} = 1 + \max\{C_{1,0}, t + C_{0,2^t}\} = 1 + t. \quad (110)$$

Use of eq. (95) on $C_{1,2^t}$ gives $1+C_{1,2^{t-1}}$. Now we can use eq. (95) repeatedly to evaluate $C_{1,2^{t-1}}$, by applying it to the points $(1, 2^{t-1})$, $(1, 2^{t-2})$, etc. until eq. (97) applies. Doing this gives $C_{1,2^{t-1}} = t$, which agrees with eq. (110), so the proposed solution is working for $m = 1$, $n = 2^t$. The case $m = 1$, $n = 2^t + 1$ is similar and is left as an exercise.

When $m = 0$, eq. (95) and eq. (96) give that $C_{0,n} = 0$, which is correct.

This finishes all the cases for which $2m + 2 \leq n$. We still have to prove the remaining cases before we have anything, because the cases we have done depend on them.

Consider the case $n = 2m$. (The case $n = 2m + 1$ is similar and is left as an exercise.) In this case $t = 1$. Applying eq. (95) to the right side of eq. (103) gives $C_{m,2m} = C_{m,m} + m$. Applying eq. (97) gives $C_{m,2m} = 3m - 1$. The right side of eq. (103) is $1 + \max\{C_{m,2m-1}, 1 + C_{m-1,2m}\}$. We can use eq. (97) to obtain $C_{m,2m-1} = 3m - 2$ and eq. (95) and eq. (97) to obtain $C_{m-1,2m} = C_{m-1,m} + m = 3m - 2$. Therefore in this case the right side of eq. (103) reduces to $3m - 1$, which is the same as the left side. So again the proposed solution is compatible.

The next case that we consider is $m < n < 2m$. In this case $t = 0$. Applying eq. (97) to the left side of eq. (103) gives $C_{m,n} = m + n - 1$. The right side of eq. (103) is $1 + \max\{C_{m,n-1}, C_{m-1,n}\}$. Applying eq. (97) gives $1 + \max\{m+n-2, m+n-2\} = m+n-1$. So again the proposed solution is compatible.

The next to last case is $m = n$. Again $t = 0$. Applying eq. (97) to the right side of eq. (103) gives $C_{m,m} = 2m - 1$. The left side of eq. (103) is $1 + \max\{C_{m,m-1}, C_{m-1,m}\}$. But $C_{m,m-1} = C_{m-1,m}$ so the right side reduces to $1 + C_{m-1,m}$ which can be simplified by eq. (97) to $2m - 1$. Thus in all cases the proposed solution is compatible.

Finally, all cases where $n < m$ are provided for by using $C_{m,n} = C_{n,m}$.

This completes the compatibility part of the proof. To finish the well-ordering part, we must go back and check that, in proving the equations are compatible for point (m_0, n_0), only points (m, n) for which $m \leq m_0$, $n \leq n_0$ and either $m \neq m_0$ or $n \neq n_0$ were used. Checking shows that this is indeed the case. The check also shows that all the values of m and n have been covered. [To verify these claims in detail, it is necessary to carefully reread the text from the paragraph of eq. (104) to here.] Since the proof for point (m_0, n_0) depends on only a finite number of previous points, no infinite descending chains are possible. Thus we know that there is a well-ordering on the pairs of points (m, n) that is compatible with the subcase dependencies present in our proof.

Notice that our proof that eqs. (95–97) satisfy eq. (94) did not define a total well-ordering. The *partial ordering* used was that element (m_0, n_0) is greater than all other points in the rectangle $0 \leq m \leq m_0$, $0 \leq n \leq n_0$. This ordering does not say whether the point $(5, 4)$ comes before or after $(4, 5)$. It is easy to extend this partial ordering to a total well-ordering. For example, lexicographic order works.

When doing a proof by induction, it is necessary to have a partial well-ordering such that all elements that occur in the proof of a given element are less than that element. If you have a total well-ordering, then you automatically know that you did not leave out any required part of the ordering. (Your proof is invalid if you leave anything out.) On the other hand, by constructing only the required part of a partial ordering, you save yourself the effort of ordering irrelevant elements.

The Binary Merge Algorithm is discussed by Knuth [11, Section 5.3.2], Baase [2, Section 2.8], Horowitz and Sahni [7, Section 10.2], and Reingold and Hansen [61, Section 8.2.2]. The algorithm was originally developed and analyzed by Hwang and Lin [98].

EXERCISES

1. Give a proof that $C_{m,n} = C_{m,(n-1)/2} + m$ when n is odd, $m \geq 2$, and $n \geq 2m+3$. Only use techniques that will permit you to use your proof as a subpart of the inductive proof that eqs. (95–97) give the solution of eq. (103).

2. Give a proof that $C_{1,2^t+1} = 1 + C_{1,2^{t-1}}$. Only use techniques that will permit you to use your proof as a subpart of the inductive proof that eqs. (95–97) give the solution of eq. (103).

3. Give a proof that the value for $C_{m,2n+1}$ given by eq. (96) is compatible with eq. (94). Use techniques that will permit you to use your proof as a subpart of the inductive proof that eqs. (95–97) give the solution of eq. (103).

1.12.2* Noetherian Induction

Up to now we have considered induction proofs where a well-ordering or partial well-ordering has been imposed on the set of cases as part of the proof technique. The important thing about proofs by induction, however, is that we show that each case can be proved in a finite number of steps. Well-orderings were just a technique to use in such proofs. If the cases that occur in the proof obey any Noetherian relation, we know that there are no infinite chains of steps in our proof; any case can be proved in a finite number of steps. Some proofs which would otherwise be long and complex become simple when Noetherian induction is used.

We need a few definitions to explain the method. Let $\rightarrow$ be a Noetherian relation on a set S (i.e., $\rightarrow$ has no infinite descending chains). For each element x in S, the set $\delta^+(x)$ is defined to be the set of elements y in S such that $x \overset{+}{\rightarrow} y$ (recall that $\overset{+}{\rightarrow}$ is the transitive closure of $\rightarrow$). For example, if S is the integers and $x \rightarrow y$ is the relation that is true when $y = x - 1$, then $\delta^+(x)$ is the set of integers that are less than x. A *predicate* is a function where the range of values is *true* and *false*. A predicate P is called $\rightarrow$-complete if, whenever $P(y)$ is *true* for all y in $\delta^+(x)$, then $P(x)$ is also *true*. The *principle of Noetherian induction* states that an $\rightarrow$-complete predicate is *true* for all x in S.

To illustrate Noetherian induction, we will use it to prove a theorem that is fundamental in the theory of term rewriting systems and of the lambda calculus. This theorem is known as the Diamond Lemma. We need several more definitions to state the theorem.

Let $\rightarrow$ be a relation on a set S. For two elements x and y in S, we write $x \uparrow y$ if there is an element a in S such that $a \overset{*}{\rightarrow} x$ and $a \overset{*}{\rightarrow} y$. That is, $x \uparrow y$ if x and y have a *common ancestor* under the relation $\rightarrow$. We write $x \downarrow y$ if x and y have a *common descendant*: if there is an element z in S such that $x \overset{*}{\rightarrow} z$ and $y \overset{*}{\rightarrow} z$.

A relation is *confluent* if $x \uparrow y$ implies $x \downarrow y$. Confluence is important because, in a confluent relation, $x \overset{*}{\leftrightarrow} y$ if and only if $x \downarrow y$; thus, equivalence can be decided by following any path of the $\rightarrow$ relation from x for as far as possible and seeing if you obtain the same result by following any path from y. (This is known as the *Church-Rosser property*; it can be proved by induction on the length of the derivation from x to y. Informally if $x \uparrow y$, $x \overset{*}{\rightarrow} w$, and $y \overset{*}{\rightarrow} z$, where $w \neq z$, then you can extend the paths from w and z until you do find a common point, because $w \uparrow z$.)

A relation is *locally confluent* if, for all a, x, y in S, $a \rightarrow x$ and $a \rightarrow y$ implies $x \downarrow y$. Figures 1.2 and 1.3 illustrate the notions of confluence and local confluence. The name Diamond Lemma comes from the shape of these diagrams.

We are now ready to state the Diamond Lemma.

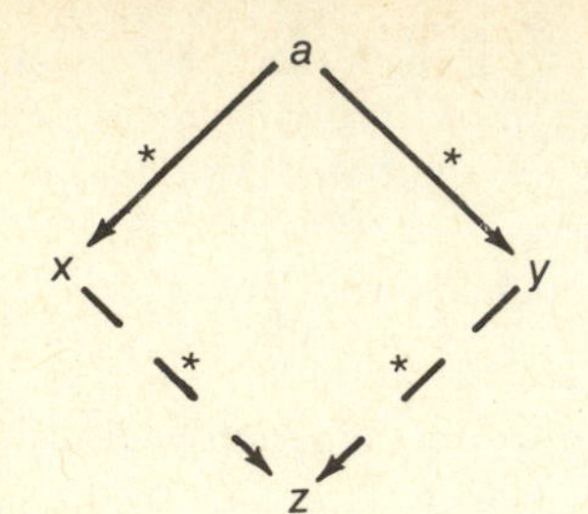

FIGURE 1.2 Confluence. For a confluent relation, if the solid paths exist, then so do the dashed paths.

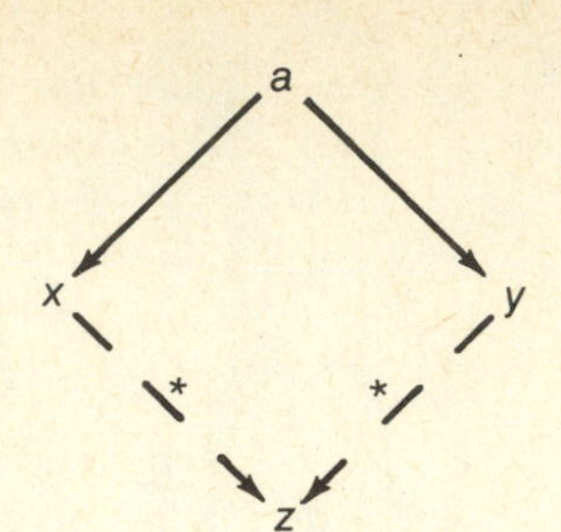

FIGURE 1.3 Local confluence. For a locally confluent relation, if the solid paths exist, then so do the dashed paths. Notice that the upper paths are one step long.

Theorem 1.5 A Noetherian relation is confluent if and only if it is locally confluent.

The only if part of the proof is simple. For a confluent relation any two elements that have a common ancestor must have a common descendant. For a locally confluent relation only those elements with a common parent must have a common descendant. Clearly those elements with a common parent have a common ancestor, so if the relation is confluent it must also be locally confluent.

Now let's do the if part of the proof. (Refer to Figure 1.4.) Assume $\rightarrow$ is a Noetherian locally confluent relation. Define $P(a)$ to be the predicate that is *true* if and only if the following condition is *true*: for all x and y such that $a \xrightarrow{*} x$ and $a \xrightarrow{*} y$ then $x \downarrow y$. In other words, $P(a)$ is *true* for just those elements a such that any pair of a's descendants have a common descendant. If $P(a)$ is *true* for all a, then the relation $\rightarrow$, which is used to define P, is confluent. We will use the assumption that $\rightarrow$ is Noetherian and locally confluent to prove that P is $\rightarrow$-complete and therefore *true* for all a.

Suppose $a \xrightarrow{*} x$ and $a \xrightarrow{*} y$. We must show that there is a z such that $x \xrightarrow{*} z$ and $y \xrightarrow{*} z$. If either $a = x$ or $a = y$, the proof is immediate. (For $a = x$ use $z = y$ and for $a = y$ use $z = x$.) This is the base case for the induction.

Our induction hypothesis is that $P(x)$ is *true* for all elements x that are proper descendants of the element a. That is, $P(x)$ is *true* for all $x \neq a$ where x is a descendant of a. If $a \xrightarrow{*} x$ with $a \neq x$ and $a \xrightarrow{*} y$ with $a \neq y$, then there exist x_1 and y_1 such that $a \rightarrow x_1 \xrightarrow{*} x$ and $a \rightarrow y_1 \xrightarrow{*} y$ (see Figure 1.4). By local confluence, there exists a u such that $x_1 \xrightarrow{*} u$ and $y_1 \xrightarrow{*} u$. The induction hypothesis applied to x_1 says that there exists a v such that $x \xrightarrow{*} v$ and $u \xrightarrow{*} v$. The induction hypothesis applied to y_1 says that there exists a z such that $y \xrightarrow{*} z$ and $v \xrightarrow{*} z$. Putting the pieces together, we have $a \xrightarrow{*} x$, $a \xrightarrow{*} y$, $x \xrightarrow{*} z$, and $y \xrightarrow{*} z$, so $P(a)$ is *true*. Since $\rightarrow$ is Noetherian, we can establish $P(a)$ for any a after a finite number of steps. This completes the induction proof.

This section is based on Huet [97].

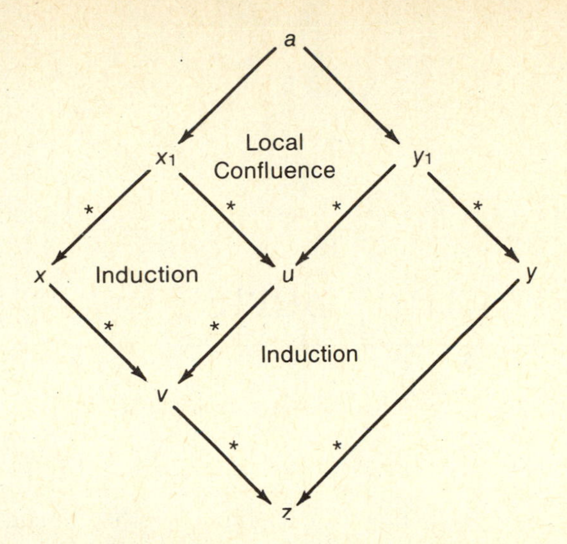

FIGURE 1.4 The induction steps for the proof of the Diamond Lemma. Since the relation is Noetherian and locally confluent, the elements x and y, which have a common ancestor, must have a common descendant.

CHAPTER 2

Summing Series

As we saw in Chapter 1, it is often necessary to sum series in order to analyze algorithms. In the previous chapter the sums of two series were given [eqs. (1.37, 1.41)]. However, one series was trivial, and for the other series we used a rather special method. This chapter gives a systematic treatment of how to sum those series that can be summed by elementary methods.

A study of this chapter will reveal a close similarity between the summation of series and integral calculus. This is not too surprising since integral calculus is concerned with summing rectangles with infinitesimal width, while summation of series is equivalent to summing rectangles with unit width. It will be helpful if you are alert for these connections so that you can apply your previous experience in evaluating integrals to the problem of summing series.

2.1 SUMMATION NOTATION

When you want to refer to the sum of a fixed number of items, you can just write the items with plus signs between them. For example, $a_1 + a_2 + a_3$ is the sum of a_1, a_2, and a_3. When you want to refer to the sum of a variable number of things, a more general notation is needed. The *dot dot dot* notation (also called *ellipsis* notation) is the most primitive extension. Thus, $a_1 + a_2 + \cdots + a_n$ is the sum of the items a_1 to a_n. The *dot dot dot* notation can be very clear for simple problems because it is so similar to the notation used for summing a fixed number of items. It is, however, not very precise. The reader must deduce from the terms given explicitly which indices are included in the sequence to be summed. Thus from $a_1 + a_3 + a_5 + \cdots + a_n$, the reader must conclude that the odd indices are included in the sum. This problem can be reduced by using the *dot dot dot* notation with an explicit formula for the ith term. Thus $a_1 + \cdots + a_{2i+1} + \cdots + a_n$ refers to the sum on the a_i with odd indices between 1 and n.

In this book we often use another notation, the standard summation notation, in which we write a large sigma with a relation on the index under the sigma and the formula for the general term after the sigma. Thus

$$S = \sum_{R(i)} x_i \tag{1}$$

means that S is the sum of those x_i such that the relation $R(i)$ is *true*. For example,

$$\sum_{1 \leq i \leq n} a_i \tag{2}$$

refers to the sum of the a_i such that $1 \leq i \leq n$. Summation notation gives an explicit formula for which terms to include in the summation.

When using summation notation you must be sure that the summation index is clearly indicated. In eq. (2) it is clear that i is the summation index because it appears as an index in the summand. For the sum $\sum_{j<k} j^k$ it is not clear whether you intend to sum over j or k. In such cases you should state which variable is the summation index.

The scope of a summation sign is the same as that of any other addition operator. Thus

$$\sum_{1 \leq i \leq n} x_i + y = \left(\sum_{1 \leq i \leq n} x_i \right) + y \tag{3}$$

and

$$\sum_{1 \leq i \leq n} x_i y = \sum_{1 \leq i \leq n} (x_i y). \tag{4}$$

In a summation, the summation index is a dummy variable, so you can give it any name without changing the value of the sum. Thus $\sum_{0 \leq i \leq n} a_i = \sum_{0 \leq j \leq n} a_j$. When one sum appears inside another, of course, you must be careful not to replace one index with an index that is already in use. Thus in $\sum_{1 \leq i \leq n} \left(\sum_{1 \leq j \leq m} a_{ij} \right)$, it would be incorrect to replace i with j or vice versa.

You can also use replacements of the form $i \leftarrow j + c$ or $i \leftarrow c - j$, where c is an integer constant. Thus

$$\sum_{0 \leq i \leq n} a_i = \sum_{0 \leq j-1 \leq n} a_{j-1}. \tag{5}$$

If j is replaced with i, eq. (5) can be rewritten as

$$\sum_{0 \leq i \leq n} a_i = \sum_{0 \leq i-1 \leq n} a_{i-1}. \tag{6}$$

Finally you can add one to each item in the relation under the second sum to obtain

$$\sum_{0 \leq i \leq n} a_i = \sum_{1 \leq i \leq n+1} a_{i-1}. \tag{7}$$

In Chapter 1 [eq. (1.38)] a replacement of the form $i \leftarrow c - j$ (with $c = n + 1$) was used to simplify a sum.

When summing a finite sum, the terms can be summed in any order. Thus $a_1 + a_2 + a_3 = a_1 + a_3 + a_2 = a_3 + a_2 + a_1$, etc. You *cannot* always change the order of the terms in an infinite sum without changing the value. (More will be said about the infinite case in Section 2.1.2.) For now, we assume that all sums are over a finite number of terms. Many interesting summation formulas are obtained by rearranging the order in a summation. For example,

$$\sum_{R(i)} (b_i + c_i) = \sum_{R(i)} b_i + \sum_{R(i)} c_i, \tag{8}$$

because the left and right sides differ only in the order in which the terms are added. On the left, you first add $b_i + c_i$ for each i and then add the results. On the right, you sum the b_i, sum the c_i, and then add the two sums. The final result is the same in either case.

Suppose you are given $S(n) = \sum_{0 \le i \le n} a_i$, and you want to solve the problem $\sum_{m<i\le n} a_i$. The problem can be solved by rearranging terms in a sum:

$$\sum_{0\le i\le n} a_i - \sum_{0\le i\le m} a_i = \sum_{0\le i\le m} (a_i - a_i) + \sum_{m<i\le n} a_i = \sum_{m<i\le n} a_i. \tag{9}$$

For example, we know from eq. (1.41) that $\sum_{1\le i\le n} i = n(n+1)/2$. Suppose we need to know $\sum_{m<i\le n} i$. Using eq. (9) we obtain

$$\sum_{m<i\le n} i = \frac{n(n+1) - m(m+1)}{2}. \tag{10}$$

Since eq. (9) gives an easy way to obtain a sum with general upper and lower limits from one with just a general upper limit, summation formulas will be given with whatever lower limit is convenient (usually zero or one).

Often a formula looks quite different after you change its index. Such differences can sometimes be used to find the value of a summation. For example, consider the sum $\sum_{1\le i\le n} i^2$. By changing the summation index you obtain the equivalent sum $\sum_{1\le i+1\le n} (i+1)^2$. This sum can be rewritten as $\sum_{0\le i\le n-1} (i^2 + 2i + 1)$. (One was subtracted from each term in the relation and the summand was multiplied out.) The difference between these two sums is zero. Thus

$$\sum_{0\le i\le n-1} (i^2 + 2i + 1) - \sum_{1\le i\le n} i^2 = 0. \tag{11}$$

Now let's rearrange the terms in the first sum of eq. (11) to simplify it.

$$\sum_{0\le i\le n-1} (i^2 + 2i + 1) = \sum_{0\le i\le n-1} i^2 + \sum_{0\le i\le n-1} 2i + \sum_{0\le i\le n-1} 1. \tag{12}$$

The last of these sums is just n. The first sum is equal to the second sum in eq. (11) minus the term for $i = n$. Thus from eq. (11) and eq. (12) we get

$$0 - n^2 + \sum_{0\le i\le n-1} 2i + n = 0. \tag{13}$$

This can be solved for the remaining summation to obtain

$$\sum_{0\le i\le n-1} i = \frac{n(n-1)}{2}. \tag{14}$$

Notice that this last result is almost the same as eq. (1.41); only the limits are different.

The same approach can be used to work out the sum of any power of i. To find the sum of i^k, first write $\sum_{1\le i\le n} i^{k+1}$. Form a second sum by replacing i with $i+1$. The difference of the two sums is zero. Simplify the second sum by multiplying out $(i+1)^{k+1}$. Now simplify the difference of the two sums. The terms of the form i^{k+1} will cancel. The remaining terms will be sums of i^k, i^{k-1}, etc. If you do not know the value of the sums of i^{k-1}, etc., you can apply the method repeatedly. For example, let's apply the method to $\sum_{1\le i\le n} i^3$. Start with

$$\sum_{0\le i\le n-1} (i+1)^3 - \sum_{1\le i\le n} i^3 = 0. \tag{15}$$

Multiplying out the $(i+1)^3$ and simplifying the sums gives

$$0 - n^3 + 3\sum_{0\le i\le n-1} i^2 + 3\sum_{0\le i\le n-1} i + \sum_{0\le i\le n-1} 1 = 0. \tag{16}$$

Simplifying further and using eq. (14) gives

$$\sum_{0\le i\le n-1} i^2 = \frac{2n^3 - 3n^2 + n}{6}. \tag{17}$$

Replacing n with $n+1$ gives

$$\sum_{0\le i\le n} i^2 = \frac{2(n+1)^3 - 3(n+1)^2 + n + 1}{6} = \frac{2n^3 + 3n^2 + n}{6}. \tag{18}$$

When more than one relation is written under a sum, the index must obey all the relations. Thus

$$\sum_{\substack{1\le i\le 4\\ i \bmod 2=0}} a_i = a_2 + a_4. \tag{19}$$

Factors that do not depend on the summation index can be factored out of sums. For example,

$$\sum_{0\le i\le n} x a_i = x \sum_{0\le i\le n} a_i. \tag{20}$$

Likewise, a factor written outside of a sum may be multiplied in [eq. (20), using the left side to replace the right side].

Much of analysis of algorithms is concerned with simplifying formulas. A complex equation is difficult to understand. It is also often difficult to compute with. The first step in simplifying a formula is to write it in closed form, if possible. A formula is in *closed form* if it is expressed in a fixed number of simple operations. For example, the formula $\sum_{0\le i\le n} a_i$ is not in closed form, because the summation sign stands for $n+1$ additions. Once the formula is expressed in closed form (or where that is not possible, when as much of the formula as possible is expressed in closed form), the formula is then simplified using the rules of algebra. There is no complete mathematical theory to characterize simple formulas. Just as people have opinions concerning what is good art, they have opinions concerning the simplest way to express some formulas. One good rule of thumb is to use as few operations as possible in the final formula. It is very important to write your results in the simplest form you can find so that they will be easier to understand. Remember that the *final* result of your analysis is not your formulas, but the understanding that the formulas convey to the people who are interested in using the analysis. *When your results are expressed simply, they will convey more information to the people who read them.*

After you learn a few formulas for solving particular summations, you need to know how to use the formulas to solve similar problems. You should not spend a lot of time trying to solve a problem from first principles if the problem is easy to solve by combining previous formulas (unless, of course, you are just practicing using the techniques for solving problems from first principles).

Consider the sum $\sum_{m\le i\le n}(ai+b)$. This is very similar to several previous sums that we have seen. First notice that the summand is the sum of two terms, so eq. (8) applies. Thus we have

$$\sum_{m\le i\le n}(ai+b)=\sum_{m\le i\le n}ai+\sum_{m\le i\le n}b. \tag{21}$$

In the first sum the a can be factored out to obtain something similar to eq. (1.41), while the second sum is already similar to eq. (1.37). The only differences are in the limits on the summations. But eq. (9) can be used to take care of this difference (using $m-1$ for the lower bound for the summation index). Thus we have

$$\sum_{m\le i\le n}(ai+b)=a\sum_{m\le i\le n}i+\sum_{m\le i\le n}b \tag{22}$$

$$=a\left(\frac{n(n+1)-m(m-1)}{2}\right)+b(n-m+1). \tag{23}$$

When trying to match a new problem against the formulas you know, look for places where it may help to break a sum into two sums or to combine two sums into one. Look for factors that can be factored out of sums. If you can get the terms being summed to match the terms in formulas that you know, then try to use a combination of change of index variable and eq. (9) to cause the limits on your problem to match the limits on your formulas. If the limits match except for a constant number of terms, using eq. (9) is the same as adjusting the limits to match by adding or subtracting the extra terms from the summation so that the net value remains the same.

It is easy to make mistakes when changing the index variable on a summation. Consider replacing i by $i+1$. When working a small problem, perhaps the best thing to do is to replace i by $j+1$ instead. This way there is no confusion between an old i and a new i, because the new i is called j. This approach can also be used on large problems, but then it tends to result in a lot of letters being used at various times. It is easy to lose track of the meanings of the various indices. Therefore many people prefer to stick with the same letters for indices as the formulas are changed. To do this without making errors, you must keep track of the old index and the new index. A useful technique is to put a prime on the new index. Once the final modified formula is written, the primes can be dropped. Thus, to make the index on a in $\sum_{1 \leq i \leq n} ia_{i-1}$ be i, write $i' = i - 1$. This can be solved for i to obtain $i = i' + 1$. Now that we know the old i in terms of the new, it is easy to write down the transformed sum as $\sum_{1 \leq i+1 \leq n} (i+1)a_i$ (dropping the primes after the transformation). Finally we simplify the relation under the sum so that the i appears by itself and the $+1$ is combined with the limits. This gives $\sum_{0 \leq i \leq n-1} (i+1)a_i$.

When working out summations, it is easy to make minor algebraic errors. There are several techniques you should use to reduce the number of such errors. First, work carefully to avoid errors in the first place. Second, plug numbers into your formulas before and after simplifying to see if the formulas produce the same answers each time. Usually these numbers should be chosen to be easy to calculate with so that you do not make errors in the checking. For sums with variable limits, values that result in there being only a few terms in the sum are particularly useful. You should try to develop a sense of how much checking a formula needs. It is very difficult (but not impossible) for a simple formula that has passed a few checks to be wrong. A complex formula needs more checks. If you find errors in your formulas after your first set of checks, then you should usually try some additional checks after you correct your formula. Otherwise you may develop a formula that is good at passing your checks rather than a correct formula. Finally, try to keep in mind the nature of your final answer. For example, if you are calculating a probability and your answer is $1 + 1/n$ (where n is positive), then you know you made an error. Answers that are wrong in this way look ridiculous to the person who remembers the original problem. Of course, you should be careful not to just come up with the answer that you want without also having correct intermediate steps. (Checked formulas are subject to a natural selection process.) A little care can do a lot to reduce the number of errors in your work.

General techniques for summing finite series are also given by Knuth [9, Section 1.2.3]. See Section 6.4 of this book for an introduction to a general theory of summing series. Many elementary summation formulas can be found in Jolley [38], but nearly all of them can be worked out rapidly by a person who has mastered the techniques for summing series.

EXERCISES

1. Write the sum $a_1 + a_4 + a_7 + \cdots + a_n$ in summation notation, where n is of the form $3k+1$ for some integer k. Do this problem two ways. Write a formula similar to eq. (19). The other way should be simpler.

2. Simplify $\sum_{1 \leq i \leq n} (a_{i+1} - a_i)$.

3. What is the value of $\sum_{1 \le i \le n} i^3$?

4. In this section the sum $\sum_i i$ was obtained by comparing $\sum_i i^2$ and $\sum_i (i+1)^2$ with the summation limits adjusted so that the two sums are the same. Using the same technique on $\sum_i x^i$ and $\sum_i x^{i+1}$, find a closed form for $\sum_{0 \le i \le n} x^i$.

2.1.1 Linear Searching

The analysis of the Linear Search Algorithm leads to some familiar summation formulas.

Algorithm 2.1 Linear Search: Input: An array X of numbers with elements x_i for $1 \le i \le n$, and a query q. Output: A number i such that $q = x_i$ if there is such an i or $i = n+1$ otherwise.

Step **1.** [Loop] For $i \leftarrow 1$ to n do:

Step **2.** [Check] If $q = x_i$, then stop; q is in the array at position i.

Step **3.** End for.

Step **4.** [Not in] Set $i \leftarrow n+1$ and stop; q is not in the array.

If q is not in the array, then Step 1 is done $n+1$ times, Step 2 is done n times, Step 3 is done n times, and Step 4 is done once. If q is in the array, then Step 1 is done the same number of times as Step 2, Step 2 is done one more time than Step 3, and Step 4 is not done at all. The worst-case time for the algorithm occurs when q is not in the array. In this case Step 1 is done $n+1$ times. When q is in the array, Step 1 is done between 1 and n times. If the elements in the array are all different, if there is no relation between q and the position of the elements in the array, and if q is known to be in the array, then the probability that $q = x_1$ is $1/n$, the probability that $q = x_2$ is $1/n$, etc. If $q = x_i$, then Step 1 is done i times. Therefore the average number of times that Step 1 is done under these assumptions is

$$\sum_{1 \le i \le n} i p_i = \sum_{1 \le i \le n} \frac{i}{n} = \frac{n+1}{2}. \tag{24}$$

The variance can be computed using eq. (1.57). First we need

$$\sum_{1 \le i \le n} i^2 p_i = \sum_{1 \le i \le n} \frac{i^2}{n} = \frac{(2n^2+3n+1)}{6}, \tag{25}$$

where the sum was simplified using eq. (18). From eq. (1.57), eq. (24), and eq. (25) we get

$$V = \frac{(2n^2+3n+1)}{6} - \left(\frac{n+1}{2}\right)^2 = \frac{n^2-1}{12}. \tag{26}$$

For large n the standard deviation is more than half the average, so it is quite large.

The Linear Search Algorithm is the first algorithm analyzed in detail in Knuth's series of books [9, Section 1.2.10].

EXERCISES

1. What is the average number of times that Step 3 of Algorithm 2.1 is done when looking for an item that is in the table?
2. What is the standard deviation of the number of times that Step 3 of Algorithm 2.1 is done when looking for an item that is in the table?

2.1.2* Infinite Sums

In analysis of algorithms we are usually concerned with finite sums. For such sums the order in which the terms are added together does not matter. When $R(i)$ specifies an infinite set, then

$$\sum_{R(i)} x_i \quad \text{means} \quad \left(\lim_{n\to\infty} \sum_{\substack{R(i)\\ 0\le i\le n}} x_i\right) + \left(\lim_{n\to\infty} \sum_{\substack{R(i)\\ 0>i\ge -n}} x_i\right). \tag{27}$$

If one or both of the limits does not exist, then the sum is divergent.

A typical example of the trouble you can have when you rearrange the order of summation in an infinite sum is illustrated by

$$\sum_{k\ge 1} \frac{(-1)^{k+1}}{k}. \tag{28}$$

Since this is an *alternating sum* (alternate terms are positive and negative), since the terms are finite, and since the terms are getting smaller in absolute value, this sum has a finite value. The value is between the sum for the first n terms and the sum for the first $n+1$ terms. The value of this sum is $\ln 2$, but the exact value is not important now. Consider what happens if you change the order of summation and sum the positive terms and negative terms separately. This produces $\sum_{k\ge 1} 1/(2k-1) - \sum_{k\ge 1} 1/(2k)$. Each of these sums *diverges* (has infinite value). The first sum, for example, is $\lim_{n\to\infty} \frac{1}{2}\sum_{1\le k\le n} 1/k$, and we will see in Section 4.5.2 that $\sum_{1\le k\le n} 1/k = \ln n + O(1)$.

It is useful to be able to change the order of terms in a summation when doing so does not cause any errors. One sufficient condition to guarantee that you can rearrange the order of summation in the sum $\sum_{R(i)} a_i$ is for the sum $\sum_{R(i)} |a_i|$ to exist (i.e., have a finite value), where $|a_i|$ is the absolute value of a_i. Such a sum is called *absolutely convergent*. The sum that caused trouble in the previous paragraph is not absolutely convergent. Very often we will be concerned with a sum of positive terms, which is automatically absolutely convergent if it is convergent at all.

The most complete reference on infinite sums is Knopp [40]. Also see Greene and Knuth [6, Section 4.1.4] and Apostol [17, Chapters 12 and 13] for problems that can arise when manipulating infinite summations. We will not cover such problems in detail.

EXERCISES

1. In Section 2.4 it is shown that $\sum_{0\le i\le n} x^i = (x^{n+1}-1)/(x-1)$. Use this result to find the value of $\sum_{i\ge 0} x^i$.

2. In Section 2.5 it is shown that

$$\sum_{0\le i<n} ix^i = \frac{(n-1)x^{n+1}-nx^n+x}{(x-1)^2}.$$

Use this result to obtain the value of $\sum_{i\ge 0} ix^i$.

2.1.3* Conventions

Much of mathematics is developed by first studying simple objects, such as positive integers. Later mathematicians discover that if they adopt suitable conventions, the results for the simple objects also apply in more complex cases. The choice of conventions can be quite important. With the proper choice of conventions it is often possible to have a single formula that covers many related cases, rather than a separate formula for each case.

One convention that needs defining for summation notation is the meaning of $\sum_{R(i)} a_i$ when there are no values of i that satisfy $R(i)$. One convention is that such a sum has the value zero. There is a very logical explanation for this convention: the way you get the value of the sum $\sum_{R(i)} a_i$ is to start with zero and add all the a_i such that $R(i)$ is true.

There is, however, a second convention that also makes sense. Consider the sum $\sum_{m\le i<n} a_i$. We have the equation

$$\sum_{m\le i<n} a_i = \sum_{m\le i<j} a_i + \sum_{j\le i<n} a_i \tag{29}$$

provided $m<j<n$. Eq. (29) is a very useful equation, and it would be nice to have a convention that makes it true *whether or not* $m<j<n$. If the convention of the previous paragraph is used, then eq. (29) is true when $m\le j\le n$, which is closer to what we want, but not quite there. The convention that will preserve eq. (29) is that

$$\sum_{m\le i<n} a_i \quad \text{means} \quad -\sum_{n\le i<m} a_i \tag{30}$$

when $m>n$ and it means zero when $m=n$. Both this convention and the one in the previous paragraph agree that $\sum_{m\le i<m} a_i$ should be zero, but the convention in this paragraph says that when $m>n$, $\sum_{m\le i<n} a_i$ should mean $-\sum_{n\le i<m} a_i$. The big advantage of this convention is that when we use it,

eq. (29) can be used for any values of j, m, and n. It also agrees with the convention in calculus that $\int_m^n f(x)\,dx = -\int_n^m f(x)\,dx$.

Notice that eq. (9) is especially powerful when used on a problem where the convention of eq. (29) is appropriate since in such cases any integer can be used for the lower limit of the sum.

Most sums that you will work with will have the form $\sum_{R(i)} a_i$ where there are some values of i such that $R(i)$ is true. In such normal cases you don't need to worry about what convention to use. In those rare cases where you do need to worry about the convention, you should proceed carefully. You can use any convention that works, but it is your responsibility to be sure that each step of your mathematics is consistent with the convention you are using.

Knuth [9, Section 1.2.3] uses the convention that sums with upper limits that are less than the lower limit are equal to zero. Karr [103] introduced the convention that is similar to the one used with integrals, and pointed out its advantages.

EXERCISES

1. Using the two conventions of this section, what value or values does the sum $\sum_{1\le i<1} i$ have?
2. Using the two conventions of this section, what value or values does the sum $\sum_{3\le i<1} i$ have?

2.2 OPERATORS

Section 2.1 gave one method for computing the sum of i^k. The method may have seemed like a special trick that was pulled out of the air. This section is an introduction to the theory of operators. This theory will be used in the following section to give a more natural method for summing i^k. As you will see, in the end the computational steps will be about the same with the new approach. The difference is that the reason for doing the steps will be more apparent after you have learned about operators.

Consider a possibly infinite sequence of numbers $\{a_n\} = \{a_0, a_1, a_2, \ldots\}$. An operator is a function on the elements of a sequence that produces a sequence as its result. The indefinite summation operator, Σ, is defined by

$$\Sigma\{a_n\} = \{a_0,\ a_0 + a_1,\ a_0 + a_1 + a_2,\ \ldots\}. \tag{31}$$

or equivalently,

$$\Sigma\{a_n\} = \Big\{\sum_{0\le i\le n} a_i\Big\}. \tag{32}$$

In other words, the sigma operator produces a sequence whose elements are the partial sums of the original sequence. For example, let $a_i = i$. Then eq. (1.41) can be expressed as $\Sigma\{0, 1, 2, 3\ldots\} = \{0, 1, 3, 6, \ldots\}$.

The simplest operators are those where the ith element of the result is obtained from the ith element of the sequence (or sequences) that the function is applied to. Let $f(x_1, \ldots, x_n)$ be an ordinary function of n variables, i.e., a function defined on elements rather than on sequences, and let $X_1, \ldots, X_n$ be n sequences. Let the elements of X_i be x_{i1}, x_{i2}, ... We use the convention that the result of $f(X_1, \ldots, X_n)$ is the sequence $\{f(x_{11}, \ldots, x_{n1}),\ f(x_{12}, \ldots, x_{n2}), \ldots\}$. For example, if the function is unary minus, we have $-\{a_1,\ a_2,\ a_3, \ldots\} = \{-a_1,\ -a_2,\ -a_3, \ldots\}$.

Two important operators are **E**, which removes the first element from a sequence, and **I**, which does not change a sequence. Thus

$$\mathbf{E}\{a_n\} = \{a_1,\ a_2,\ a_3,\ \ldots\} \tag{33}$$

and

$$\mathbf{I}\{a_n\} = \{a_0,\ a_1,\ a_2,\ \ldots\}. \tag{34}$$

It is also useful to think of **E** as an operator that, in the formula for each element in the sequence, replaces the index by the index plus one. The difference operator is defined by $\Delta\{a_n\} = \mathbf{E}\{a_n\} - \mathbf{I}\{a_n\}$. Dropping the $\{a_n\}$ gives the equation $\Delta = \mathbf{E} - \mathbf{I}$, which says that the Δ operator, when applied to any sequence, gives the same results as applying the **E** and the **I** operators to the same sequence and taking the difference of corresponding terms from the two resulting sequences. Thus

$$\Delta\{a_n\} = \{a_1 - a_0,\ a_2 - a_1,\ a_3 - a_2,\ \ldots\}. \tag{35}$$

For example, $\Delta\{i\} = \{1,\ 1,\ 1,\ \ldots\}$.

From eq. (32) and eq. (35) we have $\Delta\Sigma\{a_n\} = \{a_1,\ a_2,\ a_3,\ \ldots\}$, where $\Delta\Sigma\{a_n\}$ means $\Delta(\Sigma\{a_n\})$. Thus we have

$$\Delta\Sigma = \mathbf{E}. \tag{36}$$

Eq. (36) is quite similar to the equation in calculus that says the derivative of the integral of a function is equal to the original function. However, the right side of eq. (36) is **E**, not **I**.

Now let's consider $\Sigma\Delta$. First, though, define

$$\mathbf{P}_k\{a_n\} = \{a_k,\ a_k,\ a_k,\ \ldots\}. \tag{37}$$

Thus $\mathbf{P}_k$ forms a new sequence consisting of the kth element of the original sequence. From eq. (32) and eq. (35) we have that $\Sigma\Delta\{a_n\} = \{a_1 - a_0,\ a_2 - a_0,\ a_3 - a_0,\ \ldots\}$, so that

$$\Sigma\Delta = \mathbf{E} - \mathbf{P}_0. \tag{38}$$

Comparing eq. (36) with eq. (38) shows that the order in which operators are applied is quite important. Usually operators do not commute: changing the order in which they are applied will change the results. The $\mathbf{P}_0$ term in eq. (38) plays the same role that the constant of integration plays in the rule that the integral of the derivative of a function is equal to the original function plus a constant of integration.

In the next section we will see how to use eq. (38) to sum certain series. The approach is similar to the usual approach in calculus courses, where one first computes derivatives of many functions and then uses the results to show how to compute the related integrals.

Operator methods receive extensive coverage in classical books on finite differences, such as the one by Milne-Thomson [43, Chapters 1 and 2]. Operator methods for analyzing algorithms are discussed in detail in Greene and Knuth [6, Chapter 3]. Also see Sections 2.3 and 6.2.3 of this book.

EXERCISES

1. Rewrite eq. (18) in operator notation.
2. Let $\{a_i\}$ be the sequence $\{i^2\}$. Compute $\mathbf{I}\{a_i\}$, $\mathbf{E}\{a_i\}$, $\boldsymbol{\Delta}\{a_i\}$, $\boldsymbol{\Sigma}\{a_i\}$, $\boldsymbol{\Delta}\mathbf{E}\{a_i\}$, $\mathbf{E}\boldsymbol{\Delta}\{a_i\}$, $\boldsymbol{\Sigma}\boldsymbol{\Delta}\{a_i\}$, and $\boldsymbol{\Delta}\boldsymbol{\Sigma}\{a_i\}$.
3. Prove that $\boldsymbol{\Delta}\mathbf{E} = \mathbf{E}\boldsymbol{\Delta}$.

2.3 POLYNOMIAL SUMS

Let's now consider how to find the value of $\sum_{0\le i\le n} i^2$ by using operators. The close similarity between sums and integrals suggests that the sum may be about equal to $n^3/3$ [although we know from eq. (18) that this is not quite the right answer]. Since n^3 appears to be related to the answer, let's apply eq. (38) to the sequence $\{n^3\}$. Applying $\boldsymbol{\Delta}$ gives

$$\begin{aligned}\boldsymbol{\Delta}\{n^3\} &= \{1^3-0^3,\ 2^3-1^3,\ 3^3-2^3,\ \ldots\}\\ &= \{(n+1)^3-n^3\} = \{3n^2+3n+1\}.\end{aligned} \tag{39}$$

Therefore by eq. (38)

$$\boldsymbol{\Sigma}\{3n^2+3n+1\} = \{(n+1)^3-0^3\} = \{(n+1)^3\}. \tag{40}$$

In summation notation this says

$$\sum_{0\le i\le n}(3i^2+3i+1) = (n+1)^3. \tag{41}$$

Now, this is not quite what we wanted to sum, although it does bear a close resemblance to it. Using results from Section 1.7, we know that $\sum_{0\le i\le n} 1 = n+1$ and $\sum_{0\le i\le n} i = n(n+1)/2$. If we take 1/3 of eq. (41) minus eq. (1.41) minus 1/3 of eq. (1.37), we get [since $(3i^2+3i+1)/3 - i - 1/3 = i^2$]

$$\sum_{0\le i\le n} i^2 = \frac{(n+1)^3}{3} - \frac{n(n+1)}{2} - \frac{n+1}{3} = \frac{n(2n^2+3n+1)}{6}. \tag{42}$$

A similar approach can be used to compute $\sum_{0\le i\le n} i^k$ for any fixed value of k. First work out the sum for all smaller values of k. Then apply eq. (38) to n^{k+1}. Finally take the appropriate linear combination of the pieces to solve the problem. Although this approach works, it takes a long time for large values of k, and it leads to formulas that are difficult to remember.

The approach of the last paragraph often leads naturally to sums where the limits are $0 \le i < n$ rather than $0 \le i \le n$. For example, eq. (41) can be rewritten as

$$\sum_{0\le i<n} (3i^2 + 3i + 1) = n^3, \tag{43}$$

which is a somewhat simpler function of n.

While the sums for the powers of i are useful, it would be nice to discover a summation formula that is similar to the formula for powers of i, but which is easier to remember. A technique that often leads to nice formulas is to first work out the most basic sum in the sequence (such as the sum of i^0). Then sum the answer from the most basic sum and proceed, at each step summing the answer from the previous step. To see how this works, consider the following:

$$\sum_{0\le i<n} 1 = n \tag{44}$$

$$\sum_{0\le i<n} i = \frac{n(n-1)}{2} \tag{45}$$

$$\sum_{0\le i<n} i(i-1) = \frac{n(n-1)(n-2)}{3} \tag{46}$$

$$\sum_{0\le i<n} i(i-1)(i-2) = \frac{n(n-1)(n-2)(n-3)}{4}. \tag{47}$$

This sequence suggests the conjecture that

$$\sum_{0\le i<n} i(i-1)\cdots(i-k+1) = \frac{n(n-1)\cdots(n-k)}{k+1}. \tag{48}$$

The same method that was used to derive eq. (40) can be used to prove the conjecture. First apply Δ to the right side to obtain

$$\Delta\left\{\frac{n(n-1)\cdots(n-k)}{k+1}\right\}$$

$$= \left\{\frac{(n+1)n\cdots(n-k+1)}{k+1} - \frac{n(n-1)\cdots(n-k)}{k+1}\right\} \tag{49}$$

$$= \left\{\frac{n(n-1)\cdots(n-k+1)}{k+1}(n+1-n+k)\right\} \tag{50}$$

$$= \{n(n-1)\cdots(n-k+1)\}. \tag{51}$$

Therefore by eq. (38) we have

$$\sum_{0\le i\le n} i(i-1)\cdots(i-k+1) = \frac{(n+1)n\cdots(n-k+1)}{k+1} \tag{52}$$

or

$$\sum_{0 \le i < n} i(i-1)\cdots(i-k+1) = \frac{n(n-1)\cdots(n-k)}{k+1}. \tag{53}$$

The product on the right side of eq. (53) occurs frequently enough to warrant a special notation:

$$n^{\underline{k}} = n(n-1)\cdots(n-k+1). \tag{54}$$

The notation

$$n^{\overline{k}} = n(n+1)\cdots(n+k-1) \tag{55}$$

is also useful. We call $n^{\underline{k}}$ the *decreasing* k^{th} *power* of n and $n^{\overline{k}}$ the *increasing* k^{th} *power* of n. Using this notation, we can rewrite eq. (53) as

$$\sum_{0 \le i < n} i^{\underline{k}} = \frac{n^{\underline{k+1}}}{k+1}, \tag{56}$$

which is probably the easiest form of the formula to remember. Notice that this form is quite similar to the corresponding formula from calculus:

$$\int_0^n x^k\,dx = \frac{n^{k+1}}{k+1}. \tag{57}$$

From this section you can see that $\sum_{0 \le i \le n} i^k$ can be expressed in the form $\sum_{0 \le i \le k+1} a_i(k)n^i$. An alternate approach to finding the value of $\sum_{0 \le i \le n} i^k$ is to use the method of undetermined coefficients (Section 5.4.2) to find the values of the $a_i(k)$.

EXERCISES

1. Simplify $\sum_{1 \le i \le n}(i+3)^2$.
2. Suppose you need to multiply triangular arrays x_{ij} and y_{ij} where the elements can be nonzero only when $1 \le j \le i \le n$. A lot of time can be saved by not multiplying by the elements known to be zero. Thus one can compute

$$z_{ik} = \sum_{k \le j \le i} x_{ij}y_{jk} \quad \text{for} \quad 1 \le k \le i \le n.$$

How many multiplies will a program based on this method use to multiply the triangular arrays? For large n how does the time for this approach compare with the time required to multiply $n \times n$ matrices, using the traditional method of matrix multiplication:

$$z_{ik} = \sum_{1 \le j \le n} x_{ij}y_{jk} \quad \text{for} \quad 1 \le i,k \le n?$$

2.3.1 Quicksort

One of the quicker ways (on the average) to sort numbers is Quicksort, which takes a "splitting element" from an array and divides the array into those elements smaller than the splitting element and those bigger than it. It proceeds recursively until the entire array is sorted. The algorithm is:

Algorithm 2.2 Quicksort (X, l, r)**:** Input parameters: Bounds l and r, and an array X with elements x_{l-1}, ..., x_{r+1} where $l \le r$ and for $l \le i \le r$, $x_{l-1} \le x_i \le x_{r+1}$. Output parameters: A sorted permutation of the array X so that for $l - 1 \le i \le r$, $x_i \le x_{i+1}$. Functions: Split[X,i,l,r], which rearranges the elements x_l, x_{l+1}, ..., x_r of X so that the elements that are smaller than x_i are at the low index end and those that are larger are at the high index end.

Step **1.** If $l \ge r$, then return; the array segment is sorted.

Step **2.** Call Split[X,i,l,r], call Quicksort[X,l,$i - 1$], and call Quicksort[X,$i + 1$,r]. Then return; the array segment is sorted.

To sort an entire array X with elements x_1, x_2, ..., x_n, first set x_0 to a value that is at least as small as any element of X and set x_{n+1} to a value that is at least as large as any element of X. Then call Quicksort[X,1,n]. The extra elements x_0 and x_{n+1} are called *sentinels.* The use of sentinels eliminates the need to test whether array indices are in bounds in Steps 2 and 5 of Split. Since these steps are among the most frequently executed, and since they are so simple, the savings obtained by avoiding the extra testing makes the algorithm run about one-third faster than an algorithm that does not use sentinels.

Conceptually, the Split Algorithm works by using a moving "hole" into which elements are shifted until the desired arrangement is achieved. Originally the hole is created by putting the first element into a temporary variable x. This element is used as the splitting element. Two pointers are established: an upward moving one, i, that initially points at the hole, and a downward moving one, j, that initially points at the upper end of the segment. While the hole is at i, the j pointer is moved down past any elements that are larger than the splitting element x. If a smaller element is encountered, it is moved into the hole. The hole is now at position j, and the i pointer moves up until it finds an element larger than x. At that point the element is moved into the hole at j, and the hole is again at i. When the pointers meet, the algorithm is almost done. The element in x is put into the hole, and the algorithm returns.

Algorithm 2.3 Split (X, i, l, r)**:** Input parameters: An array X, a lower limit l, and an upper limit r, with $l \le r$. The array X is partly sorted so that, for $l \le j \le r$, $x_{l-1} \le x_j \le x_{r+1}$. Local variables: The splitting element x, a pointer i into the lower (small index) region, and a pointer j into the upper (high index) region. Output parameters: A splitting index i such that $l \le i \le r$, and a permutation of the elements $x_l, \ldots, x_r$ such that for $l \le j \le i$, $x_j \le x_i$, and for $i \le j \le r$, $x_i \le x_j$ (i.e., after Split is called, the upper region contains elements that are greater than or equal to the splitting element, while the lower region contains elements that are less than or equal to the splitting element).

Step **1.** [Initialize] Set $i \leftarrow l$, $j \leftarrow r$, and $x \leftarrow x_i$. (This step creates a hole at position i which does not contain any useful information — the information is moved into the variable x).

Step **2.** [Loop] Repeat the following steps until a return is done:

Step **3.** [Top compare] If $x_j > x$, then set $j \leftarrow j - 1$ and repeat this step. (Look for an element that should be moved out of the top part.)

Step **4.** [Done?] If $j \leq i$, then set $x_i \leftarrow x$ and return. (All elements are in place, so store x into the hole.)

Step **5.** [Move out of top] Set $x_i \leftarrow x_j$ and $i \leftarrow i + 1$. (The misplaced element is moved out of the top. This fills up the hole at position i but creates one at position j.)

Step **6.** [Bottom compare] If $x_i < x$, then set $i \leftarrow i + 1$ and repeat this step. (Look for an element that should be moved out of the bottom part.)

Step **7.** [Done?] If $j \leq i$, then set $x_j \leftarrow x$, $i \leftarrow j$, and return. (All elements are in place, so store x into the hole.)

Step **8.** [Move out of bottom] Set $x_j \leftarrow x_i$, set $j \leftarrow j - 1$. (Move the misplaced element out of the bottom. This fills up the hole at position j and creates one at position i.)

Step **9.** End repeat.

Table 2.1 shows the operation of Quicksort on some sample alphabetic data. In describing the operation of Quicksort we refer to the initial call to Quicksort as Quicksort (level 1). The calls from Quicksort (level i) invoke Quicksort (level $i+1$). The input column in Table 2.1 shows the initial data for Quicksort (level 1) where $l = 1$ and $r = 8$. The sentinels are in positions 0 and 9. At Step 2, Quicksort calls Split. Split, at Step 1, selects G (at position 1) as the splitting element. The splitting element is moved to variable x, and position 1 becomes a *hole*. At Step 3 the test fails (The B in position 8 is out of place). At Step 4 the test fails (we are not done). At Step 5 the B in position 8 is moved to position 1 (leaving a hole at 8), and i is increased to 2. At Step 6, i is increased to 3 (the H in position 3 is out of place). At Step 7 the H is moved to position 8 (leaving a hole at 3), and j is decreased to 7. At Step 3, j is decreased to 6. At Step 4 the test fails. At Step 5 the A in position 6 is moved to position 3 (leaving a hole at 6), and i is increased to 4. At Steps 6 and 7 the tests fail. At Step 8 the I in position 4 is moved to position 6 (leaving a hole at 4), and j is decreased to 5. At Steps 3 and 4 the tests fail. At Step 5 the F in position 5 is moved to position 4 (leaving a hole at 5), and i is increased to 5. At Step 6, i

Position	*Input*	*Intermediate*					*Output*
0	A	A	A	A	A	A	A
1	**G**	**B**	A	A	A	A	A
2	C	C	B	B	B	B	B
3	H	A	**C**	C	C	C	C
4	I	F	F	F	F	F	F
5	F	G	G	**G**	G	G	G
6	A	I	I	I	**I**	H	H
7	J	J	J	J	J	I	I
8	B	H	H	H	H	J	J
9	J	J	J	J	J	J	J

TABLE 2.1. Intermediate results of Quicksort. The input and output are shown in the columns so labelled. The intermediate columns show the partially sorted array immediately following each call to Split. Horizontal lines delimit the part of the array that has just been processed by Split. Dashed lines separate the previous splitting element from those elements that are smaller and from those elements that are bigger. The splitting element for the next round is shown in boldface.

is increased to 6. At Step 7 the splitting variable is stored in the hole, i is set to its position, and Split returns. The final results of these steps are shown in the first intermediate column of Table 2.1.

Quicksort (level 1) at Step 2 proceeds with its first recursive call to Quicksort (level 2) with $l = 1$ and $r = 4$. The call to Split selects the element in position 1 (B) for the splitting element. The call to Split gives the result shown in the second intermediate column of Table 2.1. Quicksort (level 2) calls Quicksort (level 3) with $l = 1$, $r = 1$, and Quicksort (level 3) returns at Step 1. Quicksort (level 2) proceeds with its second recursive call to Quicksort (level 3) with $l = 3$, $r = 4$. Quicksort (level 3) calls Split, which selects the element at position 3 (C) as the splitting element. The result of the call to Split is shown in the third intermediate column of Table 2.1. Quicksort (level 3) calls Quicksort (level 4), first with $l = 3$, $r = 2$ and then with $l = 4$, $r = 4$. In each case Quicksort (level 4) returns from Step 1. Now Quicksort (level 3) and then Quicksort (level 2) return from Step 2.

Next Quicksort (level 1) makes its second recursive call to Quicksort (level 2); this time with $l = 5$, $r = 8$. Split selects the element in position 5 (G) as the splitting element and produces the results shown in the fourth intermediate column of Table 2.1. Quicksort (level 2) calls Quicksort (level 3) with $l = 5$, $r = 4$ (and Quicksort (level 4) returns from Step 1), and then with $l = 6$, $r = 8$. Quicksort (level 4) calls Split to give the results in the last intermediate column. The two calls of Quicksort (level 4) to Quicksort (level 5) both result in Quicksort (level 5) returning from Step 1. Then Quicksort (level 4), (level 3), (level 2), and finally (level 1) finish to give the results in the output column of Table 2.1.

A variation of Quicksort is:

```
If l < r then
    begin call Split[X,i,l,r],
        if i - l < r - i then
            begin call Quicksort[X,l,i - 1],
                call Quicksort[X,i + 1,r]
            end else
            begin call Quicksort[X,i + 1,r],
                call Quicksort[X,l,i - 1]
            end
    end .
```

In other words, call Quicksort on the smaller part first.

A good implementation of this version needs much less stack space in the worst case than a good implementation of the previous version of Quicksort does.

The analysis of these algorithms is left for the exercises.

Quicksort algorithms are discussed by Knuth [11, Section 5.2.2], Aho et al. [1, Section 3.5], Baase [2, Section 2.3], Horowitz and Sahni [7, Section 3.5], Sedgewick [62, Chapter 9], and Reingold and Hansen [61, Section 8.1]. Quicksort was invented by Hoare [94].

EXERCISES

1. Are both sentinels, x_{l-1} and x_{r+1}, actually required by the Quicksort Algorithm when it uses the Split Algorithm as a subroutine? Justify your answer.
2. In Split let n be $r - l + 1$. What is the worst-case time for the Split Algorithm as a function of n?
3. What is the worst-case time for Quicksort? Hint: The worst case occurs when $i = l$ or $i = r$. Assume the algorithm is implemented in a way such that each recursive call can be done in constant time. This requires an implementation where the array X is not copied at each recursive call. Any algorithm for Quicksort that copies the entire array X between recursive calls will use a lot of space and time.
4. What is the the worst-case time for the variation of Quicksort? Compare the answer to this exercise with the answer to the previous one.
5. In the worst case, how large a stack is needed for the recursive calls to Quicksort? Assume that instead of actually making recursive calls, you program the algorithm so that it keeps its own stack of those items that must be saved from one recursive call, to the next. In particular, at the first recursive call your algorithm will save the current values of i and r (which will be needed by the second recursive call), and at the second recursive call it will save nothing (because the algorithm doesn't need to save any pointers for use after the second recursive call). (The technique of saving nothing at a recursive call that is the last statement of a routine is called *tail recursion.*) What is the worst-case stack size when tail recursion is not used? Compare the two worst-case stack sizes.
6. How large a stack (worst-case) does the variation on Quicksort need for the recursive calls? See the assumptions of the previous problem. What is the worst-case stack size when tail recursion is not used? Compare the two worst-case stack sizes. Compare the answer to this exercise with the answer to the previous one.

2.4 EXPONENTIAL SUMS

The sum $\sum_{0\le i\le n} x^i$ can be solved by relating it to itself and solving the resulting equation. Notice that $xx^i = x^{i+1}$, so if we let $S = \sum_{0\le i\le n} x^i$, the expression $xS - S$ can be simplified. That is,

$$xS - S = \sum_{0\le i\le n} x^{i+1} - \sum_{0\le i\le n} x^i \tag{58}$$

$$= x^1 + x^2 + \cdots + x^n + x^{n+1} - x^0 - x^1 - \cdots - x^{n-1} - x^n. \tag{59}$$

For $1 \le i \le n$, we have x^i in the front part of eq. (59) and $-x^i$ in the back part. (This is the reason we considered $xS - S$ in the first place.) These terms, of course, cancel, so we are left with

$$xS - S = x^{n+1} - x^0 = x^{n+1} - 1. \tag{60}$$

Solving for S gives

$$\sum_{0\le i\le n} x^i = \frac{x^{n+1} - 1}{x - 1}. \tag{61}$$

In the next section we study an algorithm whose analysis uses exponential sums.

2.4.1 Random Hashing

Computer programs often need to store information in such a way that it can be looked up by name. Each item to be remembered consists of a unique name or key, x_i, and a record, r_i. The problem for a table look-up program is to store this information in such a way that, if the program is later given x_i, it can look up r_i quickly. This is a very common problem and a variety of methods have been developed to handle it. The best method to use depends on the number of possible keys, M, and on the number of items that need to be remembered, n. Clearly $n \le M$. When n is small, linear search works fine. When n and M are about the same size, then you can usually use indexing, particularly if x_i is a number and $0 \le x_i < M$ for all i. If all the records are about the same size then you can use an array where position j is used to store the record whose key is j. (In this approach you also need a way to store an empty record — to indicate that there is no entry for a particular index.) This method, however, is very wasteful of memory when n is much less than M ($n \ll M$).

When $n \ll M$ and n is large, one usually uses a hashing method. Hashing requires a table size, N, where $N \ll M$, and a hashing function $f(x)$ such that $0 \le f(x_i) < N$ for all i. The hash function is used to compute an index, $f(x_i)$, for record r_i, where x_i is the key of record r_i. (The key is a portion of the record, and each record must have a unique key.) The hashing algorithm tries to put record r_i into location $f(x_i)$. There is just one problem with this method. Since $N \ll M$, there will be more than one key that produces the same value for the hash function. I.e., there exist $i \ne j$ such that $f(x_i) = f(x_j)$. This is the *collision* problem. Any hashing method must have some method of resolving collisions. The following algorithm uses a method which is not very practical, but which is much simpler to analyze than the more practical methods. It is very similar to the practical methods, however, so an analysis of its performance will help us understand the performance of the more practical methods.

The method is *random hashing.* In random hashing we have an infinite sequence of hashing functions $f_i(x)$ for $1 \le i$. The sequence of numbers generated for any x is random, but a given value x always causes the same sequence to be generated. That is, the probability that $f_i(x_a) = f_j(x_b)$ is $1/N$ unless $i = j$ and $a = b$, in which case the probability is one. The algorithm takes a key q and finds the location in the table where the key is stored. If the key is not in the table, then the algorithm returns the location appropriate for inserting the key and its associated record.

Algorithm 2.4 Random Hashing: Data structure: A key array X with entries x_i for $0 \le i \le N - 1$ and a corresponding record array R. Initial conditions: The number of entries, k, is zero and each key location, x_i, contains the value *empty*, a special value that is not the value of any key. Input: A query, q. Output: A location i such that $q = x_i$ if there is such an i (i.e., q is in the table) or i such that x_i contains *empty* and the hash algorithm would look in location x_i when looking for q (i.e., there exists an h such that $x_{f_j(q)} \ne \mathit{empty}$ for $1 \le j < h$, $x_{f_h(q)} = \mathit{empty}$, and $i = f_h(q)$). It is assumed that there is a second routine available which inserts items into the table. This second routine, when called, stores the key and corresponding record into the appropriate location in the table and increases k. The storing routine should report an error if $k = N$ after it inserts an item (otherwise, the next search for an item not in the table will not terminate).

Step **1.** [Initialize search] Set $h \leftarrow 0$.

Step **2.** [Search] Set $h \leftarrow h + 1$ and $i \leftarrow f_h(q)$. If $x_i = \mathit{empty}$, then stop (q is not in the table). If $x_i = q$, then stop (q has been found). Otherwise repeat this step.

Let's consider how long this algorithm takes when q is not in the table. The first step is done one time. The second step is done at least one time. There is no upper limit on how many times it can be done (this is one defect of the algorithm). Let's start the average time analysis by computing the probability that Step 2 is done at least once, then at least twice, etc. The average number of times that it is done is the sum of these probabilities.

The probability that Step 2 is done at least once is 1. Since we are assuming that q is not in the table, the algorithm will proceed to do Step 2 additional times unless an empty spot is found on the first execution of Step 2. The chance of finding an empty spot is $1 - k/N$, so the chance of going on is k/N. Since the search proceeds at random, whenever the algorithm has done the second step i times the chances of going on and doing the step some more times is again k/N. Thus if q_i is the probability that Step 2 is done at least i times, then

$$q_1 = 1 \quad \text{and} \quad q_i = \frac{k}{N} q_{i-1} \quad \text{for} \quad i \ge 2. \tag{62}$$

This reduces to

$$q_i = \left(\frac{k}{N}\right)^{i-1} \tag{63}$$

To calculate the average number of times Step 2 is done, we need the value of $\sum_{i\geq 1} ip_i$, where p_i is the probability that Step 2 is done *exactly* i times. We can express q_i, the probability that Step 2 is done *at least* i times, as $q_i = p_i + p_{i+1} + \cdots$. In particular,

$$q_1 = p_1 + p_2 + p_3 + \cdots \tag{64}$$

$$q_2 = \quad p_2 + p_3 + \cdots \tag{65}$$

$$q_3 = \quad\quad p_3 + \cdots \tag{66}$$

so

$$\sum_{i\geq 1} q_i = \sum_{i\geq 1} ip_i. \tag{67}$$

Thus the average number of times that Step 2 is done is

$$T = \sum_{1\leq i<\infty} \left(\frac{k}{N}\right)^{i-1} = \sum_{0\leq i<\infty} \left(\frac{k}{N}\right)^{i} \tag{68}$$

$$= \frac{(k/N)^\infty - 1}{k/N - 1} = \frac{1}{1-k/N} = \frac{N}{N-k} = 1 + \frac{k}{N-k} \tag{69}$$

The notation $(k/N)^\infty$ is short for $\lim_{i\to\infty}(k/N)^i$; since $0 \leq k/N < 1$, we have that $(k/N)^\infty = 0$.

The equation $1/(1-k/N) = 1 + k/N + (k/N)^2/(1-k/N)$ implies that when k/N is small compared to one, $T = 1 + k/N + O[(k/N)^2]$, but when k/N is close to one, T becomes very large.

Analyses of the Random Hashing Algorithm appear in Peterson [120], Morris [118], and Knuth [11, Section 6.4].

EXERCISES

1. How big does k/N have to be so that the probability that Step 2 in the Random Hashing routine is done more than once is greater than $1/2$?

2. How big does k/N have to be so that the average number of times that Step 2 in the Random Hashing routine is done is 2?

3. Suppose k items are put into a hash table in random order. Let T_i be the average time to look for an item not in the table when there are i items in the table (for example, with Random Hashing, $T_i = N/(N-i)$). Let U_k be the average time required to look up a random item that is in the table when the table contains k items. Calculate U_k in terms of T_i. Hint: When the item was first put into the table, the look-up time was the same (on the average) as the time for an item that was not in the table.

4. Sometimes one tries to minimize the product of the time a program uses and the space it uses. If we assume the time for Random Hashing is $N/(N-k)$ and the space is N, how large should the hash table be to minimize the product of space and time?

5. The assumptions in the last problem were not very realistic, in part because they ignored overhead. Suppose the running time is $a + N/(N-k)$ and the space is $b + N$. How large should N be (as a function of k, a, and b) to minimize the product of space and time?

6. The assumptions of the last problem are still not completely realistic. What is wrong? Hint: Does an algorithm with all the assumed space and time properties exist?

7. One difficulty with Random Hashing is that there is no good way to do deletion. Give a correct delete algorithm. (Be sure that when one item is deleted other items do not become inaccessible.) For this exercise you are not permitted to make any changes to the table structure. Thus you cannot use pointers or other data to facilitate deletion. Do an analysis to show that your method requires a long time to do a deletion. (A complete analysis might very well lead to mathematical difficulties that you do not yet know how to handle, but all you need to do is show that the delete time is large. That can be done with elementary techniques.) Sometimes a very simple analysis is all that is needed to show that an algorithm is unsuited for its intended purpose.

2.5 SUMMATION BY PARTS

Summation by parts is a general technique for simplifying sums, analogous to the integration by parts technique used in calculus. Summation by parts often permits replacement of one sum by a simpler one.

Consider a sum of the form $\sum_{m\le i<n} c_i b_i$: one where the summand is the product of two functions. The basic idea of summation by parts is to apply the summation operator to one of the factors and the difference operator to the other.

Suppose we have a sequence $\{a_i\}$ where $a_{i+1} - a_i = c_i$. Then

$$\sum_{m\le i<n} c_i b_i = \sum_{m\le i<n} (a_{i+1} - a_i) b_i = \sum_{m\le i<n} a_{i+1} b_i - \sum_{m\le i<n} a_i b_i \tag{70}$$

$$= \sum_{m\le i<n} a_{i+1} b_i - \sum_{m-1\le i<n-1} a_{i+1} b_{i+1} \tag{71}$$

$$= a_n b_n - a_m b_m - \sum_{m\le i<n} a_{i+1}(b_{i+1} - b_i). \tag{72}$$

This is the summation by parts formula. (Note the use of $<$ rather than $\le$ on the upper limit of the summation.)

The sequence $\{a_i\}$ can be any sequence satisfying the condition $a_{i+1} - a_i = c_i$, or, in operator notation, $\{c_i\} = \Delta\{a_i\}$. The easiest way to obtain such a sequence is to sum the c_i: the sequence $\{c_0,\ c_0 + c_1,\ c_0 + c_1 + c_2,\ \ldots\}$ has the required differences. Any sequence whose terms differ from these by a constant can also be used. In other words, to apply summation by parts to the formula $\sum_{m\le i<n} c_i b_i$, you should find the a_i using the formula

$$a_i = C + \sum_{m\le j<i} c_j, \tag{73}$$

where C is a constant. Often the value $C = 0$ is convenient, but any value is okay.

You should consider using summation by parts on any sum whose terms are the product of two factors. The first decision you must make is which factor to call b_i and which factor to call c_i. The c_i must be summed to obtain the a_i and the difference $b_{i+1} - b_i$ must be computed, so you must know how to sum $\{c_i\}$ and difference $\{b_i\}$. If the difference is simpler than b_i or the sum is simpler than c_i, then summation by parts may be helpful. Often the difference will be simpler while the sum will be more complex.

Let's consider the sum $\sum_{0\le i<n} ix^i$. There are two obvious ways to apply summation by parts to this sum: (1) with $c_i = i$ and $b_i = x^i$ or (2) with $c_i = x^i$ and $b_i = i$. Since $\sum_{0\le i<n} i = n(n-1)/2$, in the first case we may take $a_i = i(i-1)/2$. We also have $b_{i+1} - b_i = x^i(x-1)$, so the sum on the right side of eq. (72) will have terms of the form $i(i-1)x^i$ (after the constant terms are factored out). Since this term is more complicated than the term ix^i in the original sum, this approach is not very promising.

The second alternative is to use $c_i = x^i$ and $b_i = i$. Here $\sum_{0\le i<n} x^i = (x^n - 1)/(x-1) = x^n/(x-1) - 1/(x-1)$. Since $1/(x-1)$ is a constant, we can drop it and use $a_i = x^i/(x-1)$. The difference $b_{i+1} - b_i = 1$, so we have

$$\sum_{0\le i<n} ix^i = \frac{nx^n}{x-1} - \sum_{0\le i<n} \frac{x^{i+1}}{x-1} \tag{74}$$

$$= \frac{nx^n}{x-1} - x\frac{x^n-1}{(x-1)^2} = \frac{(n-1)x^{n+1} - nx^n + x}{(x-1)^2}, \tag{75}$$

which is the solution of our problem.

This approach can be used to solve any sum of the form $\sum_{0\le i<n} p(i)x^i$, where $p(i)$ is a polynomial in i. If you use $b_i = p(i)$, then $b_{i+1} - b_i$ will be a polynomial of one lower degree, while a_k will not be significantly more complex than $c_i = x^i$. Repeated application of summation by parts will therefore lead to a solution.

Summation by parts is covered in Knuth [9, exercise 1.2.7-10]. The relation between summation by parts and integration by parts is covered extensively in Milne-Thomson [43, Section 2.64] and concisely in Greene and Knuth [6, Section 4.2].

EXERCISES

1. Simplify $\sum_{i\ge 1} i^2x^i$. (The lack of an upper limit for i indicates that this is an infinite sum. The sum is over *all* integer values for i that are greater than or equal to one.)

2. Simplify $\sum_{i\ge 1} i^3x^i$.

3. Express $\sum_{0\le k<n} k^m z^k$ in terms of sums of the form $\sum_{0\le k<n} k^j z^k$, $j<m$

4. Simplify $\sum_{i\ge 1}(i^2 - i)x^i$.

2.5.1 Cumulative Probabilities

Summation by parts has an interesting application to the computation of averages. The usual way to compute the average number of times that a step is done is to use the formula

$$A = \sum_{0 \leq i < n} i p_i, \tag{76}$$

where n is large enough that $\sum_{0 \leq i < n} p_i = 1$. If we rewrite this formula using summation by parts with $b_i = i$, we get

$$A = n - \sum_{0 \leq i < n} a_{i+1}, \tag{77}$$

where $a_i = \sum_{0 \leq j < i} p_j$. Now define

$$q_i = \text{Prob(the step is done more than } i \text{ times)} = \sum_{j > i} p_j. \tag{78}$$

With this definition $q_i = 1 - a_{i+1}$. Thus

$$A = \sum_{0 \leq i < n} q_i. \tag{79}$$

If the q_i are easier to compute than the p_i (as they were is Section 2.4.1), this formula may provide a good method for finding the average time.

2.5.2 Heap Sorting

A *heap* is an array X with elements x_i for $1 \leq i \leq N$ where $x_i \geq x_{2i}$ and $x_i \geq x_{2i+1}$ for all i where the indices are in the array ($1 \leq i \leq \lfloor N/2 \rfloor$ for $x_i \geq x_{2i}$ and $1 \leq i \leq \lfloor (N-1)/2 \rfloor$ for $x_i \geq x_{2i+1}$). A heap is an example of an implicit data structure. If you think of the first element in the array as the root of a binary tree, the next two as its children, and so on (i.e., element x_i is the parent of elements x_{2i} and x_{2i+1}), then the heap is a binary tree where each element is at least as large as either of its children. The relations among elements in the tree are implied by their position in the array, so no explicit pointers are needed. The heap condition does not impose very much order on the array so a heap can be built quickly. The largest element of the array is always the first element in the heap.

The Heapsort Algorithm works by first making its input array into a heap. It then removes the first element (the largest) from the heap and replaces it with the last element of the array. The largest element is stored in the now empty last position. The new array, minus its last element, is then made into a heap, and the process is repeated. The final result is an array with the elements sorted from smallest to largest.

The Heap Algorithm, a subroutine of Heapsort, assumes that the portion of the array it is to work on is almost, but not quite, a heap. If x_l is the root element, then the implicit subtrees starting at x_{2l} and x_{2l+1}, must already be heaps. Let's call this the "almost heap" property. If x_l is not larger than x_{2l}

and x_{2l+1}, the Heap Algorithm interchanges x_l with its largest child. Now the element that started at position l is again the root of an implicit subtree with the almost heap property. Also, the elements starting at position l form a subheap, except for the one subtree that is an almost heap. The element is moved down until the elements starting at position l form a heap; the process continues until the heap property is satisfied for position one.

The initial heap is built by starting with almost heaps consisting of three elements (any three elements form an almost heap) and building larger and larger heaps by calls to Heap.

Algorithm 2.5 Heapsort: Input: an array X with elements x_i for $1 \leq i \leq N$. Output: an array X that is a sorted rearrangement of the input.

Step **1.** [Build heap] For $l \leftarrow \lfloor N/2 \rfloor$ to 1 call Heap(l,N,X). This step rearranges X into a heap. No processing is needed for elements $x_{\lfloor N/2 \rfloor +1}$ to x_N because each of these elements forms a trivial one-element subheap.

Step **2.** [Get biggest] For $r \leftarrow N-1$ to 2 do an interchange of x_1 and x_{r+1} followed by a call to Heap(1,r,X). This step repeatedly takes the biggest element out of the heap and rebuilds the heap (with one less element).

Step **3.** [Finish] Interchange x_1 and x_2.

Algorithm 2.6 Heap(l, r, X): Input Parameters: A starting place l, an ending place r, and an array X, where X obeys the "almost heap" property for the subtree starting at position l. Output Parameters: An array X rearranged so that the heap condition, $x_i \geq x_{2i}$ and $x_i \geq x_{2i+1}$, is satisfied for $l = i$ and for any elements that are rearranged.

Step **1.** [Initialize] Set $x \leftarrow x_l$ and $j \leftarrow l$.

Step **2.** [Find children] Set $i \leftarrow j$ and $j \leftarrow 2j$. If $j > r$, then go to Step 5 (there are no children).

Step **3.** [Find larger child] If $j < r$ and if $x_j < x_{j+1}$, then set $j \leftarrow j+1$.

Step **4.** [Move?] If $x < x_j$, then set $x_i \leftarrow x_j$ and go to Step 2. (Otherwise we have found where x goes.)

Step **5.** [Store] Set $x_i \leftarrow x$ and return.

Step 2 of the Heap subroutine is done as often as any other step in the subroutine. Since j increases by at least a factor of 2 each time around the loop, since the loop stops when $j > r$, and since $j = 2l$ the first time it is compared with r, Step 2 is done no more than $1 + \lfloor \lg(r/l) \rfloor$ times.

Now let's compute how often Step 2 of Heap is done as a result of calls from Step 2 of the Heapsort Algorithm. Step 2 of Heapsort calls Heap with $l = 1$ and $r = N - 1, \ldots, 2$, so in the worst case Step 2 of Heap is done

$$N - 2 + \sum_{2 \le r < N} \lfloor \lg r \rfloor \tag{80}$$

times. To simplify this, use summation by parts with $b_r = \lfloor \lg r \rfloor$ and $a_r = r$ (notice that here we are using summation by parts in a case where one of the factors is 1). Now $b_{r+1} - b_r$ is quite simple; it is zero unless $r + 1$ is a power of 2, since the logarithm base 2 increases slowly and goes up to the next integer at each power of 2. So we have

$$N - 2 + \sum_{2 \le r < N} \lfloor \lg r \rfloor = N \lfloor \lg N \rfloor + N - 4 - \sum_{2 \le r < N} (r+1)(\lfloor \lg(r+1) \rfloor - \lfloor \lg r \rfloor) \tag{81}$$

$$= N \lfloor \lg N \rfloor + N - 4 - \sum_{2 \le i < \lfloor \lg N \rfloor + 1} 2^i \tag{82}$$

$$= N \lfloor \lg N \rfloor + N - 2^{\lfloor \lg N \rfloor + 1}. \tag{83}$$

Now consider just how big the $2^{\lfloor \lg N \rfloor + 1}$ term is. We have

$$2^{\lg N} < 2^{\lfloor \lg N \rfloor + 1} \le 2^{\lg N + 1} \tag{84}$$

because rounding down to the next integer decreases the value by no more than one. Since $2^{\lg N} = N$ and $2^{\lg N + 1} = 2N$ and since $N \lg N - N < N \lfloor \lg N \rfloor \le N \lg N$, we have that the number of times that Step 2 of Heapsort causes Step 2 of Heap to be executed is

$$N \lg N + O(N). \tag{85}$$

Heapsort is covered in Knuth [11, Section 5.2.3], Aho et al. [1, Section 3.4], Baase [2, Section 2.5], Horowitz and Sahni [7, Section 2.3], Sedgewick [62, Chapter 11], Reingold and Hansen [61, Section 8.1.3], and Standish [63, Section 3.7.1]. Heapsort was invented by Williams [138].

EXERCISES

1. Give a simple proof that the total worst-case time for the Heapsort Algorithm is $O(N \ln N)$. Notice how much easier it is to obtain a big O result than to obtain an exact result.

2. Show that the number of times that Step 2 of the Heap Algorithm is done because of calls from Step 1 of the Heapsort Algorithm is at most

$$\left\lfloor \frac{N}{2} \right\rfloor + \sum_{1 \le i \le N/2} \left\lfloor \lg \left(\frac{N}{i} \right) \right\rfloor .$$

Show that when N is a positive power of 2, the number of times is equal to $\frac{3}{2}N - 1$. Show that for any value of N, the number of times is equal to $O(N)$.

3. [59] Develop an efficient algorithm for finding the k largest elements of a set of N elements, and show that the time for your algorithm is $O(N + k \lg N)$. How large can k be, as a function of N, and still have this time be $O(N)$? Hint: Consider having your algorithm work like the first two and a half steps of Heapsort.

2.5.3 Random Hashing (Variance)

The average time for random hashing was computed in Section 2.4.1. Now we compute the variance of the time. Once again we will concentrate on Step 2 and compute the variance for how often Step 2 is done. We continue to use q_i for the probability that Step 2 is done at least i times. The probability that Step 2 is done exactly i times is

$$p_i = q_i - q_{i+1} = \left(\frac{k}{N} \right)^{i-1} \left(1 - \frac{k}{N} \right) = \left(\frac{k}{N} \right)^{i-1} \left(\frac{N-k}{N} \right). \tag{86}$$

Let $x = k/N$. To compute the variance with eq. (1.57), we first need

$$\sum_{i \ge 1} i^2 p_i = \frac{(1-x)}{x} \frac{x + x^2}{(1-x)^3} = \frac{1+x}{(1-x)^2} \tag{87}$$

by Exercise 2.5–1. So the variance is

$$V = \frac{1+x}{(1-x)^2} - \frac{1}{(1-x)^2} = \frac{x}{(1-x)^2}. \tag{88}$$

The variance for random hashing is small when x is much less than 1.

2.6 FRACTIONAL SUMS

So far we have learned how to sum polynomials, simple exponentials, and their products. To complete the discussion of summing elementary functions, we will now consider sums where the summation index appears to a negative power. In the previous cases it was always possible to find an exact answer in closed form. This is not always possible when the summation index appears to a negative power. For example, there is no exact closed-form answer for the most basic such sum:

$$\sum_{1 \le i \le n} \frac{1}{i}. \tag{89}$$

This sum occurs often in the analysis of algorithms; we study how to approximate it in Sections 4.3, 4.5.1, and 4.5.4. Likewise, there is no exact value for

$$\sum_{1\le i\le n} \frac{1}{i^2}. \tag{90}$$

On the other hand, we have

$$\sum_{1\le i\le n} \frac{1}{i(i+1)} = 1 - \frac{1}{n+1}. \tag{91}$$

Our general plan for simplifying sums in which the index appears on the bottom of fractions is to convert the problem to a standard form and then to see if the summations of the fractions can be made to cancel. If they do, we obtain a simple answer. Otherwise the problem does not have a simple exact answer.

The first step is to reexpress the fractions by using partial fraction decomposition. The first step in partial fraction decomposition is to express the fraction in proper form with a quotient and a remainder. Thus if the fraction is $N(i)/D(i)$, where $N(i)$ and $D(i)$ are polynominals in i, then by long division we have

$$\frac{N(i)}{D(i)} = Q(i) + \frac{R(i)}{D(i)}, \tag{92}$$

where $Q(i)$ is the quotient of $N(i)$ divided by $D(i)$ and $R(i)$ is the remainder [the degree of $R(i)$ is less than than of $D(i)$]. For example,

$$\frac{i^3 - i + 1}{i(i+1)} = i - 1 + \frac{1}{i(i+1)}. \tag{93}$$

When the degree of $N(i)$ is less than the degree of $D(i)$, of course, $Q(i) = 0$. If $R(i)$ is zero, then you are done. Your problem does not really involve fractions.

The next step in partial fraction decomposition is to factor the denominator. After $D(i)$ has been factored it can be written as

$$D(i) = P_1^{a_1}(i)P_2^{a_2}(i)\cdots P_n^{a_n}(i), \tag{94}$$

where $P_j(i)$ is the jth factor of $D(i)$, a_j is the power to which it occurs, and n is the number of factors. In the above example with $1/(i(i+1))$, $P_1(i) = i$, $a_1 = 1$, $P_2(i) = i + 1$, $a_2 = 1$, and $n = 2$.

The partial fraction decomposition theorem says that the fractional part, $R(i)/D(i)$, can be written as

$$\frac{R(i)}{D(i)} = \sum_{1\le j\le n} \frac{T_j(i)}{P_j^{a_j}(i)}, \tag{95}$$

where each T_j is a polynomial with degree less than that of $P_j^{a_j}(i)$. For example,

$$\frac{1}{i(i+1)} = \frac{a}{i} + \frac{b}{i+1}, \tag{96}$$

for some a and b. The a and b are constants (zero-degree polynomials).

The last step of partial fraction decomposition is to find the $T_j(i)$ that make eq. (95) true. The most straightforward way to do this is to generate a set of simultaneous linear equations with the coefficients of the $T_j(i)$ as unknowns. The equations are obtained by plugging in values for i in eq. (95). Distinct values of i will usually result in independent equations; enough independent equations are needed to equal the number of unknowns.

For example, in eq. (96) we can plug in $i = 1$ and $i = 2$. With $i = 1$ we get

$$\frac{1}{2} = a + \frac{b}{2}, \tag{97}$$

while with $i = 2$ we get

$$\frac{1}{6} = \frac{a}{2} + \frac{b}{3}. \tag{98}$$

The solution of eqs. (97, 98) is

$$a = 1, \quad b = -1. \tag{99}$$

Thus the complete partial fraction decomposition of $(i^3 - i + 1)/(i(i+1))$ is

$$\frac{i^3 - i + 1}{i(i+1)} = i - 1 + \frac{1}{i} - \frac{1}{i+1}. \tag{100}$$

The partial fraction decomposition gets the summand into the simplest form for summing. If all the fractional parts cancel out in the sum, then the summation problem has a closed-form answer. To continue the example,

$$\begin{aligned}
\sum_{1 \le i \le n} \frac{i^3 - i + 1}{i(i+1)} &= \sum_{1 \le i \le n} \left(i - 1 + \frac{1}{i} - \frac{1}{i+1} \right) && (101) \\
&= \sum_{1 \le i \le n} i - \sum_{1 \le i \le n} 1 + \sum_{1 \le i \le n} \frac{1}{i} - \sum_{1 \le i \le n} \frac{1}{i+1} && (102) \\
&= \frac{n(n+1)}{2} - n + \sum_{1 \le i \le n} \frac{1}{i} - \sum_{2 \le i \le n+1} \frac{1}{i} && (103) \\
&= \frac{n^2 - n}{2} + 1 - \frac{1}{n+1}. && (104)
\end{aligned}$$

If the fractional parts do not cancel, then the sum does not have a closed form, and you must use the approximate techniques that are developed in Chapter 4.

As an example with a factor to a power higher than 1, consider the partial fraction decomposition of $1/(i^2(i+1))$:

$$\frac{1}{i^2(i+1)} = \frac{ai + b}{i^2} + \frac{c}{i+1} \tag{105}$$

Substituting $i = 1$, 2, and 3 into eq. (105) yields

$$\frac{1}{2} = a + b + \frac{c}{2}, \tag{106}$$

$$\frac{1}{12} = \frac{2a+b}{4} + \frac{c}{3}, \tag{107}$$

$$\frac{1}{36} = \frac{3a+b}{9} + \frac{c}{4}. \tag{108}$$

Solving for a, b, and c yields $a = -1$, $b = 1$, and $c = 1$, so

$$\frac{1}{i^2(i+1)} = \frac{-i+1}{i^2} + \frac{1}{i+1}. \tag{109}$$

The results in this section on the impossibility of summing some series directly come from the work of Karr [103]. Section 6.4 of this book has an introduction to his work, although you will have to finally consult his paper to find out why some sums do not have simple closed-form answers.

2.6.1* Finding Partial Fraction Decompositions

Although the method of solving partial fractions by reducing them to linear equations is simple in principle, it often leads to long, error-prone calculations. In this section we give a method that leads to simpler computations. There is one particular value of i that is good for finding a $T_j(i)$ in eq. (95), and this is the value that causes $P_j(i)$ to go to zero! With this value for i, the $T_j(i)$ of interest is being divided by an arbitrarily small number, while the other $T_j(i)$ are being divided by numbers that are not going to zero. Thus in the limit only the one $T_j(i)$ is relevant. To avoid dealing with limits of terms that have division by zero, you should first multiply eq. (95) by $P_j^{a_j}(i)$, reduce the left and right sides to lowest terms by cancelling $P_j^{a_j}(i)$ wherever it occurs on both the top and bottom of a fraction, and then let i approach a root of $P_j(i)$. This produces an equation where just one term on the right side is nonzero. If $P_j^{a_j}(i)$ is a polynomial of degree 1, then $T_j(i)$ is a constant, and you can solve for its value. Even when $T_j(i)$ is not a constant, this method gives useful information about $T_j(i)$.

Let's see how this works on an example. Starting with eq. (96) multiplied by i, we get

$$\lim_{i\to 0} \frac{i}{i(i+1)} = \lim_{i\to 0} i\frac{a}{i} + \lim_{i\to 0} i\frac{b}{i+1}, \tag{110}$$

$$\lim_{i\to 0} \frac{1}{i+1} = \lim_{i\to 0} a + \lim_{i\to 0} b\frac{i}{i+1}. \tag{111}$$

Since $\lim_{i\to 0} 1/(i+1) = 1$, $\lim_{i\to 0} a = a$, and $\lim_{i\to 0} bi/(i+1) = 0$, eq. (111) gives

$$a = 1. \tag{112}$$

Likewise, multiplying eq. (96) by $i+1$ and taking the limit as i goes to -1 gives

$$\lim_{i\to-1}\frac{i+1}{i(i+1)} = \lim_{i\to-1}\frac{a(i+1)}{i}+\frac{b(i+1)}{i+1}, \tag{113}$$

$$-1 = b, \tag{114}$$

so $b=-1$.

Let's now use this method on eq. (105) to see what partial information this method gives you when $T_j(i)$ has degree greater than 1. Multiplying eq. (105) by i^2 and taking the limit as i goes to zero gives

$$\lim_{i\to 0}\frac{i^2}{i^2(i+1)} = \lim_{i\to 0}\frac{i^2(ai+b)}{i^2}+\lim_{i\to 0}\frac{i^2c}{i+1}, \tag{115}$$

$$1 = b. \tag{116}$$

In the same way, multiplying eq. (105) by $i+1$ and taking the limit as i goes to zero gives $c=1$. The method is this section gives no information about the value of a, but using $b=1$ and $c=1$ in eq. (106) gives $a=-1$.

The method in this section can be used to compute partial fraction decompositions with less work than the method given in the previous section. If, however, you find this method too complex, you can always use the method in the previous section.

Partial fraction decomposition is covered by Johnson and Kiokemeister [37, Section 12.3] and by Apostol [18, Section 3.18].

EXERCISES

1. Simplify $\displaystyle\sum_{2\le i\le n}\frac{1}{i^2-1}$.

2. Simplify $\displaystyle\sum_{1\le i\le n}\frac{1}{i(i+1)(i+2)}$.

3. Why is it that the method in this section does not give a value for $\displaystyle\sum_{1\le i\le n}\frac{1}{i}$?

2.7* DERIVATIVES AND INTEGRALS OF SUMS

If $S(x)$ is a finite sum $\sum_{1\le i\le n} a_i(x)$ that depends on a parameter x, then

$$\frac{dS(x)}{dx} = \sum_{1\le i\le n}\frac{da_i(x)}{dx} \tag{117}$$

and

$$\int S(x)\,dx = \sum_{1\le i\le n}\int a_i(x)\,dx. \tag{118}$$

Eqs. (117, 118) also apply to infinite sums when the functions are uniformly convergent (see Knopp [40]).

Eqs. (117, 118) are useful for many of the same sums that summation by parts is useful for. Often, however, summation by parts requires more algebra.

We will illustrate the use of eq. (118) by simplifying $S(x) = \sum_{0\le i\le n}(i+1)x^i$. From eq. (118) we have

$$\int S(x)\,dx = \sum_{0\le i\le n} x^{i+1} = x\sum_{0\le i\le n} x^i = x\frac{x^{n+1}-1}{(x-1)}. \tag{119}$$

Since $(d/dx)\int S(x)\,dx = S(x)$, differentiating eq. (119) gives

$$S(x) = \frac{d}{dx}\left(x\frac{x^{n+1}-1}{x-1}\right) = \frac{(n+1)x^{n+2}-(n+2)x^{n+1}+1}{(x-1)^2}. \tag{120}$$

When you differentiate a sum term by term, you get the derivative of the answer. To obtain the answer you must integrate, which introduces a constant of integration. To find the constant of integration, it is necessary to evaluate the sum for some value of x by some other means. Often there is some value of x, such as zero, for which it is easy to obtain the value of the sum directly.

The conditions under which you can interchange the order of integration or differentiation and summation are discussed by Apostol [17, Sections 13.10 and 13.11, 18, Section 9.30].

EXERCISE

1. Simplify $\sum_{i\ge 0} i^2x^i$.

CHAPTER 3

Products and Binomials

The product of a fixed number of items can be indicated by writing the factors next to each other. Thus $a_1a_2a_3$ is the product of a_1, a_2, and a_3. When there is a variable number of items, you can use the *dot dot dot* notation in situations where the notation is clear. Thus $a_1a_2\cdots a_n$ is the product of the numbers from a_1 to a_n. You can also use product notation. Thus

$$\prod_{R(i)} a_i \tag{1}$$

is the product of those a_i such that $R(i)$ is *true*. For example,

$$\prod_{1\le i\le n} i \tag{2}$$

is the product of all the numbers from 1 to n.

Some products are so common that they receive a special notation. Thus

$$n^k = \prod_{1\le i\le k} n. \tag{3}$$

The product notation has two useful conventions for the meaning of eq. (1) when $R(i)$ is not satisfied by any i. The first convention is that the product begins with the value 1, and if there are any terms in the product, they multiply that value. When this convention is used, the empty product has the value 1. The other convention is that if the product has an upper and lower limit,

$$\prod_{m\le i<n} a_i \quad \text{means} \quad \prod_{n\le i<m} a_i^{-1} \tag{4}$$

when $n < m$, and it is 1 when $m = n$. This second convention is similar to the second convention for summation, and it has the same advantages. For example, with this convention eq. (4) gives the correct definition of n^k even when $k \le 0$ (provided $n \ne 0$). Usually you will work with products where the upper limit is above the lower limit, so you will not have to worry about these conventions. When you need one of these conventions, be sure that you are using one that applies to your situation. In those cases where it applies, we use the second convention unless we state otherwise.

There is a close relationship between sums and products. Since the logarithm of a product is the sum of the logarithms of the factors, we have

$$\prod_{R(i)} a_i = \exp\left(\sum_{R(i)} \ln a_i\right). \tag{5}$$

Thus computing the product of a series of factors is of the same difficulty as computing the *sum* of the logarithm of the factors.

EXERCISES

1. Simplify $\prod_{1\le i\le n} a^i$.

3.1 FACTORIALS

The notation $n!$ (n factorial) is used for the product

$$n! = \prod_{1\le i\le n} i. \tag{6}$$

Factorials are quite common in analysis of algorithms.

The ratio of two factorials can be used to represent the product of any sequence of consecutive integers. Thus

$$\frac{n!}{(n-k)!} = \prod_{n-k+1\le i\le n} i. \tag{7}$$

In Section 2.3 decreasing and increasing powers were defined. These definitions can be rewritten with factorials as

$$n^{\underline{k}} = \frac{n!}{(n-k)!} \tag{8}$$

and

$$n^{\overline{k}} = \frac{(n+k-1)!}{(n-1)!}. \tag{9}$$

It is desirable to design conventions so that formulas will work in as many cases as possible. Eq. (6) defines the factorial function for $n \ge 1$. Using either of our conventions for empty products, it also defines

$$0! = 1. \tag{10}$$

Notice that for positive n, the factorial function obeys the equation

$$n! = n(n-1)!; \tag{11}$$

that is, you can obtain $n!$ by multiplying $(n-1)!$ by n. If eq. (11) is to work for $n \leq 0$, then it is necessary to define

$$n! = \infty \quad \text{for} \quad n < 0. \tag{12}$$

This is consistent with our second convention for products.

It is convenient to be able to express $n!$ in terms of more elementary operations. We will show in Section 4.5.5 that for $n \geq 1$

$$n! = \sqrt{2\pi n}\left(\frac{n}{e}\right)^n \left(1 + O\left(\frac{1}{n}\right)\right), \tag{13}$$

which is quite accurate (in a relative sense) when n is large.

Factorials are discussed by Knuth [9, Section 1.2.5].

EXERCISE

1. What is the approximate value of $15!$?

3.1.1 Permutations

Suppose you have n distinct objects, and you want a linear arrangement of k of them. How many different arrangements can you have? For example, if your objects are a, b, and c, then the arrangements of the three objects taken two at a time are ab, ac, ba, bc, ca, and cb. There are six arrangements. The number of linear arrangements of n objects taken k at a time is called the number of *permutations of n objects taken k at a time*. There are n possible choices for the first item. For any particular first item, there are $n-1$ choices for the second item. Once the first two items are selected, there are $n-2$ choices for the third item, etc. Thus there are n permutations of n objects taken one at a time, $n(n-1)$ for n taken two at a time, $n(n-1)(n-2)$ for n taken three at a time, etc. In general, the number of permutations of n objects taken k at a time is

$$\prod_{0 \leq i < k} (n-i) = \frac{n!}{(n-k)!}. \tag{14}$$

The number of permutations of n objects taken n at a time is called the number of *permutations of n objects*, and is $n!$.

The most elementary way to write a permutation is to write the elements in their original order on one line and to write the elements in the new order on the line below. Thus if the original order is $abcdef$ and the new order is $ceabdf$, we can write $\begin{Bmatrix} a & b & c & d & e & f \\ c & e & a & b & d & f \end{Bmatrix}$. This notation suggests a second way to think of a permutation. Instead of thinking of c as becoming the first object, e becoming the second object, etc., we can equally well think of a changing into c, b changing into e, etc. The result is the same whichever way we think of the permutation.

The product of two permutations is the permutation that is obtained by doing the first permutation and then doing the second one. For example,

$$\left\{\begin{matrix} a & b & c & d & e \\ c & e & a & b & d \end{matrix}\right\}\left\{\begin{matrix} a & b & c & d & e \\ b & c & d & e & a \end{matrix}\right\} = \left\{\begin{matrix} a & b & c & d & e \\ d & a & b & c & e \end{matrix}\right\}. \tag{15}$$

A *cycle of a permutation* is written in the form $(a_1 a_2 \ldots a_k)$. It represents the permutation where a_1 is replaced by a_2, a_2 by a_3, and so on, with a_k replaced by a_1. Any permutation can be written as the product of a set of disjoint cycles.

To convert from the two-line notation to cycle notation, do the following:

1. If all the symbols (from the first line of the two-line representation) have been output, then stop. Otherwise write a beginning parenthesis.

2. Write the first symbol (from the first line of the two-line representation) that has not yet been written. Call this the starting symbol. Also call it the current symbol.

3. Set the current symbol to the symbol below the current symbol (in the two-line representation). If this symbol is not the start symbol, then output the current symbol and repeat this step.

4. Output a closing parenthesis and go back to Step 1.

If we apply this procedure to the permutation $\left\{\begin{matrix} a & b & c & d & e & f \\ c & e & a & b & d & f \end{matrix}\right\}$, we get $(ac)(bed)(f)$. The procedure produces each cycle with its smallest element first, and the cycles are ordered by the size of their smallest element. There is a unique way to write each permutation in the form of disjoint cycles such that each cycle has its smallest element first and such that the cycles are ordered by their smallest element.

It is customary to omit the 1-cycles when writing permutations in cycle form. With this shortened notation, the above permutation is $(ac)(bed)$.

EXERCISES

1. How many permutations are there of n objects taken zero at a time?

2. What are the permutations of a, b, c, d, e taken three at a time?

3.1.2 Anagrams

Let's consider how to design an efficient algorithm to find all the classes of anagrams contained in a dictionary of English words. (Two words are anagrams of each other if one can be obtained from the other by rearranging letters. For example, "deposit", "dopiest", "posited", and "topside" are anagrams of each other.) This problem is interesting because it has an inefficient straightforward solution and a fairly simple efficient solution. Also the analysis required to tell which method is best is easy.

The first approach is to form all the permutations of a word and compare each permutation against each word. The process is then repeated for every word. We can speed up this naïve approach by treating separately each word length. (Two words that are anagrams have the same length.) The number of permutations of an i letter word is $i!$. Let n_i be the number of words with i letters. To compare one permutation of one word with all the other words requires $n_i - 1$ comparisons. This approach uses $n_i(n_i - 1)i!$ comparisons when processing all the words of length i, so the total time is proportional to

$$\sum_{1 \le i \le l} n_i(n_i - 1)i! \tag{16}$$

where l is the length of the longest word.

How bad is this time? If we assume that there are 1000 words with 7 letters, then about 7×10^9 comparisons are done for 7-letter words. This is not too good, but for longer words the time is much worse. If we assume that there are 20 words with 15 letters, then about 5×10^{14} comparisons are done for them. The longer words take an inordinate amount of time because $i!$ grows very rapidly.

An obvious improvement in the algorithm is to use binary search when comparing the permutations of a word with the other words in the dictionary. This reduces the number of comparisons per permutation from $n_i - 1$ to $\lg n_i$. Removing the matched words from the dictionary (as they are found) would also help. But these improvements do not get at the main part of the problem.

The problem comes from the $i!$. A new approach is needed to get around it. Thinking more carefully about the problem reminds us that the objective is to find out which words contain the same letters. Sorting the letters of each word gives a "signature" which is the same for every word in a set of anagrams. Comparing signatures means that each word is compared against each other word only once. The $i!$ is eliminated, giving a remarkable improvement. If we produce the signatures with Insertion Sort, which requires time $O(n^2)$, then we need time $n_i O(i^2)$ to find all the signatures. If we then compare each signature against each other signature, we do $O(n_i^2)$ comparisons while comparing signatures. The total time with this approach is

$$\sum_{1 \le i \le l} [n_i O(i^2) + O(n_i^2)]. \tag{17}$$

The second term in this sum is the more important term since the number of words is *much* larger than the length of any word. If you calculate the time required to process 1000 words with 7 letters using a reasonable value for the constant in the big O, you will see that this algorithm is thousands of times faster than the first one. The improvement on 20 words with 15 letters each is even more remarkable. There is, however, still room for improvement. The $O(n_i^2)$ term can be reduced by sorting the signatures rather than comparing each signature against each other signature. After the list of signatures is sorted, any signatures that are the same will be adjacent, so identical signatures on the list for words of length i can be found in time $O(n_i)$ plus the time for sorting. It will not do to use just any sorting algorithm for the sort. If Insertion Sort is used, time $O(n_i^2)$ will be used for sorting the list for words of length i, so we will have little or no improvement over the method in the previous paragraph. There are, however, fast sorting methods that can sort items in time $O(n \lg n)$ (for example, Quicksort and Heapsort). If we use one of these methods for the sorting, then the time for our algorithm becomes

$$\sum_{1 \le i \le l} [n_i O(i^2) + O(n_i \lg n_i)]. \tag{18}$$

The time can be improved even further by using radix sort for the short words (see Knuth [11, Section 5.2.5]).

Notice the interaction between analysis, design, and common knowledge in the above discussion. *Analysis* permits us to identify which parts of an algorithm may be slow. *Knowledge* is needed to interpret the analysis. (In the above discussion it was important to know something about words, such as long words do exist.) Once the analysis focuses our attention on the weaknesses of an algorithm, *design* of a new algorithm is needed to overcome the weaknesses. The new algorithm may still have weaknesses which can be uncovered by further *analysis*.

This section is based on Bentley [69]. His column gives many examples of ways to mix simple analysis with design to produce efficient algorithms.

3.1.3 Nonrepeating Random Hashing

The Nonrepeating Random Hashing Algorithm is the same as the Random Hashing Algorithm discussed in Section 2.4.1, except that the hashing function is different. In Section 2.4.1, there is an infinite sequence of hashing functions such that the probability that $f_i(x_a) = f_j(x_b)$ is $1/N$ unless $i = j$ and $a = b$. In this section we will assume that there is a sequence of N hashing functions $f_i(x)$ that produces a nonrepeating sequence of indices for any x. In particular we have (1) $0 \le f_i(x) < N$ for all x and for $1 \le i \le N$, (2) $f_i(x) \ne f_j(x)$ for $i \ne j$, and (3) $\text{Prob}[f_i(x_a) = f_j(x_b)] = 1/N$ for $i \ne j$ and $a \ne b$.

Let's consider, for a nonrepeating random hash function, how many times Step 2 of the Random Hashing Algorithm is done when looking up an item not in the table. It is always done at least once. If there are k items in the table, the chance of finding an empty spot on the first attempt is $1 - k/N$, so the chance of going on and looking a second time is k/N. If we need to look a second time, then the chance of finding an empty spot is $1 - (k-1)/(N-1)$. The top of the fraction has been reduced by 1 because one of the k items is known to be in the place we looked the first time. There are only $k - 1$ other items that are in the table. The bottom of the fraction has been reduced by one because there are only $N - 1$ values that the second hash function might return. Thus if we go past Step 2 the first time, then the conditional probability that we go past Step 2 a second time is $(k-1)/(N-1)$. Since the probability that we go past Step 2 the first time is k/N, the probability that we go past Step 2 a second time is

$k(k-1)/(N(N-1))$. More generally, if q_i is the probability that Step 2 is done at least i times, then

$$q_1 = 1 \quad \text{and} \quad q_i = \frac{k-i+2}{N-i+2} q_{i-1} \quad \text{for} \quad 2 \le i \le N. \tag{19}$$

This reduces to

$$q_i = \frac{\prod_{k-i+2 \le j \le k} j}{\prod_{N-i+2 \le j \le N} j} = \frac{k!}{(k-i+1)!} \frac{(N-i+1)!}{N!}. \tag{20}$$

In Section 3.2.4 we will study how to finish computing the average time required for this algorithm.

Nonrepeating Random Hashing was first analyzed by Morris [118].

EXERCISES

1. What is the probability for a nonrepeating random hash function that Step 2 of the Random Hashing Algorithm is done exactly i times?
2. How big does k/N have to be so that the probability that Step 2 is done more than once, in the Random Hashing Algorithm with a nonrepeating hash function, is greater than $1/2$?

3.1.4* The Gamma Function

The *Gamma function* is a generalization of the factorial function to a function of a real variable, except that the argument is changed by 1:

$$n! = \Gamma(n+1). \tag{21}$$

In general, the Gamma function is defined by the limit

$$\Gamma(x) = \lim_{m\to\infty} \frac{m^x m!}{x(x+1)(x+2)\cdots(x+m)}. \tag{22}$$

For $x = 1$ most of the factors on the bottom cancel with the $m!$. The limit reduces to $\lim_{m\to\infty} m/(m+1) = 1$ in this case, so $\Gamma(1) = 1$, in agreement with our convention that $0! = 1$. Also,

$$\Gamma(x+1) = \lim_{m\to\infty} \frac{m^{x+1} m!}{(x+1)(x+2)(x+3)\cdots(x+m+1)} \tag{23}$$

$$= x \lim_{m\to\infty} \frac{m^x m!}{x(x+1)(x+2)\cdots(x+m)} \frac{m}{x+m+1} \tag{24}$$

$$= x\Gamma(x) \tag{25}$$

since $\lim_{m\to\infty} m/(x+m+1) = 1$. Since $\Gamma(1) = 1$ and $\Gamma(n+1) = n\Gamma(n)$, we have that $n! = \Gamma(n+1)$, as desired.

It is usually more convenient to work with integrals than to work with limits. The right side of eq. (22) can be transformed into an integral in the following way. Use $\Gamma_m(x)$ for $m^x m!/(x(x+1)(x+2)\cdots(x+m))$. Now consider the integral

$\int_0^1 (1-t)^m t^{x-1}\,dt$. If we evaluate this integral using integration by parts with $du = t^{x-1}dt$, $u = t^x/x$, $v = (1-t)^m$, $dv = -m(1-t)^{m-1}dt$, we get

$$\int_0^1 (1-t)^m t^{x-1}\,dt = (1-t)^m \frac{t^x}{x}\bigg|_0^1 + \frac{m}{x}\int_0^1 (1-t)^{m-1} t^x\,dt \tag{26}$$

$$= \frac{m}{x}\int_0^1 (1-t)^{m-1} t^x\,dt. \tag{27}$$

Notice that the exponent on $(1-t)$ went down by 1. Repeating this process gives

$$\int_0^1 (1-t)^m t^{x-1}\,dt = \frac{m!}{x(x+1)(x+2)\cdots(x+m-1)}\int_0^1 t^{x+m-1}\,dt \tag{28}$$

$$= \frac{m!}{x(x+1)(x+2)\cdots(x+m)} \tag{29}$$

so for integer m

$$\Gamma_m(x) = m^x \int_0^1 (1-t)^m t^{x-1}\,dt. \tag{30}$$

Now replace t in eq. (30) with t/m to get

$$\Gamma_m(x) = \int_0^m \left(1-\frac{t}{m}\right)^m t^{x-1}\,dt. \tag{31}$$

Taking the limit as m goes to infinity gives (see Exercise 3)

$$\Gamma(x) = \int_0^\infty e^{-t} t^{x-1}\,dt \tag{32}$$

for integer x. The formula is also valid for noninteger x (see Exercise 2).

Later we will need the value of $\Gamma(\frac{1}{2})$. From eq. (32) we have

$$\Gamma\left(\tfrac{1}{2}\right) = \int_0^\infty \frac{e^{-t}}{t^{1/2}}\,dt. \tag{33}$$

The integral looks somewhat better if we change variables using $t = u^2$ ($dt = 2u\,du$) to get

$$\Gamma\left(\tfrac{1}{2}\right) = 2\int_0^\infty e^{-u^2}\,du. \tag{34}$$

To evaluate the integral, it is now best to square it to obtain

$$\left(\Gamma\left(\tfrac{1}{2}\right)\right)^2 = 4\int_0^\infty \int_0^\infty e^{-(u^2+v^2)}\,du\,dv. \tag{35}$$

Now by converting to polar coordinates with $r^2 = u^2 + v^2$ and $du\,dv = r\,dr\,d\theta$, we get

$$\left(\Gamma\left(\tfrac{1}{2}\right)\right)^2 = 4\int_0^{\pi/2}\int_0^\infty re^{-r^2}\,dr\,d\theta = 2\pi\int_0^\infty re^{-r^2}\,dr. \tag{36}$$

With one last change of variable ($x = r^2$, $dx = 2r\,dr$) we get

$$\left(\Gamma\left(\tfrac{1}{2}\right)\right)^2 = \pi \int_0^\infty e^{-x}\,dx = \pi \tag{37}$$

so

$$\Gamma\left(\tfrac{1}{2}\right) = \sqrt{\pi}. \tag{38}$$

[Occasionally we will use $x!$ to mean $\Gamma(x+1)$ for noninteger x.]

The Gamma function is discussed by Artin [19], by Knuth [9, Section 1.2.5], and by Olver [46, Section 2.1].

EXERCISES

1. Using $0! = 1$, $n! = n(n-1)!$, $\Gamma(1) = 1$, and $\Gamma(n+1) = n\Gamma(n)$, give a simple proof by induction that $n! = \Gamma(n+1)$ for every nonnegative integer n.

2. [See 19] Let $f(x)$ be any function that obeys the following three conditions: (1) $f(x+1) = xf(x)$, (2) $f(1) = 1$, and (3) $f(x)$ is is defined for $x \geq 1$ and is log convex (i.e.,

$$\frac{\ln f(a) - \ln f(c)}{a - c} \geq \frac{\ln f(b) - \ln f(c)}{b - c}$$

for $a \leq b \leq c$). Prove that $f(x)$ must be the Gamma function. Hint: First prove that $f(x) = \Gamma(x)$ for integer x. This follows from the first two conditions (see the previous problem). Next bound the value of $f(n+x)$ for $0 \leq x \leq 1$ using the values of $f(n-1)$, $f(n)$, and $f(n+1)$ combined with log convexity. Take the limit as n goes to infinity and show that you get the same limit that occurs in the definition of the Gamma function. Then prove that $f(x)$ must equal the Gamma function for all nonnegative x.

3. Prove eq. (32) from eq. (31) by carefully taking the limit. Show that $0 \leq e^{-t} - (1 - t/m)^m \leq t^2 e^{-t}/m$ if $0 \leq t \leq m$. Show that

$$\int_0^\infty e^{-t} t^{x-1}\,dt - \Gamma_m(x) = \int_m^\infty e^{-t}\,dt + \int_0^m \left(e^t - \left(1 - \frac{t}{m}\right)^m\right) t^{x-1}\,dt$$

and that the limit of this as m goes to infinity is zero.

4. Express $\Gamma(p + \frac{1}{2})$ in terms of $\Gamma(\frac{1}{2})$, powers of 2, and factorials in the case where p is an integer. Hint: Use eq. (25) repeatedly. Then use $p - \frac{1}{2} = (2p-1)/2$. Show that the product of the odd integers from 1 to $2p$ is equal to $(2p)!/(2^p p!)$. (The top of this fraction is the product of all the integers up to $2p+1$, while the bottom divides out the even ones.)

n	$\binom{n}{0}$	$\binom{n}{1}$	$\binom{n}{2}$	$\binom{n}{3}$	$\binom{n}{4}$	$\binom{n}{5}$	$\binom{n}{6}$	$\binom{n}{7}$	$\binom{n}{8}$
0	1	0	0	0	0	0	0	0	0
1	1	1	0	0	0	0	0	0	0
2	1	2	1	0	0	0	0	0	0
3	1	3	3	1	0	0	0	0	0
4	1	4	6	4	1	0	0	0	0
5	1	5	10	10	5	1	0	0	0
6	1	6	15	20	15	6	1	0	0
7	1	7	21	35	35	21	7	1	0
8	1	8	28	56	70	56	28	8	1

TABLE 3.1. Binomial coefficients.

3.2 BINOMIALS

Suppose we have n distinct objects, and we want to select k of them. In contrast to the situation that led to the definition of the number of permutations of n objects taken k at a time, in this case we have no concern about what order the objects may be put in. The number of ways that we can select k of n objects without regard to order is called the number of *combinations of n objects taken k at a time.* For example, if the objects are a, b, c, then the combinations of the three objects taken two at a time are ab, ac, and bc. If you compare the combinations of n items taken k at a time with the permutations of n items taken k at a time, you will notice that for each combination there are $k!$ permutations, one for each way that you can arrange the k items into linear order. For example, associated with the combination ab you have the permutations ab and ba. Therefore the number of combinations is equal to the number of permutations divided by $k!$. Thus the number of combinations of n items taken k at a time is

$$\frac{n!}{k!(n-k)!}. \tag{39}$$

Problems involving combinations are so common that there is a special notation for the ratio of factorials in eq. (39). This ratio is called the *binomial coefficient* and is written

$$\binom{n}{k} = \frac{n!}{k!(n-k)!}, \quad \text{integer } n \geq k \geq 0. \tag{40}$$

Table 3.1 gives the first few binomial coefficients.

The following formula is equivalent to eq. (39) when r is a nonnegative integer:

$$\binom{r}{k} = \prod_{1 \leq i \leq k} \left(\frac{r+1-i}{i}\right), \quad \text{integer } k \geq 0, \tag{41}$$

and it provides the appropriate generalization when r is noninteger or negative. For $k < 0$ it is appropriate to define the binomial coefficient as

$$\binom{r}{k} = 0 \quad \text{for integer } k < 0. \tag{42}$$

Also

$$\binom{n}{k} = 0 \quad \text{for integer } n, k, \quad k > n \geq 0. \tag{43}$$

Some basic binomial coefficients are

$$\binom{r}{0} = 1, \quad \binom{r}{1} = r, \quad \binom{r}{2} = \frac{r(r-1)}{2}. \tag{44}$$

Binomial coefficients get their name from the *binomial theorem*:

$$(x+y)^n = \sum_{0 \leq i \leq n} \binom{n}{i} x^i y^{n-i}, \quad \text{integer } n \geq 0. \tag{45}$$

To see why this formula holds, consider what happens when you multiply out $(x+y)^2$. You get $x^2 + xy + yx + y^2$. For each way that you can select an item for the first factor and an item for the second factor, you get a term in the answer. Since $yx = xy$, you can combine $xy + yx$ into $2xy$. In general, when you multiply out $(x+y)^n$, you get 2^n terms (one for each way of selecting n items, where each is an x or y), but you can combine together those terms that have the same number of x's (such terms also have the same number of y's). The number of ways that you can obtain i x's when selecting n items is just $\binom{n}{i}$: it is the number of ways to select i out of n positions in a term to contain an x. Notice that, with our conventions, eq. (45) implies that anything to the zero power (even 0^0) is 1. This will usually be appropriate in our applications, but each case should be given individual consideration.

Often sums include all nonzero values of the summand. In such cases it is convenient to omit the limits on the summation. Thus eq. (45) can be rewritten as

$$(x+y)^n = \sum_{i} \binom{n}{i} x^i y^{n-i}, \quad \text{integer } n \geq 0, \tag{46}$$

because $\binom{n}{i}$ is zero for positive integer n when $i < 0$ and when $i > n$. Having the limits implied by the summand greatly reduces the amount of work required when changing variables in a summation formula. Since errors in the limits are among the most common errors that people make when manipulating summations, this notation with *suppressed limits* is very convenient. Of course, you do have the added burden of examining the summand to determine which terms are nonzero, that is, which terms actually contribute to the sum.

The binomial theorem generalizes to powers which are not positive integers:

$$(x+y)^r = \sum_{i \geq 0} \binom{r}{i} x^i y^{r-i}, \quad \text{when } |x/y| < 1. \tag{47}$$

When r is negative or noninteger, the sum in eq. (47) is an infinite sum. We still have $i \geq 0$, but we no longer have an upper bound on i. The sum converges if $|x/y| < 1$, so when using eq. (47) with negative or noninteger r, it is necessary to call the smaller (in absolute value) number x and the larger one y.

When you use the binomial theorem to simplify formulas, remember that if y is 1, then it will probably not show up in the formula you are trying to simplify. Thus

$$(1+x)^n = \sum_i \binom{n}{i} x^i, \quad \text{integer } n \geq 0. \tag{48}$$

There is another sum involving binomials that we already know how to simplify. Using eq. (8) we can rewrite eq. (2.56) as

$$\sum_{0 \leq i < n} \frac{i!}{(i-k)!} = \frac{n!}{(k+1)(n-k-1)!}. \tag{49}$$

If you divide each side by $k!$ and rewrite the ratio of factorials as binomial coefficients, then you get

$$\sum_{0 \leq i < n} \binom{i}{k} = \binom{n}{k+1}, \quad \text{integer } k \geq 0. \tag{50}$$

Abel developed the following generalization of the binomial theorem:

$$\frac{(x+y+n)^n}{x} = \sum_i \binom{n}{i} (x+i)^{i-1}(y+n-i)^{n-i} \qquad \text{for } x \neq 0, \quad \text{integer } n \geq 0. \tag{51}$$

Riordan [49, Section 1.5] gives values for the sum

$$\sum_i \binom{n}{i} (x+i)^{i+p}(y+n-i)^{n-i+q} \qquad \text{for } x \neq 0, \quad \text{integer } n \geq 0 \tag{52}$$

for integer p and q near zero, and Broder [78] has a systematic study of such sums.

Binomial coefficients are carefully considered by Cohen [23, Chapter 2] and Knuth [9, Section 1.2.6]. Much of this chapter is based on the presentation given in Knuth.

EXERCISES

1. Show that $\binom{\frac{1}{2}}{n+1} = \frac{(-1)^n 2^{-2n-1}}{n+1}\binom{2n}{n}$. Hint: Use eq. (41).

2. Show that $\binom{n}{k} = \frac{n^{\underline{k}}}{k!}$.

3.2.1 Relations among Binomials

Most basic properties of binomial coefficients with a nonnegative upper index come directly from eq. (40). When the upper index is negative or noninteger, the properties come from eq. (41).

The symmetry condition on binomial coefficients is

$$\binom{n}{i} = \binom{n}{n-i}, \quad \text{integer } n,\ i, \quad n \geq 0. \tag{53}$$

From eq. (40) one can factor out r and i to obtain

$$\binom{r}{i} = \frac{r}{i}\binom{r-1}{i-1}, \quad \text{integer } i \neq 0. \tag{54}$$

Eq. (54) can be written in the following algebraically equivalent way:

$$i\binom{r}{i} = r\binom{r-1}{i-1}, \quad \text{integer } i. \tag{55}$$

Eq. (55) is useful for moving factors into and out of a binomial. Sometimes eq. (55) is used to replace the right side with the left and sometimes it is used to replace the left side with the right. For example, suppose we want to simplify $\sum_{1\leq i<n} i\binom{i}{k}$, where $k \geq 0$. We would like to use eq. (50), but the factor of i outside the binomial gets in the way. Now, the right side of eq. (55) has the top index outside the binomial, except it has one more than the top index outside. So the first step is to turn the i outside the binomial into $i+1$. This can be done by adding and subtracting $\sum_{1\leq i<n}\binom{i}{k}$ to the sum to obtain

$$\sum_{1\leq i<n} i\binom{i}{k} = \sum_{1\leq i<n}(i+1)\binom{i}{k} - \sum_{1\leq i<n}\binom{i}{k}. \tag{56}$$

Now the first term on the right matches the right side of eq. (55) and the second term matches eq. (50), so we have

$$\sum_{1\leq i<n} i\binom{i}{k} = (k+1)\sum_{1\leq i<n}\binom{i+1}{k+1} - \binom{n}{k+1} + \binom{0}{k}. \tag{57}$$

We need to shift the summation index so that the sum on the right side of eq. (57) will match eq. (50). We get

$$\sum_{1\leq i<n} i\binom{i}{k} = (k+1)\sum_{2\leq i<n+1}\binom{i}{k+1} - \binom{n}{k+1} + \binom{0}{k}. \tag{58}$$

Now we can apply eq. (50), once we use eq. (2.9) to adjust for the fact that eq. (50) has a sum with a lower limit of zero while our sum has a lower limit of 2. We get

$$\sum_{1\leq i<n} i\binom{i}{k} = (k+1)\binom{n+1}{k+2} - (k+1)\binom{2}{k+2} - \binom{n}{k+1} + \binom{0}{k}. \tag{59}$$

If the left side of eq. (55) is matched against the first term on the right side of eq. (59), we get (with a little algebra)

$$\sum_{1\le i<n} i\binom{i}{k} = n\binom{n}{k+1} - \binom{n+1}{k+2} - (k+1)\binom{2}{k+2} + \binom{0}{k}. \tag{60}$$

Now eq. (60) can be further simplified by noticing that $\binom{0}{k}$ and $\binom{2}{k+2}$ are both zero, except when $k = 0$, when they are both 1. This gives

$$\sum_{1\le i<n} i\binom{i}{k} = n\binom{n}{k+1} - \binom{n+1}{k+2}. \tag{61}$$

If you have trouble with the derivation of eq. (61), or if you don't see how you could discover such a derivation on your own, you will want to read Section 3.2.2 with care. The derivation is discussed in great detail there to illustrate the general approaches to simplifying summations.

The *Kronecker delta function* is useful for expressing the value of functions such as $\binom{0}{k}$. The *Kronecker delta function* is defined as

$$\delta_{ij} = \begin{cases} 0 \text{ if } i \neq j, \\ 1 \text{ if } i = j. \end{cases} \tag{62}$$

Thus,

$$\binom{0}{k} = \delta_{k0}. \tag{63}$$

In eq. (41) we can factor out r and multiply by $(r-i)/(r-i)$ to obtain

$$\binom{r}{i} = \frac{r}{r-i}\binom{r-1}{i}, \quad \text{integer } i \neq r. \tag{64}$$

Adding eq. (55) times $1/r$ to eq. (64) times $(r-i)/r$ gives the addition formula for binomial coefficients:

$$\binom{r}{i} = \binom{r-1}{i} + \binom{r-1}{i-1}, \quad \text{integer } i. \tag{65}$$

If you need to compute a table of binomial coefficients, then eq. (65) combined with

$$\binom{0}{i} = 0 \quad \text{for integer } i \geq 1 \quad \text{and} \quad \binom{n}{0} = 1 \tag{66}$$

gives you one of the best ways to do it. Notice that eq. (65) can be used to compute any binomial from the two binomials that have an upper index that is one less and lower indices that are the same and one less than those of the one being calculated. For example, $\binom{6}{3} = \binom{5}{3} + \binom{5}{2} = 20$. The boundary conditions given in eq. (66) permit eq. (65) to be used recursively to calculate the binomial coefficient for any pair of positive indices.

Applying eq. (65) to its own rightmost term gives $\binom{r}{i} = \binom{r-1}{i} + \binom{r-2}{i-1} + \binom{r-2}{i-2}$. Repeated application of eq. (65) to itself (with r replaced with $r+n+1$) gives the following summation formula:

$$\sum_{0\le i\le n} \binom{r+i}{i} = \binom{r+n+1}{n}, \quad \text{integer } n \ge 0. \tag{67}$$

If we replace the r in eq. (41) with $-r$, then we obtain

$$\binom{-r}{k} = \prod_{1\le i\le k} \left(\frac{-r+1-i}{i}\right) = (-1)^k \frac{\prod_{1\le i\le k}(r-1+i)}{\prod_{1\le i\le k} i}. \tag{68}$$

In the upper product in eq. (68) we can now replace i with $k+1-i$ to obtain

$$\binom{-r}{k} = (-1)^k \prod_{1\le i\le k} \left(\frac{r+k-i}{i}\right) = (-1)^k \binom{r+k-1}{k}. \tag{69}$$

Notice that eq. (69) can be used to make the bottom index of a binomial also appear as part of the top index or to eliminate the bottom index from the top. Suppose, for example, that you need to simplify $\sum_i \binom{n+i}{i} x^i$, where $x < 1$. This can be done by negating the upper index and then using the binomial theorem:

$$\sum_i \binom{n+i}{i} x^i = \sum_i \binom{-n-1}{i} (-x)^i = (1-x)^{-n-1}. \tag{70}$$

Formulas often look quite different after making a substitution of variables. This has had two effects on the formulas in this book. First, it led us to convert formulas into simple forms. With a simple form one often sees more quickly what transformation is needed to make the formula match a more complex formula that is equivalent. Second, at times we give more than one form of the same formula. In eq. (69), if you replace $-r$ with n and k with $n-m$, where n and m are integers and $n \ge 0$, you get

$$\binom{n}{m} = (-1)^{n-m} \binom{-m-1}{n-m}, \quad \text{integer } n,\ m \quad n \ge 0. \tag{71}$$

One more useful formula with negative indices is

$$\binom{-n}{i} = (-1)^{n+i+1} \binom{-i-1}{n-1} \quad \text{for integer } n,\ i, \quad n \ge 1, \quad i \ge 0. \tag{72}$$

This formula interchanges the top and bottom indices.

Whenever you have the product (or ratio) of several binomials that contain common indices (or if the difference in indices of one matches an index of the other), then it may be useful to rewrite the product in factorial form and to see what you can obtain by cancelling common factors and by regrouping factors. For example,

$$\binom{r}{m}\binom{m}{k} = \frac{r!m!}{m!(r-m)!(m-k)!k!} = \frac{r!(r-k)!}{k!(r-k)!(m-k)!(r-m)!}$$
$$= \binom{r}{k}\binom{r-k}{m-k}, \quad \text{integer } m,\ k. \qquad \textbf{(73)}$$

One often wants to rearrange the binomial coefficients that appear in summation formulas to reduce the number of occurrences of the summation index. Notice that in eq. (73) m appears twice on the left side and only once on the right. Eq. (55) is a special case of eq. (73) with $k = 1$.

So far we have two formulas for summing expressions that use binomial coefficients. Eq. (50) shows how to sum expressions with a single binomial coefficient, where the summation index appears at the top of the binomial coefficient. The limits on the summation are arbitrary. Eq. (45) shows how to sum the product of a binomial coefficient times a power that depends linearly on the summation index. The summation must extend over all nonzero terms.

Most of the remaining formulas are for combining factors outside of the binomial with the binomial [eq. (54) and eq. (55)] or for changing the place where the indices occur [eqs. (53, 54, 55, 69, 73)].

You should not conclude that all sums with binomial coefficients can be reduced to simple exact closed-form answers. One of the most basic sums that cannot be simplified is

$$\sum_{0 \le i \le m} \binom{n}{i}. \qquad (74)$$

This sum can be simplified for some special values of m, but not for general m. For $m < 0$ the sum is zero. For $m \ge n$ the sum is 2^n. It is also simple for $m = \lfloor n/2 \rfloor$, but there is no closed form that works for all the values of m between zero and n.

EXERCISES

1. Simplify $\sum_i i\binom{n}{i}$.
2. Simplify $\sum_{0 \le i \le n/2} i\binom{n}{i}$.
3. Simplify $\sum_i \binom{n}{i}\binom{i}{k}x^i$.
4. Simplify $\sum_i \binom{n+i}{n}x^i$.
5. Simplify $\sum_i \binom{n-i}{n-m}\binom{n}{i}$.
6. Prove eq. (72). Hint: Look at some particular cases, such as $n = 5$, $i = 2$ and $n = 2$, $i = 5$, and evaluate both sides with eq. (73), before trying to construct a general proof.
7. Simplify $\sum_{0 \le i < n} \binom{i}{m}x^i$. Hint: Use summation by parts.

3.2.2 Strategies for Summation Formulas

When simplifying summations, you need two types of thinking. Strategy is concerned with what route you take to get from a summation to a simple answer. Tactics is concerned with selecting which transformations to apply at each step of the route. For easy problems the tactical considerations are usually most important. As soon as you find the right transformation, your problem is nearly solved. For difficult problems strategic considerations become more important. A large number of transformations are needed, and much time can be wasted unless a correct approach is selected.

Let's consider in more detail some of the strategic and tactical considerations that arise during the simplification of

$$\sum_{1\le i\le n} i\binom{i}{k}. \tag{75}$$

The strategy for this problem is rather simple: get rid of the i multiplying the binomial. As is usually the case when selecting a strategy, the thinking that leads to this strategy requires a good bit of tactical knowledge. Getting rid of the i is a good strategy because: (1) there is a formula [eq. (55)] for moving indices in and out of binomials, and (2) there is a formula [eq. (50)] for summing binomials on the top index. Thus this strategy has some promise of success. If you disregard the tactical considerations, you might arrive at the strategy of eliminating the binomial from the problem. After all i is easy to sum. Unfortunately there is no convenient way to eliminate the binomial, so this alternate strategy is not good.

Notice the amount of detail in the last paragraph. You must pay some attention to tactics when developing your strategy, but you must not pay too much attention if you want to develop your strategy at reasonable speed. How do you go about becoming good at strategy? One of the most important things is to ask yourself, each time you see a new formula, "How can this formula be put to good use?" Notice the characterization two paragraphs above of formulas [eqs. (50, 55)] according to what they are good for.

Now let's consider the tactical problems. We want to eliminate the i multiplying the binomial in eq. (75). Eq. (55) is the obvious equation to use, but it does not quite match. Using primes for the variables in eq. (55), we can match the variables in the binomial in eq. (75) with those in the binomial on the right side of eq. (55) using the substitution of variables $r'-1=i$ and $i'-1=k$. This gives an $r'-1$ multiplying the binomial rather than the r' that eq. (55) calls for.

This shows that to obtain a complete match, we need to convert the i multiplying the binomial in eq. (75) to an $i+1$. To prevent the value of the sum from changing, we need to simultaneously subtract $\sum_{1\le k\le n}\binom{i}{k}$, which compensates for the $+1$ of the $i+1$. This gives eq. (56).

Now we have three obvious steps. Eq. (55) applies to the first sum on the right side of eq. (56). The use of eq. (55) introduces a factor of $k+1$. Since the factor does not depend on the summation index, it can be factored out of the sum. Eq. (50) applies to the last sum in eq. (56) except that the lower summation limit is off by one, but eq. (2.29) can take care of the mismatch. Applying these steps converts eq. (56) into eq. (57).

Now we are nearing the end. We want to apply eq. (50) to the sum on the right side of eq. (57), but in eq. (57) the summation index appears as $i+1$ in the binomial rather than as i. Since i is a dummy variable (an index of summation in this case), we can cure the problem by changing all i's to $i-1$. This does not change the value of the sum (be careful to include the i under the summation sign when changing i to $i-1$). This gives us eq. (58). Finally, a few more transformations lead to eq. (61).

In order to develop a long series of transformations like this on your own, you not only need to remember which transformations can be applied, but you must also organize your knowledge so that for each transformation you know which kinds of problems it can be used to solve. This will make it possible to solve many long problems rapidly.

There are a few general principles for transforming the summand so that a formula may be applied. The most significant factor is the pattern of occurrences of the summation index. It can be uniformly adjusted up and down by linear transformation, and any constant number of missing or extra terms can be added in with compensating subtractions outside the sum. Eq. (54) can be used to move the summation index into or out of a binomial. Adding a constant to a linear factor requires subtracting a compensating summation (as in eq. (56)).

Matching the parts of the formula that do not involve the summation index is usually simpler. If they do not depend on each other in any way, there is usually no problem: a quantity like $a+3b+5$, for example, matches $k+1$ with $k=a+3b+4$. If a factor independent of the summation index is needed, it can simply be multiplied in and divided out of the sum.

The above guidelines are useful in discovering when an individual formula can be applied. Often it is necessary to apply more than one formula in order to find a closed form for a sum. In the example considered in this section, you might begin by noticing that the left side of eq. (56) is very nearly like the left side of eq. (50), except that an extra factor of i, the summation index, is present. Using eq. (55) gives a way to absorb an extra factor into a binomial. These observations provide an overall strategy; experimentation reveals the details of the necessary procedure. In this case, the strategy was successful. If no overall strategy is apparent, you can try transforming the sum so that various formulas can be applied. After a certain amount of practice you will be able to recognize quickly when a particular approach is likely to lead to a solution.

EXERCISES

1. [See 123] Consider the formula $T_{n,m} = \binom{n}{m} \sum_{0\le i\le n-m} m^i T_{n-m,i}$ with $T_{n,0}=0$ for $n>0$ and $T_{0,0}=1$. Calculate the numbers generated by this rule for small values of n and m. Determine the pattern in the numbers and prove by induction that you have found the correct pattern.

3.2.3 Coin Flipping

A *true coin* is a coin that, when flipped, lands with the head side up half the time and with the tail side up half the time. Suppose you flip a true coin n times. What is the probability that the first k times will result in heads and the remaining $n-k$ times will result in tails? The chance of heads the first time is $1/2$; the chance of heads the first two times is $1/4$; etc. The chance of heads the first k times is 2^{-k}. Likewise the chance of tails the $(k+1)^{\text{th}}$ time is $1/2$, etc. The chance of tails the $(k+1)^{\text{th}}$ time through the nth time is $2^{-(n-k)}$. The probability of heads the first k times and tails the remaining times is $2^{-k}2^{-n+k} = 2^{-n}$. Whenever you have two independent events (such as the probability of heads on one coin toss and tails on the next), you can obtain the probability that both events occur by multiplying the probabilities.

Suppose you flip a true coin 4 times. What is the probability that you obtain heads exactly twice? The probability of the pattern HHTT is $1/16$. Likewise the probability for each of HTHT, HTTH, THHT, THTH, and TTHH is also $1/16$. The number of ways one can obtain 2 heads out of 4 flips is exactly the same as the number of ways you can chose 2 out of 4 items, and each way occurs with the same probability, 2^{-4}. Thus the probability of exactly 2 heads is $\binom{4}{2}2^{-4} = 3/8$. In general, the probability of k heads from n flips is

$$\binom{n}{k}2^{-n}. \tag{76}$$

Suppose you flip a coin for which the probability of obtaining heads is p. What is the probability of obtaining k heads when flipping the coin n times? This time the probability of obtaining heads on the first k times is p^k. The probability of not obtaining heads the remaining $n-k$ times is $(1-p)^{n-k}$. There are a total of $\binom{n}{k}$ patterns that result in exactly k heads, and each pattern has the same probability. Therefore the probability of k heads from n flips is

$$\binom{n}{k}p^k(1-p)^{n-k}. \tag{77}$$

The average number of heads for n flips of a coin with probability p of heads is

$$\begin{aligned}
\sum_i i\binom{n}{i}p^i(1-p)^{n-i} &= \sum_i n\binom{n-1}{i-1}p^i(1-p)^{n-i} && (78)\\
&= \sum_{i+1} n\binom{n-1}{i}p^{i+1}(1-p)^{n-i-1} && (79)\\
&= pn\sum_i \binom{n-1}{i}p^i(1-p)^{n-i-1} && (80)\\
&= pn. && (81)
\end{aligned}$$

To compute the variance of the number of heads for n flips of a coin with probability p of heads, we first need to compute

$$\sum_i i^2 \binom{n}{i} p^i (1-p)^{n-i} = \sum_i [i(i-1)+i] \binom{n}{i} p^i (1-p)^{n-i} \tag{82}$$

$$= \sum_i i(i-1) \binom{n}{i} p^i (1-p)^{n-i} + \sum_i i \binom{n}{i} p^i (1-p)^{n-i} \tag{83}$$

$$= \sum_i n(n-1) \binom{n-2}{i-2} p^i (1-p)^{n-i} + pn. \tag{84}$$

Replacing i by $i+2$ on the right side of eq. (84), we get

$$\sum_i i^2 \binom{n}{i} p^i (1-p)^{n-i} = \sum_{i+2} n(n-1) \binom{n-2}{i} p^{i+2} (1-p)^{n-i-2} + pn \tag{85}$$

$$= n(n-1)p^2 \sum_i \binom{n-2}{i} p^i (1-p)^{n-i-2} + pn \tag{86}$$

$$= n(n-1)p^2 + pn. \tag{87}$$

The variance is

$$V = n(n-1)p^2 + pn - (pn)^2 = p(1-p)n. \tag{88}$$

Coin flipping is covered by Feller [27, Chapter 3] and Knuth [9, Section 1.2.10].

3.2.4 Nonrepeating Hashing

In Section 3.1.3 the probability that Nonrepeating Random Hashing does the second step i or more times was computed [eq. (20)]. Using this result and the formula for computing the average from cumulative probabilities [eq. (2.79)], we have that the average number of times Step 2 is done is

$$\sum_{i \geq 1} \frac{k!(N-i+1)!}{(k-i+1)!N!}. \tag{89}$$

The first step in simplifying this summation is to notice that the summation index occurs one time on the top and one time on the bottom of the fraction, so the factorials that contain the summation index can be rewritten as a binomial coefficient. We have

$$\frac{(N-i+1)!}{(k-i+1)!} = (N-k)! \frac{(N-i+1)!}{(k-i+1)!(N-k)!} = (N-k)! \binom{N-i+1}{N-k}. \tag{90}$$

After applying eq. (90) to eq. (89) we can apply eq. (50) (with a linear transformation of the summation index) to obtain

$$\sum_{i\geq 1} \frac{k!(N-i+1)!}{(k-i+1)!N!} = \frac{k!(N-k)!}{N!} \sum_{i\geq 1} \binom{N-i+1}{N-k} \tag{91}$$

$$= \binom{N+1}{N-k+1} \Big/ \binom{N}{N-k} \tag{92}$$

$$= \frac{N+1}{N-k+1} = 1 + \frac{k}{N-k+1} \tag{93}$$

It is interesting to compare eq. (93) with eq. (2.69). As far as the average time for looking up an item in the table is concerned, the effect of using Nonrepeating Hashing is the same as that of doing Repeating Hashing with a hash table of size one larger.

The variance of the time for Nonrepeating Random Hashing can be computed in much the same way that the variance was computed for ordinary random hashing. Again we will concentrate on the variance for Step 2. The probability that Step 2 is done i times is given by

$$p_i = q_i - q_{i+1} = \frac{k!(N-i+1)!}{(k-i+1)!N!} - \frac{k!(N-i)!}{(k-i)!N!} \tag{94}$$

$$= \frac{k!(N-i)!}{N!(k-i)!} \left(\frac{N-i+1}{k-i+1} - 1 \right) = \frac{k!(N-i)!(N-k)}{N!(k-i+1)!}. \tag{95}$$

To compute the variance in the number of times Step 2 is done when looking up an item not in the table, we first need

$$\sum_i i^2 p_i = \frac{k!(N-k)}{N!} \sum_{i\geq 1} \frac{(N-i)!}{(k-i+1)!} i^2 \tag{96}$$

$$= \frac{k!(N-k)!}{N!} \sum_{i\geq 1} \binom{N-i}{N-k-1} i^2. \tag{97}$$

To simplify this further, we would like to combine the i^2 with the binomial. How do we do this? First notice that we do have eq. (55) for combining $N-i+1$ into the binomial. If we apply eq. (55) twice, we can combine $(N-i+1)(N-i+2)$ with the binomial. So the first step is to express i^2 in terms of $(N-i+1)(N-i+2)$, $N-i+1$, and constants. Notice that $(N-i+1)(N-i+2) = N^2 - 2iN + i^2 + 3N - 3i + 2$, so $(N-i+1)(N-i+2)$ has the i^2 we need plus some extra terms of degrees 1 and zero in i. Also we can write $iN = -(N-i+1)N + N^2 + N$ and $i = -(N-i+1) + N + 1$. Putting all the pieces together, we have $i^2 = (N-i+1)(N-i+2) - (N-i+1)(2N+3) + (N+1)^2$. Substituting into eq. (97) gives

$$\frac{k!(N-k)!}{N!}\sum_{i\geq 1}\binom{N-i}{N-k-1}i^2$$

$$= \frac{k!(N-k)!}{N!}\left[\sum_{i\geq 1}(N-i+1)(N-i+2)\binom{N-i}{N-k-1}\right.$$

$$- (2N+3)\sum_{i\geq 1}(N-i+1)\binom{N-i}{N-k-1}$$

$$\left.+(N+1)^2\sum_{i\geq 1}\binom{N-i}{N-k-1}\right], \tag{98}$$

$$= \frac{k!(N-k)!}{N!}\left[(N-k)(N-k+1)\sum_{i\geq 1}\binom{N-i+2}{N-k+1}\right.$$

$$\left.- (2N+3)(N-k)\sum_{i\geq 1}\binom{N-i+1}{N-k} + (N+1)^2\binom{N}{N-k}\right] \tag{99}$$

$$= \frac{k!(N-k)!}{N!}\left[(N-k)(N-k+1)\binom{N+2}{N-k+2}\right.$$

$$\left.-(2N+3)(N-k)\binom{N+1}{N-k+1} + (N+1)^2\binom{N}{N-k}\right] \tag{100}$$

$$= (N+1)\left[\frac{(N-k)(N+2)}{N-k+2} - \frac{(N-k)(2N+3)}{N-k+1} + N+1\right] \tag{101}$$

$$= \frac{(N+1)(N+k+2)}{(N-k+2)(N-k+1)}. \tag{102}$$

Now the variance can be computed by subtracting the square of the average value of i [eq. (93)] from the average value of i^2 [eq. (102)] to obtain

$$V = \frac{(N+1)(N+k+2)}{(N-k+2)(N-k+1)} - \frac{(N+1)^2}{(N-k+1)^2} \tag{103}$$

$$= \frac{N+1}{N-k+1}\left[\frac{N+k+2}{N-k+2} - \frac{N+1}{N-k+1}\right] \tag{104}$$

$$= \frac{k(N+1)(N-k)}{(N-k+1)^2(N-k+2)}. \tag{105}$$

When $N-k$ is large, so is N, and we can write

$$V = \frac{kN}{(N-k)^2}\left(1 + O\left(\frac{1}{N-k}\right)\right). \tag{106}$$

The average and variance for Nonrepeating Random Hashing become quite large when k is almost equal to N. For $k = 0.8N$ and large N, for example, the average is about 5, while the variance is about 20.

The random hashing algorithms are reasonably simple to analyze, and their properties are similar to those of practical hashing algorithms. The nonrepeating algorithm is not practical, however, because of the difficulty of producing hash functions with the required properties. The Repeating Algorithm is not often used because there are simpler ways to obtain slightly better performance.

When doing a long calculation, it is common to make errors. If you do not take care, the errors can greatly increase the effort required to obtain the correct answer. When doing the above calculation of the variance, the authors used two types of checks to find errors. At the first sign of error the sum of the p_i was computed. When the p_i did not at first sum to 1, we knew we had not correctly computed the p_i. After the correct p_i were found and the initial calculations were still looking suspicious, it was noticed that the formulas were particularly simple for $k = 0$. From then on each step was checked for the value $k = 0$. This uncovered many errors. Finally, the sum of ip_i was computed the same way that the variance was being computed. When this did not agree with eq. (93), it was discovered that one of the binomial formulas was not exactly as we had remembered it. By frequently checking for errors, it was possible to correctly compute the variance in about 2 hours. Without constant checks it would have taken much longer to obtain the correct answer.

Errors in long calculations can also be avoided by using symbolic algebra systems, such as MACSYMA or REDUCE. Their use can save much time and frustration.

EXERCISES

1. Show that eq. (106) follows from eq. (105).
2. Why is no upper limited needed on the sum in eq. (89)? Why is a lower limit needed?

3.3† MULTIPLE SUMS

In the last section we saw a sum over two indices. In general, if we write

$$\sum_{\substack{R_1(i_1,i_2,\ldots,i_n)\\ R_2(i_1,i_2,\ldots,i_n)\\ \vdots \\ R_m(i_1,i_2,\ldots,i_n)}} F(i_1,i_2,\ldots,i_n), \tag{107}$$

we mean the sum of $F(i_1,i_2,\ldots,i_n)$ over all sets of integer values for the i_1, i_2, ..., i_n such that *all* the relations $R_1(i_1,i_2,\ldots,i_n)$, ..., $R_m(i_1,i_2,\ldots,i_n)$ are satisfied. For example,

$$\sum_{i,j} jq_{ij} \tag{108}$$

is the sum of jq_{ij} over all values of i and j.

The usual way to simplify such sums is to first represent them as a series of nested sums:

$$\sum_{R(i_1,i_2,\ldots,i_n)} a_{i_1,i_2,\ldots,i_n} =$$

$$\sum_{R_1(i_1)}\Bigg(\sum_{R_2(i_1,i_2)}\Bigg(\cdots\Bigg(\sum_{R_n(i_1,i_2,\ldots,i_n)} a_{i_1,i_2,\ldots,i_n}\Bigg)\cdots\Bigg)\Bigg), \qquad \textbf{(109)}$$

where $R_j(i_1,\ldots,i_j)$ is a relation that is *true* for the set of $i_1,\ldots,i_j$ such that there exist some $i_{j+1},\ldots,i_n$ making $R(i_1,\ldots,i_n)$ *true.* (Our scoping rules permit us to drop the parentheses.) Notice that R_j depends on only the first j variables while R depends on all n variables.

For example, the sum

$$\sum_{1\le j\le i\le k} a_{ij} = \sum_{1\le i\le k}\ \sum_{1\le j\le i} a_{ij}. \qquad (110)$$

The inner sum on the right has the range $1 \le j \le i$, because this is the range for j implied by the sum on the left. The outer sum on the right has the range $1 \le i \le k$, because this is the entire range that i can have in the sum on the left as j varies over its range.

In the next section we consider the sum

$$\sum_{i,j} j q_{ij}, \qquad (111)$$

where

$$q_{ij} = \begin{cases} \dfrac{Np_i}{k} & \text{for } 1 \le j \le i \\ 0 & \text{otherwise.} \end{cases} \qquad (112)$$

To simplify this sum, we can first break it into a sum over i and one over j to obtain

$$\sum_{i,j} j q_{ij} = \sum_i \sum_j j q_{ij}. \qquad (113)$$

No limit is needed on either sum here because q_{ij} is zero outside the region where we wish to sum. Since q_{ij} can be repaced by Np_i/k when $1 \le j \le i$ and since it is zero for other values of j, our sum becomes

$$\sum_{i,j} j q_{ij} = \sum_i \sum_{1\le j\le i} \frac{jNp_i}{k}. \qquad (114)$$

[Explicit limits *are* needed for the rightmost sum of eq. (114).] Now we can do the sum over j, obtaining a closed-form expression with no j's in it, and then do the sum over i.

Multiple sums may be simplified using the techniques for single sums. In addition, there are several special techniques for multiple sums. The *distributive law* says that

$$\Bigl(\sum_{R(i)} a_i\Bigr)\Bigl(\sum_{S(j)} b_j\Bigr) = \sum_{R(i)}\sum_{S(j)} a_i b_j. \tag{115}$$

Usually eq. (115) is used on sums that match its right side, since the left side is simpler if you can simplify either the sum over i or the sum over j.

For example,

$$\sum_{1\le i\le n}\sum_{1\le j\le n} ij = \Bigl(\sum_{1\le i\le n} i\Bigr)\Bigl(\sum_{1\le j\le n} j\Bigr) \tag{116}$$

$$= \left(\frac{n(n+1)}{2}\right)\left(\frac{n(n+1)}{2}\right) = \frac{n^2(n+1)^2}{4} \tag{117}$$

Since a multiple sum is just the sum over all the indices that satisfy the relations under the summation signs, you can interchange the order of summation freely for finite sums. (For infinite sums some care is needed; interchanging the order changes the value in some cases.) Interchanging the order of summation is simplest when each relation depends on only one variable. In this case

$$\sum_{R(i)}\sum_{S(j)} a_{ij} = \sum_{S(j)}\sum_{R(i)} a_{ij}. \tag{118}$$

If the inner relation depends on both variables, then the equation is

$$\sum_{R(i)}\sum_{S(i,j)} a_{ij} = \sum_{S'(j)}\sum_{R'(i,j)} a_{ij}, \tag{119}$$

where $S'(j)$ is the relation that there exists an i such that $R(i)$ and $S(i,j)$ are both *true*, and $R'(i,j)$ is the relation that both $R(i)$ and $S(i,j)$ are *true*. Eq. (119) is easy to use when it is easy to compute the primed relations, and hard to use otherwise. One reason to include the limits on summations only when necessary (and to use the convention that we only sum over indices where the summand is not zero) is that eq. (118) is so much simpler than eq. (119). [If the sums do not have explicit limits, then eq. (118) can be used.] It is simple to tell when you should use eq. (119). If you can do the sum over i, but not the one over j, then use eq. (119) [or eq. (118) if it applies]. Whenever you have a multiple sum, see if there is any index for which you can do the summation, and interchange the order of summation to make it the innermost index.

The techniques we have considered so far for multiple sums are as straightforward to apply as the corresponding techniques for single sums. There is, however, one more technique for multiple sums, one that can be applied in a large number of ways, and that is to change the summation indices. For a double sum, so long as $R(i,j)$ is a relation that is satisfied by a finite set of indices, we have

$$\sum_{R(i,j)} a_{ij} = \sum_{R'(i',j')} a_{i'j'}; \tag{120}$$

where $i' = f(i,j)$ and $j' = g(i,j)$ for some functions f and g; where $R'(i',j')$ is *true* if and only if there exist i and j such that $i' = f(i,j)$, $j' = g(i,j)$, and $R(i,j)$ is *true*. In addition the functions f and g must be chosen so that there is a one-to-one correspondence between each pair of (i,j) such that $R(i,j)$ is *true* and each pair (i',j') such that $R'(i',j')$ is *true*. When all the conditions are satisfied, then eq. (120) is just a change in the order of summation.

Eq. (118) and eq. (119) are just special cases of eq. (120) where $i' = j$, $j' = i$. Usually one uses linear functions for f and g, because otherwise it is difficult to have the transformation of indices be one-to-one.

For example, in Section 5.4 we will consider the sum

$$\sum_{0\le i\le k} a_i \sum_{0\le n<k-i} F_n z^{n+i}, \tag{121}$$

and we will want to write it in the form

$$\sum_{0\le n<k} z^n C_n, \tag{122}$$

where C_n is a function of the a's and F's. To do this, we need to make the substitution $i' = i$, $n' = n + i$ $(i = i',\ n = n' - i')$ in eq. (120). This gives

$$\sum_{0\le i\le k} a_i \sum_{0\le n<k-i} F_n z^{n+i} = \sum_{0\le i'\le k} a_{i'} \sum_{0\le n'-i'<k-i'} F_{n'-i'} z^{n'} \tag{123}$$

$$= \sum_{0\le i'\le k} a_{i'} \sum_{i'\le n'<k} F_{n'-i'} z^{n'}, \tag{124}$$

or, dropping the primes,

$$\sum_{0\le i\le k} a_i \sum_{0\le n<k-i} F_n z^{n+i} = \sum_{0\le i\le k} a_i \sum_{i\le n<k} F_{n-i} z^n. \tag{125}$$

Now we need to interchange the order of summation. The range for n is from $i_{\min} = 0$ up to $k - 1$. The range for i is zero up to n. So

$$\sum_{0\le i\le k} a_i \sum_{0\le n<k-i} F_n z^{n+i} = \sum_{0\le n<k} z^n \sum_{0\le i\le n} a_i F_{n-i}, \tag{126}$$

which has the form of eq. (122) with $C_n = \sum_{0\le i\le n} a_i F_{n-i}$.

Now let's consider in more detail the nature of an allowable transformation of the indices in the case where the new indices are a linear combination of the old indices. Let

$$\sum_{j_1,\ldots,j_n} a_{j_1,\ldots,j_n} = \sum_{j'_1,\ldots,j'_n} a_{j'_1,\ldots,j'_n}, \tag{127}$$

where

$$\begin{aligned} j'_1 &= k_{11}j_1 + k_{12}j_2 + \cdots + k_{1n}j_n \\ j'_2 &= k_{21}j_1 + k_{22}j_2 + \cdots + k_{2n}j_n \\ &\ \ \vdots \\ j'_n &= k_{n1}j_1 + k_{n2}j_2 + \cdots + k_{nn}j_n. \end{aligned} \tag{128}$$

Using matrix notation, eq. (128) can be written as $J' = KJ$, where J is the vector $j_1, j_2, \ldots, j_n$, J' is $j'_1, j'_2, \ldots, j'_n$, and K is the matrix of the k_{pq}. What conditions are necessary and sufficient for K to be a one-to-one transformation from an integer vector J to an integer vector J'? It is necessary for K to have all integer entries; otherwise for some integer vector J a noninteger vector J' will result. If K has all integer entries, then any integer vector J is mapped into an integer vector J'. The only question remaining is whether the mapping is one-to-one on the integers. If the matrix inverse of K (K^{-1}) also has integer entries, then each integer vector J' can be transformed (using K^{-1}) back into an integer vector J, but if K^{-1} has a noninteger entry (or if the inverse does not exist), then this cannot be done. So a necessary and sufficient condition is that K and K^{-1} have integer entries.

This condition can be stated in a simpler way, however. The matrix product K times K^{-1} is, by definition of K^{-1}, equal to the identity matrix. So the determinant of K times the determinant of K^{-1} is 1. Since K must be an integer matrix, its determinant must be an integer. If K^{-1} is also integer, then its determinant must be an integer. Therefore it is necessary that K have a determinant that is ± 1. If K is an integer matrix with determinant ± 1, then by Cramer's rule K^{-1} is also an integer matrix. Thus K is a suitable transformation matrix if and only if it has integer entries and it has a determinant that is ± 1.

Normally to change several variables you use a series of transformations where only one variable is changed at a time. The one variable is replaced by an integer linear combination of the original variables, while the remaining variables are left unchanged. The determinant of such a transformation equals the coefficient that relates the old variable to the new one. If this coefficient is ± 1, there is no trouble. For example $i' \leftarrow i + 5j$, $j' \leftarrow j$ is an acceptable transformation because the coefficient relating i to i' is 1, but $i' \leftarrow 2i + 5j$, $j' \leftarrow j$ is not an acceptable transformation because the coefficient relating i to i' is 2.

Interchange of the order of summation is discussed by Knuth [9, Section 1.2.3]. See Apostol [17, Section 12.13] for a discussion of the convergence problems that can arise with infinite sums. The linear algebra used in this section is covered in Birkhoff and Mac Lane [22, Chapters 7–10], Noble [45, Chapter 3], and Strang [56, Chapters 2 and 3].

EXERCISES

1. Sum $\displaystyle\sum_{i \geq 1} \sum_{0 \leq j \leq k-1} \left(\frac{j}{N}\right)^{i-1} \left(1 - \frac{j}{N}\right)$.

2. Show that

$$\sum_{n\geq 0} z^n \sum_{0\leq i\leq n-1} C_i C_{n-i-1} = z\sum_{i\geq 0} z^i C_i \sum_{n\geq i+1} z^{n-i-1} C_{n-i-1}$$
$$= z\sum_{i\geq 0} z^i C_i \sum_{j\geq 0} z^j C_j.$$

3. Show that for any number y and any vector X

$$\sum_{0\leq i<n} y^{-ki} \sum_{0\leq j<n} y^{ij} x_j = \sum_{0\leq j<n} x_j \sum_{0\leq i<n} y^{(j-k)i} = \sum_{0\leq j<n} x_j \frac{y^{(j-k)n}-1}{y^{(j-k)}-1}$$

when $y^{(j-k)} \neq 1$. What is the value of the sum when $y^{(j-k)} = 1$?

4. Write $\sum_m \sum_{t_1} \sum_n \sum_{t_2} p_{i-1,t_1}(m) p_{i-1,t_2}(n)$ as the product of a pair of double sums.

5. The time to access a track on a disk drive can be well approximated by the formula

$$t = t_0(1-\delta_{ij}) + t_1|i-j|$$

where i is the number of the current track, j is the number of the desired track, t_0 is the time for the disk head to start moving, and t_1 is the time required for the head to go from one track to the next. Under the assumption that i is equally likely to have any value in the range $1 \leq i \leq n$ and that j is equally likely to have any value in the same range $(1 \leq j \leq n)$, what is the average value of t? Hint:

$$\sum_{1\leq j\leq n} |i-j| = \sum_{1\leq j<i} (i-j) + \sum_{i<j\leq n} (j-i).$$

6. Simplify $\displaystyle\sum_{m_1,m_2,\ldots,m_k} \prod_{1\leq i\leq k} (-1)^{m_i} \binom{n}{m_i} x^{m_i}.$

3.3.1 Hashing with Chaining

Hashing with Chaining is an example of a hashing algorithm that is practical and that can also be analyzed with the techniques that we have developed so far.

The basic idea behind hashing with chaining is that we will let each of the N slots in the hash table hold as many items as necessary. The hash function, $h(q_i)$, will be computed one time for the ith item, q_i, and the item will be put into slot $h(q_i)$. A linked-list data structure permits each slot to hold as many items as necessary. There is an array, L, of links, and an array, X, for storing the data. The first N locations of L are used as list heads; the remaining locations in L are paired with locations in X and are used for storing linked lists of the items stored in the hash table. This time we will give a complete hashing algorithm that looks up items and stores them in the table if they are not already there.

Algorithm 3.1 Hashing with Chaining: Data structures: An array L of pointers with positions l_0 through l_{N+M-1} and an array X of data with positions x_N through x_{N+M-1}, where N is the size of the hash table and M is the maximum number of items that may be stored in the table. The hash function h returns a value between 0 and $N-1$. Initial conditions: The algorithm starts with $l_i = 0$ (empty) for $0 \leq i < N$ and with $Avail = N - 1$. The algorithm remembers the values of these local variables from one call to the next. Input: A query q to insert into a hash table if it is not already there. Output: The location i of the item in the hash table.

Step **1.** [Hash] Set $i \leftarrow h(q)$.

Step **2.** [Loop] While $l_i \neq 0$ do:

Step **3.** [Found?] Set $i \leftarrow l_i$. If $x_i = q$, then return (q is at position i).

Step **4.** End while.

Step **5.** [Insert] Set $Avail \leftarrow Avail + 1$. If $Avail \geq N + M$, then stop; the table is already full. Otherwise set $l_i \leftarrow Avail$, $l_{Avail} \leftarrow 0$, and $x_{Avail} \leftarrow q$. The item q has now been inserted into position i.

Let's now consider how often Step 2 is done when the algorithm is used to look up an item that is not in the table. Assume that the table already has k items in it. The number of times that Step 2 is done is one more than the length of the linked list that the item hashes to.

We will start by computing the probability that the first list has i of the k items. As a preliminary step let's compute the probability that the first i items were put on the first list (later we will generalize to any i items and any list). The hash function assigns each item to a list at random. The probability that the first item goes on the first list is $1/N$, the probability that the first two items go on the first list is $1/N^2$, etc. The probability that the first i items go on the first list is N^{-i}.

Even when the first i items go on the first list, the first list will not end up with exactly i items unless the remaining $k-i$ items go onto other lists. For each of the remaining items, the probability of going onto another list is $(N-1)/N$; the probability that all of them go onto other lists is $[(N-1)/N]^{k-i}$. So the probability that the first list has the first i items and no others is $(N-1)^{k-i}N^{-k}$.

We originally set out to compute the probability that the first list has exactly i items, without regard to which i items are on the list. For any combination of i of the k items the probability is $(N-1)^{k-i}N^{-k}$ that the first list contains those items and no others. There are $\binom{k}{i}$ ways to choose i of k items. Therefore the probability that the first list has exactly i items is $\binom{k}{i}(N-1)^{k-i}N^{-k}$.

The first list is not special. Under our assumptions, the probability that any list has exactly i of the k items is $\binom{k}{i}(N-1)^{k-i}N^{-k}$. The hash function is equally likely to have to search any of the lists for our item that is not in the table. Thus the probability that we search a list of length i is

$$p_i = \binom{k}{i}(N-1)^{k-i}N^{-k}. \tag{129}$$

Since Step 2 is done 1 more time than the length of the list that is hashed to, the average number of times that Step 2 is done is given by

$$A = \sum_i (i+1)p_i = \sum_i \binom{k}{i}(i+1)(N-1)^{k-i}N^{-k} \tag{130}$$

$$= \sum_i \binom{k}{i} i(N-1)^{k-i}N^{-k} + \sum_i \binom{k}{i}(N-1)^{k-i}N^{-k} \tag{131}$$

$$= \sum_i k\binom{k-1}{i-1}(N-1)^{k-i}N^{-k} + 1 \tag{132}$$

$$= kN^{-k}\sum_i \binom{k-1}{i}(N-1)^{k-i-1} + 1 \tag{133}$$

$$= kN^{-k}N^{k-1} + 1 = 1 + k/N. \tag{134}$$

In deriving eq. (132), the fact that probabilities sum to 1 was used. In deriving eq. (134), the binomial theorem [eq. (45)] was used.

Whenever the average number of searches is large compared to 1, Hashing with Chaining cuts the search time down by about a factor of N. Although this is a big improvement over linear search, the table length is $N + M$, so one can do better by using a more complex hashing method.

To calculate the variance, we first need

$$\sum_i p_i(i+1)^2 = \sum_i [i(i-1) + 3i + 1]\binom{k}{i}(N-1)^{k-i}N^{-k} \tag{135}$$

$$= \sum_i k(k-1)\binom{k-2}{i-2}(N-1)^{k-i}N^{-k} + \sum_i 3ip_i + 1. \tag{136}$$

Replacing i by $i+2$ gives

$$\sum_i p_i(i+1)^2 = k(k-1)N^{-k}\sum_i \binom{k-2}{i}(N-1)^{k-i-2} + 3k/N + 1 \tag{137}$$

$$= k(k-1)N^{-k}N^{k-2} + 3k/N + 1 \tag{138}$$

$$= 1 + 3\frac{k}{N} + \frac{k(k-1)}{N^2} \tag{139}$$

Therefore, the variance is

$$V = \sum_i (i+1)^2 p_i - \left(\sum_i p_i(i+1)\right)^2 = \frac{k}{N}\left(1 - \frac{1}{N}\right). \tag{140}$$

Notice that the variance is small when k is small compared to N.

Now let's calculate the probability that Step 2 is done i times when looking up a random item that is known to be in the table. Most people have more trouble calculating this probability than the previous one. The first thing to notice is that this probability is surely not the same as the probability that Step 2 is done i times when looking up a random item that is not in the table. When looking up an item not in the table, you sometimes look on a list that has no items on it. According to eq. (129) this happens with probability $(N-1)^k N^{-k}$. On the other hand, when looking up an item that is in the table, we always look on the list that contains the item, so we never look on an empty list.

The correct way to compute the probability is as follows. The probability that the first list has i items is given by eq. (129). If the first list has i items, then the probability that the item is on the first list is just i/k, because the item we are looking up has the same chance of being any of the items in the table. Therefore the probability that the first list has i items and that the item we are looking up is on the first list is $(i/k)p_i$. If the item we are looking up is on a list of i items, the probability that it is j from the beginning is $1/i$ for any value of j between 1 and i. There are N lists, and the item we are looking for has the same chance of being on any of them. Therefore if we let q_{ij} be the probability that our item is the jth item on a list of length i, we have

$$q_{ij} = \begin{cases} \dfrac{Np_i}{k} & \text{for } 1 \le j \le i \\ 0 & \text{otherwise.} \end{cases} \tag{141}$$

To obtain the average time, we need to sum jq_{ij} over i and j. We must sum over both i and j because q_{ij} is defined in terms of both of them, and we need to include all the values when computing the final answer. Therefore the average time is

$$A = \sum_{i,j} jq_{ij} = \sum_i \sum_{1\le j\le i} j\frac{Np_i}{k} = \sum_i \frac{Np_i}{k} \sum_{1\le j\le i} j \tag{142}$$

$$= \sum_i \frac{Np_i}{k}\frac{i(i+1)}{2} = \frac{N}{2k}\left(\sum_i (i+1)^2 p_i - \sum_i (i+1)p_i\right) \tag{143}$$

$$= \frac{N}{2k}\left(1 + 3\frac{k}{N} + \frac{k(k-1)}{N^2} - 1 - \frac{k}{N}\right) \tag{144}$$

$$= \frac{N}{2k}\left(2\frac{k}{N} + \frac{k(k-1)}{N^2}\right) = 1 + \frac{k}{2N} - \frac{1}{2N}. \tag{145}$$

So when k/N is large, we can look up items in the table about twice as fast as we can determine that items are not in the table. When k/N is small, either type of look-up is fast.

We leave the computation of the variance as an exercise.

There is a second way to compute the average time for looking up a random item that is known to be in the hash table. When the ith item was put into the table, Step 2 was done some number of times, say x. Each time the item is looked up, Step 2 will be done x times. Therefore, when there are k items in the table, the average time to look up an item that is in the table is the same as the average of the times to put the k items into the table. This gives

$$A = \frac{1}{k} \sum_{0 \le i \le k-1} \left(1 + \frac{i}{N}\right) = 1 + \frac{k(k-1)}{2kN} = 1 + \frac{k}{2N} - \frac{1}{2N}, \tag{146}$$

which is the same as the last answer.

In Step 5 of Hashing with Chaining the new item was added to the end of the list. If (as is common in practice) the new item is particularly likely to be reference in the near future, it is better to insert the new item at the front of the list. The analysis of this version is essentially the same. (If you make the assumption that all ordering of input are equally likely. Of course, when all orderings are equally likely, the new item is not particularly likely to be reference in the near future.) A much more complicated analysis is needed if new items are more likely to be referenced.

Hashing with Chaining is covered by Knuth [11, Section 6.4], Horowitz and Sahni [7, Section 2.6], Sedgewick [62, Chapter 16], Reingold and Hansen [61, Section 7.4.2], and Standish [63, Section 4.3.1]. See Knuth [11, pp. 540–541] for a history of the method.

An interesting variation of hashing is Coalesced Hashing where the overflow table partly shares space with the main hash table. This algorithm has been showed to have rather good performance by Vitter [136]. His long analysis illustrates many of the techniques of analysis of algorithms.

EXERCISES

1. What is the variance of the number of times that Step 2 of the Hashing with Chaining Algorithm is done when looking up an item in the table?
2. The average number of times that Step 2 of the Hashing with Chaining Algorithm was computed two different ways. Why does the second way of computing the average not lead to a method for computing the variance?

3.4 PERIODIC TERMS

Sometimes a formula is similar to one you know how to do, except that it is missing some of the terms. Previously we saw that if the missing terms were few in number, then you could add or subtract a few correction terms to obtain the sum that you knew how to work out. In this section we will show that if the missing terms have a regular pattern, you can often simplify the sum.

Let's first consider sums of the type

$$\sum_{j \text{ even}} a_j. \tag{147}$$

Assume a_j is zero except for a finite number of indices so that we do not have to worry about the complications that arise with infinite sums. The thing that makes this sum so different from the sums that we have seen before is the condition that j must be even. One approach is to replace j with $2k$, giving

$$\sum_k a_{2k}, \tag{148}$$

which is just fine if you know how to sum a_{2k}. Otherwise consider the sum

$$\sum_j a_j + \sum_j (-1)^j a_j. \tag{149}$$

Since $(-1)^j$ is 1 when j is even and -1 when j is odd, eq. (149) is almost the same as eq. (147). In eq. (149), for odd j corresponding terms from the first and second sums cancel, while for even j they have the same sign. Thus

$$\sum_{j \text{ even}} a_j = \frac{1}{2}\sum_j a_j + \frac{1}{2}\sum_j (-1)^j a_j. \tag{150}$$

Likewise,

$$\sum_{j \text{ odd}} a_j = \frac{1}{2}\sum_j a_j - \frac{1}{2}\sum_j (-1)^j a_j. \tag{151}$$

For example, consider $\sum_j \binom{n}{2j}$. The $2j$ in the bottom of the binomial prevents us from using our previous formulas for summing binomial coefficients. By applying eq. (150), however, we obtain

$$\sum_j \binom{n}{2j} = \sum_{j \text{ even}} \binom{n}{j} = \frac{1}{2}\sum_j \binom{n}{j} + \frac{1}{2}\sum_j (-1)^j \binom{n}{j} \tag{152}$$

$$= \frac{1}{2}(2^n + 0^n) = \frac{1}{2}(2^n + \delta_{n0}), \tag{153}$$

where δ_{jk} is the Kronecker delta function. Thus $n = 0$ is a special case for eq. (153), but the $n = 0$ case of the sum is easy to work out.

Now consider

$$\sum_{j \bmod 4 = 0} a_j. \tag{154}$$

Here the sum includes every fourth term. In this case we can use

$$\sum_{j \bmod 4=0} a_j = \frac{1}{4}\sum_j (1)^j a_j + \frac{1}{4}\sum_j (i)^j a_j + \frac{1}{4}\sum_j (-1)^j a_j + \frac{1}{4}\sum_j (-i)^j a_j, \tag{155}$$

where i is the square root of -1. Notice that $1^4 = i^4 = (-1)^4 = (-i)^4 = 1$. We found some numbers such that every fourth power of each number is 1, but the other powers cancel out when they are added together. What is going on here? Can this be done for powers other than 2 and 4? The next two sections give the mathematics for solving this problem in general.

EXERCISES

1. Prove eq. (151).

2. Simplify $\sum_j \binom{n}{2j} x^j$.

3. Simplify $\sum_j \binom{n}{2j} x^{2j}$.

4. Simplify $\sum_{0 \le i < n/2} \binom{n}{2i+1} x^{2i+1}$.

5. Simplify $\sum_j \binom{n}{4j}$.

6. Simplify $\sum_j \binom{n}{4j+1}$.

7. Give an equation similar to eq. (155) for the sum $\sum_{j \bmod 4 = 1} a_j$.

3.4.1 Complex Numbers

There are two common ways to represent complex numbers. One can give their real and imaginary parts to represent a number in the form $z = x + iy$, where z is a complex number and x and y are real numbers. Numbers in this form are easy to add: you just add the real and imaginary parts separately. Thus if $z_3 = z_1 + z_2$, where $z_1 = x_1 + iy_1$, $z_2 = x_2 + iy_2$, and $z_3 = x_3 + iy_3$, then $x_3 = x_1 + x_2$ and $y_3 = y_1 + y_2$. Subtraction is similar: you subtract corresponding components. Multiplication is more difficult, but it is equivalent to treating each number as a polynomial in i and using the fact that $i^2 = -1$. If $z_3 = z_1 z_2$, where $z_1 = x_1 + iy_1$, $z_2 = x_2 + iy_2$, and $z_3 = x_3 + iy_3$, then $x_3 = x_1 x_2 - y_1 y_2$ and $y_3 = x_1 y_2 + x_2 y_1$.

The second way to represent complex numbers is as a distance and angle in the complex plane. For the number $z = x + iy$, the distance $r = \sqrt{x^2 + y^2}$ is called the *magnitude* of z. The magnitude of z may be written as $|z|$. The angle is given by $\theta = \arctan(y/x)$. If the point (x, y) is graphed on the x, y plane, then r gives its distance from the origin, and θ gives the angle between the positive x axis and a line from the origin to the point. In this form complex numbers are not so easy to add, but they are easy to multiply. If z_1 has magnitude r_1 and angle θ_1, z_2 has r_2 and θ_2, and z_3 has r_3 and θ_3 with $z_3 = z_1 z_2$, then $r_3 = r_1 r_2$ and $\theta_3 = \theta_1 + \theta_2$.

It is convenient to measure angles in radians. One *radian* is the angle formed by an arc of a circle with a length equal to the radius, so the angular measure of a complete circle is 2π radians. The angle $90°$ is $\pi/2$ radians.

An nth *root of unity* is any number that, when multiplied by itself n times, gives 1. With real numbers, there is just one n^{th} root of unity when n is odd $(+1)$ and two roots when n is even (± 1). With complex numbers, however, there are n n^{th} roots of unity. The number $+1$ is always one of the roots.

The nth roots of unity all have magnitude 1, but they have different angles. Consider the point ω with magnitude 1 and angle $2\pi/n$; ω^n has magnitude 1 and angle 2π, so $\omega^n = 1$ and ω is an nth root of unity. Likewise ω^k, for integer k, is an nth root of unity, because $(\omega^k)^n = \omega^{kn} = (\omega^n)^k = 1^k = 1$. A number ω is a *principal* n^{th} root of unity if $\omega^n = 1$, but $\omega^k \neq 1$ for $1 \le k < n$. For the sake of definiteness we will call the point ω with magnitude 1 and angle $2\pi/n$

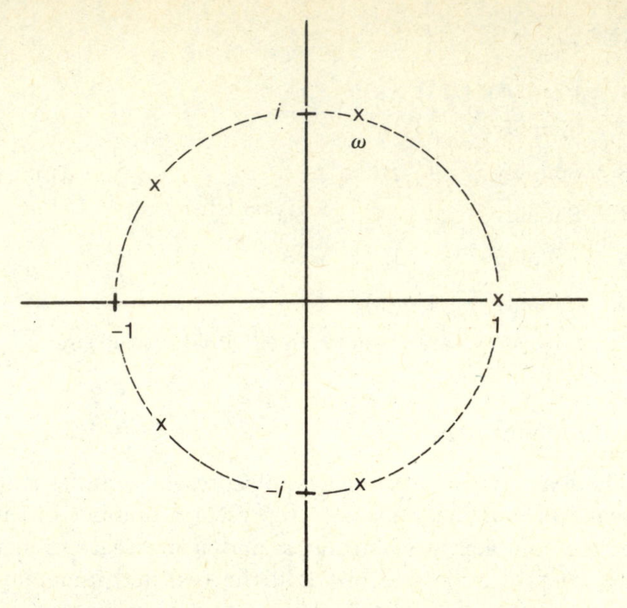

FIGURE 3.1 The five fifth roots of unity marked with crosses. All the roots are powers of the point ω with magnitude 1 and angle $2\pi/5$.

the *main* n^{th} root of unity. It is a principal n^{th} root of unity; for this value of ω the points ω^0, ω^1, ω^2, ..., ω^{n-1} are all different. They each have magnitude 1, but ω^k has angle $2\pi k/n$. Figure 3.1 shows the five 5th roots of unity. The point ω^n is the same as ω^0 (namely, +1). The points ω^0, ω^1, ω^2, ..., ω^{n-1} are evenly arranged around a circle of radius 1 centered around the origin, so the sum of these points is zero. This circle is called the *unit circle*.

Now consider the points $(\omega^0)^k$, $(\omega^1)^k$, $(\omega^2)^k$, ..., $(\omega^{n-1})^k$. We just considered how these points are arranged for $k = 1$. For $k = 0$ the points are all 1, so they sum to n. In general, the points are arranged uniformly around the unit circle with angle $2\pi k/n$. Starting at +1 and following the points in order you go around the origin k times. If d is the greatest common divisor of n and k, then the points occur at n/d distinct places, and each place has d points. If d is n (i.e., k is a multiple of n, perhaps zero), then there is just one distinct point (namely, +1), and the points $(\omega^0)^k$, $(\omega^1)^k$, $(\omega^2)^k$, ..., $(\omega^{n-1})^k$ sum to n; otherwise there is more than one distinct point, and the sequence $(\omega^0)^k$, $(\omega^1)^k$, $(\omega^2)^k$, ..., $(\omega^{n-1})^k$ sums to zero.

If you do not like the above geometrical argument for why $\sum_{0\le j\le n-1}\omega^{kj}$ is zero when ω^k is not 1, you can sum the series using eq. (2.61) to obtain

$$\sum_{0\le j\le n-1}\omega^{kj} = \frac{\omega^{kn}-1}{\omega^k-1} = 0 \quad \text{if } \omega^k \ne 1, \tag{156}$$

because ω^k is an nth root of unity, making $\omega^{kn} = 1$. When ω^k is 1, you cannot use eq. (2.61), but then all the terms in the sum are 1, so the sum is n.

The formula

$$e^{x+iy} = e^x(\cos y + i \sin y) \tag{157}$$

is the appropriate way to generalize the definition of exponent to complex numbers. With this definition the main n^{th} root of unity is $e^{2\pi i/n}$.

Confusion will result if you use i both for a summation index and for the square root of -1 in the same calculations. In this book we normally use i as a summation index, but there are a few sections where we deal with complex numbers. In those sections we only use i for the square root of -1.

Complex numbers are covered in Apostol [17, Sections 1.10 to 1.20] and Birkhoff and Mac Lane [22, Chapter 5].

EXERCISES

1. The *complex conjugate* of the number $z = x + iy$ is the number $z^* = x - iy$. In other words, the complex conjugate is formed by reversing the sign of the imaginary part. Show that $r^2 = zz^*$, where r is the magnitude of z.
2. Show that if ω is a principal n^{th} root of unity, then for any vector X

$$x_m = \frac{1}{n} \sum_{0 \le j < n} \omega^{-mj} \sum_{0 \le k} \omega^{jk} x_k$$

Hint: Consider Exercise 3.3–3.

3.4.2* General Formula

From the above one can prove the following:

$$\sum_{j \bmod n = 0} a_j = \frac{1}{n} \sum_{0 \le k < n} \sum_j (\omega^k)^j a_j, \tag{158}$$

where ω is a principal n^{th} root of unity. If the a_j are real, then the sum is also real, even though ω is complex. This observation can either be used as a check on your algebra, or as a justification for many shortcuts for computing real quantities with complex numbers.

To convert eq. (158) to a form containing only real numbers, you can first use the fact that $\omega = \cos(2\pi/n) + i \sin(2\pi/n)$. If your algebra is correct, the imaginary parts will cancel out. Notice that the nth roots of unity [except for $+1$ and -1 (if it is a root)] come in pairs with the same real parts and opposite imaginary parts. Thus the roots can be combined into pairs whose imaginary parts sum to zero.

When planning to apply eq. (158), remember to first consider replacing j with a new variable, such as nh. With such a replacement $\sum_{j \bmod n = 0} a_j$ becomes $\sum_h a_{nh}$. If you see how to do this sum, then do it; otherwise consider using eq. (158).

Let's consider the sum $\sum_j \binom{n}{3j}$. Using $\omega = \cos(2\pi/3) + i\sin(2\pi/3)$, we get

$$\sum_j \binom{n}{3j} = \frac{1}{3}\sum_j \left[\binom{n}{j} + \omega^j\binom{n}{j} + \omega^{2j}\binom{n}{j}\right] \tag{159}$$

$$= \frac{1}{3}\left(2^n + (1+\omega)^n + (1+\omega^2)^n\right). \tag{160}$$

This can now be written in a way that does not involve complex numbers, although some care is needed if you want to avoid doing a huge amount of algebra. The basic idea is to try to combine things in pairs so that the imaginary parts will cancel. We can write $(1+\omega)^n$ as $\omega^{n/2}(\omega^{-1/2}+\omega^{1/2})^n$. This can be simplified using $\omega^j = \cos(2\pi j/n) + i\sin(2\pi j/n)$. The imaginary part for the term in each pair of parentheses cancels out. For the first such term we get $\omega^{n/2}(\omega^{-1/2}+\omega^{1/2})^{n/2} = \omega^{n/2}(2\cos(\pi/3))^n$. Likewise, we can write $(1+\omega^2)^n$ as $\omega^n(\omega^{-1}+\omega)^n = \omega^n(2\cos(2\pi/3))^n$. This gives

$$\sum_j \binom{n}{3j} = \frac{1}{3}\left[2^n + \omega^{n/2}\left(2\cos\frac{\pi}{3}\right)^n + \omega^n\left(2\cos\frac{2\pi}{3}\right)^n\right]. \tag{161}$$

Now $\cos(\pi/3) = 1/2$ and $\cos(2\pi/3) = -1/2$, so

$$\sum_j \binom{n}{3j} = \frac{1}{3}\left[2^n + \omega^{n/2} + (-1)^n\omega^n\right]. \tag{162}$$

Now $\omega^{1/2} = \cos(\pi/3) + i\sin(\pi/3)$, while $-\omega = -\cos(2\pi/3) - i\sin(2\pi/3) = \cos(\pi/3) - i\sin(\pi/3)$. Thus $\omega^{1/2}$ and $-\omega$ have the same real parts, but opposite imaginary parts. Therefore $\omega^{n/2}$ and $(-1)^n\omega^n$ also have the same real parts and opposite imaginary parts. We have $\omega^{n/2} = \cos(\pi n/3) + i\sin(\pi n/3)$ and $(-1)^n\omega^n = \cos(\pi n/3) - i\sin(\pi n/3)$. This gives

$$\sum_j \binom{n}{3j} = \frac{1}{3}\left[2^n + \left(\cos\frac{\pi n}{3} + i\sin\frac{\pi n}{3}\right) + \left(\cos\frac{\pi n}{3} - i\sin\frac{\pi n}{3}\right)\right] \tag{163}$$

$$= \frac{1}{3}\left(2^n + 2\cos\frac{\pi n}{3}\right). \tag{164}$$

According to Knuth [9, exercise 1.2.6-38], the technique in this section was developed by Ramus in 1834.

EXERCISES

1. Simplify $\sum_j \binom{n}{5j}$.

2. Prove that

$$\sum_j \binom{n}{mj+k} = \frac{1}{m}\sum_{0\le j<m}\left(2\cos\frac{\pi j}{m}\right)^n \cos\frac{\pi j(n-2k)}{m}.$$

3. Obtain an equation for $\sum_j \binom{n}{mj+k}x^j$ that is similar to the result of the previous exercise.

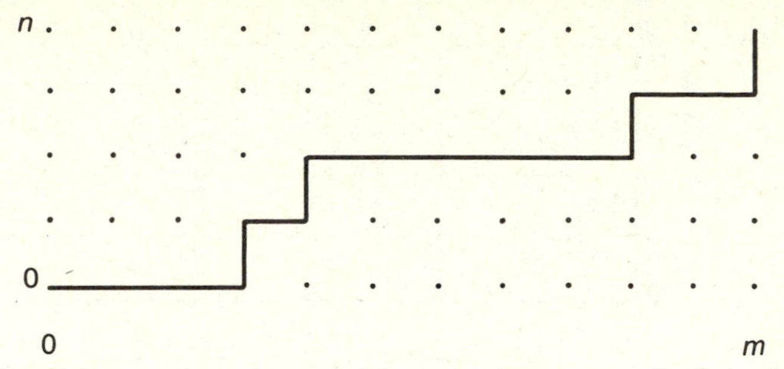

FIGURE 3.2 The number of increasing paths from point $(0, 0)$ to point (m, n) is $\binom{m+n}{m} = \binom{m+n}{n}$.

3.5 PRODUCTS OF BINOMIALS

Consider a directed increasing path (on an integer grid) from the origin to point (m, n) where each segment is either horizontal and directed to the right or vertical and directed upward. Figure 3.2 shows such a path. If you label the horizontal segments with H and the vertical segments with V, then as you traverse any such path you will use m H's and n V's, in some order. There are $\binom{m+n}{m}$ sequences of m H's and n V's, because you must choose m horizontal steps out of a total of $m+n$ steps. Each such sequence corresponds to a unique path. Therefore there is a one-to-one correspondence between the number of sequences with m H's and n V's, and the number of paths from the origin to the point (m, n). The number of such paths from the origin to point (m, n) is $\binom{m+n}{m} = \binom{m+n}{n}$.

Many of our previous formulas for binomial coefficients can be proved easily using the correspondence between increasing paths and binomial coefficients. Here we will use the correspondence to derive formulas for summing products of binomials. The general idea behind all the derivations is as follows. Draw a diagonal line with negative slope through an appropriately chosen rectangle. Since the increasing path has positive slope, it must cross the diagonal line exactly once. Now consider all the paths from the origin to point (m, n) that cross the diagonal line at a particular place, say, (i, j). Such paths are made up of a part to the left of the diagonal line and a part to the right. There is a binomial coefficient that says how many right parts are possible and another binomial coefficient that says how many left parts are possible. Since each right part can be combined with each left part to obtain a unique path, the total number of paths through (i, j) is the product of the two binomial coefficients. The total number of paths from the origin is just the sum of such products, with the sum including all the points on the diagonal line where the paths can cross. This technique also works for horizontal and vertical dividing lines provided the dividing line does not go through the grid points.

Figure 3.3 shows how this technique can be used to derive the most basic of the formulas for summing products of binomial coefficients. Consider the number of paths from the origin to $(r + s - n, n)$. There are $\binom{r+s}{n}$ such paths. Now consider a line with slope -1 that goes through the point $(r, 0)$, with

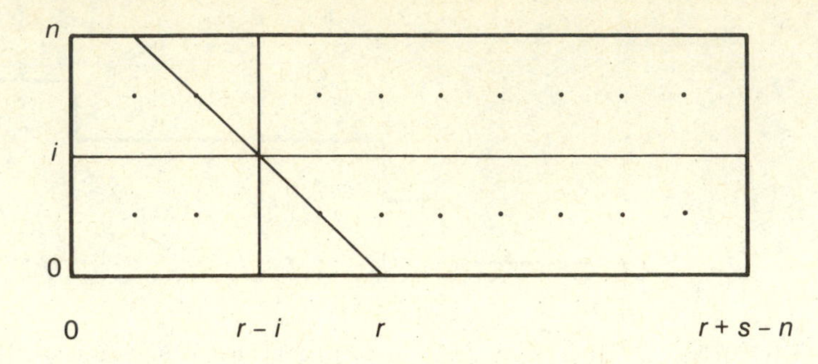

FIGURE 3.3 The number of increasing paths from point $(0,0)$ to point $(r+s-n,n)$ is $\binom{r+s}{n}$. Since each path must cross the diagonal line, the number of increasing paths is also $\sum_i \binom{r}{i}\binom{s}{n-i}$.

$0 \leq n \leq r \leq r+s-n$. All increasing lines from the origin to $(r+s-n,n)$ must cross this line exactly once. Let $(r-i,i)$ be such a crossing point. All lines through $(r-i,i)$ stay below i and to the left of $r-i$ until they get to the crossing point, i.e., they remain in the lower left rectangle of Figure 3.3. There are $\binom{r}{i}$ paths in the lower left rectangle. After the crossing point, all the paths stay above i and to the right of $r-i$, i.e., in the upper right rectangle of Figure 3.3. The number of such paths is $\binom{s}{n-i}$. The total number of paths from the origin through $(r-i,i)$ to $(r+s-n,n)$ is $\binom{r}{i}\binom{s}{n-i}$. The total number of paths from the origin to $(r+s-n,n)$ is $\sum_i \binom{r}{i}\binom{s}{n-i}$. Thus, we have

$$\sum_i \binom{r}{i}\binom{s}{n-i} = \binom{r+s}{n}, \quad \text{integer } r,\ s,\ n, \quad 0 \leq n \leq r,\ n \leq s. \tag{165}$$

Eq. (165) can be extended in the following ways. The left and right sides of eq. (165) are polynomials in r of degree n, and they are equal for all integer $r \geq n$. Two polynomials of degree less than or equal to n that are equal at $n+1$ or more points *must be identical.* Therefore, eq. (165) holds for all r. For similar reasons it holds for all s. If n is an integer below zero, then both sides of eq. (165) are zero, so again the equation is true. Thus we have

$$\sum_i \binom{r}{i}\binom{s}{n-i} = \binom{r+s}{n}, \qquad \text{integer } n, \tag{166}$$

which is the most important of the formulas for summing products of binomial coefficients.

In the derivation of eq. (166), we already knew which formula we wanted for our final answer. When deriving a formula for the first time, the process is usually a little more roundabout. One approach is to choose simple variable names for each point and see what result you can obtain. Then you can change the variables in the final formula, so that the final formula has an easy to use form. When deriving equations the first time, one usually favors methods that require no more thinking than necessary. In books and papers one often uses instead methods that lead quickly to the desired answer. This approach leads to easy-to-read books, but sometimes leaves the students wondering why they have so much more trouble obtaining the results than the book does.

By applying eq. (53) to eq. (166) twice, you can obtain $\sum_i \binom{r}{i}\binom{s}{s-n+i} = \binom{r+s}{r+s-n}$. The substitution $n' = s-n$ $(n = s-n')$ gives

$$\sum_i \binom{r}{i}\binom{s}{n+i} = \binom{r+s}{r+n}, \qquad \text{integer } n,\ s, \quad s \geq 0. \qquad \textbf{(167)}$$

So now we see that we can sum products of binomials when the summation index appears on the bottom of each coefficient, provided the summation index occurs linearly with a multiplier of ± 1.

Figure 3.4 leads to a formula with the summation index on the top. The number of increasing paths from the origin to the point $(r+s-m-n, m+n+1)$ is $\binom{r+s+1}{m+n+1}$. Consider the horizontal dashed line with height $n+1/2$. Each increasing path from the origin to $(r+s-m-n, m+n+1)$ crosses the dashed line exactly once. Let's call the place of crossing $(s-n+i, n+1/2)$. The paths that cross at this point must stay in the lower left rectangle (bounded in the x direction by 0 and $s-n+i$, and bounded in the y direction by 0 and n) until they come to the vertical crossing line (shown as a heavy line in Figure 3.4.), then follow the crossing line to the origin of the upper right rectangle (bounded in the x direction by $s-n+i$ and $r+s-n-m$, bounded in the y direction by $n+1$ and $m+n+1$). The number of such increasing paths is $\binom{s+i}{n}\binom{r-i}{m}$. The total number of increasing paths from the origin to $(r+s-m-n, m+n+1)$ is the sum of this over i with $0 \leq s-n+i \leq r+s-n-m$. Also, for the technique of Figure 3.4 to work, we need $0 \leq n \leq n+m$. This gives us

$$\sum_{-(s-n)\leq i \leq r-m} \binom{r-i}{m}\binom{s+i}{n} = \binom{r+s+1}{m+n+1}, \qquad \text{integer } m,\ n,\ s,\ r,$$
$$n \geq 0, \quad 0 \leq m \leq r. \qquad \textbf{(168)}$$

The limits on the summation and the restriction $m \leq r$ can be modified by noticing cases where the binomials are zero (See Exercise 5). Notice that unlike eq. (166) and eq. (167), eq. (168) *does* require limits on the summation.

The remaining basic formulas for products of binomials involve one index on the top and one on the bottom. They also have -1 raised to a power that depends on the summation index. They are obtained from the previous formulas by negating an index with eq. (69). The formulas are

$$\sum_i \binom{r}{i}\binom{s+i}{n}(-1)^i = (-1)^r\binom{s}{n-r}, \qquad \text{integer } n,\ r \quad r \geq 0, \qquad \textbf{(169)}$$

and
$$\sum_{0\leq i\leq r} \binom{r-i}{m}\binom{s}{i-t}(-1)^i = (-1)^t\binom{r-t-s}{r-t-m}, \qquad \text{integer } t,\ r,\ m,$$
$$t \geq 0, \ r \geq 0, \ m \geq 0. \qquad \textbf{(170)}$$

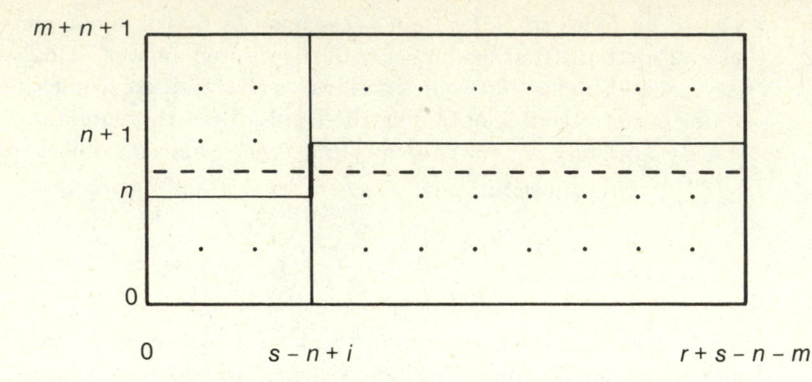

FIGURE 3.4 The number of increasing paths from point $(0, 0)$ to point $(r + s - m - n, m + n + 1)$ is $\binom{r+s+1}{m+n+1}$. Since each path must cross the horizontal line, the number of increasing paths is also $\sum_i \binom{r-i}{m}\binom{s+i}{n}$.

The general approach that will permit you to simplify many formulas that involve products of binomials is to use the elementary transformations given in Section 3.2 to reduce the summand to one of the forms given in this section. When looking for a transformation, concentrate on getting the summation index into the appropriate places. Section 3.8 gives a more powerful method of simplifying the sums of products of binomial coefficients.

Formulas for sums of products of binomial coefficients are given in Knuth [9, Section 1.2.6]; it was a source for much of this section. The use of diagrams to derive such formulas was adapted from Cohen [23, Chapter 2]. Also see Greene and Knuth [6, Sections 1.1 and 1.2].

EXERCISES

1. Use the equation $(1+x)^{r+s} = (1+x)^r(1+x)^s$ and the binomial theorem to derive eq. (166). Hint: Look at the x^i term on both sides. This exercise and the next one give elementary examples of the use of generating functions. Generating functions are discussed in much more detail in Section 5.3.
2. Consider the x^j term in $(1 + x)^{2m} = (1 + 2x + x^2)^m$ to find the value of
$$\sum_i \binom{m}{i}\binom{m-i}{j-2i}2^{j-2i}.$$
3. Why doesn't s need to be a nonnegative integer in eq. (167)? The use of eq. (53) that was made in the proof did require that s be a nonnegative integer. Hint: Think about polynomials.
4. Does eq. (167) always work for integer r, integer s, when $r + s \geq 0$? What about for noninteger negative s?
5. Show that eq. (168) works for the summation limit $0 \leq i \leq r$ when all the parameters are integers, $n \geq s \geq 0$, $m \geq 0$, and $r \geq 0$.
6. Derive eq. (169).

7. Derive eq. (170).

8. [See 9] Show that $\sum_i (-1)^i \binom{m}{i}\binom{i}{n} = (-1)^m \delta_{mn}$ for integer m and n, where $m \geq 0$.

9. Simplify $\displaystyle\sum_{m_1+m_2+\cdots+m_k=p} \prod_{1\leq i\leq k} (-1)^{m_i}\binom{n}{m_i} x^{m_i}$, where p is a positive integer. (The sum is over the k indices $m_1, m_2, \ldots, m_k$.)

3.6 MULTINOMIAL COEFFICIENTS

The number of ways that you can divide n distinct objects into m distinct piles so that the first pile has i_1 objects, the second has i_2 objects, etc., and the last (m^{th}) has i_m objects (where $i_1 + i_2 + \cdots + i_m = n$) defines the *multinomial coefficient*

$$\binom{n}{i_1, i_2, \ldots, i_m}. \tag{171}$$

Remember that the sum of the bottom indices of a multinomial coefficient must equal the top index.

The multinomial coefficient is closely related to the binomial coefficient, as you can see by considering the number of ways you can divide up n objects into m piles where the first pile has i_1 objects, the second pile has i_2, etc. This number is equal to the number of ways that you can select i_1 of the n objects for the first pile and then divide up the remaining $n-i_1$ objects into $m-1$ piles. (This observation applies even when there are restrictions on how the remaining objects are divided into piles.) This leads to the formula

$$\binom{i_1+i_2+\cdots+i_m}{i_1, i_2, \ldots, i_m} = \binom{i_1+i_2+\cdots+i_m}{i_1}\binom{i_2+\cdots+i_m}{i_2, \ldots, i_m}. \tag{172}$$

By applying eq. (172) repeatedly, we obtain

$$\binom{i_1+i_2+\cdots+i_m}{i_1, i_2, \ldots, i_m} = \binom{i_1+i_2+\cdots+i_m}{i_1}\binom{i_2+\cdots+i_m}{i_2}\cdots\binom{i_{m-1}+i_m}{i_{m-1}}. \tag{173}$$

Therefore sums with multinomial coefficients are usually simplified using the techniques for binomial coefficients.

We can rewrite eq. (173) to express multinomials in terms of factorials as follows:

$$\binom{i_1+i_2+\cdots+i_m}{i_1, i_2, \ldots, i_m} = \frac{(i_1+i_2+\cdots+i_m)!}{i_1!i_2!\cdots i_m!}. \tag{174}$$

The *multinomial theorem* is

$$(x_1+x_2+\cdots+x_m)^n = \sum_{i_1,i_2,\ldots,i_m} \binom{n}{i_1, i_2, \ldots, i_m} x_1^{i_1} x_2^{i_2} \cdots x_m^{i_m}. \tag{175}$$

Eq. (175) has been written as a sum over all values of the i's. For the multinomial to be nonzero, it is necessary for the i's to be nonnegative and to sum to n. For example, if $n = 3$ and $m = 3$, then the only nonzero terms in eq. (175) are the following triples for (i_1, i_2, i_3): $(0,0,3)$, $(0,1,2)$, $(0,2,1)$, $(0,3,0)$, $(1,0,2)$, $(1,1,1)$, $(1,2,0)$, $(2,0,1)$, $(2,1,0)$, $(3,0,0)$.

Knuth [9, Section 1.2.6] is a source for this section.

EXERCISES

1. Prove that $\displaystyle\sum_{n=n_1+n_2+\cdots+n_i} \binom{n_1+n_2+\cdots+n_i}{n_1, n_2, \ldots, n_i} = i^n$. (The expression under the summation sign means that the sum is over all combinations of values for n_1, n_2, ..., n_i that sum to n.)

3.7* STIRLING NUMBERS

The Stirling numbers are the coefficients that arise when you relate binomial coefficients to power series and powers to series of binomial coefficients. Different authors have different conventions concerning the signs and other details of the definitions of these numbers. You should check their conventions before you use their formulas and make whatever adjustments are necessary. We follow the conventions of Knuth.

The expression $n!\binom{x}{n} = x(x-1)\cdots(x-n+1)$ is a polynomial in x with integer coefficients. [Recall that we have seen $n!\binom{x}{n}$ before. It is $x^{\underline{n}}$, the decreasing n^{th} power of x.] *Stirling numbers of the first kind*, written $\left[{n \atop i}\right]$, are defined by

$$n!\binom{x}{n} = \left[{n \atop n}\right]x^n - \left[{n \atop n-1}\right]x^{n-1} + \cdots + (-1)^n\left[{n \atop 0}\right] = \sum_i (-1)^{n-i}\left[{n \atop i}\right]x^i. \quad \textbf{(176)}$$

With this definition the Stirling numbers of the first kind are all nonnegative integers. For example,

$$5!\binom{x}{5} = x^5 - 10x^4 + 35x^3 - 50x^2 + 24x \qquad (177)$$

so $\left[{5 \atop 5}\right] = 1$, $\left[{5 \atop 4}\right] = 10$, $\left[{5 \atop 3}\right] = 35$, $\left[{5 \atop 2}\right] = 50$, $\left[{5 \atop 1}\right] = 24$, and $\left[{5 \atop i}\right] = 0$ for $i < 1$ and for $i > 5$. Table 3.2 gives the value of the first few Stirling numbers of the first kind.

Notice that $n!\binom{x}{n}$ is a polynomial of degree n (we are assuming $n \geq 0$). The constant term is zero unless $n = 0$. The coefficient of the x^n term is 1. The coefficient of the x^{n-1} term is the sum of all the integers from 1 to $n-1$. The coefficient of the x term is $(n-1)!$ (for $n > 0$). These observations lead to the following special values:

$$\left[{n \atop n}\right] = 1, \quad \left[{n \atop n-1}\right] = \binom{n}{2}, \qquad \textbf{(178)}$$

n	$\left[{n \atop 0}\right]$	$\left[{n \atop 1}\right]$	$\left[{n \atop 2}\right]$	$\left[{n \atop 3}\right]$	$\left[{n \atop 4}\right]$	$\left[{n \atop 5}\right]$	$\left[{n \atop 6}\right]$	$\left[{n \atop 7}\right]$	$\left[{n \atop 8}\right]$
0	1	0	0	0	0	0	0	0	0
1	0	1	0	0	0	0	0	0	0
2	0	1	1	0	0	0	0	0	0
3	0	2	3	1	0	0	0	0	0
4	0	6	11	6	1	0	0	0	0
5	0	24	50	35	10	1	0	0	0
6	0	120	274	225	85	15	1	0	0
7	0	720	1764	1624	735	175	21	1	0
8	0	5040	13068	13132	6769	1960	322	28	1

TABLE 3.2. Stirling numbers of the first kind.

$$\left[{n \atop 1}\right] = (n-1)! \quad (\text{for } n > 0), \qquad \left[{n \atop 0}\right] = \delta_{n0}. \tag{179}$$

In general, the Stirling number $\left[{n \atop i}\right]$ is the sum of all the products of $(n-i)$ different integers taken from 1 to $n-1$; that is,

$$\left[{n \atop n-k}\right] = \sum_{1 \le i_1 < i_2 < \cdots < i_k < n} i_1 i_2 \cdots i_k. \tag{180}$$

Since $\left[{n \atop i}\right]$ is a coefficient of a term in $x(x-1)\cdots(x-n+1)$ and $\left[{n-1 \atop i}\right]$ is a coefficient of a term in $x(x-1)\cdots(x-n+2)$, you can obtain an interesting relation among Stirling numbers of the first kind as follows:

$$\begin{aligned}
&x(x-1)\cdots(x-n+1) \\
&\quad = \sum_i (-1)^{n-i} \left[{n \atop i}\right] x^i = x(x-1)\cdots(x-n+2)[x-(n-1)] && (181) \\
&\quad = x[x(x-1)\cdots(x-n+2)] - (n-1)[x(x-1)\cdots(x-n+2)] && (182) \\
&\quad = \sum_i (-1)^{n-i-1} \left[{n-1 \atop i}\right] x^{i+1} - (n-1) \sum_i (-1)^{n-i-1} \left[{n-1 \atop i}\right] x^i && (183) \\
&\quad = \sum_i (-1)^{n-i} \left[{n-1 \atop i-1}\right] x^i + (n-1) \sum_i (-1)^{n-i} \left[{n-1 \atop i}\right] x^i. && (184)
\end{aligned}$$

Since the right sides of lines 181 and 184 are polynomials in x and since they are equal for all x, each side must have the same coefficient for x^i. Therefore

$$\left[{n \atop i}\right] = (n-1)\left[{n-1 \atop i}\right] + \left[{n-1 \atop i-1}\right] \quad \text{for } n > 0. \tag{185}$$

Stirling numbers of the second kind, $\left\{{n \atop i}\right\}$, count the number of ways of partitioning a set of n elements into exactly i nonempty subsets. The number of ways you can put n labelled balls into m nonempty labelled boxes is $m!\left\{{n \atop m}\right\}$,

n	$\left\{ {n \atop 0} \right\}$	$\left\{ {n \atop 1} \right\}$	$\left\{ {n \atop 2} \right\}$	$\left\{ {n \atop 3} \right\}$	$\left\{ {n \atop 4} \right\}$	$\left\{ {n \atop 5} \right\}$	$\left\{ {n \atop 6} \right\}$	$\left\{ {n \atop 7} \right\}$	$\left\{ {n \atop 8} \right\}$
0	1	0	0	0	0	0	0	0	0
1	0	1	0	0	0	0	0	0	0
2	0	1	1	0	0	0	0	0	0
3	0	1	3	1	0	0	0	0	0
4	0	1	7	6	1	0	0	0	0
5	0	1	15	25	10	1	0	0	0
6	0	1	31	90	65	15	1	0	0
7	0	1	63	301	350	140	21	1	0
8	0	1	127	966	1701	1050	266	28	1

TABLE 3.3. Stirling numbers of the second kind.

because there are $\left\{ {n \atop m} \right\}$ ways to partition the balls into m subsets, and for each partition there are $m!$ ways to label the m sets.

In general, the Stirling number $\left\{ {n \atop i} \right\}$ is the sum of all the products of $(n-i)$ integers taken from 1 to $n-1$; that is,

$$\left\{ {n \atop n-k} \right\} = \sum_{1 \le i_1 \le i_2 \le \cdots \le i_k < n} i_1 i_2 \cdots i_k. \qquad \textbf{(186)}$$

Stirling numbers of the second kind can be used to convert powers into binomial coefficients, because

$$x^n = \sum_i \left\{ {n \atop i} \right\} \binom{x}{i} i!, \qquad \textbf{(187)}$$

where, of course, $\binom{x}{i} i! = x(x-1)\cdots(x-i+1) = x^{\underline{i}}$. For example, consider expressing x^5 with $x^{\underline{5}}$, $x^{\underline{4}}$, $x^{\underline{3}}$, $x^{\underline{2}}$, and $x^{\underline{1}}$. Now $x^{\underline{5}} = x^5 - 10x^4 + 35x^3 - 50x^2 + 24x$ is the only one of these that has an x^5, so we must take one unit of $x^{\underline{5}}$. This gives us -10 units of x^4 so far, so we must take 10 units of $x^{\underline{4}} = x^4 - 6x^3 + 11x^2 - 6x$. This gives us a total of -25 units of x^3 so far, so we must take 25 units of $x^{\underline{3}}$, and so on. We obtain

$$x^5 = x^{\underline{5}} + 10x^{\underline{4}} + 25x^{\underline{3}} + 15x^{\underline{2}} + x^{\underline{1}}. \qquad (188)$$

The first few Stirling numbers of the second kind are given in Table 3.3.

To prove eq. (187), notice that the number of ways of partitioning a set of n elements into m labelled (and possibly empty) boxes is m^n; the first item can go m different places (any box), and so can each of the remaining elements. On the other hand, the number of ways of dividing n elements into i nonempty sets is $\left\{ {n \atop i} \right\}$. The number of ways of assigning i of m labels to the i boxes is just the number of permutations of m items taken i at a time, $i!\binom{m}{i}$. Therefore the number of ways of partitioning n items into i boxes with labels from 1 to m is

$i!\binom{m}{i}\left\{ {n \atop i} \right\}$. Summing over i gives eq. (187) for all integer x. Since both sides are polynomial in x, it must be true for all x.

The following observations lead to some special values for Stirling numbers of the second kind. We cannot put n objects into zero sets (unless $n = 0$), so $\left\{ {n \atop 0} \right\} = 0$ for $n > 0$. We can put n objects into one set one way (for $n > 0$), so $\left\{ {n \atop 1} \right\} = 1$ for $n > 1$. If we divide a set into two parts, then we can put any subset into the first part so long as we put its complement into the second part. Since a set of n elements has 2^n subsets, there would be 2^n ways to divide a set into some subset and its complement, *if* we were using labelled parts. Since we are using unlabelled parts, this approach counts every division method twice; there are 2^{n-1} ways to divide a set into two unlabelled parts. One of these ways, however has one part empty. Therefore $\left\{ {n \atop 2} \right\} = 2^{n-1} - 1$ for $n > 0$. If we divide n objects into $n - 1$ nonempty sets, then each set but one has one object and the remaining set has two. There are $\binom{n}{2}$ ways to choose the two objects for that set, so $\left\{ {n \atop n-1} \right\} = \binom{n}{2}$. Finally, there is only one way to divide n objects into n sets, so $\left\{ {n \atop n} \right\} = 1$.

Consider splitting n objects into i nonempty subsets. When we do this either we get a subset that contains only the last item or we do not. If we have a set with just the last item, then the remaining $n - 1$ items must be divided into $i - 1$ nonempty sets. This can be done $\left\{ {n-1 \atop i-1} \right\}$ ways. If the last item does not occur in a set by itself, then we can remove it and still have $n - 1$ items divided into i nonempty sets. Now $n - 1$ items can be divided into i nonempty sets $\left\{ {n-1 \atop i} \right\}$ ways. If we have $n - 1$ items divided into i nonempty sets, there are i ways we can put the nth item in (it can go into any one of the i sets). Each way gives a different division of n items into i nonempty sets where the last item does not occur alone. We have now accounted for all the ways that n items can be divided into i nonempty sets, so

$$\left\{ {n \atop i} \right\} = i \left\{ {n-1 \atop i} \right\} + \left\{ {n-1 \atop i-1} \right\}. \qquad \textbf{(189)}$$

We give a few more formulas involving Stirling numbers without comment:

$$\sum_i \left[{n \atop i} \right] \left\{ {i \atop m} \right\} (-1)^i = (-1)^n \delta_{mn}, \qquad \textbf{(190)}$$

$$\sum_i \left\{ {n \atop i} \right\} \left[{i \atop m} \right] (-1)^i = (-1)^n \delta_{mn}, \qquad \textbf{(191)}$$

$$\sum_i \left[{n \atop i} \right] \binom{i}{m} = \left[{n+1 \atop m+1} \right], \qquad \textbf{(192)}$$

$$\sum_i \left\{ {n \atop i} \right\} \binom{i}{m} = \left\{ {n+1 \atop m+1} \right\}. \qquad \textbf{(193)}$$

The reader is referred to Knuth [9, Section 1.2.6] and its references for additional information on Stirling numbers. It was a source for this section. More formulas and larger tables are given in Abramowitz and Stegun [16, Chapter 24].

EXERCISES

1. [See 9] Show that

$$\sum_i (-1)^i \binom{m}{i} (b_0 + b_1 i + \cdots + b_m i^m) = (-1)^m m! b_m$$

for integer $m \geq 0$.

2. Show that

$$\sum_i \binom{n}{i} (-1)^i i^m = (-1)^n n! \left\{ {m \atop n} \right\}$$

3. Show that the number of permutations of n objects with k cycles obeys the equation $c_{nk} = (n-1)c_{n-1,k} + c_{n-1,k-1}$. Hint: Consider the last element. Either it forms a unit cycle by itself, in which case the remaining $n-1$ elements must form $k-1$ cycles, or it is part of some cycle. Continue by explaining the situation when the last element is not in a unit cycle.

4. Show that the number of permutations of n objects with k cycles is $\left[{n \atop k} \right]$. Hint: Use the result of the previous exercise and the recurrence for Stirling numbers of the first kind.

3.7.1* Find the Maximum

Consider the problem of finding the maximum of n numbers. The obvious algorithm for this problem follows.

Algorithm 3.2 Maximum: Input: an array X with elements x_i for $1 \leq i \leq n$. Output: k, the index of the largest number in X. Intermediate variable: y, the largest number so far.

Step **1.** [Initialize] Set $k \leftarrow n$ and $y \leftarrow x_n$.

Step **2.** [Loop] For $i \leftarrow n-1$ down to 1 do:

Step **3.** [Compare] If $x_i > y$, then

Step **4.** [New max.] Set $k \leftarrow i$ and $y \leftarrow x_i$.

Step **5.** End for.

The first step of this algorithm is done once, the second step n times, and the third and fifth steps are done $n-1$ times. The fourth step is done between zero and $n-1$ times depending on the input data. This section considers the probability that Step 4 is done i times when the elements of X are n different numbers in random order (any ordering equally likely). Step 4 does not have a major effect on the running time of this algorithm, but this is a simple algorithm whose analysis involves Stirling numbers.

Let p_{ni} be the probability that Step 4 is done i times when there are n items. To obtain an equation for p_{ni}, notice that when the first element is the largest, the number of times that Step 4 is done is one more than the number of times it was done for all but the first element. This case occurs $1/n$ of the time. If the first element is not the largest [which happens $(n-1)/n$ of the time], then Step 4 is done the same number of times for the entire set as it is done for all but the first element. This gives the relation

$$p_{ni} = \frac{1}{n}p_{n-1,i-1} + \frac{n-1}{n}p_{n-1,i}. \tag{194}$$

Using eq. (194) by itself, we cannot calculate the probabilities. However, we have the following special values:

$$p_{1i} = \delta_{0i} \quad \text{for } i \geq 0 \quad \text{and} \quad p_{ni} = 0 \quad \text{for } i < 0. \tag{195}$$

These special values are usually called *boundary values* because they occur at the edge of the region of interest. From eq. (194) and the boundary conditions, we can calculate p_{ni} for $n \geq 1$ and $i \geq 0$.

Let's consider how we could calculate a table of values from eq. (194) and eq. (195). Notice that to calculate p_{ni} using eq. (194) we need two values, each with a smaller first index. One has the same second index and the other has a second index that is one smaller. So to get started we need the value of p_{ni} for some starting value of n and all i from some lower limit up to the highest value of interest. To go from one row (value of n) to the next, we also need for each n the value of p_{ni} for the smallest value of i of interest. Eq. (195) gives p_{1i} for all i, so we can start our table with $n = 1$. It also gives p_{ni} for all $i < 0$. The negative values of i are not very interesting, so we will start with the largest such value, $i = -1$. So for the first row of our table we have $p_{1,-1} = 0$, $p_{1,0} = 1$, $p_{1,1} = 0$, $p_{1,2} = 0$, etc. Now we can start the second row. We get $p_{2,-1} = 0$ by eq. (195). The remaining values we get from eq. (194); $p_{2,0} = 1/2$, $p_{2,1} = 1/2$, $p_{2,3} = 0$, etc. Notice that we must be careful to calculate the values in an appropriate order so that the values we need for a calculation are available in time. If you try to calculate values from a recurrence equation [one like eq. (194), where functions are defined in terms of their values at other points], and you get stuck in your calculations because you need an item that you cannot calculate, then you do not have a complete set of boundary conditions.

Later we will study some general techniques for solving eq. (194) together with eq. (195). For now, however, let's show how to solve it using a technique similar to proof by induction combined with some lucky (i.e., carefully chosen) guesses. Notice that if you multiply eq. (194) by n, it looks similar to eq. (185), except that the left side is too big by a factor of n. This suggests guessing an answer that is related to Stirling numbers of the first kind. Let's try the general form $p_{ni} = f(n)\left[{n \atop i+a}\right]$, where $f(n)$ is a yet to be determined function of n and a is a yet to be determined constant. This is one of the more general guesses that has the right form for applying eq. (185). If we plug this guess into eq. (194), we obtain

$$f(n)\left[{n \atop i+a}\right] \quad \text{vs.} \quad \frac{1}{n}f(n-1)\left[{n-1 \atop i+a-1}\right] + \frac{n-1}{n}f(n-1)\left[{n-1 \atop i+a}\right] \tag{196}$$

$$\text{vs.} \quad \frac{f(n-1)}{n}\left(\left[{n-1 \atop i+a-1}\right] + (n-1)\left[{n-1 \atop i+a}\right]\right) \tag{197}$$

$$\text{vs.} \quad \frac{f(n-1)}{n}\left[{n \atop i+a}\right], \tag{198}$$

where we use vs. as a relation that is *true* if the two sides are equal and *false* otherwise. This is, of course, what the relation = means, but = is used so often for relations that must be *true*, that perhaps it is better to have special notation for a relation that we just hope will be *true*. Informally, vs. indicates that we want the two sides to be equal. When we write a series of lines with vs., then each vs. is *true* only if the vs. in the line below it is *true*. Eq. (198) is *true* if and only if

$$f(n) = \frac{f(n-1)}{n}, \tag{199}$$

which is *true* provided $f(n) = c/n!$, where c is a constant. Therefore our guess will satisfy eq. (194) provided it has the form

$$p_{ni} = \frac{c}{n!}\left[{n \atop i+a}\right]. \tag{200}$$

We are making definite progress. We have satisfied the recursive equation, and we have two constants left that we can adjust as we try to satisfy the boundary conditions.

To satisfy $p_{ni} = 0$ for $i \le -1$ and $n \ge 1$, we need to have $a = 1$. To satisfy $p_{1i} = \delta_{0i}$, we need $c = 1$. This simplifies our guess to

$$p_{ni} = \frac{1}{n!}\left[{n \atop i+1}\right]. \tag{201}$$

This guess satisfies all the boundary conditions as well as eq. (194), so it is the solution of our problem.

Notice that our technique for this problem consisted of first guessing a general form for the answer, then applying the technique of proof by induction, and finally taking account of the conditions that are necessary for the proof by induction to go through. This is similar to a technique that researchers often use on new problems — with one exception. The researcher's first guess is usually not correct, so the first few attempts at applying the technique lead to improved guesses rather than to the solution.

The average number of times Step 4 is done is given by $\sum_i ip_{ni}$, which can be simplified using eq. (192) (with $m = 1$) to obtain

$$\sum_i ip_{ni} = \frac{1}{n!}\left[{n+1 \atop 2}\right]. \tag{202}$$

Exercise 3 combined with Section 4.5.3 leads to a more useful form for this answer. A direct analysis of the Maximum Algorithm is given in Section 6.3.1.

EXERCISES

1. Calculate p_{ni} for $1 \le n \le 5$, $1 \le i \le 5$ directly from the recurrence equation [eq. (194)]. Give the order in which you calculate the p's.
2. Give a complete proof by induction that p_{ni} for the Maximum algorithm is given by $p_{ni} = \frac{1}{n!}\left[{n \atop i+1}\right]$.
3. Show that $\left[{n+1 \atop 2}\right]/n! = \sum_{1\le i\le n} 1/i$. Hint: Consider eq. (180).

3.8* HYPERGEOMETRIC FUNCTIONS

In solving summation problems with binomial coefficients, one difficulty is that a few transformations of the summation index can convert a problem into a form that has no obvious relation to the original problem. Therefore there are many problems which are basically the same (if you look at them with the techniques that will be given in this section), but which require quite different techniques to solve (if you use only the techniques given earlier in this chapter).

The *generalized hypergeometric function* is

$$ {}_mF_n\left[{a_1, \ldots, a_m; z \atop b_1, \ldots, b_n}\right] = \sum_{i\ge 0} \frac{a_1^{\overline{i}} \cdots a_m^{\overline{i}} z^i}{b_1^{\overline{i}} \cdots b_n^{\overline{i}} i!}. \tag{203}$$

Notice that the a's and b's are to increasing powers (see Section 2.3) while the z is to an ordinary power. The factor $i!$ is equal to $1^{\overline{i}}$.

To save space, it is common to write

$$ {}_mF_n\left[{a_1, \ldots, a_m; z \atop b_1, \ldots, b_n}\right] \quad \text{as} \quad {}_mF_n[a_1, \ldots, a_m; b_1, \ldots, b_n; z]. \tag{204}$$

In eq. (203) the a's and b's are arbitrary, except that the b's must be chosen so that no divisions by zero occur. If none of the a's are nonpositive integers, then the sum has an infinite number of terms. In this case no b can be a nonpositive integer. If some a's are nonpositive integers, *then the sum is a polynomial in z*. In this case some of the b's can also be nonpositive integers, but you must be able to pair each nonpositive integer b with a nonpositive integer a in such a way that no a is used twice and such that the a is larger (closer to zero) than the matched b.

The sum on the right side of eq. (203) converges whenever it has a finite number of terms. There are also several cases where the infinite sum converges.

$$\begin{array}{|ll|}
\hline
{}_0F_0[;\,;z] & = e^z \\
{}_1F_0[a;\,;z] & = (1-z)^{-a} \\
{}_2F_1[1,1;2;z] & = \ln(1-z) \\
{}_2F_1\left[\begin{matrix}\frac{1}{2},\frac{1}{2};z^2\\ \frac{3}{2}\end{matrix}\right] & = \dfrac{\arcsin z}{z} \\
{}_2F_1\left[\begin{matrix}\frac{1}{2},1;z^2\\ \frac{3}{2}\end{matrix}\right] & = \dfrac{\arctan z}{z} \\
\hline
\end{array}$$

TABLE 3.4. Some common functions represented as generalized hypergeometric functions.

When $m \le n$, it converges for all z. (The ratio of successive terms in the series goes to zero in this case.)

When $m = n+1$, the sum is convergent for $|z| < 1$. It is also convergent for $z = 1$ if the real part of the sum of the b's is greater than the real part of the sum of the a's, i.e., if

$$\textbf{Real}\Bigg(\sum_{1\le i\le n} b_i\Bigg) > \textbf{Real}\Bigg(\sum_{1\le i\le m} a_i\Bigg). \tag{205}$$

It is convergent for $z = -1$ when

$$\textbf{Real}\Bigg(\sum_{1\le i\le n} b_i\Bigg) > \textbf{Real}\Bigg(\sum_{1\le i\le m} a_i - 1\Bigg). \tag{206}$$

When $m > n+1$ the sum converges only at $z = 0$ unless it has a finite number of terms.

The hypergeometric function is the $m = 2$, $n = 1$ case of the generalized hypergeometric function. This case already describes a rather general function of four parameters, and many of the well-known functions of mathematics are special cases of the hypergeometric function. Table 3.4 gives some examples.

Most simple binomial summation identities reduce to the equation

$${}_2F_1\left[\begin{matrix}a,b;1\\ c\end{matrix}\right] = \frac{\Gamma(c)\Gamma(c-a-b)}{\Gamma(c-a)\Gamma(c-b)}, \tag{207}$$

which was discovered by C. F. Gauss. To illustrate how powerful this equation is, every summation formula in Section 3.5 which does not have explicit limits (including the exercises) is a special case of this formula. We will use some of these earlier formulas to illustrate the technique.

Eq. (203) can be generalized slightly. Define the *bilateral function* as

$${}_mH_n\left[\begin{matrix}a_1,\ldots,a_m;z\\ b_1,\ldots,b_n\end{matrix}\right] = \sum_i \frac{a_1^{\overline{i}}\cdots a_m^{\overline{i}} z^i}{b_1^{\overline{i}}\cdots b_n^{\overline{i}}}, \tag{208}$$

where for positive i

$$x^{-\overline{i}} \quad \text{means} \quad \frac{(-1)^i}{(1-x)^{\overline{i}}}. \tag{209}$$

Eq. (208) is similar to eq. (203), but there is no explicit limit on the summation and there is no $i!$ on the bottom.

If the a's and b's are all noninteger, then the sum for the bilateral function is defined for $|z| = 1$. If any a is a negative integer, then there is an upper limit on the values of i that contribute to the sum. If any b is a positive integer, then there is a lower limit on the values of i that contribute to the sum. If both conditions are true, then the sum contains a finite number of terms. Whenever any a is a nonnegative integer or any b is a nonpositive integer, it is necessary for the values to be chosen so that no divisions by zero result. (The conditions on nonpositive b are the same as those given above for generalized hypergeometric functions, while the condition on nonnegative integer a is that each such a can be paired with a smaller nonnegative integer b.)

When the series is infinite and $m = n$, the sum for the bilateral function converges at $z = 1$ when

$$\textbf{Real}\left(\sum_{1\le i\le n} b_i\right) > \textbf{Real}\left(\sum_{1\le i\le m} a_i\right). \tag{210}$$

With the same conditions, it is convergent for $z = -1$ when

$$\textbf{Real}\left(\sum_{1\le i\le n} b_i\right) > \textbf{Real}\left(\sum_{1\le i\le m} a_i - 1\right). \tag{211}$$

Using the bilateral function, we can obtain a result that is slightly more general than eq. (207):

$${}_2H_2\begin{bmatrix} a, b; 1 \\ c, d \end{bmatrix} = \frac{\Gamma(c)\Gamma(d)\Gamma(1-a)\Gamma(1-b)\Gamma(c+d-a-b-1)}{\Gamma(c-a)\Gamma(d-a)\Gamma(c-b)\Gamma(d-b)}, \tag{212}$$

When using eq. (207) or eq. (212), it is often the case that some of the Gamma functions will have a nonpositive integer argument. Since the Gamma function is infinite for such an argument, the answer is infinite if all such Gamma functions occur on the top of the fraction, and the answer is zero if all such Gamma functions occur on the bottom. If there are some such Gamma functions on the top and some on the bottom, then you should use the equation

$$\frac{\Gamma(-a+b)}{\Gamma(-a)} = (-1)^b \frac{\Gamma(a+1)}{\Gamma(a-b+1)} \tag{213}$$

to eliminate an indeterminate form. [Informally, this equation corresponds to viewing $\Gamma(n)$ as $\prod_{1\le i<n} i$, with the convention that $\prod_{1\le i<-n} i$ means $\prod_{-n\le i<1} 1/i$, and cancelling common factors on top and bottom (including the zeros).]

Generalized hypergeometric functions are important for the summation of binomial coefficients because problems that look wildly different (such as the various summation formulas of Section 3.5 which do not have explicit limits) transform into a single formula when expressed as generalized hypergeometric functions.

To sum a formula involving the product of binomial coefficients with no explicit limits, we recommend the following approach (oversimplified slightly):

1. Express your problem as a generalized hypergeometric function (or a bilateral function). If your problem cannot be converted into the right form, then use some other method. (The details on how to express your problem as a generalized hypergeometric function are given below.)
2. Compare your hypergeometric function with eq. (207) (or eq. (212)). If it matches, your problem is solved.
3. Compare your function with the 32 cases given in Slater [53, appendix 3]. If it matches one of them, then your problem is solved.
4. If you have not solved it in one of the previous steps, decide if you really want to solve it. Expect to do a lot of work in most cases.

As we go through the procedure, we will rederive eq. (166) to show how the method works. Our sample problem is to sum $\binom{r}{i}\binom{s}{n-i}$ over i. First we write the problem in terms of factorials:

$$\binom{r}{i}\binom{s}{n-i} = \frac{r!s!}{i!(r-i)!(n-i)!(s-n+i)!}. \tag{214}$$

In general we would next use the techniques of Section 3.2 (and its subsections) to combine any terms that are polynomial in the summation index in with the binomials. Then we would cancel out any factorials that are the same on the top and bottom. In this case there is nothing to do for these two steps. Your problem should now have the form of a ratio of factors, where each factor is the factorial of a linear function of the summation index or is a base which does not contain the summation index raised to a power that is the summation index. For all the factorial terms the coefficient of the summation index must be an integer. If your problem is not reduced to this form, then the method does not apply. The sample problem is in the correct form.

Next we see if there is a factor on the bottom of the fraction that is $i!$. If the bottom does not contain such a factor, but it does contain a factor of the form $(k+i)!$ [or $(k-i)!$], then replace i by $i-k$ (or $i+k$) so that the transformed problem does contain an $i!$. We will assume that the problem now contains an $i!$ on the bottom and continue converting the problem to a hypergeometric problem. (If we cannot obtain an $i!$, then we will use a similar process to convert the problem to a bilateral function.) In the steps given below, do not change the $i!$ factor. Save it to help form the generalized hypergeometric function.

For each factorial factor where the coefficient of the summation index is positive, apply the following transformation (remember, don't change the $i!$ factor). Let the term have the form $(ai+ad+c)!$ where $a > c \geq 0$. Then

$$(ai+ad+c)! = (ad+c)!(a^a)^i \left(d+\frac{c+1}{a}\right)^{\overline{i}} \left(d+\frac{c+2}{a}\right)^{\overline{i}} \cdots \left(d+1+\frac{c}{a}\right)^{\overline{i}}. \tag{215}$$

This transformation converts a factorial containing the summation index with multiplicity a into a increasing powers, where the power is the summation index (with multiplicity 1) and the base does not contain the summation index. A factorial which does not contain the summation index and a power (if $a \neq 1$) are also produced.

For our sample problem we can apply this transformation to the $(s-n+i)!$ factor (leave the $i!$ alone!). We have $a=1$, $d=s-n$, and $c=0$. We get

$$(s-n+i)! = (s-n)!(s-n+1)^{\overline{i}}, \tag{216}$$

so our problem becomes

$$\sum_i \frac{r!s!}{i!(r-i)!(n-i)!(s-n)!(s-n+1)^{\overline{i}}}. \tag{217}$$

Next, for each factorial factor where the summation index occurs with negative multiplicity, apply the following transformation (where $a > 0$):

$$(b-ai)! = \frac{(-1)^{ai}b!}{(a^a)^i \left(-\frac{b}{a}\right)^{\overline{i}} \left(-\frac{b-1}{a}\right)^{\overline{i}} \cdots \left(-\frac{b-a+1}{a}\right)^{\overline{i}}}. \tag{218}$$

Notice that this transformation converts factors on the bottom of the fraction into increasing powers on the top of the fraction (and factors on the top into increasing powers on the bottom).

This transformation can be applied twice to the sample problem [to $(r-i)!$ and to $(n-i)!$]. In both cases $a=1$. We get

$$(r-i)! = \frac{(-1)^i r!}{(-r)^{\overline{i}}} \tag{219}$$

and a similar result for $(n-i)!$.

Our problem becomes

$$\sum_i \frac{s!(-r)^{\overline{i}}(-n)^{\overline{i}}}{i!n!(s-n)!(s-n+1)^{\overline{i}}}. \tag{220}$$

Now we can factor out $s!/[n!(s-n)!]$, and what remains is the hypergeometric function ${}_2F_1[-r,-n;s-n+1;1]$, so our problem is to simplify

$$\frac{s!}{n!(s-n)!}\,{}_2F_1\left[\begin{matrix}-r,-n;1\\ s-n+1\end{matrix}\right], \tag{221}$$

which by eq. (207) is

$$\frac{s!\Gamma(s-n+1)\Gamma(s+r+1)}{n!(s-n)!\Gamma(s+r-n+1)\Gamma(s+1)} = \frac{(s+r)!}{n!(s+r-n)!} = \binom{s+r}{n}, \tag{222}$$

where we have used $\Gamma(x+1) = x!$ and cancelled some factorials. This is eq. (166). The approach takes a fair amount of manipulation (so be careful about errors), but it requires very little thinking once you have learned the technique.

For practice you should try the technique on eq. (167). Each step is essentially the same as the corresponding step in working out eq. (166).

When we apply the technique to eq. (169), we start out the same way. We get

$$\sum_i \binom{r}{i}\binom{s+i}{n}(-1)^i = \frac{s!(s+1)^{\bar{i}}(-r)^{\bar{i}}}{i!n!(s-n)!(s-n+i)^{\bar{i}}} \tag{223}$$

$$= \frac{s!}{n!(s-n)!}\,{}_2F_1\begin{bmatrix}-r, s+1; 1\\ s-n+1\end{bmatrix} \tag{224}$$

$$= \frac{s!\Gamma(s-n+1)\Gamma(r-n)}{n!(s-n)!\Gamma(-n)\Gamma(s+r-n+1)} \tag{225}$$

$$= \frac{s!\Gamma(r-n)}{n!\Gamma(-n)\Gamma(s+r-n+1)}. \tag{226}$$

Now $\Gamma(-n)$ is infinite (for integer $n \leq 0$), so the answer is zero unless one of the Gamma functions on the top has a nonpositive argument. We will consider the case where $r-n$ is a nonpositive integer.

Using eq. (213), we have

$$\frac{\Gamma(r-n)}{\Gamma(-n)} = \frac{(-1)^r\Gamma(n+1)}{\Gamma(n-r+1)}. \tag{227}$$

Plugging this into eq. (226) gives

$$\sum_i \binom{r}{i}\binom{s+i}{n}(-1)^i = \frac{(-1)^r s!\Gamma(n+1)}{n!\Gamma(n-r+1)\Gamma(s+r-n+1)} \tag{228}$$

$$= \frac{(-1)^r s!}{(n-r)!(s+r-n)!} = (-1)^r\binom{s}{n-r}. \tag{229}$$

Again little thinking is needed to apply the method.

Next, let's consider using hypergeometric functions to find the value of

$$\sum_i \binom{2n}{2p+2i}\binom{p+i}{i}. \tag{230}$$

The occurrence of the $2i$ makes this sum difficult to do with the techniques from the previous sections. Writing the problem with factorials gives

$$\sum_i \binom{2n}{2p+2i}\binom{p+i}{i} = \sum_i \frac{(2n)!(p+i)!}{(2p+2i)!(2n-2p-2i)!i!p!}. \tag{231}$$

Applying eq. (215), we get

$$(p+i)! = p!(p+1)^{\bar{i}}, \tag{232}$$

$$(2p+2i)! = (2p)!4^i(p+\tfrac{1}{2})^{\bar{i}}(p+1)^{\bar{i}}. \tag{233}$$

Applying eq. (218), we get

$$(2n-2p-2i)! = \frac{(-1)^{2i}(2n-2p)!}{4^i(p-n)^{\overline{i}}(p-n+\frac{1}{2})^{\overline{i}}}. \tag{234}$$

Plugging these results into eq. (231) gives

$$\sum_i \binom{2n}{2p+2i}\binom{p+i}{i} = \sum_i \frac{(2n)!(p-n)^{\overline{i}}(p-n+\frac{1}{2})^{\overline{i}}}{(2p)!(p+\frac{1}{2})^{\overline{i}}(2n-2p)!i!} \tag{235}$$

$$= \frac{(2n)!}{(2p)!(2n-2p)!}\,{}_2F_1\left[\begin{matrix} p-n, p-n+\frac{1}{2}; 1 \\ p+\frac{1}{2} \end{matrix}\right]. \tag{236}$$

Applying eq. (207) gives

$$\sum_i \binom{2n}{2p+2i}\binom{p+i}{i} = \frac{(2n)!}{(2p)!(2n-2p)!}\frac{\Gamma(p+\frac{1}{2})\Gamma(2n-p)}{\Gamma(n+\frac{1}{2})\Gamma(n)}. \tag{237}$$

Now

$$\Gamma(p+\tfrac{1}{2}) = \frac{(2p)!}{4^p p!}\Gamma(\tfrac{1}{2}) \tag{238}$$

(see exercise 3.1.4–4), so we have

$$\sum_i \binom{2n}{2p+2i}\binom{p+i}{i} = \frac{4^n(2n-p-1)!n!}{4^p p!(2n-2p)!(n-1)!}. \tag{239}$$

$$= 4^{n-p}\binom{2n-p}{p}\frac{n}{(2n-p)}. \tag{240}$$

Although the calculation is long, it is straightforward.

The use of generalized hypergeometric functions to solve binomial summations is suggested by Andrews [66, Section 5]. Slater [53] has the best introduction to generalized hypergeometric functions for people interested in summing series. She also has a table [53, appendix 3] of the identities that are likely to be useful for summing series. Minton [116] has one additional identity. Also see Abramowitz and Stegun [16, Section 15].

Gosper [90] gives a fairly simple method for deciding whether the first k terms of a hypergeometric function have a closed-form sum and for determining the sum when it has a closed form. Karr [103] has a more general, but much more complicated method for deciding whether a much wider class of sums has a closed form. Some parts of Karr's method are discussed in Section 6.4.

Many results on hypergeometric functions can be generalized to *basic hypergeometric functions*, which are obtain by replacing the increasing powers in eq. (203) with q powers, where a to the nth q power is given by

$$\prod_{0\le i<n} (1-aq^i). \tag{241}$$

These functions have applications in the study of *partitions* of a number n into m parts, none of which exceeds size N (see Andrews [66, Section 2]).

EXERCISES

1. Derive eq. (167) using hypergeometric functions.

2. Sum $\displaystyle\sum_{v\geq 0} \frac{w\binom{n}{w}\binom{m}{v}}{(n+m)\binom{n+m-1}{w+v-1}}$.

3. Sum $\displaystyle\sum_{v\geq 0} \frac{vw\binom{n}{w}\binom{m}{v}}{(n+m)\binom{n+m-1}{w+v-1}}$.

4. Sum $\displaystyle\sum_{i} i\binom{n}{i}^2$.

5. Sum $\displaystyle\sum_{i\geq 0} \frac{(2n)!}{(i!)^2[(n-i)!]^2}$.

6. Sum $\displaystyle\sum_{k} \binom{2n+1}{2k+1}\binom{j+k}{2n}$.

3.9* INCLUSION AND EXCLUSION

Consider three properties, A, B, and C. Let N be the total number of objects, $N(\emptyset)$ be the number of objects that have none of the properties, $N(A)$ be the number with property A, $N(AB)$ be the number with property A and property B, etc. The principle of *inclusion and exclusion* says that

$$N(\emptyset) = N - N(A) - N(B) - N(C) + N(AB) + N(AC) + N(BC) - N(ABC). \quad (242)$$

In other words, to obtain the number of objects with none of the properties, we can take the total number of objects, subtract the number with property A, subtract the number with property B, subtract the number with property C, add back on the number with both properties A and B (because we subtracted these objects off twice, once when subtracting those with property A and once when subtracting those with property B), add the number with both properties A and C (similar reason), add the number with both properties B and C, and finally subtract the number with all three properties (because we added back one too many times for these objects).

The general statement of inclusion and exclusion for properties A_1, A_2, ..., A_k is

$$N = \sum_{0\leq i\leq k} (-1)^i \sum_{1\leq j_1<j_2<\cdots<j_i\leq k} N(A_{j_1}A_{j_2}\ldots A_{j_i}). \quad \mathbf{(243)}$$

When the properties are independent, we have

$$N(A_1A_2\ldots A_i) = N(A_1)N(A_2)\cdots N(A_i), \quad (244)$$

and further simplifications are possible, as we see in the next example.

To illustrate inclusion and exclusion, let's compute the *Euler ϕ function.* Euler defined the function $\phi(n)$ to be the number of integers in the range $\{1, 2, \ldots, n\}$ that are not divisible by any of the prime divisors of n. Let the prime factors of n be p_1, p_2, ..., p_k. The ith property for this problem is divisibility by p_i. The total number of objects(numbers) is n, the number that are divisible by p_1 is n/p_1, the number that are divisible by both p_1 and p_2 is $n/(p_1p_2)$, etc. Thus we have

$$\phi(n) = n - \sum_{1 \le i \le k} (-1)^i \sum_{1 \le j_1 < j_2 < \cdots < j_i \le k} \frac{n}{p_{j1}p_{j2} \cdots p_{ji}}, \qquad \textbf{(245)}$$

which can be factored to give

$$\phi(n) = n\left(1 - \frac{1}{p_1}\right)\left(1 - \frac{1}{p_2}\right) \cdots \left(1 - \frac{1}{p_i}\right). \qquad \textbf{(246)}$$

Cohen [23, Chapter 5] was a source for this section.

EXERCISES

1. Write a formula for the number of permutations that have no cycles of length 1 (unit cycles). (Such permutations are called *derangements.*)
2. Show that if you put $a_1 + a_2 + \cdots + a_n + 1$ pigeons into n holes, then either the first hole has more than a_1 pigeons, or the second hole has more than a_2 pigeons, or etc.. (That is, there exists an i such that the ith hole has more than a_i pigeons.) This result is known as the *pigeon-hole principle.*
3. Define $W(i) = \sum_{1 \le j_1 < j_2 < \cdots < j_i \le k} N(A_{j_1} A_{j_2} \ldots A_{j_i})$, the number of items with at least i of k properties. Show that the number of items with exactly i properties is given by $\sum_{i \le j \le k} (-1)^{i+j} \binom{j}{i} W(j)$.
4. Prove that

$$\sum_{\substack{n_1 \ge 1, n_2 \ge 1, \ldots, n_i \ge 1 \\ n = n_1 + n_2 + \cdots + n_i}} \binom{n_1 + n_2 + \cdots + n_i}{n_1, n_2, \ldots, n_i} = i! \left\{ {n \atop i} \right\}.$$

CHAPTER 4

Asymptotic Approximation

It is often difficult or impossible to obtain an exact answer to a problem. Even when it is possible, the exact answer may be too complex to be very useful. In such cases one is usually interested in approximate answers. The two most extreme ways of approximating a function $f(n)$ are (1) to say that $f(n) \approx f_a(n)$ [$f(n)$ is approximately equal to $f_a(n)$] and (2) to say that $f_b(n) \leq f(n) \leq f_c(n)$ [$f(n)$ is between $f_b(n)$ and $f_c(n)$]. The first approach is easy to work with but not very precise. There is no statement concerning how much error might exist in the approximation. The second method is quite precise but often requires a lot of detailed calculations. Usually we want a method that is more precise than the first approach and that requires less work than the second approach.

4.1 BOUNDING FUNCTIONS

Short formulas for upper and lower bounds on a function will, when they give values that are close together, contain almost as much information about the size of the function as an exact formula for the value of the function. In fact, since short formulas are much easier for people to comprehend than long ones, simple formulas for upper and lower bounds will often convey *more* information to another person than a long complicated exact formula will.

Most upper and lower bounds calculations can be done using four groups of principles. The first group of principles is:

1. In an upper bound you can replace any quantity by another quantity which is known not to be smaller.

2. In a lower bound you can replace any quantity by another quantity which is known not to be larger.

For example, in Section 2.1.1 we found that the average number of times that Step 2 of the Linear Search Algorithm is done is $\frac{1}{2}n + \frac{1}{2}$. We can obtain a lower bound by dropping the term $\frac{1}{2}$. We can obtain an upper bound for the case $n \geq 1$ by replacing $\frac{1}{2}$ with $\frac{1}{2}n$. Thus,

$$\tfrac{1}{2}n \leq \tfrac{1}{2}n + \tfrac{1}{2} \leq n \qquad \text{for } n \geq 1. \tag{1}$$

The second group of principles tells how to combine upper and lower bounds arithmetically. Let $f(x)$ and $g(x)$ be two functions with upper bounds $f_U(x)$ and $g_U(x)$ and with lower bounds $f_L(x)$ and $g_L(x)$ in the range $x_0 \leq x \leq x_1$:

$$f_L(x) \leq f(x) \leq f_U(x) \text{ and } g_L(x) \leq g(x) \leq g_U(x) \qquad \text{for } x_0 \leq x \leq x_1. \tag{2}$$

3. The sum of upper (lower) bounds is an upper (lower) bound on the sum:

$$f_L(x) + g_L(x) \leq f(x) + g(x) \leq f_U(x) + g_U(x) \qquad \text{for } x_0 \leq x \leq x_1. \tag{3}$$

4. Multiplying an upper (lower) bound by a nonnegative number gives an upper (lower) bound on the product of the number with the function:

$$cf_L(x) \leq cf(x) \leq cf_U(x) \qquad \text{for } x_0 \leq x \leq x_1 \text{ where } c \geq 0. \tag{4}$$

5. Multiplying an upper (lower) bound by a nonpositive number gives a lower (upper) bound on the product of the number with the function:

$$cf_U(x) \leq cf(x) \leq cf_L(x) \qquad \text{for } x_0 \leq x \leq x_1 \text{ where } c \leq 0. \tag{5}$$

(Subtraction is the combination of multiplication by -1 and addition.)

6. Multiplying a nonnegative upper (lower) bound by a nonnegative upper (lower) bound gives an upper (lower) bound on the product of the functions:

$$f_L(x)g_L(x) \leq f(x)g(x) \leq f_U(x)g_U(x) \qquad \text{for } x_0 \leq x \leq x_1, \tag{6}$$

when all the functions are positive.

The next group of principles tells how upper and lower bounds behave with increasing and decreasing functions. Important increasing functions are the exponential (for all real values of its argument) and the logarithm (for positive values of its argument). An important decreasing function is the reciprocal, $1/x$, for positive and also for negative values of its argument (the reciprocal is discontinuous at zero). Although the principles are stated for functions of one variable, they can be generalized to multi-variable functions, just so long as the function is increasing (or decreasing) for the variable under consideration (with the other variables being held fixed. (The previous four principles can be viewed as cases of this group.)

7. An increasing function of an upper (lower) bound is an upper (lower) bound on the increasing function of the original function:

$$f(g_L(x)) \le f(g(x)) \le f(g_U(x)) \qquad \text{for } x_0 \le x \le x_1, \tag{7}$$

where $f(y)$ is an increasing function in the range

$$\min_{x_0 \le x \le x_1} \{g_L(x)\} \le y \le \max_{x_0 \le x \le x_1} \{g_U(x)\}.$$

8. A decreasing function of an upper (lower) bound is a lower (upper) bound on the decreasing function of the original function:

$$f(g_U(x)) \le f(g(x)) \le f(g_L(x)) \qquad \text{for } x_0 \le x \le x_1, \tag{8}$$

where $f(y)$ is a decreasing function in the range

$$\max_{x_0 \le x \le x_1} \{g_U(x)\} \le y \le \min_{x_0 \le x \le x_1} \{g_L(x)\}.$$

(Division is the combination of reciprocal and multiplication.)

For example, in Section 2.4.1 we showed that the average number of times that Step 2 of the Random Hashing Algorithm is done is $1 + k/(N-k)$, where $N > k$. If we just consider k in the range $0 \le k \le N/2$, we have $N/2 \le N-k \le N$ by using Principles 3 and 5. We have that $k/N \le k/(N-k) \le 2k/N$ by Principle 8. Finally, using Principle 1, we get

$$1 + \frac{k}{N} \le 1 + \frac{k}{N-k} \le 1 + \frac{2k}{N}. \tag{9}$$

The first eight principles are adequate for most simple problems. For the more complicated problems, we also need Principle 9, which is based on Taylor's theorem with a remainder:

Theorem 4.1 (Taylor's theorem *with a remainder*) If $f(x)$ is a function with a continuous n^{th} derivative on the closed interval $[a,b]$, and x and x_0 are two points of $[a,b]$ with $x \ne x_0$, then there exists a point c between x and x_0 such that

$$f(x) = \sum_{0 \le i \le n} \frac{f^{(i)}(x_0)}{i!}(x-x_0)^i + \frac{f^{(n+1)}(c)}{(n+1)!}(x-x_0)^{n+1}. \tag{10}$$

The following principle shows how Taylor's theorem applies to upper and lower bounds.

9. Taylor's theorem gives upper and lower bounds on the value of a function:

$$f(x) \ge \sum_{0 \le i \le n} \frac{f^{(i)}(x_0)}{i!}(x-x_0)^i + L(x-x_0)^{n+1}, \tag{11}$$

$$f(x) \le \sum_{0 \le i \le n} \frac{f^{(i)}(x_0)}{i!}(x-x_0)^i + U(x-x_0)^{n+1}, \tag{12}$$

where x is in the range $x_0 \le x \le x_1$ and U (alternately L) is an upper (lower) bound on $f^{(n+1)}(x)/(n+1)!$ in the range of x.

For example, applying Taylor's theorem with a remainder to the function e^x gives

$$e^x = \sum_{0\le i\le n} \frac{x^i}{i!} + \frac{e^c x^{n+1}}{(n+1)!}, \tag{13}$$

for some c in the range $0 \le c \le x$. Setting $n = 0$ and applying Principle 9 gives

$$1 + x \le e^x \le 1 + xe^x, \tag{14}$$

which leads to

$$e^x \ge 1 + x \quad \text{for} \quad x \ge 0 \qquad \text{and} \qquad e^x \le \frac{1}{1-x} \quad \text{for} \quad 0 \le x \le 1. \tag{15}$$

More precise bounds can be obtained by using larger values of n.

Applying Taylor's theorem with a remainder to the function $\ln(1+x)$ gives

$$\ln(1+x) = -\sum_{1\le i\le n} (-1)^i \frac{x^i}{i} + \frac{(-1)^n}{(n+1)} \left(\frac{x}{1+c}\right)^{n+1} \qquad \text{for } x \ge 0, \tag{16}$$

for some c in the range $0 \le c \le x$. Using Principle 9 (with $n = 0$ for the lower bound and $n = 1$ for the upper bound) gives

$$\frac{x}{1+x} \le \ln(1+x) \le x. \tag{17}$$

Eq. (17) also works for x below 0; for such values of x, the condition on c is $x \le c \le 0$.

Let's consider the problem of approximating $(1 - 1/100)^{10}$. Since we have an exponential function, it is useful to consider the exponential of the logarithm:

$$\left(1 - \frac{1}{100}\right)^{10} = \exp\left(10 \ln\left(1 - \frac{1}{100}\right)\right). \tag{18}$$

Using eq. (17), we obtain

$$-\frac{1}{99} \le \ln\left(1 - \frac{1}{100}\right) \le -\frac{1}{100}. \tag{19}$$

Since $-1/99 \ge -0.0101011$, this gives $\exp(-0.101011) \le (1 - 1/100)^{10} \le \exp(-0.1000000)$, or

$$0.9039 \le (1 - 1/100)^{10} \le 0.9409. \tag{20}$$

Taylor's theorem is discussed in Apostol [18, Section 7.6 and 7.7, and 17, section 5.11], and in Johnson and Kiokemeister [37, sections 14.6 and 14.7].

EXERCISES

1. Show that
$$1 + x + \tfrac{1}{2}x^2 \le e^x \le \frac{1+x}{1-x^2/2},$$
for $0 \le x < \sqrt{2}$. What are the corresponding limits on e^x for $x < 0$?

2. Show that
$$\sum_{0\le i\le n} \frac{x^i}{i!} \le e^x \le \frac{1}{1-x^n/n!} \sum_{0\le i\le n-1} \frac{x^i}{i!},$$
for $x \ge 0$. What are the corresponding limits on e^x for $x < 0$?

3. Show that
$$x - \tfrac{1}{2}x^2 \le \ln(1+x) \le x - \frac{x^2}{2(1+x)^2}$$
for $x \ge 0$. What are the corresponding limits on $\ln(1+x)$ for $x < 0$?

4. Show that $\ln(1+x)$ is between
$$-\sum_{1\le i\le n} (-1)^i \frac{x^i}{i} + \frac{(-1)^n}{(n+1)} \left(\frac{x}{1+x}\right)^{n+1} \quad \text{and} \quad -\sum_{1\le i\le n+1} (-1)^i \frac{x^i}{i}.$$

5. Show that $\dfrac{x}{1+x^2/2} \le \sin x \le x$, for $0 \le x \le \pi/2$.

4.1.1 Resource Trade-offs

It often happens that an algorithm can be speeded up by saving extra information, or made to use less memory by recalculating intermediate values. This type of situation, where the use of one resource (such as memory space) can be improved by using more of another (such as time) is called a resource trade-off. In this section we will illustrate this concept by examining a class of resource trade-offs discovered by Bentley and Brown [70].

Consider the problem of *destructive testing* with limited resources. In this problem, we assume some product is being manufactured to a uniform standard, so that all the individual items are essentially identical. We want to establish how much force is needed to break one of the items and we already know some upper bound on this force. If the items are cheap (pencils, for example), then the correct strategy is clearly binary search. If the upper bound is 100, start at 50; if the pencil breaks, try 25; and so on. If the items are extremely expensive (space shuttle engines), we may be able to afford to break only one for testing purposes. In that case the only alternative is to start with a very low force and work our way up through the possible values until the breaking point is found.

Now suppose the situation is between these two extremes. Say we are given two items to use in the testing process, and we want to find the first integer valued force larger than the breaking point. The upper bound is 100. One approach would be to start on the binary search strategy, and revert to linear search when the first item breaks. A better strategy is to start testing values 10, 20, 30, and so on; when the first item breaks at $10n$ do a linear search using the second item on the numbers $10(n-1)$, ..., $10n-1$ to find the true breaking point. For an upper limit of N, this scheme can be generalized by first testing

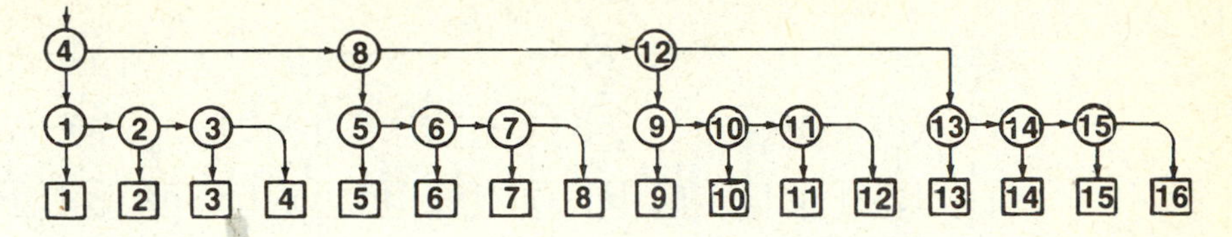

FIGURE 4.1 The decision tree for $N = 16$, $k = 2$, using the algorithm based on $\sqrt{N}$.

the values $\sqrt{N}$, $2\sqrt{N}$, ..., $\sqrt{N}\sqrt{N} = N$, and then, if the breaking point is $n\sqrt{N}$, testing $1 + (n-1)\sqrt{N}$, ..., $n\sqrt{N} - 1$. This strategy requires at most $2(\sqrt{N} - 1)$ low guesses and at most two high guesses.

This scheme for $k = 2$ can be extended to any $k < \lceil \lg N \rceil$. (For larger k we should use binary search.) The first item is used to test values that are multiples of $N^{(k-1)/k}$: $N^{(k-1)/k}$, $2N^{(k-1)/k}$, ..., N. If the item breaks on the test for $iN^{(k-1)/k}$, then the next item is used to test values from $(i-1)N^{(k-1)/k} + N^{(k-2)/k}$ to $iN^{(k-1)/k} - N^{(k-2)/k}$ in increments of $N^{(k-2)/k}$, and so on. As a concrete example, consider $k = 4$ and $N = 10,000$. We first test 1000, 2000, and so on. If the first item breaks at 4000, we next test values 3100, 3200, etc. If the second item breaks at 3300, we use the third item to test the values 3210, 3220, etc. If the third item breaks at 3290, we use the last item to test the values 3281, 3282, etc. This fourth and last test gives the answer. By increasing k, we can reduce the number of tests required — at the expense of increasing the number of items destroyed.

Does this scheme give the best possible "trade-off" between items used and the worst-case number of tests done? No, it does not. To see how the scheme can be improved, let's return to the $k = 2$ case and construct the decision tree for the algorithm. Figure 4.1 shows the tree for $N = 16$ and $k = 2$. Tests are indicated by round nodes. The final values chosen by the algorithm are shown by square nodes. Downward arcs indicate breakage of the item. Nonbreakage is indicated by horizontal arcs. Thus the algorithm takes a horizontal arc when the breaking strength is greater than the value given in the circular node, and it takes a downward arc when the strength is less than or equal to the value.

Define $d(T)$, the *rank* of tree T, as the maximum number of branches between the root of T and any of T's leaves. Define $h(T)$, the *height* of T, as the maximum number of downward pointing branches between the root and any leaf. It is clear that $d(T)$ gives the maximum number of tests that are done, while $h(T)$ gives the maximum number of test items destroyed.

Examination of Figure 4.1 shows why the $\sqrt{N}$ algorithm does not have the optimum worst-case behavior: the number of tests for an item with a high value (such as 16) is much larger than that required by a small value (such as 4). A better algorithm (from the worst-case viewpoint) would have some of the good cases take longer so that the worst case would not be so bad.

How do you build the best possible tree for this problem? A little thought indicates that the right subtree from any node should have a rank equal to that

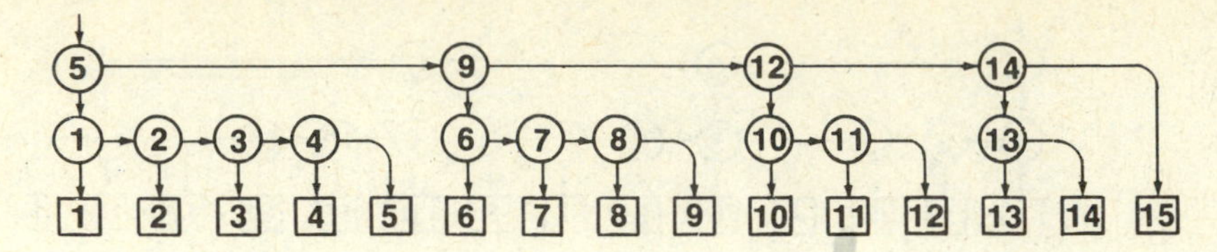

FIGURE 4.2 The decision tree for $N = 15$, $k = 2$, using the Binomial Tree Algorithm.

of the down subtree and height one more than the down subtree. Some slight adjustments are needed for cases where the total number of nodes does not have the correct form for a tree built on these principles to "come out even", and special care is needed for boundary cases, but that is the general idea. These consideration lead to the idea of a binomial tree.

A *binomial tree* is associated with each binomial coefficient $\binom{d}{h}$ for $d \ge h \ge 0$. The trees for $\binom{d}{0}$ and $\binom{d}{d}$ each contain a single node. A $\binom{d}{h}$ tree for $d > h > 0$ consists of a root, a down subtree that is a $\binom{d-1}{h-1}$ tree, and a right subtree that is a $\binom{d-1}{h}$ tree. The name binomial tree is justified by the close similarity between the addition formula for binomial coefficients (eq. (3.65)) and the recursive principle used to construct binomial trees. The number of leaves on a binomial $\binom{d}{h}$ tree is $\binom{d}{h}$.

Figure 4.2 shows a $\binom{6}{2}$ tree. With 15 leaves, it can serve as the optimum decision tree for $N = 15$, $k = 2$. Comparing Figure 4.2 with Figure 4.1, you can see that the tree in Figure 4.2 is better balanced for testing purposes: the rank of the subtrees gets progressively less as the number of tests needed to reach the subtrees increases.

For testing values up to N we need a tree with at least N nodes. The bottom index of the binomial is determined by the number of items we are willing to destroy in testing. The top index must be large enough so that the tree has at least N nodes. (If N is between two binomial coefficients, then we build a tree of the next larger size and omit the irrelevant nodes.)

A binomial tree will have at least N nodes if the top index is at least $(h!N)^{1/h}+h-1$ because the number of leaves is $\binom{d}{h}$, and for $d = (h!N)^{1/h}+h-1$

$$\binom{d}{h} = \binom{(h!N)^{1/h} + h - 1}{h} \ge \frac{((h!N)^{1/h})^h}{h!} = N. \tag{21}$$

Thus we see that we can test for one of N values while destroying no more than h items by doing at most $\lceil (h!N)^{1/h} \rceil$ tests.

This section is based on Bentley and Brown [70].

EXERCISES

1. Draw a $\binom{7}{3}$ binomial tree.
2. Show that the number of leaves in a $\binom{d}{h}$ tree is $\binom{d}{h}$.

3. How is the inequality in eq. (21) obtained?
4. Show that if T is a $\binom{d}{h}$ tree with $d > h > 1$, then $d(T) = d - 1$ and $h(T) = h$.
5. Show that the algorithm based on binomial trees is optimal for destructive testing. Hint: Consider the set of all possible decision trees for fixed N and h. Show that for all trees of height $h + 1$ with N leaves, the binomial trees have the minimum rank. It may help to first consider the case where $N = \binom{d}{h}$ for some d.

4.2† ASYMPTOTIC NOTATION

Asymptotic analysis is concerned with providing approximations that are useful when some parameter becomes large. The Θ, O, and Ω notations lead to results that are intermediate between approximating with bounds and approximating by saying one function is approximately equal to another. Using these notations requires less work than computing explicit tight upper and lower bounds on the answer does, but it does not lead to a value of the error at any point. Instead it gives the functional form of a bound on the error. When necessary you can convert an asymptotic error analysis into explicit upper and lower bounds, although doing so requires additional work.

The following are the definitions for big O, big Θ, big Ω, and little o:

$O(f(x))$ is the set of all functions $g(x)$ such that there exist positive constants C and x_0 with $|g(x)| \le Cf(x)$ for all $x \ge x_0$, **(22)**

$\Theta(f(x))$ is the set of all functions $g(x)$ such that there exist positive constants C, C', and x_0 with $Cf(x) \le g(x) \le C'f(x)$ for all $x \ge x_0$, **(23)**

$\Omega(f(x))$ is the set of all functions $g(x)$ such that there exist positive constants C and x_0 with $g(x) \ge Cf(x)$ for all $x \ge x_0$, **(24)**

$o(f(x))$ is the set of all functions $g(x)$ such that $\lim_{x\to\infty} \dfrac{g(x)}{f(x)} = 0$. **(25)**

We use the term *error notation* in this section to refer to any one of the four notations. Frequently one uses only one of the notations at a time. In error notation the equal sign means that the left side of an "equation" is a subset of the right side, i.e., the right side is a less precise representation of the left side. If you forget this convention you can obtain some amazing (wrong) results.

These notations are often used in very simple situations. For example,

$$\begin{array}{cccc} 2n^2 = O(n^2) & 2n^2 = \Theta(n^2) & 2n^2 = \Omega(n^2) & 2n^2 = o(n^2 \ln n) \\ 2n^2 = O(n^3) & & 2n^2 = \Omega(n) & 2n^2 = o(n^3) \\ 2n^2 + n = O(n^2) & 2n^2 + n = \Theta(n^2) & 2n^2 + n = \Omega(n^2) & 2n^2 + n = o(n^2 \ln n). \end{array}$$

There is usually an interaction between the best values for the implied constants for the error and the range of x. For example, $1/(x-1) = \Theta(1/x)$ for $x > 1$. For $x > 2$ we have $1/x < 1/(x-1) < 2/x$, so eq. (23) is satisfied with $C = 1$ and $C' = 2$. For $x > 11$ we have $1/x < 1/(x-1) < 1.1/x$, so eq. (23) is satisfied with $C = 1$ and $C' = 1.1$. The big Θ notation permits us to say that

$1/x$ and $1/(x-1)$ behave roughly the same way for large x without going into the details of calculating C and C' for any range of x.

Error notation is good for saying how a function behaves for large x, but it does not say how large x must be. Thus $1/(x-1000) = \Theta(1/x)$, but x must be larger than 1000. Occasionally we want to be a little more specific. If we write

$$\Theta(f(x)) \qquad \text{for } x \geq x_0, \tag{26}$$

we mean the set of functions $g(x)$ such that

$$Cf(x) \leq g(x) \leq C'f(x) \qquad \text{for } x \geq x_0, \tag{27}$$

where C and C' are unspecified positive constants (but x_0 is a specified constant).

We also use error notation when x approaches values other than infinity. For example, we write

$$\Theta(f(x)) \qquad \text{as } x \to c \tag{28}$$

to mean the set of functions $g(x)$ such that

$$Cf(x) \leq g(x) \leq C'f(x) \qquad \text{for } |x-c| \leq r \tag{29}$$

where r, C, and C' are unspecified positive constants. The other three error notations can also be extended in the ways suggested by eqs. (26–29).

The range of an approximation can affect the functional form of the approximation. For example,

$$x + \frac{1}{x} = O(x) \qquad \text{for } x \geq 1, \tag{30}$$

but

$$x + \frac{1}{x} = O\left(\frac{1}{x}\right) \qquad \text{for } 0 < x \leq 1. \tag{31}$$

Although error notation permits you to ignore finite constants, you should pay some attention to the size of the suppressed constants. If they are huge, your result may not be of much use even though it is technically correct.

Error notation gives a way to represent the functional behavior of an approximation while avoiding the large amount of calculation that is often required to obtain explicit accurate bounds. However, for any derivation with an error notation, there is a closely parallel derivation that gives explicit bounds. In the derivation with error notation you make simplifications using facts such as: a constant plus a constant is a constant, the product of two constants is a constant, and a function of a constant is a constant; whereas when finding an explicit bound, you must actually do the arithmetic to find the value of the constants.

The following rules can be used to simplify expressions with big Θ. There are similar rules for the other notations; comments will indicate when there are significant differences between the notations.

The basic rule for introducing big Θ is

$$f(x) = \Theta(f(x)). \tag{32}$$

In other words, if you have $f(x)$ in your formula, you can replace it on the right side with $\Theta(f(x))$. The most basic simplification rule is

$$cf(x) = \Theta(f(x)), \tag{33}$$

where c is a positive finite constant. With big O we have the corresponding rule:

$$g(x)f(x) = O(f(x)) \quad \text{provided } |g(x)| \leq c \quad \text{for } x \geq x_0, \tag{34}$$

where c is a positive constant. For big Ω we have

$$g(x)f(x) = \Omega(f(x)) \quad \text{provided } g(x) \geq c \quad \text{for } x \geq x_0, \tag{35}$$

where c is a positive constant. There are several formulas for replacing multiple big Θ's with single ones:

$$\Theta(\Theta(f(x))) = \Theta(f(x)), \tag{36}$$

$$\Theta(f(x))\Theta(g(x)) = \Theta(f(x)g(x)), \tag{37}$$

$$\Theta(f(x))\Theta(g(x)) = f(x)\Theta(g(x)). \tag{38}$$

Part of the simplifying effect of using the error notations comes from the fact that constants disappear [eq. (33)]. More of the effect is due to the disappearance of small functions. We have

$$\Theta(f(x)) \pm \Theta(g(x)) = \Theta(f(x)), \quad \text{when } g(x) = o(f(x)). \tag{39}$$

Also we have

$$\Theta(f(x)) + \Theta(f(x)) = \Theta(f(x)). \tag{40}$$

For big O we can replace eq. (40) with

$$O(f(x)) \pm O(f(x)) = O(f(x)). \tag{41}$$

Eq. (41) is valid with the minus sign because of the absolute value in the definition of big O and because big O is associated with upper limits.

The equations that we have given all work when applied a fixed number of times. *Care is needed, however, when they are applied an unbounded number of times.* A large number of small errors can add up to a large error. When using error notation to simplify away an unbounded number of terms, you must usually go back to the definitions to justify your method of simplification. Such problems arise frequently when simplifying summations.

Most of the details of the notation used in this section are from Knuth [107], where he discusses the history of these notations and proposes the specific notations that are usually used today. Elementary uses of these notations are discussed in Knuth [9, Section 1.2.11.1], Aho et al. [1, Section 1.1], Baase [2, Section 1.3], and Horowitz and Sahni [7, Section 1.4]. One of the most complete treatments is given in de Bruijn [25, Chapter 1]. Advanced topics are covered in de Bruijn [25] and Greene and Knuth [6, Chapter 4].

Asymptotics is necessary for the analysis of algorithms, even at the fairly elementary level, so this book covers it without going into advanced mathematics. Yet for some of the more delicate problems of convergence that sometimes arise when doing asymptotics, advanced mathematics is quite helpful and relevant. A good treatment of the appropriate topics is given in Apostol [17, Chapters 12 and 13]. A more elementary but quite precise treatment is given in Apostol's calculus book [18, Chapter 9].

EXERCISES

1. Find an asymptotic expression for $\dfrac{x}{x-1}$ which has accuracy $O(x^{-3})$. Hint: Use polynomial division.

2. Show that
$$\frac{1}{1+O(x)} = 1 + O(x) \qquad \text{as } |x| \to 0.$$
Is it true that
$$\frac{1}{1+O(x)} = 1 + O(x) \qquad \text{for } |x| < 1?$$

3. Show that eq. (33) is true: for any positive constant c and for any function $f(x)$, $cf(x) = \Theta(f(x))$.

4. Show that $\log_b x = O(\ln x)$ for any base $b > 1$. This exercise justifies writing $O(\log x)$ for $O(\log_b x)$ (i.e., omitting the base of the logarithm in big O expressions).

5. Show that eq. (34) is true.

6. Show that eq. (35) is true.

7. Prove eqs. (36–38).

8. Show that $\Theta(f(x)) + \Theta(f(x)) = \Theta(f(x))$ [eq. (40)]. Why is it incorrect to say $\Theta(f(x)) - \Theta(f(x)) = \Theta(f(x))$? Why are both signs correct in eq. (41)?

9. Show that if $f(x) = \Omega(g(x))$, then $g(x) = O(f(x))$.

4.2.1† Power Series

Any function that has derivatives of all orders has a *power series* expansion. This includes most functions in common use. That is, we can nearly always write a function $f(x)$ as
$$f(x) = \sum_{i \ge 0} a_i x^i. \tag{42}$$

This is the power series for $f(x)$. The existence of a power series for $f(x)$ depends on *Taylor's theorem* in calculus (see Section 4.1), which says that differentiable functions can be written in the form of eq. (42) with
$$a_i = \frac{f^{(i)}(0)}{i!}, \tag{43}$$
where $f^{(i)}(0)$ is the ith derivative of $f(x)$ evaluated at $x = 0$. Most common functions satisfy eq. (42) for all complex x such that $|x| < r$ for some real r. The largest such r is called the *radius of convergence* of the power series. The

radius of convergence is the same as the smallest absolute value of a complex number x such that the function goes to infinity or has some other such unusual behavior when evaluated at x.

The error from approximating a function by the first m terms of its power series is $\Theta(x^{m+1})$ provided x is not too large. To prove this, we first show that the error is no more than $O(x^{m+1})$ when x is bounded. Then we show that the error is at least $\Omega(x^{m+1})$ when $|x|$ is small enough.

Suppose $f(x)$ satisfies eq. (42) and $\sum_{k\geq 0}|a_k r^k|$ is finite for some $r > 0$. Let $K = \sum_{k\geq 0}|a_k r^k|$. Then $|f(x)| \leq K$ for all x such that $|x| \leq r$. Also,

$$f(x) = \sum_{0\leq i\leq m} a_i x^i + O(x^{m+1}) \quad \text{for } |x| \leq r. \tag{44}$$

This big O result is for small x. To prove eq. (44) let's calculate the implied constant for the big O. First $f(x) = \sum_{0\leq i\leq m} a_i x^i + x^{m+1}(a_{m+1} + a_{m+2}x + \cdots)$, so any finite upper limit on the maximum value (as a function of x) of $|a_{m+1} + a_{m+2}x + \cdots|$ will serve as the constant C in eq. (22). Now $|a_{m+1} + a_{m+2}x + \cdots| \leq |a_{m+1}| + |a_{m+2}r| + \cdots$, and this last sum is finite because it is equal to $(K - \sum_{0\leq i\leq m}|a_i r^i|)/r^{m+1}$, K is finite, and the sum over i is finite (because it has a fixed number, $m+1$, of finite terms). Therefore, we have a finite value $(|a_{m+1}| + |a_{m+2}r| + \cdots)$ for the implied constant in the big O definition. (This result is similar to Abel's theorem, which states that if power series converges for $x = c$, then it converges absolutely for $|x| < c$.)

There is a similar result for big Θ notation. Suppose $f(x)$ satisfies eq. (42) and $\sum_{k\geq 0}|a_k r^k|$ is finite for some $r > 0$. Let $K = \sum_{k\geq 0}|a_k r^k|$. Then provided $a_{m+1} \neq 0$, we have

$$f(x) = \sum_{0\leq i\leq m} a_i x^i + \Theta(x^{m+1}) \quad \text{for } |x| \leq r', \tag{45}$$

where r' is some constant greater than zero. The results of the previous paragraph imply that the upper limit part of this result is true. We will finish proving eq. (45) by finding an r' suitable for the lower bound. We still have $f(x) = \sum_{0\leq i\leq m} a_i x^i + x^{m+1}(a_{m+1} + a_{m+2}x + \cdots)$, so any nonzero lower limit on the minimum value (as a function of x) of $a_{m+1} + a_{m+2}x + \cdots$ will serve as the constant C in eq. (23). Let's rewrite $a_{m+1} + a_{m+2}x + \cdots$ as $a_{m+1} + x(a_{m+2} + a_{m+3}x + \cdots)$ and let U be an upper limit on the part in parentheses. Then for $|x| < a_{m+1}/(2U)$, $a_{m+1} + a_{m+2}x + \cdots$ is not zero, because it differs by no more than $a_{m+1}/2$ from a_{m+1}. Following arguments similar to the ones in the previous paragraph, we can use $U = a_{m+2} + a_{m+3}r + \cdots = f(r) - \sum_{0\leq i\leq m} a_i x^i$, which is finite. This establishes eq. (45) with the C for eq. (23) equal to $a_{m+1}/2$ and $r' = \min(r, a_{m+1}/(2U))$.

The arguments used in the last two paragraphs are typical of the arguments you must make to use error notation with sums when you cannot reduce your problem to a case you have worked before. The formulas we have developed so far, however, have many applications. Often you will be able to easily reduce your problems to one of the cases treated above.

For example, since

$$e^x = \sum_{i\geq 0} \frac{x^i}{i!}, \tag{46}$$

and the series converges for all finite x, we have

$$e^x = \sum_{0\leq i\leq m} \frac{x^i}{i!} + \Theta(x^{m+1}) \quad \text{for } |x| \leq r, \tag{47}$$

where r is any fixed bound. Likewise,

$$\ln(1+x) = -\sum_{1\leq i\leq m} (-1)^i \frac{x^i}{i} + (-1)^m \Theta(x^{m+1}) \quad \text{for } |x| \leq r < 1, \tag{48}$$

where r is fixed.

Some of our previous formulas can be considered as power series. For example, the binomial theorem [eq. (3.47)] gives

$$(1+x)^r = \sum_{0\leq i<m} \binom{r}{i} x^i + \Theta(x^m), \quad \text{as } x \to 0, \tag{49}$$

where r is fixed. Replacing x by $-x$, r by $-r$, and using eq. (3.69), eq. (49) can be rewritten as

$$\frac{1}{(1-x)^r} = \sum_{0\leq i<m} \binom{i+r-1}{i} x^i + \Theta(x^m), \tag{50}$$

which is useful for negative exponents. The technique of Section 3.4.2 can be used for power series with periodic terms.

Algorithms for quickly manipulating large power series are given in Knuth [10, Section 4.7]. The mathematics of power series is extensively discussed in Knopp [40, Chapter 5].

EXERCISES

1. Prove eq. (47). Hint: Use Taylor's theorem with a remainder.
2. Prove eq. (48). Hint: Use Taylor's theorem with a remainder.
3. Give an asymptotic approximation for $(1 - a/x)^x$ that is accurate to $O(x^{-2})$. Why is it much more difficult to obtain accuracy $\Theta(x^{-2})$ than it is to obtain accuracy $O(x^{-2})$? Hint: Start doing the problem as though you were asked to obtain accuracy $\Theta(x^{-2})$.
4. Approximate $(1 - a/x)^x$ to accuracy $O(x^{-3})$.
5. Show that

$$(e^x - 1)^m = \sum_{i\geq 0} \left\{ {m+i \atop m} \right\} \frac{m!}{(m+i)!} x^{m+i}.$$

4.2.2† Techniques for Asymptotic Expansions

The previous two sections gave the basic principles needed for asymptotic expansions of closed-form formulas. In this section we will illustrate the techniques for applying the principles.

Consider the expression $1/(1+O(1/x))$ (where $1/x$ can be replaced by any quantity that asymptotically approaches zero). Long division gives

$$\frac{1}{1+O(1/x)} = 1 + \frac{O(1/x)}{1+O(1/x)}, \tag{51}$$

so

$$\frac{1}{1+O(1/x)} = 1 + O(1/x). \tag{52}$$

This formula is quite useful in simplifying big O expressions.

Suppose we are interested in the expression $(1 - p(v))^v$, in the case where $\lim_{v\to\infty} p(v) = 0$. A constant fraction raised to the v power would approach zero as $v \to \infty$, but a larger fraction goes to zero more slowly than a smaller one. The value of this expression thus depends on how fast $p(v)$ approaches zero. We first consider some special values for $p(v)$.

When $p = a/v$, our expression becomes $(1 - a/v)^v$. The expression can be rewritten as

$$\left(1 - \frac{a}{v}\right)^v = \exp\left(v \ln\left(1 - \frac{a}{v}\right)\right). \tag{53}$$

Since $\lim_{v\to\infty} a/v = 0$, we can use the power series expansion for the logarithm: $\ln(1 - a/v) = -a/v + O(1/v^2)$. So we have

$$\exp\left(v \ln\left(1 - \frac{a}{v}\right)\right) = \exp\left[v\left(-\frac{a}{v} + O\left(\frac{1}{v^2}\right)\right)\right] = \exp\left[-a + O\left(\frac{1}{v}\right)\right] \tag{54}$$

$$= e^{-a} e^{O(1/v)} = e^{-a}\left(1 + O\left(\frac{1}{v}\right)\right), \tag{55}$$

since $e^{O(1/v)} = 1 + O(O(1/v)) = 1 + O(1/v)$.

Similarly, when $p = a/v^2$, we have

$$\left(1 - \frac{a}{v^2}\right)^v = \exp\left[v \ln\left(1 - \frac{a}{v^2}\right)\right] = \exp\left[v\left(-\frac{a}{v^2} + O\left(\frac{1}{v^4}\right)\right)\right] \tag{56}$$

$$= \exp\left[-\frac{a}{v} + O\left(\frac{1}{v^3}\right)\right] = e^{-a/v} e^{O(1/v^3)} \tag{57}$$

$$= e^{-a/v}\left(1 + O\left(\frac{1}{v^3}\right)\right). \tag{58}$$

This time we can also expand the remaining exponential in a power series to obtain

$$\left(1 - \frac{a}{v^2}\right)^v = \left(1 - \frac{a}{v} + \frac{a^2}{2v^2} + O\left(\frac{1}{v^3}\right)\right)\left(1 + O\left(\frac{1}{v^3}\right)\right) \tag{59}$$

$$= 1 - \frac{a}{v} + \frac{a^2}{2v^2} + O\left(\frac{1}{v^3}\right). \tag{60}$$

If we want less accuracy, we can say

$$\left(1 - \frac{a}{v^2}\right)^v = \exp\left[v \ln\left(1 - \frac{a}{v^2}\right)\right] = \exp\left[vO\left(\frac{1}{v^2}\right)\right] \tag{61}$$

$$= e^{O(1/v)} = 1 + O\left(\frac{1}{v}\right). \tag{62}$$

Now consider what happens when $p = a/v^{1/2}$.

$$\left(1 - \frac{a}{v^{1/2}}\right)^v = \exp\left[v\left(-\frac{a}{v^{1/2}} + O\left(\frac{1}{v}\right)\right)\right] = e^{-av^{1/2}+O(1)} \tag{63}$$

$$= e^{-av^{1/2}} e^{O(1)} = e^{-av^{1/2}} O(1) = O(e^{-av^{1/2}}). \tag{64}$$

This time we did not get our answer in the form of a term times 1 plus a big O term. To get more accuracy, we need more terms from the power series expansion of the logarithm:

$$\exp\left[v \ln\left(1 - \frac{a}{v^{1/2}}\right)\right] = \exp\left[v\left(-\frac{a}{v^{1/2}} - \frac{a^2}{2v} + O\left(\frac{1}{v^{3/2}}\right)\right)\right] \tag{65}$$

$$= e^{-av^{1/2} - a^2/2 + O(v^{-1/2})} \tag{66}$$

$$= e^{-av^{1/2} - a^2/2}(1 + O(v^{-1/2})). \tag{67}$$

Let's now consider what to do if we want an asymptotic expansion for general $p(v)$. Since the various equations we use apply only for certain ranges of $p(v)$, our derivation will only work for a certain range of $p(v)$. Depending on what use we plan to make of the expansions, these restrictions may or may not be a problem. For this example we will just note the restrictions as they arise. If we wanted to remove the restrictions, we would need to obtain a series of formulas, each one suitable for certain ranges of the function $p(v)$.

Provided $p(v) \le r < 1$ for some fixed r, we have

$$(1 - p(v))^v = \exp(v \ln(1 - p(v))) = \exp[v(-p(v) + O(p(v)^2))] \tag{68}$$

$$= \exp[-p(v)v + O(p(v)^2 v)] = e^{-vp(v)} e^{O(vp(v)^2)}. \tag{69}$$

If $vp(v)^2$ is bounded [i.e., $|\lim_{v\to\infty} vp(v)^2| \le K$ for some finite constant K], then we can expand the rightmost exponential in eq. (69) to obtain

$$[1 - p(v)]^v = e^{-vp(v)}[1 + O(vp(v)^2)]. \tag{70}$$

If $\lim_{v\to\infty} vp(v)^2 = 0$, the big O term goes to zero, and $e^{-vp(v)}$ is a good asymptotic approximation. In deriving eq. (70) we made two assumptions: (1) $p(v) \le r < 1$ and (2) $\lim_{v\to\infty} vp(v)^2$ is bounded. The second assumption, however, implies the first, so eq. (70) applies when $\lim_{v\to\infty} vp(v)^2$ is bounded.

You must be careful not to misuse asymptotic expressions. Suppose you want the value of $(1 - 1/100)^{10}$. If you apply eq. (58) with $a = 1$, you get

$$\left(1 - \frac{1}{100}\right)^{10} = e^{1/10}\left[1 + O\left(\frac{1}{10000}\right)\right] \tag{71}$$

which is true but not very useful. Eq. (71) says that our answer is bounded by some number, because $O(1/1000)$ is bounded by some number. We do not know how big that number is. In other words eq. (71) is equivalent to

$$\left(1 - \frac{1}{100}\right)^{10} = O(1). \tag{72}$$

Big O notation just gives the functional form of the error, not the size of the error. We showed how to find a correct bound for $(1 - 1/100)^{10}$ in Section 4.1.

Let's now consider one last example:

$$\ln\left(1 + \frac{1}{x^2}\right)e^{1/x^3}. \tag{73}$$

Suppose we want to develop an asymptotic expansion for this expression where the error is $O(x^{-8})$. Our strategy will be to expand both factors in a power series. We need to decide how many terms to use in each power series expansion. If we use too few terms, the end result will be an approximation that is not as accurate as we wish; we will have to repeat the derivation with extra terms added. If we use too many terms, we will do a lot of unnecessary work. In general, it is better to start with an underestimate rather than an overestimate of the number of terms needed. Each successive term in the final result is usually much more difficult to obtain than the preceding one. Thus, little work is wasted; compared to the final calculation, a less accurate preliminary calculation requires very little time. Also each successive term in the final result is much less important than the ones that come before it. Repeating the calculation with increasing accuracy helps ensure that the important early terms in the answer are done correctly. Finally, a less accurate preliminary calculation can serve as a guide to the more accurate one. For an unfamiliar problem it is worth doing a rough preliminary calculation just for this purpose.

The above advice not withstanding, we will use

$$\ln\left(1 + \frac{1}{x^2}\right) = \frac{1}{x^2} - \frac{1}{2x^4} + \frac{1}{3x^6} + O\left(\frac{1}{x^8}\right) \tag{74}$$

and

$$e^{1/x^3} = 1 + \frac{1}{x^3} + \frac{1}{2x^6} + O\left(\frac{1}{x^9}\right). \tag{75}$$

This gives

$$\ln\left(1+\frac{1}{x^2}\right)e^{1/x^3} = \left(\frac{1}{x^2}-\frac{1}{2x^4}+\frac{1}{3x^6}+O\left(\frac{1}{x^8}\right)\right)$$
$$\times\left(1+\frac{1}{x^3}+\frac{1}{2x^6}+O\left(\frac{1}{x^9}\right)\right) \tag{76}$$
$$= \frac{1}{x^2}-\frac{1}{2x^4}+\frac{1}{3x^6}+O\left(\frac{1}{x^8}\right)$$
$$+\frac{1}{x^5}-\frac{1}{2x^7}+\left[\frac{1}{3x^9}\right]$$
$$+\left[\frac{1}{2x^8}\right] \tag{77}$$
$$= \frac{1}{x^2}-\frac{1}{2x^4}+\frac{1}{x^5}+\frac{1}{3x^6}-\frac{1}{2x^7}+O\left(\frac{1}{x^8}\right). \tag{78}$$

Consider the computation that led to eq. (77). We started by multiplying the largest term of eq. (75) by the terms of eq. (74), keeping in mind that terms smaller than $O(1/x^8)$ cannot contribute to the answer. The second row was obtained by multiplying $1/x^3$ [the second term of eq. (75)] by the terms of eq. (74). Only the first two terms contribute to the answer; the third term, $1/(3x^9)$, is too small since it can be absorbed into $O(1/x^8)$. If you notice in advance that such a term is going to be too small, there is no need to actually write it down; we indicate this by writing the term in square brackets in eq. (77). The first term of the product of $1/(2x^6)$ and eq. (74) falls in the same category, so we need not bother with any others; and the whole product of $O(1/x^9)$ with eq. (77) is too small to matter. In fact, we could have ended eq. (75) with $O(1/x^6)$. Noticing which terms to throw away as you compute them saves at least half the work in many asymptotic calculations.

EXERCISES

1. Prove that eq. (51) implies eq. (52).
2. Why can we expand the $e^{-a/v}$ in eq. (58) in a power series, but not the e^{-a} in eq. (55)?
3. Give the asymptotic expansion for $e^{1/x}(1-1/x)$ with accuracy $O(1/x^3)$.
4. Show that for any constant a, $e^{-a}-\Theta(1/v)=e^{-a}+O(1/v)$ and $e^{-a}+O(1/v)= e^{-a}(1+O(1/v))$.
5. What is the asymptotic expansion of $(1-a/v^3)^{v^2}$? Give your answer in the form of a function of v times 1 plus a big O term where the big O term goes to zero as v goes to infinity.
6. Find the asymptotic expansion of $\ln(1+1/x^2)e^{1/x^3}$ to accuracy $O(1/x^9)$.
7. Show that
$$\left[1-\frac{t^2}{2n}+o\left(\frac{t^2}{n}\right)\right]^n = \exp\left(\tfrac{1}{2}t^2+o(t^2)\right).$$
Hint: First consider the logarithm of the problem.

4.2.3* Asymptotics and Convergence

An *asymptotic expansion* for a function $f(x)$ at x_0 is a series of terms $s_n(x)$, such that for each n, $\lim_{x \to x_0}[f(x) - s_n(x)]/f(x) = 0$. The more rapidly this limit approaches zero, the better the asymptotic expansion. A series is *convergent* at x_0 if $\lim_{n \to \infty}[f(x) - s_n(x)] = 0$. In other words, an asymptotic expansion relates to the limit as x varies for fixed n, while convergence depends on the limit for fixed x as n varies.

An asymptotic series is not necessarily convergent. For example, consider the integral

$$e^{-x} \int_1^x \frac{e^t}{t}\, dt. \tag{79}$$

An asymptotic expansion can be obtained using integration by parts with $u = t^{-n}$ and $dv = e^t\, dt$ on the following integral:

$$\int_1^x \frac{e^t}{t^n}\, dt = \frac{e^x}{x^n} - e^1 + n \int_1^x \frac{e^t}{t^{n+1}}\, dt. \tag{80}$$

Repeated application of eq. (80) to eq. (79) gives

$$e^{-x} \int_1^x \frac{e^t}{t}\, dt = \sum_{1 \le i \le n} \frac{(i-1)!}{x^i} - e^{-x+1} \sum_{1 \le i \le n} (i-1)! + n! e^{-x} \int_1^x \frac{e^t}{t^{n+1}}\, dt. \tag{81}$$

For large x and fixed n, the last two terms on the right side of eq. (81) become small faster than the first one, so we have the asymptotic expansion

$$e^{-x} \int_1^x \frac{e^t}{t}\, dt = \frac{0!}{x} + \frac{1!}{x^2} + \frac{2!}{x^3} + \cdots \qquad \text{as } x \to \infty. \tag{82}$$

This expansion does not converge for any finite x. (For $n > x$, $n!$ grows more rapidly (as a function of n) than x^n does.)

For a nonconvergent asymptotic expansion and fixed x, there is a best value for n. Increasing the number of terms past the optimum point results in a worse approximation. (Usually the closer x is to the asymptotic value, the larger the optimum number of terms is.) When working with such expansions, you should increase the number of terms until you have a low enough error or until the size of the error begins to increase. In the second case, you will need to decide whether you can tolerate the resulting error.

4.3 BOUNDING SUMMATIONS

The *asymptotic analysis* of summations is usually concerned with approximating the value of the sum when the number of terms becomes very large. When finding an asymptotic value for a summation, you should start by observing several basic properties of the summand. First determine which term has the largest absolute value. If the summand is a monotonic function of the summation index, then the biggest term will be either the first term or the last term. If the summand is not monotonic or if it is not clear whether or not the sum is monotonic, then you can take the derivative of the summand with respect to the summation index. Any value of the summation index that makes the derivative equal to zero will correspond to a local maximum or a local minimum of the summand. If the derivative of the summand goes to infinity, such points must also be considered. You can compare the value of the summand at the endpoints of the summation range and at the places where the derivative goes to zero (or infinity) to determine the global maximum. There is one complication. The derivative often goes to zero at noninteger values of the summation index. In such cases you need to consider the value of the summand at the next lower integer and at the next higher integer. (We are assuming here that the summand varies slowly enough that the function changes smoothly between integers. If the function oscillates too much between integer values, then derivatives will be of little use.)

When looking for the maximum term, factors of the form $(-1)^i$, where i is the summation index, are dropped. Such factors have an absolute value of one, and they oscillate too rapidly for methods that assume that the summand changes smoothly from one integer value of the summation index to the next.

If the summation has n terms and the largest term has value L, then the sum is less than or equal to nL. If the terms are all about the same size, this may be a fairly good approximation. If the terms are positive and vary rapidly, this estimate usually will be too large by a factor of about n. If the terms vary in sign the estimate may be *way too large* in absolute value.

While noticing which term has the largest absolute value, you should also notice whether all the terms have the same sign and whether the value of a term changes rapidly or slowly with the summation index. If the terms are all positive, the sum is larger than the largest term. The size of the largest term is a pretty good estimate of the value of the sum when the summand has one peak and the terms rapidly decrease away from the peak.

From these preliminary observations you can often get a rough idea of the value of the sum. They will also suggest which method to try when a more accurate estimate of the value of the sum is needed. Since the more accurate methods require much more calculation, compare your initial estimates with your final estimates. This will save you much embarrassment if you made a major error in your "accurate" estimate.

If you want to compute the sum $\sum_{R(i)} a_i$ and you can find sequences b_i

and c_i such that $b_i \geq a_i \geq c_i$ for i such that $R(i)$ is *true*, then

$$\sum_{R(i)} b_i \geq \sum_{R(i)} a_i \geq \sum_{R(i)} c_i. \tag{83}$$

It is always possible to find a sequence whose elements are smaller than the corresponding a_i and one whose elements are larger than the corresponding a_i. It is necessary that the sequences be quite close to a_i for this approach to be very accurate.

Consider the sum $\sum_{1 \leq i \leq n} 1/i$. We know that $i = 2^{\lg i}$, so $i \geq 2^{\lfloor \lg i \rfloor}$ and $1/i \leq 2^{-\lfloor \lg i \rfloor}$. In addition, $i \leq 2^{\lceil \lg i \rceil}$ and $1/i \geq 2^{-\lceil \lg i \rceil}$. This gives

$$\begin{aligned} &1 + \tfrac{1}{2} + \tfrac{1}{4} + \tfrac{1}{4} + \tfrac{1}{8} + \tfrac{1}{8} + \tfrac{1}{8} + \tfrac{1}{8} + \cdots \\ &\leq 1 + \tfrac{1}{2} + \tfrac{1}{3} + \tfrac{1}{4} + \tfrac{1}{5} + \tfrac{1}{6} + \tfrac{1}{7} + \tfrac{1}{8} + \cdots \\ &\leq 1 + \tfrac{1}{2} + \tfrac{1}{2} + \tfrac{1}{4} + \tfrac{1}{4} + \tfrac{1}{4} + \tfrac{1}{4} + \tfrac{1}{8} + \cdots. \end{aligned} \tag{84}$$

Thus by eq. (83) we have

$$\sum_{1 \leq i \leq n} 2^{-\lceil \lg i \rceil} \leq \sum_{1 \leq i \leq n} \frac{1}{i} \leq \sum_{1 \leq i \leq n} 2^{-\lfloor \lg i \rfloor}. \tag{85}$$

Now the right and left sums in eq. (85) are not so hard to work out; they have a lot of terms that are the same. We have $2^{-\lceil \lg i \rceil} = 2^{-j}$ for $2^{j-1} < i \leq 2^j$ (2^{j-1} cases for $j \geq 1$) and $2^{-\lfloor \lg i \rfloor} = 2^{-j}$ for $2^j \leq i < 2^{j+1}$ (2^j cases). We can also drop some positive terms from the lower limit and add some positive terms to the upper limit and still have upper and lower limits. Thus we have

$$1 + \sum_{1 \leq j \leq \lfloor \lg n \rfloor} \tfrac{1}{2} \leq \sum_{1 \leq i \leq n} \frac{1}{i} \leq \sum_{0 \leq j \leq \lceil \lg n \rceil} 1 \tag{86}$$

or

$$1 + \tfrac{1}{2} \lfloor \lg n \rfloor \leq \sum_{1 \leq i \leq n} \frac{1}{i} \leq 1 + \lceil \lg n \rceil, \tag{87}$$

so

$$\sum_{1 \leq i \leq n} 1/i = \Theta(\ln n). \tag{88}$$

We will obtain a more accurate approximation to the value of this sum in Sections 4.5.1 and 4.5.4.

Bender [68, Section 3] gives several techniques for estimating the sum of positive terms.

EXERCISES

1. In Section 2.5.2 we showed that

$$\sum_{1\le i\le N} \lfloor \lg i \rfloor = (N+1)\lfloor \lg N \rfloor - 2^{\lfloor \lg N \rfloor + 1} + 2 + N.$$

Use this result to show that

$$N \lg N - 2N + \lg N + 1 \le \lg(N!) \le N \lg N + N + \lg N + 3.$$

From this deduce that

$$2N\left(\frac{N}{4}\right)^N \le N! < 8N(2N)^N.$$

This gives us a *rough* estimate on the size of $N!$; the upper and lower limits differ by a factor of over 8^N. A much better result is obtained in Section 4.5.5.

2. Use the result of Exercise 1.5–8 and the approach of the last exercise to obtain a better limit on the value of $N!$. (The factor of 2^N in the upper limit should go away. The limit is still rather rough. The limits differ by a factor of more than 4^N.)

4.3.1* Backtracking

For many problems there is a strategy for gradually obtaining a solution. After a certain amount of work has been done, definite progress has been made. More work brings more progress, until finally a complete answer is obtained. All the problems we have considered so far are like this. When you have an algorithm where each step brings you closer to a solution, you usually have a fast algorithm.

For some problems, however, no such approach has been discovered. You may start on one attempt at a solution, only to discover after much work that the problem has no solutions that start out like your first attempt. You have to start over with a second attempt. The only information that you gain from the first attempt is that it did not lead to a solution. To solve such problems you must search through a space of possible solutions. Since you have no guarantee that the initial directions you search will lead to a solution, such searching algorithms are usually very slow for large problems.

There are many interesting problems, some of great practical importance, where the best known solution method is to search a space of possible solutions. One example is the register allocation problem. Suppose you have a complex assembly language program for a computer with several registers. If you want the program to run as fast as possible, you must carefully consider which intermediate results to leave in registers and which results to store in memory. Some general principles, such as putting the variables that are used in inner loops into registers, will help, but the general principles will not completely solve the problem. If you want to find the very best way to use the registers, you will need to consider a large number of possibilities.

A *puzzle* is a problem where it is much easier to check that a solution is correct than it is to find a solution in the first place. Many puzzles are best

FIGURE 4.3 A solution of the four-queens problem.

solved by searching a space of possible solutions. One such puzzle is the *n-queens problem.* You are given an $n \times n$ chess board with rows 1 to n and columns 1 to n. The problem is to place n-queens on the board so that no queen can capture another. That is, no two queens can be placed on the same row, no two queens can be placed on the same column, and no two queens can be placed on the same diagonal. A lot of ways of arranging the queens must be considered if you want to solve this problem, particularly if you want to find all the solutions. Figure 4.3 shows a solution for $n = 4$. The problem, of course, is not very difficult for $n = 4$, but it rapidly becomes more difficult as n increases.

When you are solving a problem by searching through a space of possible solutions, it helps to be clever and search no more of the space than is necessary. *Backtracking* is a method of organizing a search though a solution space in such a way that at times you can eliminate large regions of the space from further consideration. By disposing of entire groups of possibilities at once, much time can be saved. To solve problems with backtracking, you need a method of determining that some of the partial solutions of the problem cannot be extended to complete solutions. A *partial solution* is a way of setting some of the variables of the problem so that none of the constraints of the problem are violated. If you have a perfect method for eliminating partial solutions that cannot be extended to complete solutions, then you can develop a fast algorithm for your problem. Backtracking works even when only some of the partial solutions that cannot be extended to complete solutions can be identified. It is essential, however, that the method for rejecting partial solutions never reject partial solutions that do lead to complete solutions.

Let's suppose the original problem is to satisfy some predicate P. For example, in the n-queens problem, since we will need to have exactly one queen per row (if we have more they can take each other and if we have less we cannot get all n queens on the board) let's use the variable x_i to give the column for the queen on the ith row. We have a solution if three sets of conditions are satisfied. First, we need $x_i \neq x_j$ for all i and j such that $i \neq j$ and $1 \leq i, j \leq n$. In other words, no two queens can be in the same column. Second, we need $x_i - x_j \neq i - j$ for $i \neq j$, $1 \leq i, j \leq n$ (no two queens can be on the same diagonal that goes from the lower left to the upper right). Third, we need $x_i - x_j \neq -i + j$ for $i \neq j$, $1 \leq i, j \leq n$ (this takes care of the other kind of diagonal). Using $\wedge$ to mean logical *and*, our predicate P is

$$\bigwedge_{\substack{1 \leq i \leq n \\ 1 \leq j \leq n \\ i \neq j}} \Big((x_i \neq x_j) \wedge (x_i - x_j \neq i - j) \wedge (x_i - x_j \neq -i + j) \Big). \tag{89}$$

As shown in Figure 4.3, one way to make $P(x_1, x_2, x_3, x_4)$ *true* is to set $x_1 = 2$, $x_2 = 4$, $x_3 = 1$, and $x_4 = 3$.

Suppose your original problem is defined by a predicate P that is a function of a set S of variables. (For the n-queens problem the set of variables is $x_1, \ldots, x_n$.) To do backtracking you need, in addition to P, a set of *intermediate predicates* P_A that are defined for various sets A that are subsets of S. The predicate P_A is a function of the variables in A, and if it is *false* for some set of values for the variables in A, then there is no hope of a solution of the complete predicate (P) using the same values for the variables that are in A (and any value for the variables that are not in A). When P_A is *true*, there may or may not be a solution using those values for the variables in A. In other words, using a_j to represent a value for variable x_j, P_A must be defined so that if $A = \{x_1, \ldots, x_k\}$ and $P_A(a_1, \ldots, a_k)$ is *false*, then there do not exist *any* values for the variables $x_{k+1}, \ldots, x_n$ such that $P(a_1, \ldots, a_n)$ is *true*. When P_A is *false*, it is time to give up on the current attempt at a partial solution. This eliminates the entire class of potential solutions that begins with those values for the variables in A. It is convenient to require that P_S, the intermediate predicate for the entire set of variables, equal P, the predicate for the original problem.

When P is the conjunction (logical *and*) of several relations (which is often the case), an obvious choice for P_A is the conjunction of all the relations that depend only on the variables in A. For example, for the four-queens problem, this approach gives $P_\emptyset = \textit{true}$, where $\emptyset$ is the empty set, $P_{\{1\}} = \textit{true}$, $P_{\{1,2\}} = (x_1 \neq x_2) \wedge (x_1 - x_2 \neq 1) \wedge (x_1 - x_2 \neq -1)$, etc.

There are many variations on backtracking. Here we give the simplest one. We assume that the ith variable, x_i, has the range one to v_i. This makes it easier to state the algorithm, but it is in no way essential, as you will see below.

Algorithm 4.1 Backtracking: Input: A set $\{x_1, \ldots, x_n\}$ of variables, where the values for the ith variable are $1, \ldots, v_i$, and a set of intermediate predicates P_i, where P_i is an intermediate predicate for the set $\{x_1, \ldots, x_i\}$. Output: All the sets of values for the variables such that P_S is *true*.

Step **1.** [Initialize] Set $i \leftarrow 0$.

Step **2.** [Solution?] If $i \neq n$, then go to Step 3. Otherwise, the current set of values is a solution. Output it and go to Step 5.

Step **3.** [Search deeper] Set $i \leftarrow i + 1$.

Step **4.** [Zeroth value] Set $x_i \leftarrow 0$.

Step **5.** [Next value] If $x_i \geq v_i$, then go to Step 7. (All values of x_i have been tried.) Otherwise set $x_i \leftarrow x_i + 1$.

Step **6.** [Test] If $P_i(x_1, \ldots, x_i)$, is *true*, then go to Step 2. Otherwise go to Step 5.

Step **7.** [Backtrack] Set $i \leftarrow i - 1$. If $i = 0$, then stop. Otherwise go to Step 5.

To modify the algorithm to work with any finite set of values for x_i, define an array $V_i[j]$ for each i such that $V_i[j]$ is the jth value for the ith variable of the original problem. In Step 6, replace $P_i(x_1, \ldots, x_i)$ with $P_i(V_1[x_1], \ldots, V_i[x_i])$. This change results in x_i being the index of the value for the ith problem variable rather than being the value for the problem variable itself. Introducing an extra variable to serve as an index to the values of the problem variables makes the statement of backtracking algorithms more general but harder to follow.

4.3.1.1 Backtracking (Average Time)*

The running time for this algorithm depends on the nature of the intermediate predicates. If they always evaluate to *false*, then the algorithm runs very rapidly (in time $\Theta(v_1)$), while if they always evaluate to *true* the algorithm runs very slowly (in time $\Theta(v_1 \ldots v_n)$). Of course, neither of these extreme cases corresponds to very interesting problems. For most problems the intermediate predicates sometimes evaluate to *true* and other times to *false*. Computing the average running time is rather complex.

The running time of the Backtracking Algorithm is a constant plus big O of the number of times that Step 6 is done. The constant is not important since it never leads to a large running time, and it will be ignored for the rest of the analysis. The sets of values for the x_i in Step 6 can be represented as a tree. Each level in the tree corresponds to a value of i. Each node on level i in the tree corresponds to a particular setting of values for the variables $x_1, x_2, \ldots, x_i$. (The path from the root to the node determines the values.) The parent of a node on level i is the node on level $i - 1$ that corresponds to the same set of

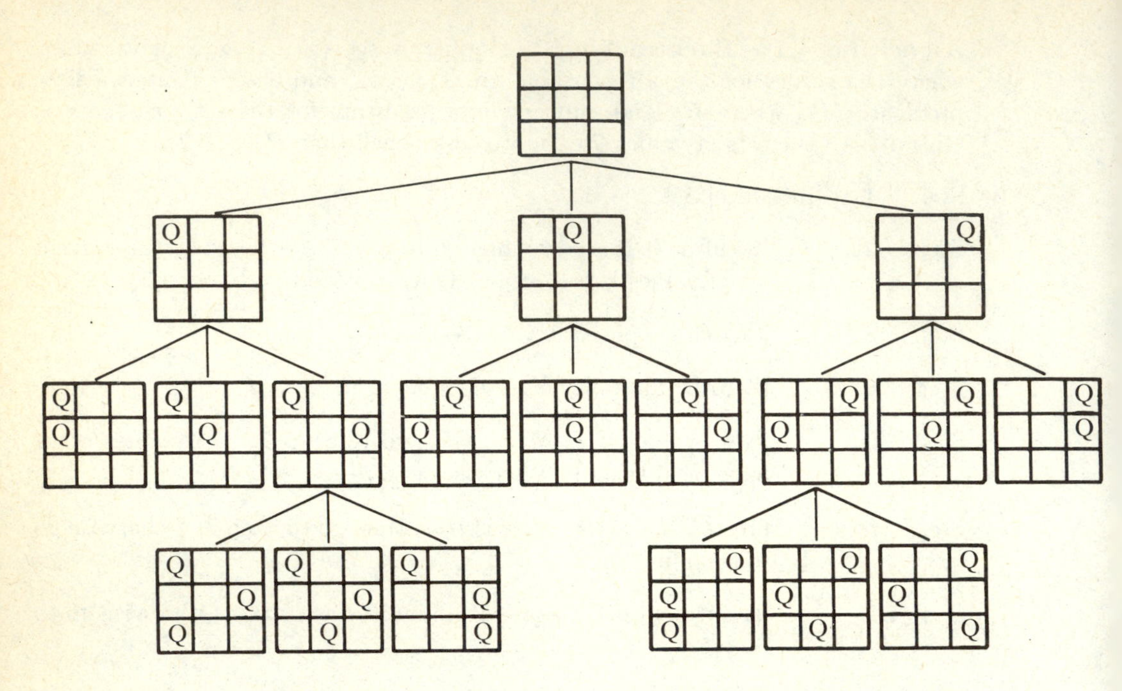

FIGURE 4.4 The backtrack tree for the three-queens problem.

values for the first $i - 1$ variables. Figure 4.4 shows the backtrack tree for the three-queens problem. From the backtrack tree you can see that the three-queens problem has no solution.

We now consider a random set of problems for which the average running time for backtracking is not too difficult to compute. The set of problems has been selected because it leads to a relatively straightforward analysis. For some values of the parameters it contains, with high probability, many problems that are definitely unlike typical problems that are solved by backtracking. Care is therefore needed in interpreting the results of the analysis.

The problems are made up of simple predicates over logical variables; the problem is to find values for the variables that make the predicate *true*. There are v variables. Each variable has two possible values: *true* and *false*. A *literal* is a variable, x_i, or the negation of a variable, $\neg x_i$. Thus there are $2v$ possible literals. A *clause* is the disjunction (logical *or*) of a series of literals. Thus $x_1 \vee \neg x_3 \vee x_4$ is a possible clause. There are 2^{2v} possible clauses, since each clause can contain or not contain each of the $2v$ literals. A *predicate* in our random problem set is the conjunction of a series of clauses. Our predicates contain t clauses. A possible predicate, with $v = 4$ and $t = 2$, is $(x_1 \vee \neg x_3 \vee x_4) \wedge (x_1 \vee x_2 \vee x_3)$. We have no requirement that the clauses in a predicate all be different. There thus are 2^{2tv} possible predicates. Determining whether such a predicate has a solution (an assignment of values to variables that causes the predicate to evaluate to *true*) is called the *satisfiability* problem.

We assume the predicates are numbered from 1 to 2^{2tv} (the particular way in which we assign numbers is not important). The ith predicate is called $P^{(i)}$. Intermediate predicates are formed by taking the conjunction of all the clauses in the predicate that contain only the variables that have values. The symbol $P_j^{(i)}$ stands for the intermediate predicate which is associated with the ith predicate and which uses the first j variables.

To have a random set of problems, we also need a rule for assigning a probability to each problem in the set. We do this in two steps. First, we choose a probability p, and we form random clauses by independently selecting each possible literal with probability p. Then we form random predicates by taking the conjunction of t independently generated random clauses. This random process for generating clauses depends on v, t, and p. Once you are given these three numbers, you can calculate the probability that any particular predicate is generated.

Let's first consider the nature of the random clauses that are generated. The probability that the clause contains both x_1 and $\neg x_1$ is p^2. Such clauses are not very interesting since they are always *true*. If you do not want such clauses to be generated very often, then p must be small. (You should be wondering how small; see the exercises.) The Backtracking Algorithm as we have stated it will not notice that such clauses are always *true*. A more sophisticated algorithm would test for such clauses and eliminate them.

The probability that a clause does not contain variable x_1 is $1-p$. The probability that it does not contain either x_1 or $\neg x_1$ is $(1-p)^2$. The probability that the clause is completely empty is $(1-p)^{2v}$ (there are $2v$ independent chances for a literal to get into the clause). A clause with no literals is *false*. If any clause in the predicate is *false*, then the entire predicate is also *false*. The probability that a predicate with one clause is **not** *false* is $1-(1-p)^{2v}$. The probability that a predicate with t clauses does not contain the empty clause is $\left(1-(1-p)^{2v}\right)^t$. (A predicate with more than one clause can evaluate to *false* for every value of the variables even if it does not contain the empty clause.) The probability that the predicate does contain the empty clause is $1-\left(1-(1-p)^{2v}\right)^t$.

Let's now compute the average time required by the Backtracking Algorithm for these random problem sets. To compute the average, you take the time for each problem, multiply by the probability of the problem, and sum over all the problems. The time required by each problem is big O of the number of nodes in its backtrack tree, assuming that all nodes take about the same time to process. For the problems we are considering, all nodes take the same time to process within a factor of v. Since we will eventually be interested in whether the time is a polynomial or an exponential function of v, the assumption of a constant time per node is good enough for our purposes. (This assumption would not be good enough for someone who was interested in whether backtracking took 1 day or 1 week of computer time for some type of problem.) We now compute how many nodes the average backtrack tree has, since the number of nodes is closely related to the running time.

The expected number of nodes can be calculated using the formula

$$N_{\text{Average}} = \sum_i p_i M_i, \tag{90}$$

where the sum is over all problems in the set, p_i is the probability for the ith problem, and M_i is the number of nodes for the ith problem. This straightforward way of calculating the average, however, is too difficult to carry out. Although p_i is easy enough to compute, there is no reasonable way to compute M_i. So instead we will compute the average with an even more primitive formula:

$$N_{\text{Average}} = \sum_i p_i \sum_j C_{ij}, \tag{91}$$

where the sum over j is over all the nodes that the backtrack tree for a problem in the set might have, and $C_{ij} = 1$ if the ith problem produces a backtrack tree that contains node j and $C_{ij} = 0$ otherwise. The sum over j, of course, just counts the number of nodes in the backtrack tree for problem i, that is, $M_i = \sum_j C_{ij}$. Now if we were to use eq. (91) directly, we would be making no progress; we would just get eq. (90) back again. The next step is to change the order of summation in eq. (91) to obtain

$$N_{\text{Average}} = \sum_j \sum_i p_i C_{ij}. \tag{92}$$

With this equation we will be able to make progress, because $\sum_i p_i C_{ij}$ is not all that difficult to compute.

The first thing to notice about $\sum_i p_i C_{ij}$ is that its value depends only on the level of node j. Consider two nodes, j and j', that are on the same level in backtrack trees. Each node corresponds to a way to set some of the variables in the problem. Since the two nodes are on the same level, the same variables are set for each node; only the values assigned to the variables are different. Let's call the variables that have been assigned values $x_1, \ldots, x_k$. Let $a_1 \ldots, a_k$ be the values that lead to node j, and let $a'_1 \ldots, a'_k$ be the values that lead to node j'. Now consider $\sum_i p_i C_{ij}$ and $\sum_i p_i C_{ij'}$. The index i is over all possible predicates. We will show that for each value of i in the first sum there is a corresponding value i' in the second sum such that $p_i C_{ij} = p_{i'} C_{i'j'}$. Furthermore the mapping between i and i' is one to one. Thus, the two sums have the same values; the terms in each sum are the same, just the order is different.

To find the correspondence, we must show that there is a predicate $P^{(i')}$ that leads to a node for values $a'_1, \ldots, a'_k$. To obtain $P^{(i')}$ from $P^{(i)}$, compare $a_1 \ldots, a_k$ with $a'_1 \ldots, a'_k$. For each l such that $a_l \neq a'_l$, replace variable x_l with $\neg x_l$ in $P^{(i)}$. (This replacement may generate some literals of the form $\neg\neg x_l$, which should be replaced with x_l.) Clearly now $p_i C_{ij} = p_{i'} C_{i'j'}$ because each literal in $P^{(i')}$ has the same value as the corresponding literal in $P^{(i)}$ (where $P^{(i')}$ is evaluated with $x_1 = a'_1, \ldots, x_l = a'_l$, while $P^{(i)}$ is evaluated with $x_1 = a_1, \ldots, x_l = a_l$). The transformation is one to one because $P^{(i)}$ can be obtained from $P^{(i')}$ by applying the same transformation to $P^{(i')}$.

Since the value of $\sum_i p_i C_{ij}$ depends only on the level of node j, we can replace eq. (92) with

$$N_{\text{Average}} = \sum_{0 \le k \le v} \sum_i N_k p_i C'_{ik} = \sum_{0 \le k \le v} N_k \sum_i p_i C'_{ik}, \tag{93}$$

where N_k is the number of nodes on level k, C'_{ik} is C_{in_k}, and n_k is the node on level k with all of the values of the variables equal to *false*. The choice of *false* for the values of the variables is arbitrary—it just happens to be a set of values that is easy to work with.

To summarize what we have shown so far, there are two ways we could compute the average number of nodes in a backtrack tree. The standard way is to compute the number of nodes in the tree for each problem and take the average of the number of nodes. An alternate way is to consider each node that might be in a backtrack tree, compute the probability for each node that the node is in the backtrack tree and sum over the number of nodes. Either approach leads to the same answer. The first approach, however, is difficult to carry out, whereas the second approach is not so difficult. This method of transforming an average time calculation is useful in general (see Section 7.1.2.1 for an another example).

The value of N_k is easy to compute. There are 2^k possible nodes on level k (where the root is level zero). Thus $N_k = 2^k$.

The next step is to compute $\sum_i p_i C'_{ik}$. This sum is the probability that a random predicate leads to a backtrack tree that has a node corresponding to setting the first k variables to *false*. The ith predicate produces such a tree if and only if the corresponding intermediate predicate for $k-1$ variables ($k \ge 1$), $P^{(i)}_{k-1}(\mathit{false}, \ldots, \mathit{false})$, is *true*. (For $k = 0$, there is always a root node). So $\sum_i p_i C'_{ik}$ is the probability that $P^{(i)}_{k-1}(\mathit{false}, \ldots, \mathit{false}) = \mathit{true}$. Now $P^{(i)}_{k-1}(\mathit{false}, \ldots, \mathit{false}) = \mathit{true}$ unless $P^{(i)}_{k-1}$ contains a clause where every literal is positive (i.e., does not contain a *not* sign). Let's use r_k for this probability. Thus we have

$$r_k = \sum_i p_i C'_{ik} \qquad \text{for } k \ge 0, \tag{94}$$

$$r_0 = 1, \tag{95}$$

and eq. (93) becomes

$$N_{\text{Average}} = \sum_{0 \le k \le v} N_k r_k = \sum_{0 \le k \le v} 2^k r_k. \tag{96}$$

Now, each clause in our random predicate is selected independently. A clause in a predicate leads to a clause in the intermediate predicate which contains only positive literals if and only if the clause in the original predicate consists only of positive literals and only literals for the first k variables, where k is the number of variables for the intermediate predicate. Let u_k be the probability that a clause has this form. Since the clauses are independent, we have

$$r_k = u_{k-1}^t. \tag{97}$$

We still need to compute u_k. A clause consists only of positive literals for variables from 1 to k if it does *not* contain a positive literal for a variable with index greater than k or a negative literal for any variable, so u_k is 1 minus the probability that a random clause has either a positive literal with index greater than k or a negative literal with any index. There are v negative literals and $v-k$ positive literals with index greater than k, so $2v-k$ distinct literals lead to the clause having a negative literal or a literal with index greater than k. The probability that each such literal does not occur in the clause is $1-p$, and the literals are selected independently, so

$$u_k = 1-(1-p)^{2v-k} \qquad \text{for } k \geq 1. \tag{98}$$

To illustrate the previous derivation, let's calculate the probabilities for the case $v=3$ and $t=2$. To calculate u_0, we must calculate the probability that a random clause does have some literals. This is 1 minus the probability that it does not have any literals. The probability that it has x_1 is p, so the probability that it does not have it is $1-p$. Likewise for $\neg x_1$, x_2, $\neg x_2$, x_3, and $\neg x_3$. The probability that it has no literals is thus $(1-p)^6$. So $u_0 = 1-(1-p)^6$, as given by eq. (98). To calculate r_1, we notice that we want the probability that each clause has some literals. Since the clauses are generated independently, $r_1 = u_0^2 = (1-(1-p)^6)^2$.

For u_1, we must calculate the probability that a random clause has some literal besides x_1. This is 1 minus the probability that it has no literal other than x_1 (it may or may not have x_1). We just calculated that the probability that a random clause has no literals is $(1-p)^6$. The probability that it has x_1 and no other literals is $p(1-p)^5$. The probability that it has only x_1 (either zero or one time) is $(1-p)^6 + p(1-p)^5 = (1-p)^5$. So $u_1 = 1-(1-p)^5$ and $r_2 = u_1^2$.

For u_2, we must calculate the probability that a random clause has some literals besides x_1 and x_2. The probability that it has no literals is $(1-p)^6$ and the probability that it has only x_1 is $p(1-p)^5$, as we have already calculated. Likewise the probability that it has only x_2 is $p(1-p)^5$. The probability that it has literals x_1 and x_2 but no others is $p^2(1-p)^4$. Therefore the probability that it has only literals from the set $\{x_1, x_2\}$ is $(1-p)^6 + 2p(1-p)^5 + p^2(1-p)^4 = (1-p)^4$, and the probability that it has literals not from that set is $1-(1-p)^4$. Thus $r_3 = (1-(1-p)^4)^2$.

Combining the results so far, we have that the average number of nodes in a backtrack tree for a random predicate is

$$1 + \sum_{1\leq i\leq v} 2^i \left(1-(1-p)^{2v-i+1}\right)^t. \tag{99}$$

To continue the example, for $v = 3$ and $t = 2$, the average number of nodes in a backtrack tree is $1 + (1-(1-p)^6)^2 + 2(1-(1-p)^5)^2 + 4(1-(1-p)^4)^2$.

There is no obvious way to find the exact value of eq. (99), so let's consider finding an approximate value. Since the sum contains a 2^i, it might get to be very large. On the other hand $\left(1-(1-p)^{2v-i+1}\right)^t$ decreases with increasing i, and for some values of i, p, t, and v, it becomes extremely small. The first step in obtaining our rough approximation is to find the value of i in the range 1 to v that maximizes $2^i\left(1-(1-p)^{2v-i+1}\right)^t$. The value of the sum is no more than v times the maximum value of $2^i\left(1-(1-p)^{2v-i+1}\right)^t$.

Let's use the notation

$$N_i = 2^i\left(1-(1-p)^{2v-i+1}\right)^t; \tag{100}$$

let $i_{\max}$ be the value of i that maximizes N_i, and i_* be the value of i in the range 1 to v that maximizes N_i. (Eq. (100) is a new definition of N_i; the previous definition was used up through eq. (96).) To compute $i_{\max}$, we could start by taking the derivative of N_i with respect to i and solving for the value of i that results in the derivative being zero. Since the logarithm is a monotonic increasing function, we can also find $i_{\max}$ by taking the derivative of $\ln N_i$ and solving for the value of i that results in the derivative of the logarithm being zero. Since N_i is an exponential function, the latter approach results in less algebra. So,

$$\ln N_i = i \ln 2 + t \ln\left[1-(1-p)^{2v-i+1}\right] \tag{101}$$

and

$$\frac{d \ln N_i}{di} = \ln 2 - \frac{t(1-p)^{2v-i+1}(-\ln(1-p))}{1-(1-p)^{2v-i+1}}. \tag{102}$$

Now eq. (102) is a decreasing function of i [because $(1-p)^{-i}$ is a decreasing function], so the value of i that results in the derivative being zero does maximize (rather than minimize) N_i.

Thus $i_{\max}$ is the solution of the equation

$$\ln 2 - \frac{t(1-p)^{2v-i_{\max}+1}(-\ln(1-p))}{1-(1-p)^{2v-i_{\max}+1}} = 0. \tag{103}$$

The first step in finding $i_{\max}$ is to solve for $(1-p)^{2v-i_{\max}+1}$, obtaining

$$(1-p)^{2v-i_{\max}+1} = \frac{\ln 2}{\ln 2 + t(-\ln(1-p))} = \frac{1}{1+t(-\ln(1-p))/\ln 2}. \tag{104}$$

Taking logarithms and solving for $i_{\max}$ gives

$$i_{\max} = 2v + 1 - \frac{\ln\left(1+\dfrac{t(-\ln(1-p))}{\ln 2}\right)}{-\ln(1-p)}. \tag{105}$$

Since the derivative [eq. (103)] is a decreasing function of i, i_*, the value of i in the range 1 to v that maximizes N_i, is

$$i_* = \begin{cases} v & \text{when} \quad i_{\max} \geq v, \\ i_{\max} & \text{when} \quad 1 \leq i_{\max} \leq v, \\ 1 & \text{when} \quad i_{\max} \leq 1. \end{cases} \tag{106}$$

Now let's consider which values for p, t, and v cause $i_{\max} > v$. We have

$$2v + 1 - \frac{\ln\left(1 + \dfrac{t(-\ln(1-p))}{\ln 2}\right)}{-\ln(1-p)} > v, \tag{107}$$

or

$$(v+1)(-\ln(1-p)) > \ln\left(1 + \frac{t(-\ln(1-p))}{\ln 2}\right). \tag{108}$$

Taking exponentials gives

$$(1-p)^{-(v+1)} > 1 + \frac{t(-\ln(1-p))}{\ln 2}. \tag{109}$$

Replacing $-\ln(1-p)/\ln 2$ with $-\lg(1-p)$ and solving for t gives

$$t < \frac{(1-p)^{-(v+1)} - 1}{-\lg(1-p)}. \tag{110}$$

When t is below the limit given by eq. (110), i_* is v, so in this case

$$N_{i_*} = 2^v \left(1 - (1-p)^{v+1}\right)^t. \tag{111}$$

The total number of nodes, N, is bounded by

$$N_{i_*} \leq N \leq vN_{i_*}. \tag{112}$$

Now let's consider the case $i_* = i_{\max}$. Plugging eq. (105) into eq. (99) gives

$$N_{i_*} = 2^{2v+1-\frac{\ln\left(1+\frac{t(-\ln(1-p))}{\ln 2}\right)}{-\ln(1-p)}} \left(1 - (1-p)^{\frac{\ln\left(1+\frac{t(-\ln(1-p))}{\ln 2}\right)}{-\ln(1-p)}}\right)^t \tag{113}$$

which can be simplified by noticing that

$$(1-p)^{\frac{\ln\left(1+\frac{t(-\ln(1-p))}{\ln 2}\right)}{-\ln(1-p)}} = \exp\left(\ln(1-p)\frac{\ln\left(1 + \dfrac{t(-\ln(1-p))}{\ln 2}\right)}{-\ln(1-p)}\right) \tag{114}$$

$$= \exp\left(-\ln\left(1 + \frac{t(-\ln(1-p))}{\ln 2}\right)\right) \tag{115}$$

$$= \frac{1}{1 + \dfrac{t(-\ln(1-p))}{\ln 2}} = \frac{\ln 2}{\ln 2 + t(-\ln(1-p))}. \tag{116}$$

Using the simplification, we get

$$N_{i_*} = 2^{2v+1-\frac{\ln\left(1+\frac{t(-\ln(1-p))}{\ln 2}\right)}{-\ln(1-p)}} \left(\frac{t(-\ln(1-p))}{\ln 2 + t(-\ln(1-p))}\right)^t. \tag{117}$$

Likewise

$$2^{2v+1-\frac{\ln\left(1+\frac{t(-\ln(1-p))}{\ln 2}\right)}{-\ln(1-p)}} = \left(\frac{t(-\ln(1-p))}{\ln 2 + t(-\ln(1-p))}\right)^{\ln 2/(-\ln(1-p))} \tag{118}$$

so

$$N_{i_*} = 2^{2v+1} \left(\frac{t(-\ln(1-p))}{\ln 2 + t(-\ln(1-p))}\right)^{t+\ln 2/(-\ln(1-p))}. \tag{119}$$

Now let's do an asymptotic analysis of eq. (111). Although eq. (112) leads to a bound on the average number of nodes that is good enough for many purposes and although it is in closed form, it is not in a form that is very easy to understand. Since eq. (111) has an exponential nature, let's consider $\ln N_{i_*}$. Using a for $(1-p)^{v+1}$, we have

$$N_{i_*} = \exp\left(v\ln 2 + t\ln\left(1-(1-p)^{v+1}\right)\right) = \exp(v\ln 2 + t\ln(1-a)). \tag{120}$$

Now let's consider the relation between v, p, and a when v becomes large, p is a function of v, and a remains constant. This is a useful way to look at eq. (120) because eq. (120) is a complicated function of p, but a simple function of a. Since $a = (1-p)^{v+1}$, we have

$$\ln a = (v+1)\ln(1-p). \tag{121}$$

Since we want a to be constant as v becomes large, it is necessary for $\ln(1-p)$ to become small, which will happen only if p approaches zero. If p approaches zero, we can use eq. (48) to obtain

$$\ln a = -vp\left(1 + \Theta\left(\frac{1}{v}\right) + \Theta(p)\right). \tag{122}$$

When a is nonzero and finite $p = \Theta(1/v)$, and eq. (122) can be simplified to

$$\ln a = -vp\left(1 + O\left(\frac{1}{v}\right)\right). \tag{123}$$

Solving eq. (123) for a and plugging into eq. (120) gives (for the case that $\lim_{v\to\infty} vp$ is constant)

$$\begin{aligned} N_{i_*} &= \exp\left\{v\ln 2 + t\ln\left[1-(1-p)^{v+1}\right]\right\} && (124)\\ &= \exp\{v\ln 2 + t\ln[1-\exp(-vp)\exp(1+O(p))]\} && (125)\\ &= \exp\left\{v\ln 2 + t\ln\left[1-\exp\left(-vp\left(1+O\left(\frac{1}{v}\right)\right)\right)\right]\right\} && (126)\\ &= \exp\left\{v\ln 2 + t\ln\left[1-\exp(-vp)\left(1+O\left(\frac{1}{v}\right)\right)\right]\right\} && (127)\\ &= \exp\left\{v\ln 2 + t\left[\ln(1-\exp(-vp)) + O\left(\frac{1}{v}\right)\right]\right\}. && (128) \end{aligned}$$

From eq. (128) we can see that, when $\lim_{v\to\infty} vp$ is constant and t is below the limit given by eq. (110), the average time used by backtracking is an exponentially increasing function of v and an exponentially decreasing function of t [notice that $\ln(1-\exp(-vp))$ is negative, since the argument of the logarithm is less than 1]. Depending on the value of $\ln(1-\exp(-vp))$ and on the rate of increase of t with v, the average time for large problems can be exponentially large (or not be exponentially large).

Many investigations of predicates concern predicates in conjunctive normal form. A predicate is in this form if it consists of a set of distinct clauses, where each clause consists of literals with distinct variables. The predicates that we just considered are similar to predicates in conjunctive normal form except there was no requirement that the clauses be distinct and there was no requirement that the variables of a clause be distinct. Clauses of the form we considered can be converted to conjunctive normal form in the following way. First, drop all but one copy of any repeated literal in each clause. Second, drop any clause that contains a variable and the negation of the same variable. Third, drop all but one copy of each repeated clause. The original predicate and the resultant conjunctive normal form predicate are *true* for exactly the same set of values of the variables.

Backtracking is covered in Horowitz and Sahni [7, Chapter 7] and in Hu [8, Chapter 4]. One of the first average time analyses of backtracking is given in Brown and Purdom [79].

EXERCISES

1. Give an equation relating v, t, p, and f so that only fraction f of the random clauses contain variable x_1 and its negation.

2. Give an equation relating v, t, p, and f so that only fraction f of the random clauses contain a variable and its negation, i.e., the clause counts in the computation if it contains any variable and the negation of that same variable.

3. What are the conditions on t such that in eq. (106) $i_* = i_{\max}$? What are the conditions on t such that in eq. (106) $i_* = 1$?

4. Simplify eq. (112) for the case $\lim_{v\to\infty} vp$ is zero.

5. Simplify eq. (112) for the case $\lim_{v\to\infty} vp$ is infinite.

6. Do an asymptotic analysis of eq. (117) similar to the one for eq. (111) done in the text.

4.4 RAPIDLY CHANGING SUMMANDS

When the summand changes rapidly, consider dividing the summation range into two parts. The first part should contain all the large terms. If you can then sum the first part and show that the second part is small, you can approximate the original sum.

Consider the sum $\sum_{1\le i\le n} i!$. There is no simple exact value for this sum, but the sum is easy to approximate. Since $i!$ is an increasing function, the largest term is the $i = n$ term. Let's first try using $i = n$ as the first part and $1 \le i \le n-1$ as the second part. This gives

$$\sum_{1 \le i \le n} i! = n! + \sum_{1 \le i \le n-1} i!. \tag{129}$$

We can approximate the sum on the right side by taking the number of terms, $n-1$, and multiplying by the largest term, $(n-1)!$, to obtain [using $n-1 = O(n)$]

$$\sum_{1 \le i \le n} i! = n! + O(n!) = O(n!), \tag{130}$$

which is an approximation, although not a very good one.

Now consider approximating the sum with the two biggest terms ($i = n$ and $i = n-1$) in the first part. This gives

$$\sum_{1 \le i \le n} i! = n! + (n-1)! + \sum_{1 \le i \le n-2} i! \tag{131}$$

$$= n! + (n-1)! + O((n-1)!) = n! + O((n-1)!) \tag{132}$$

$$= n! \left(1 + O\left(\frac{1}{n}\right)\right), \tag{133}$$

which is a fairly good approximation for large n. If you use several terms in the first part of the sum, you can obtain an even better approximation.

For a more complex example, consider the sum $\sum_{1 \le i \le n} 2^i \ln i$. Both 2^i and $\ln i$ are increasing functions of i, so the largest term occurs at $i = n$. Also 2^i is a rapidly varying function of i while $\ln i$ is a slowly varying function of i (for large i), so the sum should be about the same as $\ln n \sum_{1 \le i \le n} 2^i = (2^{n+1} - 2) \ln n$. How can we use these observations to derive an approximation to the sum? A promising approach is to expand the slowly varying function in a power series. First let's make the change of variable $j = n-i$ so that the important part of the sum will correspond to small values of the index. Then we have, for $1 \le i \le n$,

$$\ln i = \ln(n-j) = \ln\left[n\left(1 - \frac{j}{n}\right)\right] = \ln n + \ln\left(1 - \frac{j}{n}\right) \tag{134}$$

$$= \ln n - \frac{j}{n} + O\left(\frac{j^2}{n^2}\right). \tag{135}$$

Eq. (135) is a good approximation when j is small, but not when j is near n. So we will write our original problem as

$$\sum_{1 \le i \le n} 2^i \ln i = \sum_{0 \le j \le n-1} 2^{n-j} \ln(n-j) \tag{136}$$

$$= \sum_{0 \le j < \lfloor n/2 \rfloor} 2^{n-j} \ln(n-j) + \sum_{\lfloor n/2 \rfloor \le j \le n-1} 2^{n-j} \ln(n-j). \tag{137}$$

The rightmost sum in eq. (137) can be approximated by replacing $\ln(n-j)$ by the larger quantity $\ln n$, resulting in a sum that can be simplified with eq. (2.61). Doing this gives

$$\sum_{1\le i\le n} 2^i \ln i = 2^n \sum_{0\le j<\lfloor n/2\rfloor} 2^{-j}\left(\ln n - \frac{j}{n} + O\left(\frac{j^2}{n^2}\right)\right) + O\left(2^{\frac{1}{2}n}\ln n\right) \tag{138}$$

$$= 2^n\left[\ln n \sum_{0\le j<\lfloor n/2\rfloor} 2^{-j} - \frac{1}{n}\sum_{0\le j<\lfloor n/2\rfloor} j2^{-j} + \frac{1}{n^2}O\left(\sum_{0\le j<\lfloor n/2\rfloor} j^2 2^{-j}\right)\right] + O\left(2^{\frac{1}{2}n}\ln n\right). \tag{139}$$

To simplify eq. (139) use

$$\sum_{0\le j<\lfloor n/2\rfloor} 2^{-j} = \frac{1-(1/2)^{\lfloor n/2\rfloor}}{1-(1/2)} = 2 + O(2^{-\frac{1}{2}n}), \tag{140}$$

$$\sum_{0\le j<\lfloor n/2\rfloor} j2^{-j} = \frac{1/2 - \lfloor n/2\rfloor(1/2)^{\lfloor n/2\rfloor} + (\lfloor n/2\rfloor - 1)(1/2)^{\lfloor n/2\rfloor+1}}{(1-(1/2))^2} \tag{141}$$

$$= 2 + O(n2^{-\frac{1}{2}n}), \tag{142}$$

and

$$\sum_{0\le j<\lfloor n/2\rfloor} j^2 2^{-j} = O(1) \tag{143}$$

(see eq. (2.61), eq. (2.75), and Exercise 2.5–1). The result is

$$\sum_{1\le i\le n} 2^i \ln i = 2^{n+1}\left[\ln n - \frac{1}{n} + O\left(\frac{1}{n^2}\right)\right]. \tag{144}$$

The simplification starting at eq. (134) was adapted from de Bruijn [25, Section 3.2]. De Bruijn gives additional examples [25, Section 3.2 and 3.3]. Bender [68, Section 3] gives several techniques for estimating the sum of positive terms.

EXERCISES

1. Give an asymptotic formula with accuracy $O(n!/n^3)$ for $\sum_{1\le i\le n} i!$.
2. Find an approximation to: (a) $\sum_{0\le i\le 10} i^{100}$, (b) $\sum_{0\le i\le 1000} i^{1{,}000{,}000}$, (c) $\sum_{0\le i\le 1000} i^x$ for large x, and (d) $\sum_{0\le i\le n} i^x$ for large x and fixed n. Pay careful attention to whether or not a big O answer is appropriate for each part.
3. Show that $\sum_{1\le i\le n} i^m = \dfrac{n^{m+1}}{m+1} + O(n^m)$. Hint: Use $\sum_{0\le i\le n}(i+1)^{m+1} - \sum_{1\le i\le n+1} i^{m+1} = 0$.
4. [See 6] Give the answer to the previous exercise to accuracy $O(n^{m-2})$.

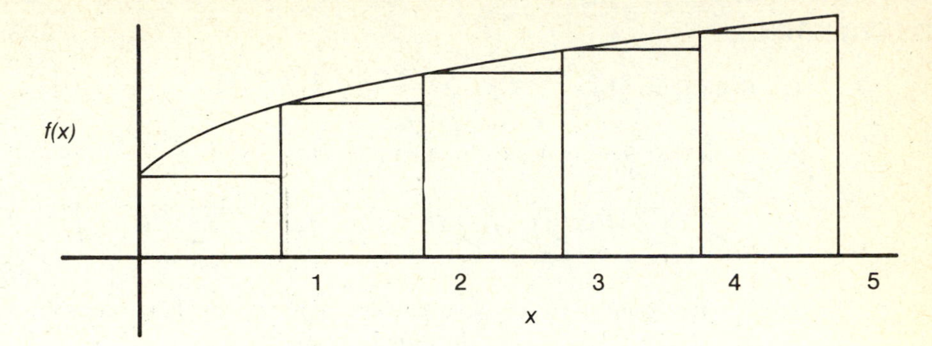

FIGURE 4.5 A comparison of $\int_0^5 f(x)\,dx$ with $\sum_{0\le i\le 4} f(i)$.

5. Approximate $\displaystyle\sum_i \binom{n}{i}\binom{n}{j-i}x^i$ for large x.

6. Approximate $\displaystyle\sum_i \binom{n}{i}\binom{n}{j-i}x^i$ for small x.

7. Give upper and lower bounds which differ by no more than a factor of n for the sum

$$\sum_{1\le j\le n} j^k(n-j)^k.$$

4.5 SLOWLY VARYING SUMMANDS

For a slowly varying function there is a close relation between the integral of the function and the sum of the function. Figure 4.5 shows the curve $\frac{1}{2}+x^{1/2}$. The integral of this function between the limits 0 and 5 gives the area under the curve between $x=0$ and $x=5$. The sum of $\frac{1}{2}+i^{1/2}$ for the limits 0 and 4 gives the area of the rectangles in Figure 4.5. The area under the curve and the area of the rectangles that go up to the curve are about the same.

For a monotonically increasing function, such as the one in Figure 4.5, the sum is smaller than the integral, but it is larger than the integral that is obtained by replacing x by $x+1$. When $f(x)$ is monotonically increasing

$$\int_{a-1}^{b-1} f(x)\,dx \le \sum_{a\le i<b} f(i) \le \int_a^b f(x)\,dx \tag{145}$$

When $f(x)$ is monotonically decreasing

$$\int_a^b f(x)\,dx \le \sum_{a\le i<b} f(i) \le \int_{a-1}^{b-1} f(x)\,dx. \tag{146}$$

EXERCISES

1. Show that $\frac{2}{3}(n-1)^{3/2} \le \sum_{1\le i<n} i^{1/2} \le \frac{2}{3}n^{3/2} - \frac{2}{3}$.

2. Show that for any decreasing function $f(x)$

$$\sum_{1\le i<m} f(i) + \int_m^n f(x)\,dx \le \sum_{1\le i<n} f(i) \le \sum_{1\le i<m} f(i) + \int_{m-1}^{n-1} f(x)\,dx.$$

3. Use the result of the previous exercise to obtain the following approximate limits on the sum $\sum_{1\le i<n} 1/i^2$ for small values of m.

m	Lower limit	Upper limit
1	$1 - 1/n$	∞
2	$1.5 - 1/n$	$2 - 1/(n-1)$
3	$1.58 - 1/n$	$1.75 - 1/(n-1)$
4	$1.61 - 1/n$	$1.69 - 1/(n-1)$

Give the exact values (as fractions) for all the coefficients in these four approximations.

4.5.1 The Euler Summation Formula

The Euler summation formula gives the difference between an integral and the corresponding sum. To derive the formula, let's start by comparing the area of one rectangle with the area under the corresponding part of the curve. Integration by parts is a technique that should be considered when trying to develop a formula relating the area under a curve to the height of the curve at the end points, because the formula contains both an integral and the value of a function at the end points. If we integrate $(x-i-1/2)f'(x)$, we get

$$\int_i^{i+1} \left(x-i-\frac{1}{2}\right) f'(x)\,dx = \left(x-i-\frac{1}{2}\right) f(x)\Big|_i^{i+1} - \int_i^{i+1} f(x)\,dx \quad (147)$$

$$= \frac{1}{2}[f(i+1)+f(i)] - \int_i^{i+1} f(x)\,dx. \quad (148)$$

That is, $\int_i^{i+1}(x-i-1/2)f'(x)\,dx$ is the difference between the area of the trapezoid with vertices $(i,0)$, $(i,f(i))$, $(i+1,f(i+1))$, and $(i+1,0)$, and the area under $f(x)$ between $x=i$ and $x=i+1$.

Next we would like to sum eq. (148) to relate the integral for the entire area to the value of the function at integer points. First, however, we need to rewrite $x-i-1/2$ in a way that removes the i dependence. Define the *sawtooth function* $\{x\}$ as

$$\{x\} = x \bmod 1 = x - \lfloor x \rfloor. \quad \mathbf{(149)}$$

Using the sawtooth function eq. (148) can be written as

$$\int_i^{i+1} \left(\{x\} - \frac{1}{2}\right) f'(x)\,dx = \frac{1}{2}[f(i+1) + f(i)] - \int_i^{i+1} f(x)\,dx. \tag{150}$$

Notice that we have removed the i dependence from the function on the left that is being integrated.

Now let's sum eq. (150) to obtain

$$\int_1^n \left(\{x\} - \frac{1}{2}\right) f'(x)\,dx = \frac{1}{2} \sum_{1 \le i < n} [f(i+1) + f(i)] - \int_1^n f(x)\,dx \tag{151}$$

$$= \sum_{1 \le i < n} f(i) + \frac{1}{2}[f(n) - f(1)] - \int_1^n f(x)\,dx. \tag{152}$$

The lower limit of 1 on the integral and summation is just a convenient lower limit. One can obtain similar results with any lower limit on the summation.

Eq. (152) can be used to compute sums from integrals or vice versa, provided you have a way to estimate the size of the left side. Since we are mainly interested in computing sums from integrals, we will rewrite eq. (152) as

$$\sum_{1 \le i < n} f(i) = \int_1^n f(x)\,dx - \frac{1}{2}[f(n) - f(1)] + \int_1^n B_1(\{x\}) f'(x)\,dx \tag{153}$$

where $B_1(x)$ is the polynomial $x - 1/2$.

To illustrate eq. (153), let's consider the sum $\sum_{1 \le i < n} 1/i$. Eq. (153) gives

$$\sum_{1 \le i < n} \frac{1}{i} = \int_1^n \frac{dx}{x} - \frac{1}{2}\left(\frac{1}{n} - 1\right) - \int_1^n B_1(\{x\}) \frac{dx}{x^2} \tag{154}$$

$$= \ln n + \frac{1}{2} - \frac{1}{2n} - \int_1^n B_1(\{x\}) \frac{dx}{x^2}. \tag{155}$$

To make further progress, we must approximate the integral in eq. (155). Notice that $B_1(\{x\})$ is between $-1/2$ and $1/2$ for any x, so we have

$$-\frac{1}{2} \int_1^n \frac{dx}{x^2} \le \int_1^n B_1(\{x\}) \frac{dx}{x^2} \le \frac{1}{2} \int_1^n \frac{dx}{x^2} \tag{156}$$

or

$$-\frac{1}{2}\left(1 - \frac{1}{n}\right) \le \int_1^n B_1(\{x\}) \frac{dx}{x^2} \le \frac{1}{2}\left(1 - \frac{1}{n}\right). \tag{157}$$

Therefore $\sum_{1 \le i < n} 1/i = \ln n + O(1)$. This is a much more accurate approximation than we got in Section 4.3, but in Section 4.5.4 we will obtain an even more accurate approximation.

Knuth [9, Section 1.2.11.2] was a source for this section. Also see Apostol [17, Section 9.8].

EXERCISES

1. Show that
$$\frac{2}{3}n^{3/2} - n^{1/2} + \frac{1}{3} \leq \sum_{1\leq i<n} i^{1/2} \leq \frac{2}{3}n^{3/2} - \frac{2}{3}.$$

2. Show that
$$\sum_{1\leq i<n} f(i) \leq \sum_{1\leq i<m} f(i) + \int_m^n f(x)\,dx - \frac{1}{2}(f(n) - f(m)) + \int_m^n B_1(\{x\})f'(x)\,dx.$$

3. Use the result of the previous exercise to obtain the following approximate limits on the sum $\sum_{1\leq i<n} 1/i^2$ for small values of m.

m	Lower limit	Upper limit
1	$1 - 1/n$	$2 - 1/n - 1/n^2$
2	$1.5 - 1/n$	$1.75 - 1/n - 1/n^2$
3	$1.58 - 1/n$	$1.69 - 1/n - 1/n^2$
4	$1.61 - 1/n$	$1.67 - 1/n - 1/n^2$

Give the exact values (as fractions) for all the coefficients in these four approximations.

4.5.2* The Generalized Euler Summation Formula

Higher-degree approximations can be obtained by integrating eq. (153) by parts. To do this we will need to integrate $B_1(x)$. As the integration by parts proceeds, it will give rise to a sequence of polynomials called *Bernoulli polynomials.* The Bernoulli polynomial $B_m(x)$ is defined recursively as $\int mB_{m-1}(x)\,dx$ with the constant of integration equal to 1. The *Bernoulli numbers* B_m, which occur as coefficients in the Bernoulli polynomials, are defined by the recurrence

$$B_0 = 1, \qquad B_n = \frac{-1}{n+1}\sum_{0\leq i\leq n-1}\binom{n+1}{i}B_i \quad \text{for } n \geq 1. \tag{158}$$

Except for B_1 all the Bernoulli numbers of odd index are zero. The first few Bernoulli numbers are given in Table 4.1.

The *Bernoulli polynomials* are given by

$$B_m(x) = \sum_i \binom{m}{i} B_i x^{m-i}. \tag{159}$$

(Notice that B_i refers to the ith Bernoulli number, while $B_i(x)$ is the ith Bernoulli polynomial.) For $m = 1$ this gives $B_1(x) = B_0x + B_1 = x - 1/2$, which is the polynomial we used in eq. (153). For $m > 1$, we have $B_m(1) = B_m = B_m(0)$, so for $m > 1$, $B_m(\{x\})$ is a continuous function of x.

$B_0 = 1$	$B_1 = -1/2$	$B_2 = 1/6$	$B_3 = 0$
$B_4 = -1/30$	$B_5 = 0$	$B_6 = 1/42$	$B_7 = 0$
$B_8 = -1/30$	$B_9 = 0$	$B_{10} = 5/66$	$B_{11} = 0$
$B_{12} = -691/2730$	$B_{13} = 0$	$B_{14} = 7/6$	$B_{15} = 0$
$B_{16} = -3617/510$	$B_{17} = 0$	$B_{18} = 43867/798$	$B_{19} = 0$

TABLE 4.1. Bernoulli numbers.

The derivative of $B_m(x)$ is given by

$$B'_m(x) = \sum_i \binom{m}{i}(m-i)B_i x^{m-i-1} = m\sum_i \binom{m-1}{i} B_i x^{m-i-1} \tag{160}$$
$$= mB_{m-1}(x). \tag{161}$$

For $m \geq 1$, we can use integration by parts to obtain

$$\frac{1}{m!}\int_1^n B_m(\{x\})f^{(m)}(x)\,dx$$
$$= \frac{1}{(m+1)!}\Big(B_{m+1}(1)f^{(m)}(n) - B_{m+1}(0)f^{(m)}(1)\Big) - \frac{1}{(m+1)!}\int_1^n B_{m+1}(\{x\})f^{(m+1)}(x)\,dx \tag{162}$$
$$= \frac{1}{(m+1)!}\Big(B_{m+1}f^{(m)}(n) - B_{m+1}f^{(m)}(1)\Big) - \frac{1}{(m+1)!}\int_1^n B_{m+1}(\{x\})f^{(m+1)}(x)\,dx. \tag{163}$$

Applying eq. (163) repeatedly to eq. (153) gives *Euler's general summation formula*:

$$\sum_{1\leq i<n} f(i) = \int_1^n f(x)\,dx + \sum_{1\leq i\leq m} \frac{B_i}{i!}\Big(f^{(i-1)}(n) - f^{(i-1)}(1)\Big) + R_m \tag{164}$$

where

$$R_m = \frac{(-1)^{m+1}}{m!}\int_1^n B_m(\{x\})f^{(m)}(x)\,dx. \tag{165}$$

The remainder R_m will be small when $B_m(\{x\})f^{(m)}(x)/m!$ is small. It is known that $|B_m(\{x\})| \leq |B_m|$ when m is even and that

$$\left|\frac{B_m(\{x\})}{m!}\right| < \left|\frac{4}{(2\pi)^m}\right| \tag{166}$$

so $B_m(\{x\})/m!$ does not cause any trouble. On the other hand for many functions $f^{(m)}(x)$ becomes very large for large m. In such cases, for fixed x, there is a best value of m at which R_m has its minimum value.

The most common way to use eq. (164) is to consider R_m as an error term. In such cases one tries to compute an upper bound on its value. When $f^{(2i+1)}(x)$ tends monotonically toward zero as x increases from 1 to n, we have

$$R_{2i} = \theta \frac{B_{2i+2}}{(2i+2)!}\left(f^{(2i+1)}(n) - f^{(2i+1)}(1)\right) \quad \text{where } 0 < \theta < 1, \tag{167}$$

so in this case the error has the same sign as, and is less than, the first discarded term (see Knopp [40]). If $f^{(2i)}(x)$ has a constant sign for $1 \le x \le n$, then

$$|R_{2i}| \le \left| \frac{B_{2i}}{(2i)!}\left(f^{(2i-1)}(n) - f^{(2i-1)}(1)\right)\right|, \tag{168}$$

so the remainder has a smaller absolute value than the last term computed.

Knuth [9, Section 1.2.11.2] is a source for this section. Also see Knopp [40, Section 64].

EXERCISES

1. Show that $\sum_{1\le i<n} i^{1/2} \ge \frac{2}{3}n^{3/2} - \frac{1}{2}n^{1/2} - \frac{601}{2880} + \frac{1}{24}n^{-1/2} + \frac{1}{2880}n^{-3/2}$ and that $\sum_{1\le i<n} i^{1/2} \le \frac{2}{3}n^{3/2} - \frac{1}{2}n^{1/2} - \frac{599}{2880} + \frac{1}{24}n^{-1/2} - \frac{1}{2880}n^{-3/2}$. Hint: Use the generalized Euler summation formula with $m = 4$.

2. Obtain the following approximate limits on the sum $\sum_{1\le i<n} 1/i^2$ by considering it as $\sum_{1\le i<k} 1/i^2 + \sum_{k\le i<n} 1/i^2$ and summing the first part exactly (for small values of k) and approximating the second part with the generalized Euler summation formula using $m = 4$.

k	Lower limit	Upper limit
1	$1.6333 - \frac{1}{2}n^{-2} - \frac{1}{6}n^{-3} + \frac{1}{30}n^{-5}$	$1.7000 - \frac{1}{2}n^{-2} - \frac{1}{6}n^{-3} - \frac{1}{30}n^{-5}$
2	$1.6448 - \frac{1}{2}n^{-2} - \frac{1}{6}n^{-3} + \frac{1}{960}n^{-5}$	$1.6469 - \frac{1}{2}n^{-2} - \frac{1}{6}n^{-3} - \frac{1}{960}n^{-5}$
3	$1.6449 - \frac{1}{2}n^{-2} - \frac{1}{6}n^{-3} + \frac{1}{7290}n^{-5}$	$1.6452 - \frac{1}{2}n^{-2} - \frac{1}{6}n^{-3} - \frac{1}{7290}n^{-5}$

Give the exact values (as fractions) for all the coefficients in these three approximations.

4.5.3* Functions with Slowly Varying Tails

When Euler's general summation formula is applied to a slowly varying function, you obtain an error term that becomes rapidly smaller as the order of the approximation is increased. Many interesting functions, however, are not quite so simple. The function may vary quite smoothly for large values of the argument, but vary rather rapidly for small values. In such cases if we try to use Euler's general summation formula directly, a problem arises. With a low-order formula we obtain a result that is moderately accurate, but as we increase the order, the accuracy does not improve much.

In such cases we have to be more clever in our application of Euler's summation formula. The error term, R_m, is actually a function of both m, the order of the summation, and n, the upper limit on the sum. We can write it as R_{mn} when we want to emphasize the dependence on both m and n. If R_{mn} does not go to zero rapidly as m increases, then it is useful to look at $\lim_{n\to\infty} R_{mn}$. For functions that become smooth for large values of n, this limit approaches a constant. If we write $R_{m\infty}$ for this limit, then $R_{mn} - R_{m\infty}$ will become small as n becomes large.

If we can find a way to calculate $R_{m\infty}$, then we are no longer troubled by the fact that R_{mn} does not become small as n becomes large; by subtracting off $R_{m\infty}$, we obtain a formula that does have a small error term.

This approach is illustrated in the next several sections.

4.5.4† Harmonic Numbers

The sum $\sum_{1\le i\le n} 1/i$ has been mentioned several times before. It occurs so frequently that it has been given a name: it is called the nth *harmonic number*:

$$H_n = \sum_{1\le i\le n} \frac{1}{i}. \tag{169}$$

We have already shown in Section 4.5.1 that $H_n = \ln n + O(1)$. Now let's apply Euler's general summation formula. We have $f(x) = 1/x$ and $f^{(m)}(x) = (-1)^m m!/x^{m+1}$, so

$$H_{n-1} = \ln n + \sum_{1\le i\le m} \frac{B_i}{i}(-1)^{i-1}\left(\frac{1}{n^i} - 1\right) + R_{mn}. \tag{170}$$

Unfortunately the error term on this formula does not become small rapidly as we increase m. Using the techniques of Section 4.5.2, we obtain that $H_n = \ln n + O(1)$ for each value of m. The constant for the big O goes down as m goes up, but no dramatic improvement is obtained.

The problem is that $1/i$ is a rapidly varying function of i for small values of i. It becomes quite smooth as i increases. We need to consider the value of the error as n becomes large and subtract off the limiting value, as suggested in Section 4.5.3. The next several paragraphs give the details of how to do this.

Since H_{n-1} is about equal to $\ln n$, let's now compute the limit as n goes to infinity of $H_{n-1} - \ln n$. The $1/n^i$ term in eq. (170) goes to zero, and R_{mn} approaches $\int_1^\infty B_m(\{x\})\,dx/x^{m+1}$. If we call this limit γ, we have

$$\gamma = \lim_{n\to\infty}(H_{n-1} - \ln n) = \sum_{1\le i\le m} \frac{B_i}{i}(-1)^i - \int_1^\infty \frac{B_m(\{x\})}{x^{m+1}}\,dx. \tag{171}$$

Notice that the value of the limit does not depend on m. The constant γ is called *Euler's constant* and its value is $0.57721\ldots$. Now

$$\sum_{1\le i\le m} \frac{B_i}{i}(-1)^i - \sum_{1\le i\le m} \frac{B_i}{i}(-1)^{i-1}\left(\frac{1}{n^i} - 1\right) = \sum_{1\le i\le m} \frac{(-1)^i B_i}{i n^i} \tag{172}$$

and

$$\int_1^\infty \frac{B_m(\{x\})}{x^{m+1}}\,dx - R_{mn} = \int_n^\infty \frac{B_m(\{x\})}{x^{m+1}}\,dx \tag{173}$$

so eq. (170) can be written as

$$H_{n-1} = \ln n + \gamma + \sum_{1\le i\le m} \frac{(-1)^{i-1}B_i}{in^i} + \int_n^\infty \frac{B_m(\{x\})}{x^{m+1}}\,dx. \tag{174}$$

The integral in eq. (174) is $O(1/n^m)$ for fixed m. Adding $1/n$ to both sides of eq. (174) and setting $m = 8$ gives

$$H_n = \ln n + \gamma + \frac{1}{2n} - \frac{1}{12n^2} + \frac{1}{120n^4} - \frac{1}{252n^6} + O\left(\frac{1}{n^8}\right). \tag{175}$$

Eq. (175) is a good illustration of our earlier remark that asymptotic equations often *cannot* be extended to form convergent infinite series. The first few B_i are successively smaller in absolute value (for even i), but for large even i, B_i is a large number [about $2i!/(2\pi)^i$] so the sum in eq. (174) is divergent when m goes to infinity (for fixed n).

The most common way that harmonic numbers arise in algorithm analysis is from finding the largest of n numbers (Algorithm 3.2). When you look at the numbers from first to last, the probability that the first number is the largest so far is 1, the probability that the second is the largest so far is $\frac{1}{2}$, etc. The sum of these numbers gives the expected number of new maximums, so the average number of new maximums is H_n.

It is no harder to sum harmonic numbers than it is to compute them. From the definition of H_i, and interchanging the order of summation, we have

$$\sum_{1\le i\le n} H_i = \sum_{1\le i\le n}\sum_{1\le j\le i}\frac{1}{j} = \sum_{1\le j\le n}\sum_{j\le i\le n}\frac{1}{j} = \sum_{1\le j\le n}\frac{n+1-j}{j} \tag{176}$$

$$= (n+1)\sum_{1\le j\le n}\frac{1}{j} - \sum_{1\le j\le n} 1 = (n+1)H_n - n. \tag{177}$$

Notice that this technique of interchanging the order of summation will work for any sum of the form

$$\sum_{1\le i\le n} p(i)H_i \tag{178}$$

where $p(i)$ is any polynomial in i.

Knuth [9, Sections 1.2.7 and 1.2.11.2] was a source for this section. Also see Knopp [40, Section 64].

EXERCISES

1. [See 40] Using $n = 4$ and $m = 8$ in eq. (174), evaluate γ with an accuracy of 10^{-6}. Hint: Show that the integral in eq. (174) is no more than 10^{-6} in this case. Use the numerical value of the logarithm and of the sum to solve for γ.
2. If you use the method of the last exercise but increase n, how accurate an approximation to the value of γ do you obtain?
3. Simplify $\sum_{1\le i\le n} \binom{i}{m} H_i$. Hint: Use summation by parts.
4. Find an approximate value for $\sum_{1\le i\le n} i^{-2}$ that is accurate to $O(n^{-5})$. You may use the result ([40, Section 32]) $\sum_{i\ge 1} i^{-2k} = \frac{(-1)^{k+1}(2\pi)^{2k}B_{2k}}{2(2k)!}$ for integer k. In particular $\sum_{i\ge 1} i^{-2} = \frac{1}{6}\pi^2$.
5. Use the equation $\sum_{i\ge 1} i^{-2k} = \frac{(-1)^{k+1}(2\pi)^{2k}B_{2k}}{2(2k)!}$ to obtain upper and lower bounds on the size of B_{2k} that differ by no more than a factor of 2.

4.5.4.1† *Random Hashing—Finding Items*

In Section 2.4.1 we computed the average time required by the Random Hashing Algorithm to look for items that were not in the hash table. Now we will compute the average time to look for items that are in the table.

The probability p_i that it takes i steps to look for a random item not in the table is given by eq. (2.86). To emphasize the fact that p_i is a function of k, the number of items in the table, we will call it $p_i(k)$ in this section. We now need the probability that it takes i steps to look up a random item that is in the table. Let's call this probability q_i. For each item in the table, the number of steps needed to look it up is the same as the number of steps that were needed to insert it. Let N be the table size. When we look up a random item the probability that it was the first item put into the hash table is $1/k$. Likewise the probability that it was the jth item put into the table is $1/k$ for $1 \le j \le k$. If the item was the first put in the table, then the probability that i steps were needed to look it up is $p_i(0)$, if it was the second, the probability that i steps were needed is $p_i(1)$, etc. The probability that a random item was the jth item put into the table and that it takes i steps to look it up is just $p_i(j)/k$ for $0 \le j \le k-1$ (see eq. (1.44)). Using eq. (2.86) for $p_i(k)$, the probability we need, q_i, is given by

$$q_i = \sum_{0\le j\le k-1} \frac{p_i(j)}{k} = \frac{1}{k} \sum_{0\le j\le k-1} \left(\frac{j}{N}\right)^{i-1} \left(1 - \frac{j}{N}\right). \tag{179}$$

There is no obvious way to simplify this sum.

The average number of times that the inner loop of the algorithm is done is

$$\sum_{i\ge 1} iq_i = \frac{1}{k} \sum_{i\ge 1} i \sum_{0\le j\le k-1} \left(\frac{j}{N}\right)^{i-1} \left(1 - \frac{j}{N}\right). \tag{180}$$

Although there is no obvious way to simplify the sum over j, we do know how to do the sum over i. Interchanging the order of summation (since the sum has positive terms, it is absolutely convergent, so the interchanging of the order of summation causes no problems) and using eq. (2.75) gives

$$A = \frac{1}{k} \sum_{0 \le j \le k-1} \sum_{i \ge 1} i \left(\frac{j}{N}\right)^{i-1} \left(1 - \frac{j}{N}\right) = \frac{1}{k} \sum_{0 \le j \le k-1} \frac{N}{N-j}. \tag{181}$$

Now by changing the range of the summation

$$A = \frac{N}{k} \sum_{N-k+1 \le j \le N} \frac{1}{j} = \frac{N}{k}(H_N - H_{N-k}). \tag{182}$$

This is the exact answer in terms of harmonic numbers, and it is the best equation for calculating the performance of the algorithm for small values of k and N. When N is large and $N-k$ is small, you should expand the first harmonic number into a logarithm. When both N and $N-k$ are large, you should expand both harmonic numbers.

We will now consider the case where N and $N-k$ are large more carefully. Using the approximation for harmonic numbers, we obtain

$$A = \frac{N}{k}\left[\ln N - \ln(N-k) + \frac{1}{2N} - \frac{1}{2(N-k)} + O\left(\frac{1}{N^2}\right) + O\left(\frac{1}{(N-k)^2}\right)\right] \tag{183}$$

$$= \frac{N}{k}\left[-\ln\left(1 - \frac{k}{N}\right) - \frac{k}{2N(N-k)} + O\left(\frac{1}{(N-k)^2}\right)\right]. \tag{184}$$

This is the best form of the answer when N and $N-k$ are large but k/N is not small.

If N is large and k/N is small (in which case $N-k$ is also large), we can use the power series for the logarithm [eq. (48)] to obtain

$$A = 1 + \frac{k}{N} - \frac{1}{2(N-k)} + O\left(\frac{k^2}{N^2}\right) + O\left(\frac{N}{k(N-k)^2}\right). \tag{185}$$

EXERCISES

1. If random hashing is used on a large table that is 80 percent full ($k/N = .80$), how much time, on the average, is used to look items up?

2. Going from eq. (183) to eq. (184), what happened to the $O(1/N^2)$ term?

3. Why are two big O terms needed for eq. (185)?

4. Give an asymptotic expression for the average number of times Step 2 of the Random Hashing Algorithm is done. Your expression should be suitable for the case N is large and $N-k$ is small.

5. Compute the average time required for looking up an item known to be in the table when Nonrepeating Random Hashing is used.

4.5.5† Stirling's Approximation

To approximate the factorial function it is interesting to look at the logarithm of $n!$. Since $n!$ is the product of the integers from 1 to n, $\ln n! = \ln 1 + \ln 2 + \cdots + \ln n$. Although $n!$ is a rapidly varying function of n, the sum for its logarithm is a slowly varying function of n. Therefore the Euler summation formula can be used to approximate the logarithm of $n!$. An accurate approximation to the logarithm of $n!$ leads to an approximation for $n!$ itself which has a *small relative error, but not a small absolute error.*

Using $f(x) = \ln x$, we have $f^{(m)}(x) = (-1)^{m+1}(m-1)!/x^m$ and $\int_1^x f(y)\,dy = x \ln x - x + 1$. Euler's general summation formula, eq. (164), gives

$$\ln(n-1)! = n \ln n - n + 1 - \frac{1}{2}\ln n + \sum_{1<i\le m} \frac{B_i(-1)^i}{i(i-1)}\left(\frac{1}{n^{i-1}} - 1\right) + R_{mn}, \quad (186)$$

where

$$R_{mn} = \frac{1}{m}\int_1^n \frac{B_m(\{x\})}{x^m}\,dx \quad \text{for } m \ge 2. \quad (187)$$

Again we need to look at the error in detail to obtain accurate results. The term $\ln i$ varies slowly with i for large i, but rapidly for small i. We will proceed as we did for harmonic numbers. Since $1/n^{i-1}$ approaches zero as n goes to infinity (for $i > 1$), taking the limit of eq. (186) gives

$$\lim_{n\to\infty}\left(\ln n! - n \ln n + n - \frac{1}{2}\ln n\right) = 1 + \sum_{1<i\le m} \frac{B_i(-1)^{i+1}}{i(i-1)} + \lim_{n\to\infty} R_{mn}. \quad (188)$$

[Several steps of the derivation of eq. (188) have been omitted, but they are similar to the corresponding steps in Section 4.5.4.] Examining the limit of R_{2n} as n goes to infinity makes it clear that the limit on the right exists, so the limits in eq. (188) are finite. We will compute the value of the limits in a moment, but for now let's use σ for the right side of eq. (188). We have

$$\ln n! = \left(n + \frac{1}{2}\right)\ln n - n + \sigma + \sum_{1<i\le m} \frac{B_i(-1)^i}{i(i-1)n^{i-1}} + O\left(\frac{1}{n^m}\right). \quad (189)$$

Setting $m = 3$ and taking exponentials gives

$$n! = e^{\sigma}\sqrt{n}\left(\frac{n}{e}\right)^n \exp\left(\frac{1}{12n} + O\left(\frac{1}{n^3}\right)\right). \quad (190)$$

Before simplifying eq. (190) further, let's investigate the value of σ. Consider the expression $\sqrt{n}(2n)!/(4^n n! n!)$. By dividing a large factorial ($(2n)!$) by two smaller factorials (both $n!$), we have a ratio whose value, as expressed using eq. (190), depends on σ. The other factors are carefully chosen so that the ratio will approach a finite limit as n goes to infinity. Using eq. (190) we have

$$\frac{\sqrt{n}(2n)!}{4^n n!n!} = \frac{\sqrt{n}e^\sigma\sqrt{2n}(2n/e)^{2n}\exp(O(1/n))}{4^n e^\sigma\sqrt{n}(n/e)^n\exp(O(1/n))e^\sigma\sqrt{n}(n/e)^n\exp(O(1/n))} \tag{191}$$

$$= \frac{\sqrt{2}}{e^\sigma}\left(1+O\left(\frac{1}{n}\right)\right) \tag{192}$$

so the limit as n goes to infinity of this expression is $\sqrt{2}/e^\sigma$. If we can find another way to evaluate the limit, then we can compute e^σ and finish developing Stirling's approximation.

If we multiply out the left side of eq. (191), we get

$$\frac{\sqrt{n}(2n)!}{4^n n!n!} = \sqrt{n}\,\frac{1\cdot 2\cdot 3\cdots 2n}{2\cdot 4\cdot 6\cdots 2n\cdot 2\cdot 4\cdot 6\cdots 2n} = \sqrt{n}\,\frac{1\cdot 3\cdot 5\cdots(2n-1)}{2\cdot 4\cdot 6\cdots(2n)}. \tag{193}$$

If we square eq. (193), we can write the result as

$$n\left(\frac{(2n)!}{4^n n!n!}\right)^2 = \frac{1\cdot 3}{2\cdot 2}\;\frac{3\cdot 5}{4\cdot 4}\;\frac{5\cdot 7}{6\cdot 6}\;\cdots\;\frac{(2n-1)(2n+1)}{(2n)(2n)}\;\frac{n}{(2n+1)} \tag{194}$$

$$= \frac{n}{(2n+1)}\prod_{1\le i\le n}\left(\frac{(2i-1)(2i+1)}{(2i)(2i)}\right). \tag{195}$$

Dividing out the 2's gives

$$n\left(\frac{(2n)!}{4^n n!n!}\right)^2 = \frac{n}{(2n+1)}\prod_{1\le i\le n}\left(\frac{(i-1/2)(i+1/2)}{i\cdot i}\right). \tag{196}$$

If we now divide top and bottom by a pair of n factorials and by some powers of n, we get four factors whose limits (as n goes to infinity) are Gamma functions (see eq. (3.22)) and one factor whose limit is 1/2:

$$n\left(\frac{(2n)!}{4^n n!n!}\right)^2 = \frac{n}{(2n+1)}\,\frac{\dfrac{\prod_{1\le i\le n}(i-1/2)}{n^{1/2}n!}\,\dfrac{\prod_{1\le i\le n}(i+1/2)}{n^{3/2}n!}}{\dfrac{\prod_{1\le i\le n} i}{nn!}\,\dfrac{\prod_{1\le i\le n} i}{nn!}}. \tag{197}$$

Now if we take the limit as n goes to infinity, express the limits of the products as Gamma functions, and use the value of $\Gamma(1/2)$ [eq. (3.38)], we get

$$\lim_{n\to\infty} n\left(\frac{(2n)!}{4^n n!n!}\right)^2 = 2\frac{\Gamma(1)\Gamma(1)}{\Gamma(1/2)\Gamma(3/2)} = \frac{1}{\pi}. \tag{198}$$

Comparing the square root of eq. (198) with the limit of eq. (192) gives

$$e^\sigma = \sqrt{2\pi} \tag{199}$$

which is the constant we need for Stirling's formula:

$$n! = \sqrt{2\pi n}\left(\frac{n}{e}\right)^n\exp\left(\frac{1}{12n}+O\left(\frac{1}{n^3}\right)\right). \tag{200}$$

Expanding the exponential in a power series gives

$$n! = \sqrt{2\pi n}\left(\frac{n}{e}\right)^n \left(1 + \frac{1}{12n} + O\left(\frac{1}{n^2}\right)\right). \quad \textbf{(201)}$$

Using the same approach, but carrying the expansions out to more terms, gives

$$n! = \sqrt{2\pi n}\left(\frac{n}{e}\right)^n \left(1 + \frac{1}{12n} + \frac{1}{288n^2} - \frac{139}{51840n^3} - \frac{571}{2488320n^4} + O\left(\frac{1}{n^5}\right)\right). \quad (202)$$

Knuth [9, Section 1.2.11.2] was a source for this section. Also see Knopp [40, Section 64].

4.6 ALTERNATING SUMMANDS

Many sums have terms that alternate in sign. Such sums are often more difficult to approximate than sums where all the terms have the same sign. One approach is to consider the sum of the positive and negative sums separately, sum each part, and combine the results. This often does not work well, however. The sums of the two parts are likely to be nearly the same, so unless you know the value for each part very accurately, you will learn little about the value of the difference.

There are several other approaches for such sums that merit special consideration. First, one can use the technique of the previous section and try to find a portion of the sum that is a good approximation to the entire sum. If you have a sum where the absolute value of the summand varies rapidly with the summation index, this approach will often work.

For such sums where the absolute value of the summand is monotonically decreasing, we have

$$\sum_{0\le i\le k}(-1)^i a_i > \sum_{0\le i\le n}(-1)^i a_i > \sum_{0\le i\le k+1}(-1)^i a_i$$
$$\text{where } a_i > a_{i+1}, \quad k \text{ is even}, \quad \text{and } 0 < k < n-2. \quad \textbf{(203)}$$

The way to remember eq. (203) is that the first k terms of an alternating series of terms that decrease in absolute value approximate the complete sum with an error that has the same sign as the first neglected term and with an error that has an absolute value that is less than the absolute value of the first neglected term. Notice that if the terms are decreasing rapidly, this gives an approximation with low relative error, but if the terms decrease slowly, then the relative error is large.

Consider the sum $S = \sum_{i\ge1}(-1)^i/i$. By using eq. (203) with small values of k, we get

$$S > -1, \quad S < -1/2, \quad S > -5/6, \quad S < -7/12, \quad \text{etc.} \quad (204)$$

The second approach is to add the terms in pairs (add each term with even index to the term with index one higher) and then add the sums of the pairs. When the terms vary slowly, this often leads to a better-behaved sum to work with. For example, if the original sum has terms with monotonic absolute values, then the sum of pairs will be a sum of terms which all have the same sign.

Alternating sums are discussed by Apostol [18, Section 9.16 and 17, Section 12.7], de Bruijn [25, Section 3.11], and Knopp [40, Sections 15, 34, 35]. Bender [68, Section 4] gives techniques for approximating sums that arise from inclusion and exclusion. These sums, which are important in many applications, often require more sophisticated techniques than the ones given in this section.

EXERCISES

1. How many terms do you need in order to compute the value of $\sum_{i\geq 1}(-1)^i/i$ to an accuracy of ± 0.01 using eq. (203)?

2. Show that $\displaystyle\sum_{i\geq 1}\frac{(-1)^i}{i} = -\sum_{j\geq 1}\frac{1}{2j(2j-1)} = -1+\sum_{j\geq 1}\frac{1}{2j(2j+1)}$. Hint: Combine pairs of terms in the sum over i. For the second formula do not include the first term in the pairing.

3. Use the result of the previous exercise to show that

$$-\sum_{1\leq j\leq n}\frac{1}{2j(2j-1)} \leq \sum_{i\geq 1}\frac{(-1)^i}{i} \leq -1+\sum_{1\leq j\leq n}\frac{1}{2j(2j+1)}.$$

How large does n need to be to approximate the value of the sum over i to an accuracy of ± 0.01?

4. Prove eq. (203). Hint: Use proof by induction starting with the case $2k = n$ or $2k+1 = n$.

4.7* ASYMPTOTIC ITERATION

Asymptotic iteration is concerned with solving equations of the form

$$x = f(x,t) \tag{205}$$

for x as t becomes large. The basic method consists of guessing a value for x [i.e., $x = g_0(t)$], plugging the guess into the right side to calculate a new guess for x [i.e., $x = g_1(t) = f(g_0(t),t)$] and repeating the process. If you are lucky, the guesses will converge. Once you have a good guess, then you can try to prove that it is the solution of your equation to within some accuracy.

If your equation has more than one root, you will probably be interested in one particular root. In such cases you will also need to prove that your iteration converges to the root that you are interested in.

If your iteration does not converge, then there are several things you can try. First, you can try to find a better starting value. Iterative techniques often work only when you have a good starting value. Second, you can find a new form for the equation you are solving. Iterative techniques work only when $f(x,t)$ is a slowly varying function of x.

Let's consider the equation

$$xe^x = t. \tag{206}$$

The e^x term is likely to be the most important term on the left side of the equation, so the first step is to rewrite eq. (206) in a form that emphasizes this term. Taking the logarithm of eq. (206), and rearranging terms, this gives

$$x = \ln t - \ln x. \tag{207}$$

The right side is now a slowly varying function of x, so there is some hope that asymptotic iteration will lead to a solution. Eq. (207) suggests that $x = \ln t$ may be a good first guess.

Using the guess $x_0 = \ln t$ gives

$$x_1 = \ln t - \ln\ln t. \tag{208}$$

This is similar to the first guess, so we appear to be making progress. Now let's plug this guess in. We get

$$x_2 = \ln t - \ln(\ln t - \ln\ln t). \tag{209}$$

Again the new guess is close to the old one. Since we are concerned with large t, this guess can be simplified to

$$x_2 = \ln t - \ln\ln t - \ln\left(1 - \frac{\ln\ln t}{\ln t}\right) = \ln t - \ln\ln t + O\left(\frac{\ln\ln t}{\ln t}\right). \tag{210}$$

We can iterate some more if we want additional accuracy.

Now it is time to try to prove that we have a solution. (If the iteration had not converged, it would instead be time to try to find a better way to iterate.) Let's try to show that $x = \ln t + O(\ln\ln t)$ is a solution. To do this, we plug $x = \ln t + a\ln\ln t$ into eq. (207) and compare the left and right sides as we vary a. We get

$$\ln t + a\ln\ln t \quad \text{vs.} \quad \ln t - \ln(\ln t + a\ln\ln t), \tag{211}$$

$$\ln t + a\ln\ln t \quad \text{vs.} \quad \ln t - \ln\ln t - \ln\left(1 + \frac{a\ln\ln t}{\ln t}\right), \tag{212}$$

$$\ln t + a\ln\ln t \quad \text{vs.} \quad \ln t - \ln\ln t - \left(\frac{a\ln\ln t}{\ln t}\right) + O\left(\left(\frac{a\ln\ln t}{\ln t}\right)^2\right). \tag{213}$$

If a is -1, then the left side of eq. (213) is smaller than the right side (for large t), while if a is above -1, then the left side is larger. Thus the two sides are equal for a value of a that goes to -1 as t goes to infinity. This gives

$$\ln t - \ln\ln t \le x < \ln t - \tfrac{1}{2}\ln\ln t, \tag{214}$$

for large t, which implies $x = \ln t + O(\ln\ln t)$.

Asymptotic iteration is discussed by de Bruijn [25, Chapter 8] and Greene and Knuth [6, Section 4.1.2].

EXERCISES

1. Put eq. (206) in the form $x = te^{-x}$ and try iteration with the initial guess $x = t$. Does this approach lead to a solution of eq. (206)? If not, what is the problem? In this exercise it is best to proceed fairly informally so that you can see the overall picture rather than getting bogged down in details. Consider the first few formulas of your iteration from the same point of view that was used when solving eq. (206) in the text.
2. Show that $x = \ln t - \ln\ln t - O(\ln\ln t/\ln t)$ is a solution of eq. (207).
3. Solve the equation $x^2 - \ln x = u$ for x as a function of u to accuracy $O(\ln u/u^{3/2})$.

4.8* ASYMPTOTICS AND MEASUREMENTS

Many serious uses of approximations require that bounds on the errors of the approximations be calculated. An asymptotic result does not say what the error is at any point. When necessary, the approach used to derive the asymptotic estimate can usually be refined to also provide an error bound. Usually much additional calculation is required to obtain good error bounds in this way.

Another approach is to measure the error at selected points and to deduce the error bounds from the measurements. Thus if your theory says that $f(x) = 1+O(1/x)$ and your measurement of $f(100)$ gives 1.0201, you may conclude that the constant for the big O should be about 2. This approach has an important weakness: there is no theoretical justification for the assumption that the error at $x = 100$ is representative of the error at other points. Perhaps you were lucky and your formula just happens to work well at $x = 100$. Suppose you make measurements at several points, say, $f(100) = 1.02010$, $f(200) = 1.01005$, $f(300) = 1.00668$, $f(400) = 1.00501$. Now you can observe both the size of the error and its variation. If the rate of variation fits your asymptotic formula, then you are not likely to be fooled when you use the data to deduce the size of the coefficients in the asymptotic formula. (It can still happen, however.)

With asymptotic formulas the general form of the formula is usually relatively easy to obtain, while much computational effort is needed to obtain the coefficients and error bounds. Therefore the idea of using measurements along with the analysis is attractive, although it does not replace the need for mathematical proofs of important results. One additional consideration is that mathematical analyses do sometimes contain errors (although they are never *supposed* to contain errors). Often a few measurements will clearly demonstrate the existence of such errors.

When doing an analysis for a practical application, it is important to demonstrate that the analysis applies to the intended application. For example, if your result is $f(x) = 1 + O(1/x)$ and you are interested in values of x around 100, then everything is probably okay if the constant implied by the big O for $x = 100$

is around 1, but your result may be of little importance if the implied constant is around 1000. Although these practical questions can be answered by analysis, it is usually much easier to answer them by measurement.

The role of measurements in analyzing algorithms is somewhat similar to the role of testing in developing correct programs. It is difficult to develop a correct program if testing is the only technique that you use to ensure correctness. On the other hand, it is much more difficult to develop a correct program with no testing at all than by using testing along with other methods. In this book we will concentrate on what you can learn about program performance by analysis of the program, but often combining some of these mathematical techniques with appropriate measurements is the easiest way to learn about the program's behavior.

CHAPTER 5

Simple Linear Recurrences

A *recurrence equation* is an equation that gives the value of a function at a point in terms of its value at other points. For example, the equation

$$T(n) = T(n/2) + 1 \tag{1}$$

gives the value of $T(n)$ in terms of the value of $T(n/2)$.

A recurrence equation by itself does not completely determine a function. We also need the value of the function at some specified points. Such values are called *boundary values.* If you have a boundary value, such as $T(1) = 0$, then you can use eq. (1) to calculate $T(n)$ for other values of n. Thus

$$T(2) = T(1) + 1 = 1, \tag{2}$$

$$T(4) = T(2) + 1 = 2, \tag{3}$$

$$T(8) = T(4) + 1 = 3, \tag{4}$$

$$T(16) = T(8) + 1 = 4. \tag{5}$$

Recurrence equations arise naturally in the analysis of many algorithms. Eq. (1), for example, comes from the analysis of the Binary Search Algorithm. The most common source of such equations is *divide and conquer algorithms*, which divide the original problem into smaller problems. For example, in the Binary Search Algorithm, the original problem of finding one of n numbers in a table is converted into the problem of finding one of $n/2$ numbers that are in one-half of the table.

Whenever you have a recurrence equation that describes the running time of an algorithm, it is a good idea to first use it to generate a table of running times for small values of the parameters. This will let you become familiar with the nature of the recurrence you are working with. There are several additional benefits. First, you may find that you cannot generate a set of values from your equations, in which case your equations are incomplete; they do not completely specify the solution of your problem. Second, you can compare the numbers you generate from the equations with numbers you generate by direct inspection of the algorithm. If the numbers are different, there is a mistake in the recurrence. In either of these two cases, you will want to do more work on obtaining the correct set of equations before you spend time solving them.

If your equations do generate correct answers, then you should inspect the answers to see if they have a simple pattern. (Sometimes very complex equations have very simple solutions.) If you recognize a simple pattern to the solution, then you will want to prove that the pattern gives the answer in all cases. Usually proof by induction is the best way to construct the proof. When you do not see a simple pattern, then you will want to use the techniques in this and the next three chapters. Even then your initial calculations are useful. They can be used to check your final answer and intermediate results (see Section 7.1.2.1, for example).

The rest of this chapter is concerned with systematic techniques for finding closed-form solutions and approximations to solutions of the simpler types of recurrence equations. The next section introduces the notation for recurrence equations and gives a brief discussion of the structure of their solutions. You may wish to read it quickly and then *refer back to it as you need to understand it in more detail.*

5.1 NOTATION

There are two major types of solutions of recurrence equations. A *particular solution* is a specific function with no arbitrary constants that satisfies the equation. Usually we are interested in the particular solution that satisfies some additional condition. For example, if we have the boundary value $T(1) = 0$, then the solution of eq. (1) is $T(n) = \lg n$ for $n = 2^k$, integer k. [The equation and this boundary value do not determine the value of $T(n)$ for $n \neq 2^k$.] If we are given the boundary value $T(1) = 1$, then the solution is $T(n) = \lg n + 1$ for $n = 2^k$, integer k.

The other type of solution is a *general solution*, which is actually the set of all solutions of the recurrence equation (see some technical qualifications below). If you have a general solution of a recurrence equation, then you can usually use given boundary conditions to select the particular solution that fits those boundary conditions. A general solution is usually given in the form of a function with arbitrary constants, where the function satisfies the recurrence for each set of values for the constants. For example, the general solution of eq. (1) is $T(n) = \lg n + C$, where C is a constant. The two particular solutions in the previous paragraph correspond to $C = 0$ and $C = 1$.

In what follows we will sometimes use *function notation* [as in $T(n)$], and other times *sequence notation* (T_n). We use the two notations interchangeably, according to what is convenient in each case. Often equations are easier to read in sequence notation, but when we have several solutions of an equation, we need to use the subscript position to differentiate them. Thus later we will use $B_i(n)$ to indicate the ith solution of an equation.

A recurrence equation in which the value of the function depends on its value at one previous point is called an *extended first order recurrence equation.* An extended first order recurrence has the form

$$f(n, T_n, T_{g(n)}) = 0, \tag{6}$$

where f is an arbitrary function and g is normally an integer valued function. Eq. (1) is an example of an extended first order recurrence equation. A *first order* recurrence has the form

$$f(n, T_n, T_{n-1}) = 0. \tag{7}$$

In other words, a first order equation is an extended first order equation in which $g(n)$ is the function $n-1$.

If the recurrence equation has the form

$$f(n, T_n, T_{g(n)}, T_{g(g(n))} \cdots T_{g^{[k]}(n)}) = 0 \tag{8}$$

(where f and g are as before) for some fixed k (and does not have this form for any smaller value of k), then the equation is an extended recurrence equation of *order* k. [The notation $g^{[k]}(n)$ means $g(g(\cdots g(n) \cdots))$, where there are k applications of the function g.] A kth *order* equation has the form

$$f(n, T_n, T_{n-1}, \ldots T_{n-k}) = 0. \tag{9}$$

In other words, a kth order equation is an extended k^{th} order equation where $g(n)$ is the function $n-1$.

If the recurrence equation has the form

$$f(n, T_n, T_{n-1}, \ldots, T_0) = 0 \tag{10}$$

[and does not have the form of eq. (9) for any fixed k], then it is a *full-history* recurrence equation.

A *linear recurrence equation* is a recurrence equation where the unknown function appears only to the first power. A linear recurrence equation of *order* k has the form

$$\sum_{0 \le i \le k} a_i(n) T_{n-i} = b(n), \tag{11}$$

where $a(n)$ and $b(n)$ are functions of n. If $b(n) = 0$ for all n, then the equation is *homogeneous.* Associated with any linear equation is a homogeneous equation, which is obtained by replacing $b(n)$ with 0. A *linear full-history* recurrence equation has the same form as eq. (11), but with k replaced by n.

A second form that we will use for linear recurrence equations is

$$\sum_{0 \le i \le k} a'_i(n) T_{n+i} = b'(n). \tag{12}$$

This form relates $k+1$ values with indices from n to $n+k$ where as the previous form relates $k+1$ values with indices from $n-k$ to n. To convert between the form of eq. (11) and the form of eq. (12), you use the equations

$$a'_i(n) = a_{k-i}(n+k) \tag{13}$$
$$b'(n) = b(n+k) \tag{14}$$

or

$$a_i(n) = a'_{k-i}(n-k) \tag{15}$$
$$b(n) = b'(n-k). \tag{16}$$

If you are applying formulas that give the solutions of linear recurrence equations, you must be sure to check which form the formula applies to. Otherwise you will get the wrong answer. We will usually use the first form for first order equations and with generating functions. The second form, however, is more convenient for operator methods and for forming characteristic equations.

The general solution of a kth order linear recurrence equation is a function of the form

$$T(n) = \sum_{1\le i\le k} c_i G_i(n) + P(n), \tag{17}$$

where the c_i are arbitrary constants, the G_i are k linearly independent solutions of the homogeneous recurrence equation [eq. (11) with $b(n)$ set to zero], and $P(n)$ is a particular solution of the original recurrence equation. For example, the general solution of

$$T_n = 5T_{n-1} - 6T_{n-2} + 4 \tag{18}$$

is

$$T_n = c_1 2^n + c_2 3^n + 2. \tag{19}$$

Here $k = 2$, $G_1(n) = 2^n$, $G_2(n) = 3^n$, and the particular solution is 2. Of course, the G_i are not unique when $k > 1$ (any set that works can be replaced by any independent linear combination of the original set), and $P(n)$ is not unique. We will see in Section 5.3 and following how to find such solutions.

If you have the general solution of a recurrence equation, and some boundary values, then you can use the boundary values to solve for the c_i in eq. (19). It takes k nondegenerate boundary values to solve for all the c_i. For example, if we are given eq. (18) along with the boundary conditions $T_0 = 5$ and $T_1 = 7$, then by plugging $n = 0$ and $n = 1$ into eq. (19) we obtain

$$5 = c_1 + c_2 + 2 \tag{20}$$
$$7 = 2c_1 + 3c_2 + 2, \tag{21}$$

which has the solution $c_1 = 4$, $c_2 = -1$.

A kth order nonlinear recurrence equation can have both general solutions that depend on k arbitrary constants and *singular solutions* that depend on fewer than k constants. Consider the first order nonlinear equation

$$T_{n+1} = n(T_{n+1} - T_n) - (T_{n+1} - T_n)^2. \tag{22}$$

This equation has the general solution $T_n = cn+c^2$, as you can verify by plugging it into eq. (22). In addition it has the singular solution $T_n = (n^2-1)/4$, which you can also verify by plugging into eq. (22).

The notation and general principles that we have given so far are comparable to corresponding principles and definitions for differential equations. To emphasize this close correspondence between the two fields, recurrence equations are sometimes called *difference equations.* The *first difference* of T_n is given by $\Delta T_n = T_{n+1} - T_n$, and the ith *difference* is obtained by $\Delta^i T_n = \Delta(\Delta^{i-1}T_n) = \Delta^{i-1}T_{n+1} - \Delta^{i-1}T_n$. The ith difference is analogous to the ith derivative.

In the following sections we will present a number of techniques for solving recurrences and apply them to a variety of problems. It is more important to master the techniques than to learn a catalog of solutions. The techniques usually apply to many different problems, and most of the techniques are easier to remember than the particular results. You are advised to work the problems along with the text in order to get a feeling for how the techniques are applied. Once you understand the techniques, then you will probably find it useful to start remembering the particular formulas that you use frequently.

An introduction to solving recurrence equations is given in Bender and Orszag [21, Chapter 2] and Lueker [110]. Greene and Knuth [6, Chapters 2 and 3] cover the more advanced recurrence equations that are of particular interest to computer scientists. One of the best of the advanced classical books on the subject is Milne-Thomson [43]. We drew heavily on all four of these books when writing this and the following three chapters. Knuth [9, Sections 1.2.8 and 1.2.9] has a brief introduction to the subject.

5.2 LINEAR FIRST ORDER RECURRENCES

The basic form of the linear first order recurrence equation is

$$T_n = a_n T_{n-1} + b_n. \tag{23}$$

The usual way to solve this equation is to repeatedly replace the T on the right side with the value given by eq. (23):

$$\begin{aligned}
T_n &= a_n T_{n-1} + b_n && (24)\\
&= a_n(a_{n-1}T_{n-2} + b_{n-1}) + b_n && (25)\\
&= a_n(a_{n-1}(a_{n-2}T_{n-3} + b_{n-2}) + b_{n-1}) + b_n && (26)\\
&= a_n(a_{n-1}(a_{n-2}(a_{n-3}T_{n-4} + b_{n-3}) + b_{n-2}) + b_{n-1}) + b_n && (27)\\
&\vdots\\
&= a_n(a_{n-1}(a_{n-2}(a_{n-3}(\cdots(a_1T_0 + b_1)\cdots)\\
&\qquad + b_{n-3}) + b_{n-2}) + b_{n-1}) + b_n. && (28)
\end{aligned}$$

Multiplying out eq. (28) gives

$$T_n = T_0 \prod_{1\le i\le n} a_i + \sum_{1\le i\le n} b_i \prod_{i<j\le n} a_j. \tag{29}$$

Notice that T_0 serves as the one arbitrary constant needed to specify the general solution of a first order recurrence equation. Eq. (29) gives a closed-form solution of the original recurrence, provided you can find a closed-form for the sums and products that appear.

For example, consider the recurrence

$$T_n = 2T_{n-1} + 1. \tag{30}$$

Using eq. (29), we find that the general solution is

$$T_n = 2^n T_0 + \sum_{1 \le i \le n} 2^{n-i} = 2^n T_0 + 2^n - 1. \tag{31}$$

When a_n and b_n are rational functions of n, the technique of Gosper [90] can be used to determine whether eq. (29) has a closed-form solution. The technique of Karr [103] can be used under more general conditions.

EXERCISES

1. What is the general solution of the recurrence $T_n = aT_{n-1} + b$?

2. What is the particular solution of the recurrence $T_n = aT_{n-1} + b$ with the boundary condition $T_0 = 1$?

3. What is the general solution of the recurrence $T_n = nT_{n-1}$?

4. What is the general solution of the recurrence $T_n = T_{n-1} + n$?

5. Show that if $b_n = 0$ for all n and a_n is a rational function of n, then the solution of eq. (23) can be expressed with Gamma functions.

6. Give the general solution of the second order recurrence $T_n = a_n T_{n-2} + b_n$. Hint: Define $U_n = T_{2n}$ and $V_n = T_{2n+1}$. Write first order linear equations for U_n and for V_n. Use the techniques of Section 3.4 to combine the solutions for even and odd indices.

5.2.1 Secondary Recurrences

Most first order recurrence equations that arise in the analysis of algorithms are slightly more complex than eq. (23). They are extended linear first order recurrences that have the form

$$T_n = a_n T_{f(n)} + b_n. \tag{32}$$

The usual way to solve this equation is again to repeatedly replace the T on the right side with the value given by eq. (32). If you do this k times, you get

$$
\begin{aligned}
T_n &= a_n T_{f(n)} + b_n && (33)\\
&= a_n\big(a_{f(n)} T_{f(f(n))} + b_{f(n)}\big) + b_n && (34)\\
&= a_n\big(a_{f(n)}\big(a_{f(f(n))} T_{f(f(f(n)))} + b_{f(f(n))}\big) + b_{f(n)}\big) + b_n && (35)\\
&= a_n\big(a_{f(n)}\big(a_{f(f(n))}\big(a_{f(f(f(n)))} T_{f(f(f(f(n))))} + b_{f(f(f(n)))}\big)\\
&\qquad + b_{f(f(n))}\big) + b_{f(n)}\big) + b_n && (36)\\
&\ \ \vdots\\
&= a_n\big(a_{f(n)}\big(a_{f(f(n))}\big(a_{f(f(f(n)))}\big(\cdots\big(a_{f^{[k-1]}} T_{f^{[k]}(n)} + b_{f^{[k-1]}}\big)\\
&\qquad \cdots\big) + b_{f(f(f(n)))}\big) + b_{f(f(n))}\big) + b_{f(n)}\big) + b_n, && (37)
\end{aligned}
$$

where $f^{[k]}(n)$ stands for the result of applying f to n k times [i.e., $f^{[k]}(n) = f(f(f(\cdots f(n)\cdots)))$, where f occurs k times on the right side of the equation]. Multiplying out eq. (37) gives

$$T_n = T_{f^{[k]}(n)} \prod_{0\le i<k} a_{f^{[i]}(n)} + \sum_{0\le i<k} b_{f^{[i]}(n)} \prod_{0\le j<i} a_{f^{[j]}(n)}. \qquad \textbf{(38)}$$

Here $T_{f^{[k]}(n)}$ serves as the one arbitrary constant needed to specify the general solution of a first order recurrence equation. [As we explain below, usually k is chosen to be that function of n that results in $f^{[k]}(n)$ being a particular constant, such as zero.] Eq. (38) gives you a closed-form solution of your original recurrence provided you can find a closed-form for the sums and products that appear and provided you can obtain a closed-form for $f^{[j]}(n)$.

Before continuing the development of the general technique, let's consider an example. Notice carefully the relationship between the various steps we go through while solving the example and the subscripts that appear on the formulas in the previous paragraph. Let T_n be the solution of the recurrence

$$T_n = 3T_{n/2} + n \qquad (39)$$

with the initial condition $T_1 = 1$. In the notation of the previous paragraph we have $a_n = 3$, $b_n = n$, and $f(n) = n/2$. This implies that $f^{[k]}(n) = n/2^k$. We can solve eq. (39) by repeatedly plugging in for the T on the right side of the equation to obtain

$$
\begin{aligned}
T_n &= 3T_{n/2} + n && (40)\\
&= 3(3T_{n/2^2} + n/2) + n && (41)\\
&= 3(3(3T_{n/2^3} + n/2^2)2 + n/2) + n && (42)\\
&= 3(3(3(3T_{n/2^4} + n/2^3) + n/2^2) + n/2) + n && (43)\\
&\ \ \vdots\\
&= 3(3(3(3(\cdots(3T_{n/2^k} + n/2^{k-1})\cdots) + n/2^3) + n/2^2) + n/2) + n. && (44)
\end{aligned}
$$

Multiplying out eq. (44) gives

$$T_n = 3^k T_{n/2^k} + n \sum_{0\le i<k} \left(\frac{3}{2}\right)^i = 3^k T_{n/2^k} + 2n\left[\left(\frac{3}{2}\right)^k - 1\right]. \qquad (45)$$

[You can, of course, obtain eq. (45) directly from eq. (38); it is just a question of whether you find it easier to remember the formula and how to use it or to remember the technique and how to use it.] Now we need to set k so that the index on $T_{n/2^k}$ is the same as the index on our boundary condition. Since we are given T_1 for our boundary condition, we must set k so that $n/2^k = 1$. This can be done only when n is a power of 2. For other values of n, eq. (39) and the value of T_1 do not define the value of T_n. The required value of k is $k = \lg n$. With this value for k, and 1 for T_1, eq. (45) becomes

$$T_n = 3^{\lg n} + 2n[(\tfrac{3}{2})^{\lg n} - 1]. \tag{46}$$

Equations with logarithms of variables appearing in exponents can usually be simplified. Since $a^x = b^{\log_b(a^x)}$ for any base b, we have (using $b = 2$) $3^{\lg n} = 2^{\lg n \lg 3} = n^{\lg 3}$, so

$$T_n = n^{\lg 3} + 2n^{1+\lg(3/2)} - 2n \tag{47}$$

$$= 3n^{\lg 3} - 2n \approx 3n^{1.58496} - 2n. \tag{48}$$

Now let's return to the problem of how to compute $f^{[k]}$ in general. To do this, you need to solve a *secondary recurrence*. Consider eq. (32) and assume we are trying to calculate T_n for some particular index n. Assume we have an initial condition T_m. Define k to be any integer such that $f^{[k]}(n) = m$. (If there is no such k, then the value of T_n cannot be deduced from the value of T_m.) Define the sequence d_i so that $f^{[i]}(n) = d_i$. In other words, d_i gives the value of the index after i iterations of the secondary recurrence. Since

$$f^{[i+1]}(n) = f(f^{[i]}(n)) = f(d_i) \tag{49}$$

and

$$f^{[i+1]}(n) = d_{i+1}, \tag{50}$$

we have that

$$f(d_i) = d_{i+1}. \tag{51}$$

Since

$$f^{[0]}(n) = n, \tag{52}$$

the initial condition for eq. (51) is

$$d_0 = n. \tag{53}$$

Eq. (51) is a first order recurrence equation. If f is a linear function, then it is a linear first order recurrence equation.

For example, the secondary recurrence for eq. (39) is

$$d_{i+1} = \frac{1}{2}d_i \tag{54}$$

with the boundary condition $d_0 = n$. The solution is

$$d_i = n2^{-i}. \tag{55}$$

If $k = \lg n$ is an integer, then $d_i = 2^{k-i}$ and $d_k = 1$, so eq. (39), with the initial condition $T_1 = 1$, can be solved.

Let's consider a slightly more complicated example. Let

$$T_n = 3T_{n/2+1} + n \tag{56}$$
$$T_3 = 1. \tag{57}$$

The secondary recurrence for this equation is

$$d_{i+1} = \frac{1}{2}d_i + 1, \tag{58}$$

with the boundary condition $d_0 = n$. The solution is

$$d_i = \frac{1}{2}d_{i-1} + 1 \tag{59}$$
$$= \frac{1}{2}\left(\frac{1}{2}d_{i-2} + 1\right) + 1 \tag{60}$$
$$\vdots$$
$$= \left(\frac{1}{2}\right)^i d_0 + \sum_{0\le j<i} \left(\frac{1}{2}\right)^j \tag{61}$$
$$= \frac{n}{2^i} - \left(\frac{1}{2}\right)^{i-1} + 2 = \frac{n-2}{2^i} + 2. \tag{62}$$

Since our boundary condition gives the value of T_3, we need to choose k so that $d_k = 3$. (We will call this new index k the *starting index of the secondary recurrence.*) We get

$$3 = \frac{n-2}{2^k} + 2, \qquad 2^k = n-2, \qquad k = \lg(n-2). \tag{63}$$

If k is an integer, then the original problem has a solution. Using eq. (38) with $a = 3$, $b_{f^{[i]}(n)} = d_i = 2^{k-i} + 2$, and $T_3 = 1$, we get

$$T_n = 3^k + \sum_{0\le i<k} [2^{k-i} + 2]3^i = 3^k + 2^k \left[\frac{\left(\frac{3}{2}\right)^k - 1}{\frac{1}{2}}\right] + 2\left[\frac{3^k - 1}{2}\right] \tag{64}$$
$$= 3^k + 2\cdot 3^k - 2\cdot 2^k + 3^k - 1 = 4\cdot 3^k - 2\cdot 2^k - 1. \tag{65}$$

Finally, we need to replace k with $\lg(n-2)$ to get

$$T_n = 4\cdot 3^{\lg(n-2)} - 2\cdot 2^{\lg(n-2)} - 1 \tag{66}$$
$$= 4(n-2)^{\lg 3} - 2(n-2) - 1 = 4(n-2)^{\lg 3} - 2n + 3. \tag{67}$$

The solutions of eq. (39) and eq. (56) are both $\Theta(n^{\lg 3})$. In each case the constant associated with the big Θ is determined by the boundary condition.

Sometimes it is convenient to rewrite a recurrence in terms of the index of the secondary recurrence, once the secondary recurrence has been solved. We will illustrate this technique on eqs. (56, 57). The first part of this approach is just like the previous approach. We find the secondary recurrence and solve it. Then, from the solution of the secondary recurrence, we find the index of the original recurrence in terms of the starting index of the secondary recurrence. In this case we need to find n in terms of k. From eq. (63) we get

$$n = 2^k + 2. \tag{68}$$

Next we replace all occurrences of the original index in the original recurrence equation by its value in terms of the starting index of the secondary recurrence. From eqs. (56, 57, 68) we get

$$T_{2^k+2} = 3T_{2^{k-1}+2} + 2^k + 2 \tag{69}$$

with boundary condition $T_{2^1+2} = 1$. This new equation can be thought of as a first order linear equation in the index k, so it can be solved with eq. (29). In other words, if you define the function $R(k) = T_{2^k+2}$, then eq. (69) is a linear first order equation for $R(k)$. Solving eq. (69) gives

$$T_{2^k+2} = 4.3^k - 2 \cdot 2^k - 1 \tag{70}$$

as before. Finally, we need to replace k in terms of n to obtain eq. (67).

Secondary recurrences were developed by Lueker [110].

EXERCISES

1. What is the solution of the recurrence $T_n = 3T_{n/4+1} + 2$ with boundary condition $T_2 = 1$?
2. For large n, what is the ratio of the solution of eq. (39) to the solution of eq. (56) using the given boundary conditions [$T_1 = 1$ for eq. (39) and $T_3 = 1$ for eq. (56)]?
3. For large n, what is the ratio of the solution of eq. (39) to the solution of eq. (56) using the boundary condition $T_3 = 1$ for both equations? Why does this exercise provide a better way to compare two algorithms than the previous exercise?
4. A variation of the Binary Search Algorithm follows:

Algorithm 5.1 Binary Search with Equality Checking: Input: A number q, called a *query*, a length $n > 1$, and a sorted array X of $n+1$ numbers, where $x_i \leq x_{i+1}$ for $0 \leq i \leq n-1$. It is given that $x_0 < q < x_n$. Output: The result "success" and an index m if there is an index m such that $q = x_m$ and the result "failure" otherwise.

Step **1.** [Initialize] Set the bottom pointer $b \leftarrow 0$ and the top pointer $t \leftarrow n$.

Step **2.** [Loop] While $t - b \geq 0$ do:

Step **3.** [Halve region] Compute the midpoint $m \leftarrow \left\lfloor \frac{b+t}{2} \right\rfloor$. If $q = x_m$, then report "success" and return. Otherwise, if $x_m < q$, then set $b \leftarrow m+1$. Otherwise, set $t \leftarrow m-1$. (When we change b or t, we know that element q is not at position m, unlike the corresponding case for Binary Search without equality checking.)

Step **4.** End while.

Step **5.** [Not found] Report "failure"; the item is not in the table.

Show that the average time required for this algorithm obeys the recurrence

$$T_{2n+1} = t_1 + t_2 \frac{2n}{2n+1} T_n$$

with boundary condition $T_1 = t_3$, where t_1, t_2, and t_3 are constants.

5. The algorithm of the previous exercise can be modified to work with two-way tests by replacing each three-way test with 2 two-way tests. On a computer that does not have three-way branches, would you be better off to use this modified algorithm or to use the Binary Search Algorithm(Algorithm 1.6)?

5.2.2 Divide and Conquer Algorithms

One of the most important types of extended first order recurrence equation comes from divide and conquer algorithms. Let n be a measure of your problem size. Suppose you have a method of solving a problem of size n by solving a problems of size n/c, and that the time required to divide a problem of size n into a parts is $f(n)$. Suppose you solve problems of size 1 directly in time b. Then the time for your method obeys the recurrence equation

$$T_n = aT_{n/c} + f(n) \tag{71}$$

$$T_1 = b. \tag{72}$$

The secondary recurrence for eqs. (71, 72) is

$$d_{i+1} = d_i/c \tag{73}$$

$$d_0 = n. \tag{74}$$

The solution of the secondary recurrence is $d_i = n/c^i$. The boundary condition requires that $d_k = 1$, so $k = \log_c n$. Thus $d_i = c^{k-i}$. By eq. (38) you get

$$T_n = ba^k + \sum_{0 \le i < k} f(c^{k-i})a^i. \tag{75}$$

Now let's consider eq. (75) for the case $f(n) = n^x$ for some constant x. Then eq. (75) reduces to

$$T_n = ba^k + \sum_{0 \le i < k} c^{x(k-i)}a^i = ba^k + c^{kx} \sum_{0 \le i < k} \left(\frac{a}{c^x}\right)^i \tag{76}$$

$$= ba^k + c^{kx}\frac{(a/c^x)^k - 1}{a/c^x - 1} \tag{77}$$

$$= ba^k + c^x \frac{a^k - c^{kx}}{a - c^x}. \tag{78}$$

If we replace k with $\log_c n$, we get

$$T_n = ba^{\log_c n} + c^x \frac{a^{\log_c n} - c^{x \log_c n}}{a - c^x} = bn^{\log_c a} + c^x \frac{n^{\log_c a} - n^x}{a - c^x}. \tag{79}$$

Eq. (78) has the form zero divided by zero when $a = c^x$. In this case all the terms in the sum in eq. (76) are 1, so

$$T_n = ba^k + c^{kx} \log_c n = bn^x + n^x \log_c n. \tag{80}$$

Notice that when $a = c^x$, $x = \log_c a$.

For

$$T_n = aT_{n/c} + n^x, \tag{81}$$

we have

$$\begin{aligned} T_n &= O(n^{\log_c a}) \qquad \text{when } a > c^x, &(82)\\ T_n &= O(n^x \log n) \qquad \text{when } a = c^x, &(83)\\ T_n &= O(n^x) \qquad \text{when } a < c^x. &(84) \end{aligned}$$

So the most important consideration when designing a divide and conquer algorithm is the value of $\log_c a$, as long as the overhead of the dividing part of the algorithm (n^x) is small enough that $x \leq \log_c a$. In this case, the smaller $\log_c a$ is, the faster the algorithm will run asymptotically. On the other hand, if a way is found to reduce $\log_c a$ to below x, then further reduction in the size of $\log_c a$ is relatively unimportant. Instead, reducing the overhead of the division part of the algorithm becomes more important. Later we will see more complicated examples of divide and conquer algorithms where the rate of division varies on every step.

To summarize, if $\log_c a > x$, then the way to improve the speed of your algorithm is to find a way to *reduce the number of parts* (a) or *reduce the size of the parts* (increase c). The overhead (n^x) for dividing into parts is not very important for large problems. In this case improving the time for small cases (decreasing b) will result in a proportional improvement in the time for large problems. On the other hand, if $\log_c a < x$ you need to find a way to reduce your overhead (x). The number of parts and the sizes of the parts are of limited importance in this case, particularly if $\log_c a$ is a lot less than x.

If you want to design a divide and conquer algorithm to run in time $O(n^y)$, then eq. (79) tells you that the overhead associated with dividing (n^x) must be such that $x < y$, and you must also arrange the division into parts so that $\log_c a < y$. [You can have equality on one or the other of these limits, but not both; when $x = y$ and $\log_c a = y$, the time is $O(n^y \log n)$.] Using these two conditions, you can rapidly determine approximately how fast any divide and conquer algorithm runs. This is very helpful when designing improved algorithms. In such cases, you know how good the best previous algorithm is. As you develop various alternate algorithms, you can quickly test whether or not they are faster (for large n). Once you find an approach that leads to a faster algorithm, you can then refine it and do a detailed analysis. This saves refining and carefully analyzing algorithms that are not an improvement on previous algorithms.

The material in this section is covered in Bentley et al. [71], which carries the ideas to even greater generality. Much of the book by Aho et al. [1] is devoted to analyzing divide and conquer algorithms. See particularly Section 2.6.

EXERCISES

1. Suppose you can solve a problem directly in time n^2. Suppose in time n^x you can divide the problem into two subproblems of size $n/2$. How small must x be so that it is better to divide a large problem into subproblems rather than to solve it directly?

2. Suppose you can solve a problem directly in time n^3. Suppose you can, in one unit of time, divide the problem into k subproblems of size $n/2$. How small must k be so that it is better to divide a large problem into subproblems rather than to solve it directly?

5.2.3 Faster Multiplication

The traditional algorithm (Algorithm 1.8) for multiplying two n-bit integers requires time $O(n^2)$. Several methods that are faster than this (for large numbers) have been developed. Such methods are useful for doing arithmetic on numbers that are too big to fit into one computer word. The simplest of these faster methods is based on the following idea. Suppose you have two $2n$-bit numbers. Let the first number be $U = U_1 2^n + U_2$ where $0 \leq U_1 < 2^n$ and $0 \leq U_2 < 2^n$. (In other words, U_1 is the upper n bits of the number and U_2 is the lower n bits.) Likewise let the second number be $V = V_1 2^n + V_2$ where $0 \leq V_1 < 2^n$ and $0 \leq V_2 < 2^n$. The ordinary algorithm for multiplying integers in base 2^n uses the equation

$$UV = U_1 V_1 2^{2n} + (U_1 V_2 + U_2 V_1) 2^n + U_2 V_2. \tag{85}$$

The bottom (least significant) n bits of the product are given by the bottom n bits of $U_2 V_2$. To obtain the next n bits of the product, first form the sum of $U_1 V_2 + U_2 V_1$ and the upper n bits of $U_2 V_2$. The desired n bits of the product are the bottom n bits of the sum. The top bits of this sum are added to $U_1 V_1$ to obtain the top $2n$ bits of the product. The approach in eq. (85) requires four multiplications of n-bit numbers to multiply two $2n$-bit numbers. It also requires three additions of $2n$-bit numbers and some overhead for finding the digits and doing the carrying. (See Algorithm 1.8.)

Now let's consider the following equation, which is algebraically equivalent to eq. (85):

$$UV = U_1 V_1 2^{2n} + [(U_1 + U_2)(V_1 + V_2) - U_1 V_1 - U_2 V_2] 2^n + U_2 V_2 \tag{86}$$

(multiply it out and check for yourself that it is equivalent). Although eq. (86) is algebraically equivalent to eq. (85), it suggests a much different algorithm for multiplication:

Algorithm 5.2 Faster Multiplication: Input: The $2n$-bit number $U = U_1 2^n + U_2$ where $0 \leq U_1 < 2^n$ and $0 \leq U_2 < 2^n$, and the $2n$-bit number $V = V_1 2^n + V_2$ where $0 \leq V_1 < 2^n$ and $0 \leq V_2 < 2^n$. Output: The product UV represented by $UV = W_1 2^{3n} + W_2 2^{2n} + W_3 2^n + W_4$.

Step **1.** Set $T_1 \leftarrow U_1 + U_2$.

Step **2.** Set $T_2 \leftarrow V_1 + V_2$.

Step **3.** Set $W_3 \leftarrow T_1 T_2$.

Step **4.** Set $W_2 \leftarrow U_1 V_1$.

Step **5.** Set $W_4 \leftarrow U_2 V_2$.

Step **6.** Set $W_3 \leftarrow W_3 - W_2 - W_4$.

Step **7.** Set $C \leftarrow \lfloor W_4 / 2^n \rfloor$ and $W_4 \leftarrow W_4 \bmod 2^n$.

Step **8.** Set $W_3 \leftarrow W_3 + C$, $C \leftarrow \lfloor W_3 / 2^n \rfloor$ and $W_3 \leftarrow W_3 \bmod 2^n$.

Step **9.** Set $W_2 \leftarrow W_2 + C$, $W_1 \leftarrow \lfloor W_2 / 2^n \rfloor$ and $W_2 \leftarrow W_2 \bmod 2^n$.

Divisions and mods using 2^n are actually bit extraction operations, so they can be computed rapidly. The algorithm multiplies two $2n$-bit numbers by doing two multiplications of n-bit numbers, one multiplication of $n+1$-bit numbers, no more than six additions (and subtractions) of numbers with $O(n)$ bits, and one comparison. For an upper-bound analysis of the algorithm we will use the fact that the time for multiplication increases with the number of bits, so the time is no more than the time required for three $n+1$-bit multiplications plus $O(n)$ overhead. If we now use the above algorithm recursively (down to $n = 3$), then the upper bound on the time required to multiply two numbers with this method is given by the recurrence

$$T_{2n} = 3T_{n+1} + O(n), \tag{87}$$

$$T_3 = c, \tag{88}$$

where c is the time required to multiply 3-bit numbers. This is essentially the same as eq. (56), and the solution is

$$T_n = O(n^{\log_2 3}) \approx O(n^{1.59}). \tag{89}$$

The Faster Multiplication Algorithm was developed by Karatsuba and Ofman [100]. This and other faster algorithms for multiplication of large numbers are given in Knuth [10, Sections 4.3.2 and 4.3.3].

5.2.3.1 Break-even Points

Eq. (89) shows that the Faster Multiplication Algorithm is faster than the traditional multiplication algorithm [which takes time $O(n^2)$ for large n], but it does not make the best use of the idea upon which it is based. Although the recursive algorithm is faster for large n, it is not faster for small n. For small n the overhead of the six additions is important. For large n the addition time is not important, because it grows like n while the multiplication time grows like $n^{1.59}$. Rather than using the recursive algorithm down to size 3, it is better to use the recursive algorithm down to some larger size n_* and then use the traditional algorithm.

Let's consider how to calculate the best value for n_*. Suppose we have two algorithms, A and B, that do the same calculation, but A is recursive (like Faster Multiplication) and B is not (like the traditional multiplication algorithm). What we need to do is to compute the set of values for n such that doing a problem of size n by algorithm A followed by algorithm B is slower than using algorithm B directly. Usually the set of values that results will consist of all values below some number n_*. In this case algorithm A should be used recursively until a problem with size less than n_* is obtained.

Let's apply this approach to the Faster Multiplication Algorithm. To make the problem more realistic, we will assume we are multiplying n word numbers rather than $2n$-bit numbers; the algorithm is essentially the same. To find the *exact* break-even point between two algorithms, you need to know the running time of each algorithm with more accuracy than you need for a good asymptotic analysis. On the other hand, switching algorithms at *any* point near the break-even point usually results in a running time that is quite close to that obtained by switching exactly at the break-even point. Accordingly, we will simply assume that the traditional algorithm requires time $n^2 + n$ and that the time for the Fast Multiplication Algorithm is given by

$$T_n = 3T_{n/2+1} + 3n, \tag{90}$$

where n is the number of words needed to represent each number. (A more precise approach would be to look at the assembly language code for each routine or to accurately time each part of each routine.) The problem is to choose n_* so that if the initial condition $T_{n_*} = n_*^2 + n_*$ is combined with eq. (90), the solution, T_n, is as small as possible. That is, we need to know when n^2+n (the time for using the traditional algorithm directly) is smaller than $3((n/2+1)^2 + n/2+1) + 3n$ (the time for using the recursive algorithm once and followed by using the traditional algorithm). Let n_* be the value such that

$$n_*^2 + n_* = 3\left[\left(\frac{n_*}{2}+1\right)^2 + \frac{n_*}{2}+1\right] + 3n_*, \tag{91}$$

or

$$\frac{1}{4}n_*^2 - \frac{13}{2}n_* - 6 = 0. \tag{92}$$

The potential break-even points are $n_* = 13 - \sqrt{193} \approx -0.89$ and $n_* = 13 + \sqrt{193} \approx 26.89$. (The negative root, of course is unrelated to the performance of the algorithm on real problems.) We have

$$n^2 + n > 3\left[\left(\frac{n}{2}+1\right)^2 + \frac{n}{2}+1\right] + 3n \qquad \text{for } n \geq 27 \tag{93}$$

and

$$n^2 + n < 3\left[\left(\frac{n}{2}+1\right)^2 + \frac{n}{2}+1\right] + 3n \qquad \text{for } 0 < n < 27, \tag{94}$$

so (under the assumptions of this example) it is best to use the recursive algorithm until subproblems of size 26 words or less are obtained and then to use the traditional algorithm.

EXERCISES

1. Using the assumptions about the running time of Faster Multiplication and traditional multiplication that were made in the text compare the time required if the recursion is carried down to $n = 3$ with the time required if the recursion is carried down to $n = 18$ (the largest value below 27 on the sequence that leads to $n = 3$).

2. Program Faster Multiplication and traditional multiplication in an assembly language. Do an exact analysis of the time required by each method and compute the break-even point for your pair of routines. How long does your version of Faster Multiplication take when you use the best break-even point?

3. When we analyzed the recursive multiplication algorithm, we assumed that three multiplies of size $(n/2+1)$ are needed. Actually, two of the multiplies are of size $n/2$. The original assumption simplifies the analysis and gives a useful asymptotic result. When calculating break-even points, however, more accuracy is needed. What break-even point do you get if you analyze the algorithm point using the correct size for each multiply?

5.2.4 Quicksort (Best Case)

The Quicksort Algorithm (Algorithm 2.2) spends most of its time inside Split (Algorithm 2.3). The time taken by the Split Algorithm is given by $a(r-l)+b$, where the values of a and b depend on just which pathways through Split are taken. Which path is followed through Split, of course, depends on the data being sorted.

After the analysis we do below, it will be evident that Quicksort runs most rapidly if the data is arranged so that each time Quicksort is called, exactly half of the elements are larger than the splitting element and half are smaller. The fastest path through Split with such a splitting element is as follows. Come into Step 1, go through the loop in Step 3 until j has moved over half of the elements, go to Step 4, go through the loop in Step 6 until i has moved over the other half of the elements, and exit at Step 7.

When there are $2^k - 1$ elements for some integer k, then there is one data ordering that will cause the splitting element selected by Algorithm 2.2 to divide the data in half and cause Split to take this best path. The median element of the file needs to be first so that it will be selected as the splitting element. Then the elements that are less than the median element need to be next, followed by the elements that are bigger than the median element. The internal order of the two halves of the data following the median element must be similar. The end result, for data items with distinct values from 1 to $2^k - 1$, is that the data item with value i should be in position j, where j is obtained from i by writing i as a k-bit binary number and reversing the bits.

The result of these considerations is the following. Let a and b be constants such that the time for one call of Quicksort (not including the recursive calls in Step 2) plus the resulting call of Split is $a(r-l)+b$ when $r > l$ (a will have the same value as above, but b will now be larger because it now includes the time inside of Quicksort as well as the constant part of the time inside of Split). Let c be the time used by Quicksort when $l = r$. Then the time used by Quicksort (for $2^k - 1$ elements) obeys the recurrence

$$T_n = an + b + 2T_{(n-1)/2}, \tag{95}$$
$$T_1 = c. \tag{96}$$

The secondary recurrence for this equation is

$$d_i = \tfrac{1}{2}d_{i-1} - \tfrac{1}{2}, \tag{97}$$
$$d_0 = n. \tag{98}$$

The solution is

$$d_i = \frac{n}{2^i} - \frac{1}{2}\sum_{0 \le j < i}\left(\frac{1}{2}\right)^j = \frac{n+1}{2^i} - 1. \tag{99}$$

The boundary condition gives

$$d_k = 1, \tag{100}$$
$$1 = \frac{n+1}{2^k} - 1, \qquad 2^{k+1} = n+1, \qquad k = \lg(n+1) - 1. \tag{101}$$

Thus the solution of eqs. (95–96) is

$$T_n = a(n+1)\lg\frac{n+1}{2} + \frac{b-a+c}{2}n + \frac{a-b+c}{2}. \tag{102}$$

An analysis of the best-case behavior of Quicksort is given in Baase [2, Section 2.3].

$$\left(\begin{array}{cc|cc} a_{11} & a_{12} & a_{13} & a_{14} \\ a_{21} & a_{22} & a_{23} & a_{24} \\ \hline a_{31} & a_{32} & a_{33} & a_{34} \\ a_{41} & a_{42} & a_{43} & a_{44} \end{array}\right) = \begin{pmatrix} A_{11} & A_{12} \\ A_{21} & A_{22} \end{pmatrix}$$

FIGURE 5.1 A 4×4 matrix divided into four 2×2 matrices.

EXERCISES

1. Prove that putting the data into reverse binary order results in an equal division of the data on each call to Split. (A list of integers is in reverse binary order if they are sorted with the last bit considered to be the most significant bit for the sort; all the integers should be written with the same number of bits).
2. Show that the reverse binary order results in Split exiting the first time it gets to Step 7.
3. Show that if Split exits the first time it gets to Step 7 and if the time to go around the loop in Step 3 is equal to the time to go around the loop in Step 6, then the time for Quicksort obeys the recurrence $T_n = an + b + T_{n_1} + T_{n_2}$ when the splitting element is such that Split divides the data into sets of size n_1 and n_2 (where $n = n_1 + n_2 + 1$).
4. Use the result of the previous exercise to show that the best time for Quicksort results when $n_1 = n_2$.
5. Derive eq. (102). Hint: Eq. (80) is quite useful.

5.2.5 Strassen's Algorithm

The classical algorithm for matrix multiplication (Algorithm 1.9) uses 8 multiplications and 4 additions to multiply 2×2 matrices. Strassen [133] discovered a method that uses 7 multiplications and 18 additions. Winograd [140] improved this to 7 multiplications and 15 additions. (We will call all such 7 multiplication algorithms Strassen algorithms.)

Replacing 8 multiplications and 4 additions with 7 multiplications and 15 additions is no improvement on most computers — if the elements being multiplied are numbers. However, these new matrix multiplication algorithms are designed so that the elements being multiplied can be submatrices. Since the time to multiply matrices (with the classical algorithm) is $\Theta(n^3)$, while the time to add matrices is $\Theta(n^2)$, it is faster to do 7 multiplications and 15 additions (rather than 8 multiplications and 4 additions) when the elements are moderate size matrices. Thus, if you have a pair of large $n \times n$ matrices to multiply, it is quicker to break each matrix into four $n/2 \times n/2$ matrices (see Figure 5.1) and use Strassen's algorithm rather than the classical algorithm. If the matrices for the subproblems are also large, the method should be applied recursively. Spiess [131] has determined the break-even point between the classical and the Strassen matrix multiplication algorithms for some real programs.

The 7 multiplication, 15 addition algorithm follows:

Algorithm 5.3 Strassen's Algorithm: Input: Two $n \times n$ matrices A and B with elements a_{ij} and b_{ij}. Output: The $n \times n$ matrix C with elements c_{ij}, where $C = AB$.

Let $A = \begin{pmatrix} A_{11} & A_{12} \\ A_{21} & A_{22} \end{pmatrix}$, $B = \begin{pmatrix} B_{11} & B_{12} \\ B_{21} & B_{22} \end{pmatrix}$, and $C = \begin{pmatrix} C_{11} & C_{12} \\ C_{21} & C_{22} \end{pmatrix}$.

Step **1.** Set $S_1 \leftarrow A_{21} + A_{22}$.

Step **2.** Set $S_2 \leftarrow S_1 - A_{11}$.

Step **3.** Set $S_3 \leftarrow A_{11} - A_{21}$.

Step **4.** Set $S_4 \leftarrow A_{12} - S_2$.

Step **5.** Set $S_5 \leftarrow B_{12} - B_{11}$.

Step **6.** Set $S_6 \leftarrow B_{22} - S_5$.

Step **7.** Set $S_7 \leftarrow B_{22} - B_{12}$.

Step **8.** Set $S_8 \leftarrow S_6 - B_{21}$.

Step **9.** Set $M_1 \leftarrow S_2 S_6$.

Step **10.** Set $M_2 \leftarrow A_{11} B_{11}$.

Step **11.** Set $M_3 \leftarrow A_{12} B_{21}$.

Step **12.** Set $M_4 \leftarrow S_3 S_7$.

Step **13.** Set $M_5 \leftarrow S_1 S_5$.

Step **14.** Set $M_6 \leftarrow S_4 B_{22}$.

Step **15.** Set $M_7 \leftarrow A_{22} S_8$.

Step **16.** Set $T_1 \leftarrow M_1 + M_2$.

Step **17.** Set $T_2 \leftarrow T_1 + M_4$.

Step **18.** Set $C_{11} \leftarrow M_2 + M_3$.

Step **19.** Set $C_{12} \leftarrow T_1 + M_5 + M_6$.

Step **20.** Set $C_{21} \leftarrow T_2 - M_7$.

Step **21.** Set $C_{22} \leftarrow T_2 + M_5$.

EXERCISES

1. Prove that Strassen's algorithm is correct: that it does indeed calculate the product of two matrices.

2. Find a closed-form solution for $T_n = 7T_{n/2}$ with the boundary condition $T_1 = 1$.

3. Find a closed-form solution for $T_n = 7T_{n/2} + n^2$ with the boundary condition $T_1 = 1$. This is the recurrence for the running time of the matrix multiplication method of Strassen [133], neglecting the size of some coëfficients and some lower order terms.

4. Suppose you discover a method for multiplying 3×3 matrices in k multiplications. Suppose your method can be used recursively to multiply larger matrices. How small does k need to be before your algorithm is faster than the classical algorithm for large matrices? How small does k have to be before your method is faster than Strassen's method for large matrices?

5. Strassen's algorithm takes time:

$$T_{2n} = 7T_n + cn^2 + \text{lower order terms}$$

for some constant c. The traditional algorithm for multiplying matrices takes time

$$T_n = an^3 + bn^2 + \text{lower order terms}$$

for some constants a and b. Neglecting the lower order terms, calculate the break-even point. See Spiess [131] for a discussion of the actual break-even point for real programs for this problem.

6. Write a program for Strassen's method that can be used recursively on large matrices and which uses the classical algorithm for small matrices. Use your program to determine the break-even point for switching from Strassen's algorithm to the classical algorithm experimentally.

5.3 GENERATING FUNCTIONS

Given a sequence of numbers $\{a_0, a_1, a_2, \ldots\}$, consider the function $G(z)$ defined by the infinite sum

$$G(z) = \sum_{i \geq 0} a_i z^i. \tag{103}$$

This function is called the *generating function* for the sequence. It contains complete information about the sequence. If you are given the function, then you can compute the sequence by evaluating the derivative of the function at $z = 0$:

$$a_n = \frac{1}{n!} \left. \frac{d^n f(z)}{dz^n} \right|_{z=0}. \tag{104}$$

The generating function is useful because it is often easier to work with one function than it is to work with an infinite sequence of numbers. In this section we will see how to use generating functions to solve some higher-order recurrence equations.

Let's start with a simple example. Consider the sequence defined by

$$T_n = 5T_{n-1} - 6T_{n-2} \qquad \text{for } n \geq 2, \tag{105}$$

$$T_0 = 0, \qquad T_1 = 1. \tag{106}$$

This is a second order linear recurrence equation. If we multiply eq. (105) by z^n we get

$$z^n T_n = 5z^n T_{n-1} - 6z^n T_{n-2} \qquad \text{for } n \geq 2. \tag{107}$$

Let's sum eq. (107) for all $n \geq 2$. The result is

$$\sum_{n \geq 2} z^n T_n = 5 \sum_{n \geq 2} z^n T_{n-1} - 6 \sum_{n \geq 2} z^n T_{n-2}. \tag{108}$$

Now let's try to make each sum look like the generating function for T_n. First replace the n in the second sum with $n+1$ and the n in the third sum with $n+2$. This gives

$$\sum_{n \geq 2} z^n T_n = 5 \sum_{n+1 \geq 2} z^{n+1} T_n - 6 \sum_{n+2 \geq 2} z^{n+2} T_n \tag{109}$$

$$= 5z \sum_{n \geq 1} z^n T_n - 6z^2 \sum_{n \geq 0} z^n T_n. \tag{110}$$

Now let's add and subtract some cancelling terms so that each sum can have a lower limit of zero. We get

$$T_0 + zT_1 + \sum_{n\geq 2} z^n T_n - T_0 - zT_1$$
$$= 5zT_0 + 5z \sum_{n\geq 1} z^n T_n - 5zT_0 - 6z^2 \sum_{n\geq 0} z^n T_n, \quad (111)$$

$$\sum_{n\geq 0} z^n T_n - T_0 - zT_1 = 5z \sum_{n\geq 0} z^n T_n - 5zT_0 - 6z^2 \sum_{n\geq 0} z^n T_n. \quad (112)$$

If we now define the generating function $G(z) = \sum_{n\geq 0} z^n T_n$, eq. (112) can be written as

$$G(z) - T_0 - zT_1 = 5zG(z) - 5zT_0 - 6z^2 G(z). \quad (113)$$

This is a nontrivial equation for $G(z)$ which can be solved to give

$$G(z) = \frac{T_0 + z(T_1 - 5T_0)}{1 - 5z + 6z^2}. \quad (114)$$

Using $T_0 = 0$ and $T_1 = 1$, we get

$$G(z) = \frac{z}{1 - 5z + 6z^2}. \quad (115)$$

Notice the need to be very careful with the boundary conditions and the first few terms of each sum when deriving eq. (114).

Now $G(z)$ contains information about the value of T_n for every n. If we expand $G(z)$ into a power series, we will get the values of the T_n. The best way to obtain the power series expansion is as follows. First factor the denominator. Since the denominator has degree 2, it can be written as $c(1 - z/a)(1 - z/b)$ for some constants a, b, and c (we will see why this is a particularly useful way to write the denominator in a moment). If we multiply this out, we get $c - c(1/a + 1/b)z + c/(ab)z^2$. For this to be equal to $1 - 5z + 6z^2$, we must have $c = 1$. Now $(1 - z/a)(1 - z/b)$ is equal to zero if $z = a$ or $z = b$. Therefore we can find the values of a and b by finding the roots of the equation

$$1 - 5z + 6z^2 = 0. \quad (116)$$

The roots are

$$z = \tfrac{1}{3} \quad \text{and} \quad z = \tfrac{1}{2}, \quad (117)$$

so if we let a be the smaller root, then $a = \frac{1}{3}$ and $b = \frac{1}{2}$.

We can now obtain the value of $G(z)$ by partial fraction decomposition. We have

$$G(z) = \frac{z}{1 - 5z + 6z^2} = \frac{z}{(1 - z/a)(1 - z/b)} = \frac{A}{1 - z/a} + \frac{B}{1 - z/b}. \quad (118)$$

We can obtain the value of A by multiplying by $1 - z/a$ and taking the limit as $z \to a$. The result is

$$A = \lim_{z\to a} \frac{z}{1 - z/b} = \frac{a}{1 - a/b} = \frac{ab}{b - a} = \frac{\frac{1}{6}}{\frac{1}{2} - \frac{1}{3}} = 1. \quad (119)$$

Likewise $B = -1$, so

$$G(z) = \frac{1}{1-3z} - \frac{1}{1-2z}. \tag{120}$$

Now to get a power series for $G(z)$, we use the fact that $1/(1-3z)$ equals the infinite sum $1+3z+3^2z^2+\cdots$. [You can obtain this by polynomial division or by eq. (2.61).] Using the same expansion for $1/(1-2z)$, we get

$$G(z) = (1+3z+3^2z^2+\cdots-1-2z-2^2z^2-\cdots). \tag{121}$$

Since by the definition of $G(z)$ the coefficient of z^n is T_n, we have

$$T_n = (3^n - 2^n). \tag{122}$$

Since 2 is less 3, the second term in eq. (122) has very little effect (compared to the first term) for large n. We can write eq. (122) as $T_n = 3^n\left(1-\left(\frac{2}{3}\right)^n\right)$ to emphasize the importance of the 3^n term.

If the nth coefficient in the generating function is no larger than $O(a^n)$ for some a, then the generating function converges for all small values of z. If the coefficients grow too rapidly to have such a bound, then the series converges only for $z = 0$. If the series converges, then the use of generating functions leads to correct results. Even if the series does not converge, the use of generating functions often leads to correct results. The interested reader should see the treatment of formal power series by Goulden and Jackson [33, Chapter 1].

Since any long calculation is likely to contain errors, any results obtained with generating functions should be checked by comparing the first few terms calculated using generating functions with the first few terms calculated directly. Also, unless you are familiar with the theory of formal power series, for each result you obtain using generating functions you should either (1) verify that your series converges or (2) give an independent proof of the final result. Induction is often the easiest way to produce an independent proof. The main weakness of proof by induction is that you need to know the answer before you can develop the proof. When proving a result obtained by generating functions, you already know what answer you are trying to establish, so you do not have this problem.

Cohen [23, Chapter 3] has a slow careful introduction to generating functions. Knuth [9, Section 1.2.9] also has an introduction to generating functions. A more advanced treatment is given in Riordan [48, Chapter 2].

5.3.1 Fibonacci Numbers

The *Fibonacci numbers* are defined by the recurrence

$$F_n = F_{n-1} + F_{n-2} \qquad \text{for } n \geq 2, \tag{123}$$

$$F_0 = 0, \qquad F_1 = 1. \tag{124}$$

These numbers occur in the analysis of many algorithms. We can obtain their values using the techniques that were used in the last section. The algebra is somewhat more complex this time, but the approach is quite similar. If you start to get lost in the algebra in this section, it will help to refer back to the corresponding part of the previous section, where the algebra is simpler.

The Fibonacci numbers are defined by a second order linear recurrence equation. If we multiply eq. (123) by z^n, we get

$$z^n F_n = z^n F_{n-1} + z^n F_{n-2} \qquad \text{for } n \geq 2. \tag{125}$$

Summing eq. (125) for all $n \geq 2$ gives

$$\sum_{n\geq 2} z^n F_n = \sum_{n\geq 2} z^n F_{n-1} + \sum_{n\geq 2} z^n F_{n-2}. \tag{126}$$

Now let's try to make each sum look like the generating function for F_n. First let's replace the n in the second sum with $n+1$ and the n in the third sum with $n+2$. This gives

$$\sum_{n\geq 2} z^n F_n = \sum_{n+1\geq 2} z^{n+1} F_n + \sum_{n+2\geq 2} z^{n+2} F_n \tag{127}$$

$$= z\sum_{n\geq 1} z^n F_n + z^2 \sum_{n\geq 0} z^n F_n. \tag{128}$$

Now let's add and subtract some cancelling terms so that each sum can have a lower limit of zero. We get

$$F_0 + zF_1 + \sum_{n\geq 2} z^n F_n - F_0 - zF_1$$

$$= zF_0 + z\sum_{n\geq 1} z^n F_n - zF_0 + z^2 \sum_{n\geq 0} z^n F_n. \tag{129}$$

$$\sum_{n\geq 0} z^n F_n - F_0 - zF_1 = z\sum_{n\geq 0} z^n F_n - zF_0 + z^2 \sum_{n\geq 0} z^n F_n. \tag{130}$$

If we now define the generating function $G(z) = \sum_{n\geq 0} z^n F_n$, eq. (130) can be written as

$$G(z) - F_0 - zF_1 = zG(z) - zF_0 + z^2 G(z). \tag{131}$$

This is a nontrivial equation for $G(z)$ which can be solved to give

$$G(z) = \frac{F_0 + z(F_1 - F_0)}{1 - z - z^2}. \tag{132}$$

Using $F_0 = 0$ and $F_1 = 1$, we get

$$G(z) = \frac{z}{1 - z - z^2}. \tag{\textbf{133}}$$

Now $G(z)$ contains information about the value of F_n for every n. If we expand $G(z)$ into a power series, we will get the values of the F_n. The best way to obtain the power-series expansion is to first factor the denominator. Since the denominator has degree 2, it can be written as $c(1 - z/a)(1 - z/b)$ for some constants a, b, and c. For this to be equal to $1 - z - z^2$, we must have $c = 1$.

Since $(1-z/a)(1-z/b)$ is equal to zero if $z=a$ or $z=b$, we can find the values of a and b by finding the roots of the equation

$$1-z-z^2=0. \tag{134}$$

The roots are

$$z=\frac{-1\pm\sqrt{1+4}}{2} \tag{135}$$

so if we let a be the positive root, $a=(-1+\sqrt{5})/2$ and $b=(-1-\sqrt{5})/2$. We will need $\phi=1/a=2/(-1+\sqrt{5})=2(1+\sqrt{5})/((1+\sqrt{5})(-1+\sqrt{5}))=(1+\sqrt{5})/2\approx 1.6180$ and $\hat{\phi}=1/b=2/(-1-\sqrt{5})=2(-1+\sqrt{5})/((-1+\sqrt{5})(-1-\sqrt{5}))=(1-\sqrt{5})/2\approx -0.6180$.

We can obtain the value of $G(z)$ by partial fraction decomposition. We have

$$G(z)=\frac{z}{1-z-z^2}=\frac{z}{(1-z/a)(1-z/b)}=\frac{A}{(1-z/a)}+\frac{B}{(1-z/b)}. \tag{136}$$

We can obtain the value of A by multiplying by $1-z/a$ and taking the limit as $z\to a$. The result is

$$A=\lim_{z\to a}\frac{z}{1-z/b}=\frac{a}{1-a/b}=\frac{ab}{b-a}=\frac{-1}{-\sqrt{5}}=\frac{1}{\sqrt{5}}. \tag{137}$$

Likewise $B=-1/\sqrt{5}$, so

$$G(z)=\frac{1}{\sqrt{5}}\left(\frac{1}{1-\phi z}-\frac{1}{1-\hat{\phi}z}\right). \tag{138}$$

Now to get a power series for $G(z)$, we use the fact that $1/(1-\phi z)$ equals the infinite sum $1+\phi z+\phi^2z^2+\cdots$. [You can obtain this by polynomial division or by eq. (2.61).] This gives

$$G(z)=\frac{1}{\sqrt{5}}(1+\phi z+\phi^2z^2+\cdots-1-\hat{\phi}z-\hat{\phi}^2z^2-\cdots). \tag{139}$$

Since by the definition of $G(z)$ the coefficient of z^n is F_n, we have

$$F_n=\frac{1}{\sqrt{5}}(\phi^n-\hat{\phi}^n). \tag{140}$$

Since the absolute value of $\hat{\phi}$ is less than 1, the second term in eq. (138) has very little effect for large n. For large n we have $F_n\approx\phi^n/\sqrt{5}$.

Generating functions give a very powerful method for solving both linear and nonlinear recurrence equations. The rest of this chapter contains many examples of the use of generating functions.

This material on Fibonacci numbers in this section is covered in Knuth [9, Section 1.2.8].

EXERCISES

1. What is the value of $\sum_{0 \le i \le n} F_i$?

2. Express the solution of the recurrence $T_n = T_{n-1} + T_{n-2}$ with the boundary conditions $T_0 = a$ and $T_1 = b$ in terms of Fibonacci numbers.

3. Suppose you calculate the nth Fibonacci number with the following recursive algorithm:

Algorithm 5.4 F(n):

Step **1.** If $n = 0$ then $F(n) = 0$, if $n = 1$ then $F(n) = 1$, and if $n \ge 2$ then $F(n) = F(n-1) + F(n-2)$.

How many additions are done to calculate the nth Fibonacci number? As a check on your answer, the following table gives the answer for small n.

n	0	1	2	3	4	5
additions	0	0	1	2	4	7

Hint: First show that the number of additions obeys the recurrence

$$A(n) = A(n-1) + A(n-2) + 1.$$

4. Suppose you calculate the nth Fibonacci number with the following algorithm:

Algorithm 5.5 F(n):

Step **1.** Set $F \leftarrow 1$ and $P \leftarrow 0$.

Step **2.** For $i \leftarrow 2$ to n do

Step **3.** Set $N \leftarrow F + P$, $P \leftarrow F$, and then $F \leftarrow N$.

How many additions are done to calculate the nth Fibonacci number? As a check on your answer, the following table gives the answer for small n.

n	2	3	4	5
additions	1	2	3	4

5.3.2 Properties of Fibonacci Numbers

Fibonacci numbers have many interesting properties. In this section we want to consider several properties that we will need later. The first property is obtained using the definition of Fibonacci numbers [eq. (123)] recursively to replace the larger term on the right side of itself. By replacing the F_{n-1} in eq. (123) with $F_{n-2} + F_{n-3}$, we get

$$F_n = 2F_{n-2} + F_{n-3}. \tag{141}$$

If we replace the F_{n-2} in eq. (141) by $F_{n-3} + F_{n-4}$, we get

$$F_n = 3F_{n-3} + 2F_{n-4}. \tag{142}$$

Continuing this way for $m - 1$ times gives

$$F_n = F_m F_{n-m+1} + F_{m-1} F_{n-m}. \tag{143}$$

Next we want to show that $\phi^n + \hat{\phi}^n$ can be expressed with Fibonacci numbers. Since $F_n = (1/\sqrt{5})(\phi^n - \hat{\phi}^n)$, we cannot express $\phi^n + \hat{\phi}^n$ in terms of a single Fibonacci number. (We do not have the correct dependence on n.) Let's see what we can do with two consecutive Fibonacci numbers. We write

$$\phi^n + \hat{\phi}^n = aF_n + bF_{n-1} \tag{144}$$

and see whether or not we obtain constants for a and b. Expanding the Fibonacci numbers gives

$$\phi^n + \hat{\phi}^n = \frac{a}{\sqrt{5}}(\phi^n - \hat{\phi}^n) + \frac{b}{\sqrt{5}}(\phi^{n-1} - \hat{\phi}^{n-1}) \tag{145}$$

$$= \frac{a}{\sqrt{5}}(\phi^n - \hat{\phi}^n) + \frac{b}{\phi\sqrt{5}}\phi^n - \frac{b}{\hat{\phi}\sqrt{5}}\hat{\phi}^n \tag{146}$$

$$= \left(\frac{a}{\sqrt{5}} + \frac{b}{\phi\sqrt{5}}\right)\phi^n - \left(\frac{a}{\sqrt{5}} + \frac{b}{\hat{\phi}\sqrt{5}}\right)\hat{\phi}^n. \tag{147}$$

For eq. (147) to hold for all n, the coefficient of ϕ^n must be the same on both sides. Likewise the coefficient of $\hat{\phi}^n$ must be the same on both sides. This gives

$$1 = \frac{a}{\sqrt{5}} + \frac{b}{\phi\sqrt{5}}, \tag{148}$$

$$1 = -\frac{a}{\sqrt{5}} - \frac{b}{\hat{\phi}\sqrt{5}}. \tag{149}$$

This pair of equations has the solution $a = 1$, $b = 2$. Therefore,

$$\phi^n + \hat{\phi}^n = F_n + 2F_{n-1}. \tag{150}$$

The sum $\sum_{0\le k\le n} F_k F_{n-k}$ can be expressed in terms of Fibonacci numbers. Start by expanding the Fibonacci numbers. This gives

$$\sum_{0\le k\le n} F_k F_{n-k} = \frac{1}{5}\sum_{0\le k\le n}(\phi^k - \hat{\phi}^k)(\phi^{n-k} - \hat{\phi}^{n-k}) \tag{151}$$

$$= \frac{1}{5}\sum_{0\le k\le n}(\phi^n - \hat{\phi}^k\phi^{n-k} - \phi^k\hat{\phi}^{n-k} + \hat{\phi}^n) \tag{152}$$

$$= \frac{n+1}{5}(\phi^n + \hat{\phi}^n) - \frac{2}{5}\sum_{0\le k\le n}\phi^k\hat{\phi}^{n-k} \tag{153}$$

$$= \frac{n+1}{5}(\phi^n + \hat{\phi}^n) - \frac{2}{5}\hat{\phi}^n\left(\frac{\phi^{n+1}/\hat{\phi}^{n+1} - 1}{\phi/\hat{\phi} - 1}\right). \tag{154}$$

$$= \frac{n+1}{5}(\phi^n + \hat{\phi}^n) - \frac{2}{5}\left(\frac{\phi^{n+1} - \hat{\phi}^{n+1}}{\phi - \hat{\phi}}\right). \tag{155}$$

We simplify the first term using eq. (150). We simplify the second term using $\phi - \hat{\phi} = \sqrt{5}$ and $F_{n+1} = (1/\sqrt{5})(\phi^n - \hat{\phi}^n)$. The result is

$$\sum_{0 \le k \le n} F_k F_{n-k} = \frac{n+1}{5}(F_n + 2F_{n-1}) - \frac{2}{5}F_{n+1}. \tag{156}$$

Equations with three consecutive Fibonacci numbers should be simplified to use only two Fibonacci numbers. Replacing F_{n+1} with $F_n + F_{n-1}$ gives

$$\sum_{0 \le k \le n} F_k F_{n-k} = \frac{n-1}{5}F_n + \frac{2n}{5}F_{n-1}. \tag{157}$$

Additional properties of Fibonacci numbers are given in Knuth [9, Section 1.2.8].

5.3.3* Tape Sorting: Three Tapes

Fibonacci numbers arise often during the analysis of algorithms. In this section we will consider ways to sort huge lists of numbers on a computer with three tape drives. One of the methods we will consider is based on the properties of Fibonacci numbers.

For our analysis we will make several assumptions. The analysis techniques that are being illustrated are easy to adapt to other assumptions, but the particular answers we obtain, of course, only apply to the particular set of assumptions we make. If you want to know the best way to sort numbers on a particular computer, then you should refer to a book that discusses sorting in more detail, such as Knuth [11, Section 5.4]. As we list our assumptions, we will indicate that some assumptions are quite reasonable and that others, such as neglecting rewind time, are unreasonable. This approach permits us to ignore some pesky details which would make the analysis longer but which would not introduce any new principles. In any event it is not possible to give a single analysis that covers all tape units, because various tape units differ in many details. After you understand our analysis, you can do your own analysis for any particular tape unit.

Our tape units can do three things: (1) read a record and advance to the next record, (2) write a record and advance to the next record, and (3) rewind (go back to the first record). We assume that the tape unit cannot read old records that follow a newly written record. Once you begin writing on a tape, the only useful thing to do is additional writing or rewinding followed by reading.

We assume that in one time unit the computer can read or write on all three tape units. (On some computers it would be more reasonable to assume that in one time unit the computer can read or write on only one tape unit.) We assume that the data read during one time unit cannot be written until the next time unit. We also assume that the computations can be done during the same time as the reading and writing, and that they never require any additional computer time above that used for the reading and writing. (This assumption

Input Tape 1	*Input Tape 2*	*Output Tape*
1	5	1
2		2
9		5
	6	6
	10	9
3		10
	4	3
7		4
	8	7
11		8
	12	11
		12

TABLE 5.1. Two tapes with sorted blocks of length 3 and the results of merging to produce a tape with sorted blocks of length 6.

is quite reasonable unless the computer is extremely slow or the tape units are extremely fast.) We assume no time is required for rewinding the tapes. (This assumption is definitely unreasonable. Some tape units can rewind somewhat faster than they can read or write, while others rewind at the same speed that they read and write.) Finally we assume that our operators, who have to mount and unmount each tape, are extremely slow. (Real operators are often slow, but not this slow). This means that we will only consider algorithms that do not require mounting and dismounting of tapes during their execution.

We will mount the tapes with the initial data and some blank tapes. The algorithms will overwrite the initial data during the sorting. No new tapes will be mounted during the sorting. (If you use a routine which destroys its input for important work, you should first make a copy of the input, so that you will still have your input if something goes wrong.) This completes the ground rules that our sorting programs must work under.

We will now analyze two tape-sorting algorithms. The algorithms follow the same general approach. Initially the input tapes will contain blocks of data with the items within each block already sorted. Let L be the number of items to be sorted, and let N be the number of blocks. (Typically L/N will be about twice the size of the computer memory; see Knuth [11, Section 5.4.1].) The sorting programs will use two tapes for input and one for output. They will read in items from the two input tapes and merge them to produce sorted blocks for the output tape. The merged blocks will be as long as the sum of the input blocks, so fewer, but longer, blocks are needed for the data. As this process is repeated, the number of blocks decreases until there is just one block which contains all the data.

Table 5.1 shows how two tapes, each with two sorted blocks (a total of four blocks) of three items each, can be merged to produce one tape with two blocks of six items each.

5.3.3.1* *Simple Merge Sort*

The first approach we consider will begin with half the blocks on one input tape and the other half on the other input tape. After the first merge is done, we will have one tape with all the data. To continue merging, we must then distribute the data on the one tape back to two tapes. After this, we are ready to merge again. The merging is repeated until a single block results. The details are given in the following algorithm, which is stated in a general form where k tapes are used.

Algorithm 5.6 Simple Merge Sort: Input: Tapes 1 through $k-1$ with sorted blocks. Tape k is blank. Intermediate variables: n is the number of blocks that have been put onto Tape k. The variables a_i are used as input buffers. The value $-\infty$ for a_i indicates that all the data from Tape i has been read. Output: Tape k with the input completely sorted. (The input is destroyed.)

Step **1.** Set $n \leftarrow 0$. For $1 \leq i \leq k-1$ set a_i to the first value from tape i.

Step **2.** Call Merge$((1,\ldots,k-1),k)$. Set $n \leftarrow n+1$. For $1 \leq i \leq k-1$ if any $a_i \neq -\infty$, repeat this step.

Step **3.** If $n = 1$, then stop. The sorted output is on Tape k.

Step **4.** Call Distribute$(k,(1,\ldots,k-1))$. Go to step 1.

Algorithm 5.7 Merge(A, B): Input: A list A of Input Tapes (A_1 to A_j) with sorted blocks. Constant: $-\infty$, a number smaller than any on the tapes (this constant can be avoided by making the algorithm somewhat more complex). Convention: When we read from a tape, if we are at the end of the tape, we obtain the value $-\infty$. Intermediate variables: e_i, *true* if Tape i has just read to the end of a block, *false* otherwise; b, the last number written on Tape B; a_i, the variable that has the last item read from Tape A_i. The variable a_i is available to the calling program. Output: Tape B with one more sorted block, where the length of the block on Tape B is the sum of the block lengths from Tapes A_1 through A_j.

Step **1.** [Initialize] For $1 \leq i \leq j$ set $e_i \leftarrow$ *false*.

Step **2.** [Merge] While any e_i is *false* do:

Step **3.** [Select smallest] Set m so that a_m is the smallest a_i such that e_i is *false* where $1 \leq i \leq j$.

Step **4.** [Output smallest] Set $c \leftarrow a_m$, write a_m onto Tape 3, and set a_m to the next value on Tape A_m.

Step **5.** [End of block?] If $a_m < c$, then set $e_m \leftarrow$ *true*.

Step **6.** End of while any e_i.

Algorithm 5.8 Distribute(B, A): Input: A Tape B with sorted blocks and j, the number of output tapes. Constant: $-\infty$, a value smaller than any on the tapes. (This special value can be avoided if suitable changes are made in the algorithm.) Output: A is a list of tapes (A_1 to A_j) such that the blocks from Tape B have been evenly distributed among Tapes A_1 to A_j.

Step **1.** [Initialize] Rewind Tape B. Set $a \leftarrow -\infty$, and $m \leftarrow 1$.

Step **2.** [Loop] Set b to the next value from Tape B. If $b = -\infty$, stop.

Step **3.** [End of Block?] If $b < a$, then set $m \leftarrow (m \bmod j) + 1$. (The end of a block has been found; put the next block on the next tape).

Step **4.** Write b onto Tape m and set $a \leftarrow b$. Go to Step 2.

The time for the Simple Merge Sort Algorithm is easy to analyze. We need L time units to output the results of a merge. (The input can be read as fast as the output is done.) We also need L time units to distribute the results of a merge back onto the input tapes. The time for one pass thus is $2L$. We start with N blocks. On each pass the number of blocks is decreased by a factor of $k-1$. Thus $\lceil \log_{k-1} N \rceil$ passes are needed. Finally, we don't need the time for distribute (L time units) on the last pass. Thus the total time is

$$T_{\text{Simple Merge}} = L(2\log_{k-1} N - 1) \tag{158}$$

time units. For three tape units ($k = 3$), $T_{\text{Simple Merge}} = L(2\lg N - 1)$ time units.

EXERCISES

1. Repeat the analysis of Simple Merge Sort under the assumption that one time unit is needed for each number that is read or written. Thus to read two numbers and output them takes a total of four time units.

2. Repeat the analysis of Simple Merge Sort under the assumption that rewinding a tape takes one time unit for each number on the tape.

3. Repeat the analysis of Simple Merge Sort under the assumptions from both of the last two problems.

5.3.3.2 Polyphase Merge*

Almost half of the time in Simple Merge Sorting is used by the distribution phase of the algorithm, where no merging is done. The motivation for Polyphase Merging is to try to always be merging. It is easiest to understand Polyphase Merging by first considering the last phase, then the next to last phase, and so on up to the first phase.

Phase	*Tape 1*		*Tape 2*		*Tape 3*	
	Blocks	*Length*	*Blocks*	*Length*	*Blocks*	*Length*
7	*Output*		21	1	13	1
6	13	2	8	1	*Output*	
5	5	2	*Output*		8	3
4	*Output*		5	5	3	3
3	3	8	2	5	*Output*	
2	1	8	*Output*		2	13
1	*Output*		1	21	1	13
0	1	34				

TABLE 5.2. The arrangement of blocks on tapes for polyphase merge with three tape units and seven phases. Block sizes are given for the case where the initial block size is 1.

On the last phase we want to be merging two tapes which each contain one block. (This is no different from the last phase of the Simple Merge Algorithm.) On the next to last phase we want to merge two tapes so that, when we finish, the output tape will have one block, one of the two input tapes will have one block left, and the other input tape will be empty. (This is quite different from the Simple Merge Algorithm.) The only way to obtain this result is to start the next to last phase with one tape containing two blocks and one tape containing one block. After merging, the tape that contained one block will be empty; it will serve as the output tape for the final phase. The blocks in the final phase will usually be of different lengths.

On the third to last phase we want to produce an output tape with two blocks, to have one tape with no blocks left, and to have one tape with one block left. This requires that we start with tapes of two and three blocks. Table 5.2 shows the results of continuing this *backwards analysis.* In general, if on phase i from the end we begin with tapes having a_i and b_i blocks (where $a_i \geq b_i$), then on the previous phase (phase $i+1$ from the end) we want to produce an output tape of length a_i, have one tape with zero blocks left, and one tape with b_i blocks left. This requires that we start phase $i+1$ with one tape of length a_i and one tape of length $a_i + b_i$. That is,

$$a_{i+1} = a_i + b_i \tag{159}$$

$$b_{i+1} = a_i. \tag{160}$$

Using eq. (160) to replace the b_i in eq. (159) with a_{i-1}, we get $a_{i+1} = a_i + a_{i-1}$. Replacing i with $i-1$ gives

$$a_i = a_{i-1} + a_{i-2}, \tag{161}$$

the recurrence equation for the Fibonacci numbers. Since $b_{i+1} = a_i$, b_i also obeys the recurrence for the Fibonacci numbers. If we start with $a_1 = 1$ and

Phase	*Tape 1*		*Tape 2*		*Tape 3*	
	Blocks	*Length*	*Blocks*	*Length*	*Blocks*	*Length*
7	*Output*		F_8	F_2	F_7	F_1
6	F_7	F_3	F_6	F_2	*Output*	
5	F_5	F_3	*Output*		F_6	F_4
4	*Output*		F_5	F_5	F_4	F_4
3	F_4	F_6	F_3	F_5	*Output*	
2	F_2	F_6	*Output*		F_3	F_7
1	*Output*		F_2	F_8	F_1	F_7
0	F_1	F_9				

TABLE 5.3. The arrangement of blocks on tapes for polyphase merge with three tape units and seven phases. Block sizes are given for the case where the initial block size is 1.

$b_1 = 1$, then we have

$$a_i = F_{i+1}, \tag{162}$$

$$b_i = F_i, \tag{163}$$

where F_i is the ith Fibonacci number. Table 5.3 shows the number of blocks and the block lengths for a seven-phase Polyphase Merge, with each number expressed as a Fibonacci number.

Although polyphase merging can be done with any number of items, the algorithm is particularly simple when the number of blocks is $N = F_{n+1}$ for some integer n and where each block has L/N items. We will just consider this simple case. The algorithm is stated for an arbitrary number of tapes, although in this section we are interested in the three-tape merge.

Algorithm 5.9 Polyphase Merge: Input: Tapes 1 through $k - 1$ with sorted blocks. Tape k is blank. Intermediate variables: B is the output tape, A is the *set* of input tapes, n is the number of blocks that have been put onto the output tape. The variables a_i are input buffers. The value $-\infty$ for a_i indicates that all the data from Tape i has been read. Output: Tape B with the input completely sorted. (The input is destroyed.) **Warning:** This algorithm expects to have the correct number of blocks on each input tape when it starts (see the following analysis in this section and in Section 8.1.3.1). If it ever gets to Step 5 with only one input tape where $a_i \neq -\infty$, then the algorithm will never terminate. A special test could be added to call the distribute routine in this case and thereby prevent the infinite loop. However, for the algorithm to be efficient, it is important that each input record have the appropriate number of items, so we have omitted the extra test.

Step **1.** Set $B \leftarrow k$, and set A to the set $\{1, 2, \ldots, k-1\}$. For each Tape i in A, set a_i to the first value on Tape i.

Step **2.** Set $n \leftarrow 0$.

Step **3.** Call Merge(A, B). Set $n \leftarrow n + 1$. For $1 \leq i \leq k$, $i \neq B$, if all $a_i \neq -\infty$ repeat this step.

Step **4.** If $n = 1$ and if for all j in A, $a_j = -\infty$, then stop. The sorted output is on Tape B.

Step **5.** Let i be an index such that $a_i = -\infty$. Rewind Tapes i and B. Set a_B to the first value on Tape B. Insert Tape B into the set A, remove Tape i from A, and set $B \leftarrow i$.

Step **6.** Go to Step 2.

If we have three tape units $(k = 3)$ and if we start with F_{n+1} blocks, then Polyphase Merge uses $n - 1$ phases. The following is true when we are i steps from the end in Polyphase Merge (if we want the final output to be on Tape 1):

Tape $i \bmod 3 + 1$ is the input tape with the greater number of blocks.
Tape $(i + 1) \bmod 3 + 1$ is the input tape with the lesser number of blocks.
Tape $(i + 2) \bmod 3 + 1$ is the output tape.

If the initial block length is L/N, then at phase i from the end:

Tape $i \bmod 3 + 1$ starts with F_{i+1} blocks of length LF_{n-i+1}/N.
Tape $(i + 1) \bmod 3 + 1$ starts with F_i blocks of length LF_{n-i}/N.
Tape $(i + 2) \bmod 3 + 1$ gets F_i blocks of length LF_{n-i+2}/N.

Continuing with the same assumptions, the time for the phase i from the end is LF_iF_{n-i+2}/N time units since writing the output tape determines the amount of time used. The total time therefore is

$$T_{\text{Polyphase Merge}} = \frac{L}{N} \sum_{1 \leq i \leq n-1} F_i F_{n-i+2} \tag{164}$$

$$= \frac{L}{N}\left(\sum_{0 \leq i \leq n+2} F_i F_{n-i+2} - F_n F_2 - F_{n+1} F_1 \right). \tag{165}$$

Applying eq. (143) gives

$$T_{\text{Polyphase Merge}} = \frac{(n-4)L}{5N} F_{n+2} + \frac{2(n+2)L}{5N} F_{n+1}. \tag{166}$$

Since $N = F_{n+1}$, eq. (166) reduces to

$$T_{\text{Polyphase Merge}} = \frac{(n-4)L}{5F_{n+1}} F_{n+2} + \frac{2(n+2)L}{5}. \tag{167}$$

Let's now compare this result with the result for Simple Merge when N is large. First we need to use asymptotic approximations to simplify eq. (167). We can write F_{n+2} in terms of F_{n+1} as follows:

$$F_{n+2} = \frac{1}{\sqrt{5}}(\phi^{n+2} - \hat{\phi}^{n+2}) \tag{168}$$

$$= \frac{\phi}{\sqrt{5}}\left[\phi^{n+1} - \hat{\phi}^{n+1} - \left(\frac{\hat{\phi}}{\phi}\hat{\phi}^{n+1} - \hat{\phi}^{n+1}\right)\right] \tag{169}$$

$$= \phi F_{n+1} + \frac{\phi - \hat{\phi}}{\sqrt{5}}\hat{\phi}^{n+1} \tag{170}$$

$$= \phi F_{n+1}\left[1 + O\left(\left(\frac{\hat{\phi}}{\phi}\right)^n\right)\right] = \phi F_{n+1}[1 + O(0.553^n)]. \tag{171}$$

Using this result in eq. (167) gives

$$T = (n+1)L\left(\frac{2+\phi}{5} + \frac{2}{5(n+1)} + O(0.553^n)\right). \tag{172}$$

Next we need to express $n+1$ in terms of N. We have

$$N = \frac{1}{\sqrt{5}}(\phi^{n+1} - \hat{\phi}^{n+1}), \tag{173}$$

$$\phi^{n+1} = \sqrt{5}N\left(1 - \frac{\hat{\phi}^{n+1}}{\sqrt{5}N}\right), \tag{174}$$

$$(n+1)\lg\phi = \lg N + \frac{1}{2}\lg 5 + \lg\left(1 - \frac{\hat{\phi}^{n+1}}{\sqrt{5}N}\right) \tag{175}$$

$$= \lg N + \frac{1}{2}\lg 5 + O(0.553^n), \tag{176}$$

$$n+1 = \frac{\lg N}{\lg\phi} + \frac{\lg 5}{2\lg\phi} + O(0.553^n) = \frac{\lg N}{\lg\phi} + O(1). \tag{177}$$

Using this result in eq. (172) gives

$$T = L\lg N\left(\frac{2+\phi}{5\lg\phi} + O\left(\frac{1}{\lg N}\right)\right) \approx 1.042L\lg N. \tag{178}$$

Comparing this time with that for Simple Merge Sorting [eq. (158) with $k = 3$], we see that for three tapes Polyphase Merge is almost twice as fast. (But keep in mind all the assumptions made during this analysis including the neglect of rewind time.)

EXERCISES

1. Repeat the analysis of Polyphase Merge under the assumption that one time unit is needed for each number that is read or written. Thus to read two numbers and output them takes a total of four time units.
2. Repeat the analysis of Polyphase Merge under the assumption that rewinding a tape takes one time unit for each number on the tape.
3. Repeat the analysis of Polyphase Merge under the assumptions of both of the last two problems.
4. Compare the answer to Exercise 1 of this section to the answer to Exercise 1 of the previous section when N is large.
5. Compare the answer to Exercise 2 of this section to the answer to Exercise 2 of the previous section when N is large.
6. Compare the answer to Exercise 3 of this section to the answer to Exercise 3 of the previous section when N is large.

5.4 CONSTANT COEFFICIENTS

A linear k^{th} order recurrence equation with *constant coefficients* has the form

$$\sum_{0\le i\le k} a_i T_{n-i} = b_i, \tag{179}$$

where $a_0 \neq 0$ and $a_k \neq 0$. The Fibonacci recurrence of Section 5.3.1 is an example of a linear second order recurrence. In this section we consider how to solve such equations in general and give the form of the solution in the homogeneous case. It is helpful to compare each step in this general derivation with the corresponding step in the derivation of the solution of the Fibonacci recurrence.

To solve this recurrence with generating functions, define the generating function

$$G(z) = \sum_{i\ge 0} T_i z^i. \tag{180}$$

Multiply eq. (179) by z^n and sum over $n \ge k$ to obtain

$$\sum_{0\le i\le k} a_i \sum_{n\ge k} T_{n-i} z^n = \sum_{n\ge k} b_n z^n. \tag{181}$$

Now by factoring and by change of the summation index we have

$$\sum_{n\ge k} T_{n-i} z^n = z^i \sum_{n\ge k-i} T_n z^n \tag{182}$$

$$= z^i \left[G(z) - \sum_{0\le n<k-i} T_n z^n \right]. \tag{183}$$

Using this to simplify eq. (181), we get

$$\sum_{0\le i\le k} a_i z^i \left[G(z) - \sum_{0\le n<k-i} T_n z^n \right] = \sum_{n\ge k} b_n z^n, \tag{184}$$

$$G(z) \sum_{0\le i\le k} a_i z^i = \sum_{0\le i\le k} a_i z^i \sum_{0\le n<k-i} T_n z^n + \sum_{n\ge k} b_n z^n \tag{185}$$

$$= \sum_{0\le n<k} z^k \sum_{0\le i\le n} a_i T_{n-i} + \sum_{n\ge k} b_n z^n. \tag{186}$$

[See eqs. (3.121–3.126) for the details leading from eq. (185) to eq. (186).] Now if we solve eq. (186) for $G(z)$, we get

$$G(z) = \frac{\sum_{0\le n<k} z^n \sum_{0\le i\le n} a_i T_{n-i}}{\sum_{0\le i\le k} a_i z^i} + \frac{\sum_{n\ge k} b_n z^n}{\sum_{0\le i\le k} a_i z^i}. \tag{187}$$

The first term in eq. (187) is a rational function of z, where the denominator has degree k and the numerator has degree at most $k-1$.

Linear recurrence equations with constant coefficients are covered in Bender and Orszag [21, Section 2.3], Milne-Thomson [43, Chapter 13], and Greene and Knuth [6, Section 2.1.1.1].

EXERCISE

1. Suppose that pairs of foxes have one pair of children (one child of each sex) when they are 1 year old, a second pair of children when they are 2 years old, and that they die when they are 3 years old, without producing any more children. Write a recurrence equation for the number of foxes at the end of the nth year. Solve the recurrence equation for the case where you start at year zero with one pair of foxes of age zero.

5.4.1 The Characteristic Equation

We will now consider the homogeneous case ($b_n = 0$ for all n) in more detail. In this case, the second term in eq. (187) is zero. We will use partial fraction decomposition on the first term of eq. (187).

The *characteristic equation* for a linear recurrence equation with constant coefficients in the form $\sum_{0\le i\le k} a_i T_{n-i} = b_i$ is

$$\sum_{0\le i\le k} a_{k-i} z^i = 0. \tag{188}$$

If the original equation is written in the form $\sum_{0\le i\le k} a'_i T_{n+i} = b'_i$, then the characteristic equation is

$$\sum_{0\le i\le k} a'_i z^i = 0. \tag{189}$$

The characteristic equation plays a fundamental role in the solution of linear recurrence equations with constant coefficients, as we will see in this section and the next one. Let $\lambda_1, \lambda_2, \ldots, \lambda_j$ be the roots of the characteristic equation, where the pth root has multiplicity β_p. Thus

$$\sum_{0\le i\le k} a_{k-i}z^i = a_0 \prod_{1\le p\le j} (z-\lambda_p)^{\beta_p} \tag{190}$$

and

$$\sum_{1\le p\le j} \beta_p = k. \tag{191}$$

By replacing i with $k-i$ eq. (190) can be rewritten as

$$\sum_{0\le i\le k} a_{k-i}z^i = \sum_{0\le i\le k} a_i z^{k-i}. \tag{192}$$

Substituting eq. (192) into eq. (190) gives

$$\sum_{0\le i\le k} a_i z^{k-i} = a_0 \prod_{1\le p\le j} (z-\lambda_p)^{\beta_p}. \tag{193}$$

Replacing z with $1/z$ gives

$$\sum_{0\le i\le k} a_i z^{i-k} = a_0 \prod_{1\le p\le j} \left(\frac{1}{z}-\lambda_p\right)^{\beta_p}. \tag{194}$$

Multiplying by z^k gives

$$\sum_{0\le i\le k} a_i z^i = a_0 \prod_{1\le p\le j} (1-\lambda_p z)^{\beta_p}. \tag{195}$$

Now that we have a product expression for the denominator of eq. (187), we can use partial fractions. Using eq. (195), we can define the functions $T_p(z)$ so that the generating function of eq. (187) has a partial fraction decomposition involving the roots of the characteristic equation:

$$\frac{\sum_{0\le n<k} z^n \sum_{0\le i\le n} a_i T_{n-i}}{\sum_{0\le i\le k} a_i z^i} = \sum_{1\le p\le j} \frac{T_p(z)}{(1-\lambda_p z)^{\beta_p}}. \tag{196}$$

The degree of $T_p(z)$ is less than β_p, and $T_p(z)$ can be written in the form

$$T_p(z) = \sum_{0\le q<\beta_p} c_{pq} z^q. \tag{197}$$

(See Section 2.6.)

To find the coefficient of z^n on the right side of eq. (196), we expand it in a power series, term by term. Expanding $1/(1-\lambda_p z)^{\beta_p}$ in a power series and using eq. (3.72) gives

$$(1-\lambda_p z)^{-\beta_p} = \sum_{i\ge 0} \binom{-\beta_p}{i} (-\lambda_p)^i z^i \tag{198}$$

$$= (-1)^{\beta_p+1} \sum_{i\ge 0} \binom{-i-1}{\beta_p-1} \lambda_p^i z^i. \tag{199}$$

The coefficient of z^i is an exponential λ_p^i times a binomial. The binomial is a polynomial in i of degree $\beta_p - 1$. The quotient $T_p(z)/(1-\lambda_p z)^{\beta_p}$ is given by

$$\frac{T_p(z)}{(1-\lambda_p z)^{\beta_p}} = (-1)^{\beta_p+1} \sum_{0 \le q < \beta_p} c_{pq} \sum_{i \ge 0} \binom{-i-1}{\beta_p - 1} \lambda_p^i z^{i+q}. \tag{200}$$

Using q' for $i+q$ $(q = q' - i)$ and then dropping the primes gives

$$\frac{T_p(z)}{(1-\lambda_p z)^{\beta_p}} = (-1)^{\beta_p+1} \sum_{i \ge 0} \sum_{i \le q < \beta_p + i} c_{p,q-i} \binom{-i-1}{\beta_p - 1} \lambda_p^i z^q. \tag{201}$$

Now interchanging the sums over i and q gives

$$\frac{T_p(z)}{(1-\lambda_p z)^{\beta_p}} = (-1)^{\beta_p+1} \sum_{0 \le q < \beta_p} \sum_{q-\beta_p < i \le q} c_{p,q-i} \binom{-i-1}{\beta_p - 1} \lambda_p^i z^q, \tag{202}$$

so using $d_q(p)$ for the coefficient of z^q in $T_p(z)$, we have

$$d_q(p) = (-1)^{\beta_p+1} \sum_{q-\beta_p < i \le q} c_{p,q-i} \binom{-i-1}{\beta_p - 1} \lambda_p^i \tag{203}$$

$$= (-1)^{\beta_p+1} \lambda_p^q \sum_{q-\beta_p < i \le q} c_{p,q-i} \binom{-i-1}{\beta_p - 1} \lambda_p^{-q+i}. \tag{204}$$

Now replace $i - q$ by i' $(i = q + i')$ and drop the prime to obtain

$$d_q(p) = (-1)^{\beta_p+1} \lambda_p^q \sum_{-\beta_p < i \le 0} c_{p,-i} \binom{-i-q-1}{\beta_p - 1} \lambda_p^i. \tag{205}$$

Finally replace i by $-i$ to obtain

$$d_q(p) = (-1)^{\beta_p+1} \lambda_p^q \sum_{0 \le i < \beta_p} c_{p,i} \binom{i-q-1}{\beta_p - 1} \lambda_p^{-i}. \tag{206}$$

Thus the form of $d_q(p)$ is an exponential in q (i.e., λ_p^q) times a polynomial in q of degree $\beta_p - 1$ [i.e., the rest of the right side of eq. (206)].

Continuing with the solution of the homogeneous case, the coefficient of z^n in the generating function $G(z)$ is

$$\sum_{1 \le p \le j} d_j(p). \tag{207}$$

Now this coefficient is also T_n by the definition of a generating function. Thus T_n is given by a sum over the roots of the characteristic equation of exponentials times polynomials, where the base of each exponential is the corresponding root, and where the degree of the polynomial is given by 1 less than the multiplicity of the corresponding root. We can use these observations about the form of the solution to develop a method for solving homogeneous linear recurrences, which is done in the next section.

Characteristic equations are discussed in Milne-Thomson [43, Section 13.0] and Greene and Knuth [6, Section 2.1.1.1].

EXERCISES

1. Show that if λ is a root of the equation $\sum_{0\le i\le k} a_{k-i}z^i = 0$, then $1/\lambda$ is a root of the equation $\sum_{0\le i\le k} a_i z^i = 0$.
2. Consider a divide and conquer algorithm that reduces a problem of size n to two problems, one of size $n-i$ and the other of size $n-j$. Suppose that the time to do the division is a polynomial function of n, i.e. $O(n^x)$ for some number x. Give the recurrence equation for the best upper bound on the running time of the algorithm. (Assume that all problems with size $\max\{i,j\}$ or less are solved directly.)
3. The solution of the equation for the previous exercise can be written as $T(n) = \Theta(b^n)$, where the value of b depends on i and j. Find the value of b to four decimal places for all i and j satisfying $1 \le i \le j \le 4$.

5.4.2 Undetermined Coefficients

In the previous section we saw that the general solution of an nth-degree linear homogeneous recurrence equation with constant coefficients has a simple functional form that involves the roots of the characteristic equation. In particular, let $\lambda_1, \ldots, \lambda_j$ be the roots of the characteristic equation, and let $\beta_1, \ldots, \beta_j$ be their multiplicities. Then, letting f_{pq} be unknown constants, the solution for T_k has the form

$$T_k = \sum_{1\le p\le j} \sum_{0\le q\le \beta_p - 1} f_{pq} k^q \lambda_p^k. \tag{208}$$

That is, T_k is a sum of terms, where each term is a root λ_p to the kth power, multiplied by a polynomial in k of degree $\beta_p - 1$. When λ_p (as often happens) is of multiplicity 1, the polynomial is just a constant. Once we know the values of $\lambda_1, \ldots, \lambda_j$, $\beta_1, \ldots, \beta_j$, and the n constants f_{pq}, we know T_k for all k. This suggests the following method for solving such equations.

1. Solve the characteristic equation [eq. (188)]. Let the roots be $\lambda_1, \lambda_2, \ldots, \lambda_j$ and let the corresponding multiplicities be $\beta_1, \beta_2, \ldots, \beta_j$.
2. Use the boundary conditions to solve for the unknown f_{pq} in eq. (208).

In general, a method of solving an equation by plugging in a partially specified form of the solution, where the partially specified form has some unknown coefficients, is called a method of *undetermined coefficients*.

Let's see how undetermined coefficients work for linear recurrence equations with constant coefficients. Consider the equation

$$T_n = 3T_{n-1} - 2T_{n-2} \tag{209}$$

with the boundary conditions $T_0 = 0,\ T_1 = 1$. The characteristic equation is

$$z^2 - 3z + 2 = 0, \tag{210}$$

so the roots are $\lambda_1 = 2$ and $\lambda_2 = 1$. Each root has multiplicity 1. Thus the solution of the equation has the form

$$T_k = f_{1,0}2^k + f_{2,0}1^k. \tag{211}$$

From $T_0 = 0$ and $T_1 = 1$, we get

$$f_{1,0}2^0 + f_{2,0}1^0 = 0 \tag{212}$$

$$f_{1,0}2^1 + f_{2,0}1^1 = 1, \tag{213}$$

which has the solution $f_{1,0} = 1,\ f_{2,0} = -1$, so the solution of the recurrence is

$$T_k = 2^k - 1. \tag{214}$$

The recurrence

$$T_n = 4T_{n-1} - 4T_{n-2} \tag{215}$$

has the characteristic equation

$$z^2 - 4z + 4 = 0, \tag{216}$$

which has a double root at $z = 2$. Therefore the general solution of eq. (215) has the form

$$T_k = (f_{1,0} + f_{1,1}k)2^k. \tag{217}$$

The solution for the boundary conditions $T_0 = 1,\ T_1 = 1$ is found by solving

$$f_{1,0}2^0 + f_{1,1}0 \cdot 2^0 = 1, \tag{218}$$

$$f_{1,0}2^1 + f_{1,1}1 \cdot 1^1 = 1, \tag{219}$$

to obtain $f_{1,0} = 1,\ f_{1,1} = -\frac{1}{2}$, so the solution of that matches the boundary conditions is

$$T_k = 2^k - \tfrac{1}{2}k2^k. \tag{220}$$

The method of undetermined coefficients is discussed in Bender and Orszag [21, Section 2.4] and in Greene and Knuth [6, Section 2.1.1.1].

EXERCISES

1. What is the solution of the recurrence $T_n = 2T_{n-1} - T_{n-2}$ with the boundary conditions $T_0 = 0$ and $T_1 = 1$?
2. Find a particular solution of the recurrence $F_{n+2} + F_{n+1} + F_n = n$. Hint: Try for a solution of the form $c_1 n + c_0$.
3. Find a particular solution of the recurrence

$$F_{n+3} - F_{n+2} + F_{n+1} - F_n = n^2 2^n.$$

4. If the coefficients of a linear k^{th} order recurrence equation are real, then the solution is real. The roots of the characteristic equation may, however, be complex. Solve the recurrence

$$T_n = 5T_{n-1} + 4T_{n-2}$$

with boundary conditions $T_0 = 0$, $T_1 = 1$.
5. When the answer to a problem is real, it is often best to express the answer using only real numbers. Give the solution of the previous problem using only real numbers. Hint: use the techniques of Section 3.4.2.

5.4.3* A Simple Queue

When you have a device that can service one customer at a time and you have customers arriving at random, problems can occur. At times there will be more than one customer wanting service. A common way to handle such problems is to place a queue between the device and the customers. Whenever the device is free, it begins servicing the customer at the front of the queue. For a system with a queue it is interesting to know how long the queue is likely to be.

In this section we will consider one simple queueing problem. Assume that the customers arrive at random. To be precise, if dt is a small time interval (we will soon take the limit as dt goes to zero), we will assume that the probability a customer arrives between time t and $t + dt$ is $\lambda\, dt + o(dt)$, where λ is a positive constant and $o(dt)$ stands for any function $f(dt)$ such that $\lim_{dt\to 0} f(dt)/dt = 0$. (A process that generates inputs with these properties is called a *Poisson process.*) Assume that the probability that more than one customer arrives can be neglected [i.e., it is $o(dt)$]. The probability that no customers arrive is then $1 - \lambda\, dt + o(dt)$. Assume that in a time interval dt, the probability that the server finishes with a customer is $\mu\, dt + o(dt)$, where μ is a positive constant. [This assumption about the server is equivalent to assuming that the probability that the service time is between t and $t + dt$ is given (in the limit as dt goes to zero) by $e^{-\mu t}\mu dt$.] Assume that after the server finishes with one customer, it starts servicing the next customer immediately, if there is a customer waiting.

With these assumptions, we can now derive a set of differential equations that give the probability that there are k customers in the queue at time t. Let's first consider the probability that there are zero customers waiting at time $t + dt$. Well, there are zero customers at time $t + dt$ if there were zero customers waiting at time t and no new customers arrived, or if there was one customer waiting at time t and the server finished with that customer. There are also other possibilities, but they have a small probability $[o(dt)]$. If we let $p_k(t)$ be the probability that there are k customers in the queue at time t, then we have

$$p_0(t+dt) = (1-\lambda\,dt)p_0(t) + (\mu\,dt)p_1(t) + o(dt) \tag{221}$$

$$p_0(t+dt) - p_0(t) = -\lambda p_0(t)\,dt + \mu p_1(t)\,dt + o(dt) \tag{222}$$

$$\frac{p_0(t+dt) - p_0(t)}{dt} = -\lambda p_0(t) + \mu p_1(t) + o(1). \tag{223}$$

Now if we take the limit as dt goes to zero, the left side approaches $dp_0(t)/dt$, while $o(1)$ approaches zero, so we get

$$\frac{dp_0(t)}{dt} = -\lambda p_0(t) + \mu p_1(t). \tag{224}$$

This differential equation can be used to calculate p_0 as a function of time provided $p_1(t)$ is known.

We will now obtain equations for $p_k(t)$ for $k \geq 1$. There will be k customers waiting at time $t + dt$: (1) if there were k customers waiting at time t and no customers arrived and service was finished for none, (2) if there were $k+1$ customers at time t and service was finished for one and no new ones arrived, or (3) if there were $k-1$ at time t, one new customer arrived, and service was finished for none. There are other possibilities but the other possibilities all have small probabilities. Thus we have, for $k \geq 1$,

$$p_k(t+dt) = (1-\lambda\,dt-\mu\,dt)p_k(t) + (\mu\,dt)p_{k+1}(t) + (\lambda\,dt)p_{k-1}(t) + o(dt), \tag{225}$$

$$p_k(t+dt) - p_k(t) = -(\lambda+\mu)p_k(t)\,dt + \mu p_{k+1}(t)\,dt + \lambda p_{k-1}(t)\,dt + o(dt), \tag{226}$$

$$\frac{p_k(t+dt) - p_k(t)}{dt} = -(\lambda+\mu)p_k(t) + \mu p_{k+1}(t) + \lambda p_{k-1}(t) + o(1), \tag{227}$$

$$\frac{dp_k(t)}{dt} = -(\lambda+\mu)p_k(t) + \mu p_{k+1}(t) + \lambda p_{k-1}(t). \tag{228}$$

From eqs. (224, 228), we can determine $p_k(t)$ for all k and t provided we know $p_k(0)$ for all k. We will not, however, solve eqs. (224, 228) since it would take us too far afield from the other topics in this book.

Often sets of differential equations have solutions that approach some limiting steady value. This is the case for eqs. (224, 228). The limiting steady value, if it exists, can be found by setting all the derivatives equal to zero. If the set of differential equations has a solution where all the derivatives are zero, then there surely is a *steady-state solution.* The only remaining question is whether or not you get to the steady-state solution from the initial conditions. We will show that eqs. (224, 228) does have a steady-state solution under certain conditions, and we will just state without proof that from initial conditions where the $p_k(0)$ are probabilities, the time dependent solution does approach the steady-state solution.

To find the steady-state solution, we must solve the equations

$$\lambda p_0 = \mu p_1, \tag{229}$$

$$\mu p_{k+1} - (\lambda+\mu)p_k + \lambda p_{k-1} = 0 \qquad \text{for } k \geq 1, \tag{230}$$

where p_k is the steady-state value for $p_k(t)$. If we do not obtain a solution where $0 \le p_k \le 1$ for each k, then the original problem does not have a steady-state solution. Eq. (230) is a second order linear recurrence equation with constant coefficients. To obtain a unique solution, we need two additional conditions. Eq. (229) is one such condition. The other condition is that

$$\sum_{k \ge 0} p_k = 1. \tag{231}$$

[It is quite common with problems related to probabilities to need a *normalization condition* like eq. (231).] The characteristic equation for eq. (230) is

$$\mu x^2 - (\lambda + \mu)x + \lambda = 0, \tag{232}$$

or

$$x^2 - (1 + \lambda/\mu)x + \lambda/\mu = 0. \tag{233}$$

Since λ and μ enter only in the ratio λ/μ, it is useful to define $\rho = \lambda/\mu$. This gives the characteristic equation

$$x^2 - (1 + \rho)x + \rho = 0. \tag{234}$$

The roots are $x = 1$ and $x = \rho$, so the general solution is

$$p_k = c_1 + c_2\rho^k. \tag{235}$$

To have p_k as given by eq. (235) obey eq. (229), we need

$$\rho(c_1 + c_2) = c_1 + c_2\rho, \tag{236}$$

which for $\rho \ne 1$ can be the case only if $c_1 = 0$. (For $\rho = 1$ the problem has no steady-state solution.) To have the probabilities sum to 1, we need

$$1 = c_2 \sum_{k \ge 0} \rho^k = \frac{c_2}{1 - \rho}, \tag{237}$$

or

$$c_2 = 1 - \rho. \tag{238}$$

Thus the steady-state solution is

$$p_k = (1 - \rho)\rho^k. \tag{239}$$

The p_k are positive for $0 \le \rho < 1$. For $\rho \ge 1$ the problem does not have a steady-state solution; customers arrive more quickly than they can be served, and the queue becomes longer and longer, without limit. For $\rho > 1$, the p_k given by eq. (239) are negative, but this does not have anything to do with a steady-state solution of the original problem. You can think of it as the equation's way of telling you that there is no steady-state probability in this case.

The equations for a simple queue with a single server are considered in much more detail by Saaty [50, Section 2.5a]. Also see Feller [27, Section 17.7]. Some of the earliest work in this field was done by Erlang (1878–1929) in connection with the mathematical study of telephone systems.

EXERCISES

1. What is the average length of the queue analyzed in this section as a function of ρ?
2. The probability p_0 is the fraction of the time the server is idle. What is the relation between p_0 and the average length of the queue? How long is the queue when the server is idle 20 percent of the time? What about when the server is idle 10 percent of the time? (Most people have a poor intuitive understanding of the relation between average idle time and average queue length. This can lead to designs with excessive waiting.)

5.5 DIVIDE AND CONQUER WITH UNEVEN PARTS

In Section 5.2.2 we considered divide and conquer algorithms that divide a problem of size n into a parts of size c, with an overhead of n^x for doing the division. We found that for $a > c^x$ the time required by the algorithm depended mainly on the size of $\log_c a$, while for $a < c^x$ the time depended mainly on the size of x.

We will now consider more general divide and conquer algorithms where a problem of size n is divided into parts with k different sizes. Assume that there are a_i parts of size $c_i n$, and that the overhead for the division into parts is bn^x. We require that $0 < c_i < 1$, $a_i > 0$, and $b > 0$. The time obeys the recurrence

$$T_n = \sum_{1 \le i \le k} a_i T_{c_i n} + bn^x. \tag{240}$$

The algorithm will do some small cases directly. The small cases give the boundary conditions for solving eq. (240).

Let's now determine when the time is bounded by $T_n \le \alpha n^y$ for some α and y. We are interested in bounds where y is as small as possible and where, for that value of y, α is as small as possible. Assuming that the limit is true for all indices smaller than n, plugging the limit into the right side of eq. (240) gives

$$T_n \le \alpha \sum_{1 \le i \le k} a_i (c_i n)^y + bn^x. \tag{241}$$

If this limit on T_n is below αn^y, then eq. (241) can be used as the general step in an induction proof of the limit. (And if the limit is above αn^y, we can establish a counterexample to the limit we wish to prove.) That is, we need

$$\alpha n^y \ge \alpha \sum_{1 \le i \le k} a_i (c_i n)^y + bn^x, \tag{242}$$

which can be rearranged to give

$$\alpha\Big(1 - \sum_{1\le i\le k} a_i c_i^y\Big) \ge b n^{x-y}. \tag{243}$$

When $1-\sum_{1\le i\le k} a_i c_i^y$ is zero or negative, eq. (243) has no solution for α. When the quantity is positive, the smallest value of α that obeys eq. (243) is

$$\alpha = \left(\frac{b}{1-\sum_{1\le i\le k} a_i c_i^y}\right) n^{x-y}. \tag{244}$$

From eq. (243) we obtain the following rules.

1. If $1-\sum_{1\le i\le k} a_i c_i^x > 0$, then the divide and conquer algorithm runs in time

$$\left(\frac{b}{1-\sum_{1\le i\le k} a_i c_i^x}\right) n^x \tag{245}$$

provided the time for the small cases obeys this limit. (If the small cases do not obey this limit, then you can raise the coefficient in the limit until it is large enough to provide for the small cases.)

2. For any $y > x$ such that $1 - \sum_{1\le i\le k} a_i c_i^y > 0$, the divide and conquer algorithm runs in time $O(n^y)$.

The first rule follows from setting $y = x$ in eq. (244). The second rule follows from setting y large enough so that $1 - \sum_{1\le i\le k} a_i c_i^y$ is positive.

When designing a divide and conquer algorithm to run in time n^y, these rules can save you the effort of doing a detailed analysis on each algorithm that occurs to you. You can wait until you have a fast algorithm to refine it and to do a detailed analysis. The following section shows an application of this technique.

These two rules can be considered as an application of characteristic equations (see Section 5.4.1). The appropriate characteristic equation for eq. (240) is

$$1 - \sum_{1\le i\le k} a_i c_i^z = 0. \tag{246}$$

When $a_i > 0$ and $0 < c_i < 1$ for all i, rule 2 is equivalent to saying that the largest root of eq. (246) is smaller than y. Since the largest root determines the asymptotic rate of growth, its value is of principal concern when designing an algorithm. When necessary, you can do a more detailed analysis of eq. (240) by combining the techniques of secondary recurrences (Section 5.2.1) with the techniques for solving linear recurrences with constant coefficients (Section 5.4).

5.5.1 Medians

The *rank* of an element x_i of a totally ordered set S is the number of elements of S that are less than or equal to x_i. The same idea applies if the set contains some equal elements, except that a repeated element would occupy a number of consecutive ranks if we did not adopt a more precise definition: an element x_i in an n element set has rank k if there are at least $k-1$ elements in the set that are less than x_i and at least k values that are less than or equal to x_i. The *median* of n numbers (where n is odd) is the number with rank $(n+1)/2$. In other words, the median is the middle number of the set. The traditional method of finding the median of n numbers is to sort the numbers and then take the middle element. This approach takes worst-case time $\Theta(n \ln n)$ when a fast sorting algorithm is used. For large n there is, however, a more efficient divide and conquer algorithm.

The basic idea of the divide and conquer algorithm for finding the median is to first divide the original n numbers into n/c groups of c numbers (we will use $c = 9$). Next the median of each group is found by sorting the numbers in the group. (This takes 19 compares per group if the best algorithm for sorting nine numbers is used; see Knuth [11, Section 5.3.1].) The median algorithm is then used recursively on the n/c medians of the groups to produce a grand median. Of the original n elements, at least $\frac{1}{4}n$ will be larger than the grand median, and at least $\frac{1}{4}n$ will be less than the grand median (these fractions will be considered in more detail below). The grand median is used to divide the original set of n numbers into the set S_1 of numbers smaller than the grand median, the set S_2 of numbers equal to the grand median, and the set S_3 of numbers greater than the grand median. If the number of elements in S_1 is greater than $n/2$, then the overall median is in S_1. If the number of elements in S_3 is greater than $n/2$, then the overall median is in S_3. Otherwise the grand median is the overall median. When the overall median is in S_1 or S_3, the algorithm is used recursively. (This requires that the algorithm be general enough to find elements of any rank, not just the median.) Since the sets S_1 and S_3 each contains fewer than $\frac{3}{4}n$ elements, the recursive call is somewhat quicker than the original call. From the analysis of divide and conquer algorithms given in the previous section, we can see that this approach should give us an algorithm that runs in linear time provided $c > 4$ and provided c is not too large.

We will now give the details of the algorithm and a more precise analysis. The above analysis is the type of analysis that you go through, when first inventing an algorithm, in order to determine whether you have a promising approach. The analysis that follows is the type that you go through after you have invented the algorithm and want to tune up some details.

The following algorithm finds the kth smallest element of an n-element set. It is designed to be fast for large n and for k near $n/2$.

Algorithm 5.10 Select (k, S)**:** Input: A set S that has a total order (except that S may contain equal elements) and an integer k. The number of elements in S ($|S|$) is n. Output: An element of S with rank k. The parameters c

and d are chosen to make the algorithm run rapidly. (See the analysis below; c should be odd; $c = 9$ and $d = 19{,}762$ are suitable values.) Intermediate values: m is the grand median—the median of a set of medians.

Step **1.** If the relation of n to d is:

Step **2.** $n \le d$, then sort S and return the kth smallest element of S.

Step **3.** $n > d$, then

Step **4.** Partition S into $\lfloor n/c \rfloor$ groups of c elements and, if n mod $c \ne 0$, one small group of n mod c elements.

Step **5.** Sort each group from the previous step. Form the set M by taking the element median of each of those groups. Since c is odd, all groups except perhaps the small group have an odd number of elements. If the small group has an even number of elements, we will call the element with rank $\frac{1}{2}(n \bmod c)$ the median.

Step **6.** Set $m \leftarrow \text{Select}(\lceil (|M| + 1)/2 \rceil, M)$, i.e., find the median of the set M. (When $|M|$ is even, we are calling the element of rank $n/2 + 1$ the median. By adopting the opposite convention from Step 5, we improve the worst-case performance slightly.)

Step **7.** Partition S into the sets S_1, S_2, and S_3, where S_1 contains elements less than m, S_2 contains elements equal to m, and S_3 contains elements greater than m. (The algorithm still works if some elements equal to m end up in sets S_1 and S_3.)

Step **8.** If k has the value:

Step **9.** $k \le |S_1|$, then return $\text{Select}(k, S_1)$.

Step **10.** $|S_1| < k \le |S_1| + |S_2|$, then return m.

Step **11.** $|S_1| + |S_2| < k \le n$, then return $\text{Select}(k - |S_1| - |S_2|, S_3)$.

Now let's compute an upper limit on the number of comparisons of data elements. Let $C(n)$ be the worst-case number of comparisons when the algorithm is used on a set of size n. Step 2, when it applies, uses no more than $d \lg d$ comparisons if an efficient sort algorithm is used (see Knuth [11, Section 5.3.1]). Step 5 uses no more than $n \lg c$ comparisons. (The groups of size c can be sorted with no more than $c \lg c$ comparisons.) For $c = 9$, Step 5 uses no more than $\frac{19}{9}n$ comparisons. Step 6 uses no more than $C(\lceil n/c \rceil)$ comparisons. [Here and several subsequent places we use the fact that $C(n)$ is an increasing function of n. Likewise, we will consider only approximations to $C(n)$ that are increasing functions of n.] Comparisons are also used in Step 7 and in Step 9 or 11. Before

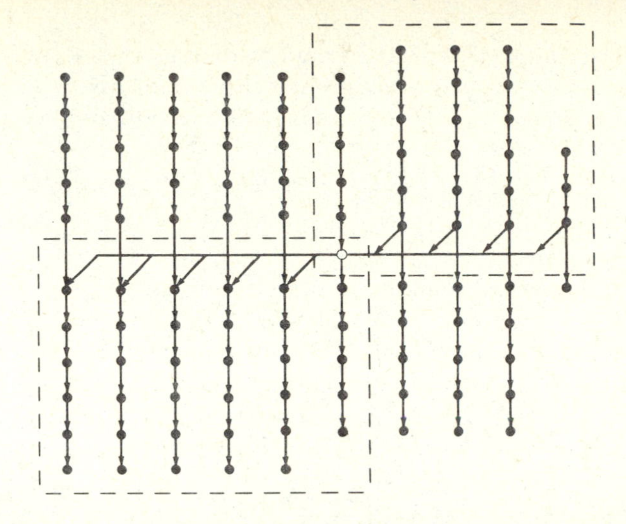

FIGURE 5.2 The grand median, indicated by a small circle, of 85 elements in 9 groups of 9 elements and one group of 4 elements is shown. Each element is known to be greater than or equal to any element that it can reach by following a chain of arrows. The grand median is known to be less than or equal to all the elements in the upper right dashed box. It is known to be greater than or equal to all the elements in the lower left box. The relation of the grand median to the remaining elements is not known.

computing how many comparisons can be used by these steps, however, we need to study the relation of various elements to the grand median.

Figure 5.2 shows the relation of various elements to the grand median. The grand median is definitely less than or equal to the elements in the upper right part of the figure, and it is definitely greater than or equal to the elements in the lower left part of the figure. Assume that c is odd. If there is no small group, then the grand median is less than or equal to $(c+1)/2$ elements in its own group, and it is less than or equal to $(c+1)/2$ elements in each of the groups with a larger (or equal) median. This is a total of $\lceil \lceil n/c \rceil /2 \rceil$ groups, for a total of $\lceil \lceil n/c \rceil /2 \rceil (c+1)/2$ elements. After considering that one of the groups may be the small group, we find that the grand median is less than or equal to at least

$$\left\lceil \frac{\lceil n/c \rceil}{2} \right\rceil \left(\frac{c+1}{2} \right) - \left\lfloor \frac{c-(n-1) \bmod c}{2} \right\rfloor \leq \frac{(c+1)n}{4c} - \left\lfloor \frac{c}{2} \right\rfloor \tag{247}$$

elements. For similar reasons the grand median is greater than or equal to at least $(c+1)n/(4c) - \lfloor c/2 \rfloor$ elements in the lower left part of Figure 5.2.

No comparisons are needed to determine the relation of the grand median to the elements in the upper right and lower left parts of Figure 5.2. (This can result in some elements equal to the grand median in sets S_1 and S_3, but this will cause no trouble.) Comparisons are needed to determine the relation between the grand median and the the remaining elements. The remaining elements come in groups of $(c-1)/2$ elements, except for a part of the small group, which can have up to $\lfloor (n \bmod c)/2 \rfloor$ elements. Using binary search, the elements in a group can be compared with the grand median using $1 + \lceil \lg[(c-1)/2] \rceil$ comparisons. The elements in the small group can be compared with the grand median using $1 + \lg\lfloor (n \bmod c)/2 \rfloor$ comparisons. Thus Step 7 uses at most

$$(1 + \lceil \lg[(c-1)/2] \rceil)\left(\left\lfloor \frac{n}{c} \right\rfloor - 1\right) + 1 + \lg\lfloor (n \bmod c)/2 \rfloor \le \frac{1 + \lceil \lg[(c-1)/2] \rceil}{c} n \tag{248}$$

comparisons.

The set S_1 contains at most $(3c+1)n/(4c) + \lfloor c/2 \rfloor$ elements, because at least $(c+1)n/(4c) + \lfloor c/2 \rfloor$ of the n elements are put into the sets S_2 and S_3 (all the elements in the upper right part of Figure 5.2). Likewise S_3 contains at most $(3c-1)n/(4c) + \lfloor c/2 \rfloor$ elements. At most one of Steps 9 and 11 is done. The number of comparisons used by these steps is at most $C((3c-1)n/(4c) + \lfloor c/2 \rfloor)$.

Putting all the pieces of the analysis together, we have

$$C(n) \le C\left(\frac{n}{c} + 1\right) + C\left(\frac{3c-1}{4c} n + \frac{c}{2}\right) + \left(\lg c + \frac{1 + \lceil \lg[(c-1)/2] \rceil}{c}\right) n. \tag{249}$$

In order for $C(n)$ to be a linear function of n, it is necessary to choose c so that the coefficients of n on the two uses of C on the right side of eq. (249) sum to less than 1 (see the previous section and the more detailed analysis that comes below); i.e.,

$$\frac{1}{c} + \frac{3c-1}{4c} = \frac{3c+3}{4c} \tag{250}$$

must be less than 1. Since c must be odd, this means that we need $c \ge 5$.

Examination of small values of c (and using the actual number of comparisons required to sort c elements, rather than the upper limit of $c \lg c$) shows that $c = 9$ is the best value of c to use in this algorithm. (To sort 9 elements, 19 comparisons are needed.) Using $c = 9$, we find that

$$C(n) \le C(\tfrac{1}{9}n + 1) + C(\tfrac{13}{18}n + 4) + \tfrac{22}{9}n. \tag{251}$$

To obtain an upper limit on the solution of eq. (251), we can consider upper limits of the form $C(n) \le \alpha n + \beta$ and try to find the smallest values of α and β that work. Proceeding as in a proof by induction, assume that the limit has been established for all values of n below n_0. To establish the limit for n_0, we plug the formula into the right side of eq. (251). This gives

$$C(n_0) \le \tfrac{1}{9}\alpha n_0 + \alpha + \beta + \tfrac{13}{18}\alpha n_0 + 4\alpha + \beta + \tfrac{22}{9} n_0. \tag{252}$$

This is consistent with the induction hypothesis if this limit is below $\alpha n_0 + \beta$, i.e.,

$$\alpha n_0 + \beta \ge \tfrac{1}{9}\alpha n_0 + \alpha + \beta + + \tfrac{13}{18} n_0 + 4\alpha + \beta + \tfrac{22}{9} n_0 \tag{253}$$

for all n_0 larger than our base cases. (We will set up and prove our base cases as soon as we see what we need for base cases. We are combining proof by induction with the development of the details of our induction hypothesis.)

For eq. (252) to be satisfied for large n_0, we need α to obey the relation

$$\alpha n_0 \ge \tfrac{1}{9}\alpha n_0 + \tfrac{13}{18} n_0 + \tfrac{22}{9} n_0. \tag{254}$$

The smallest value of α that obeys eq. (254) is $\alpha = 14\frac{2}{3}$. Having selected α to be this smallest possible value, we need β to obey the relation

$$\beta \geq \alpha + \beta + 4\alpha + \beta \tag{255}$$
$$\geq 73\tfrac{1}{3} + 2\beta \tag{256}$$

Solving eq. (256) gives

$$\beta \leq -73\tfrac{1}{3}. \tag{257}$$

We will now consider how small we can make β and still have the limit satisfied by the boundary conditions. (We could at this point just set $\beta = -73\frac{1}{3}$ and notice that everything works for $d = 26,000$.) Since the recursive part of the algorithm generates problems as small as $(d+1)/9+1$, and since we plan to use a sorting algorithm that uses (at most) $n \lg n$ comparisons, we need

$$14\tfrac{2}{3}\,n + \beta \geq n \lg n \tag{258}$$

for n in the range $(d+1)/9+1 \leq n \leq d$. By trying various values for d, we find that $d = 19,762$ permits β to be as small as -7830. With $\beta = -7830$ and $d = 19,762$, our induction hypothesis is true for the required base cases $(2196 \leq n \leq 19,762)$. From the derivations of eq. (253) and eq. (257) we know that the hypothesis is true in general provided it is true for the base cases. Thus the Select Algorithm with $c = 9$ and $d = 19,762$ makes no more than

$$14\tfrac{1}{3}\,n - 7830 \tag{259}$$

comparisons.

The upper-bound analysis on the worst-case time for the Select Algorithm suggests that the algorithm is better than sorting only for rather large n. (A lower-bound analysis is needed to prove that it is not good for small n.) One reason that Select is not more efficient is that it does not make full use of the comparisons that it does. After it cuts the problem size down, it starts on the subproblem from scratch. Divide and conquer approaches often give the first indication that an improved algorithm can be developed. Once the divide and conquer algorithm is discovered, it is common for another algorithm to be developed which combines some of the ideas of the divide and conquer algorithm with some additional ideas to given an even more efficient (but complicated) algorithm.

The original paper on the linear-time median algorithm is Blum et al. [73]. They give a version of the algorithm in the text, and then they give an improved algorithm that does fewer than $5.44n + o(n)$ comparisons. Schönhage et al. [129] give an improved algorithm that does $3n + o(n)$ comparisons. Their algorithm goes to great lengths to use information from many of the comparisons more than once. None of these algorithms has been shown to be practical for small values of n. Elementary discussions of the Select Algorithm are given in Aho et al. [1, Section 3.6] and in Knuth [11, Section 5.3.3].

EXERCISES

1. Show that $c = 7$ and $c = 11$ do not work as well in the Select Algorithm as $c = 9$. (To sort 7 elements requires 13 compares; to sort 11 elements requires 26 compares.)

2. Analyze the Select Algorithm more carefully to find the smallest value of n that results in fewer than $n \lg n$ comparisons being used. Assume that $c = 9$ and that d has the best possible value.

CHAPTER 6

General Linear Recurrences

In the previous chapter we considered some of the simpler types of linear recurrence equations. We now continue with the study of more complex and more general linear recurrences.

6.1* POLYNOMIAL COEFFICIENTS

Consider the generating function

$$G(z) = \sum_{i \geq 0} a_i z^i = a_0 + \sum_{i \geq 1} a_i z^i \tag{1}$$

and its derivative

$$\frac{dG(z)}{dz} = \sum_{i \geq 1} i a_i z^{i-1}. \tag{2}$$

Multiplying the derivative by z gives

$$z\frac{dG(z)}{dz} = \sum_{i \geq 1} i a_i z^i = \sum_{i \geq 0} i a_i z^i. \tag{3}$$

Eq. (3) is useful for dealing with recurrence equations that contain coefficients that are linear functions of the index. There are similar formulas that use higher derivatives to handle higher powers of the index.

Let's consider the equation

$$(k+1)\rho p_{k+1} - (1+\rho k)p_k + p_{k-1} = 0 \qquad \text{for } k \geq 0 \tag{4}$$

with the boundary condition $p_{-1} = 0$ (i.e., $\rho p_1 = p_0$). If we multiply eq. (4) by z^k and sum over k, we get

$$\sum_{k\geq 1}(k+1)\rho p_{k+1}z^k - \sum_{k\geq 1}(1+\rho k)p_k z^k + \sum_{k\geq 1} p_{k-1}z^k = 0. \tag{5}$$

Now let's consider how to make the various pieces look like the generating function $G(z) = \sum_{k\geq 0} p_k z^k$ and its derivatives. Using $k' = k+1$ ($k = k'-1$), we have

$$\sum_{k\geq 1}(k+1)p_{k+1}z^k = \sum_{k\geq 2} kp_k z^{k-1} = \sum_{k\geq 0} kp_k z^{k-1} - p_1 \tag{6}$$

$$= \frac{dG(z)}{dz} - p_1. \tag{7}$$

We also have

$$\sum_{k\geq 1} p_k z^k = \sum_{k\geq 0} kp_k z^k - p_0 = G(z) - p_0, \tag{8}$$

and

$$\sum_{k\geq 1} kp_k z^k = z\sum_{k\geq 0} kp_k z^{k-1} = z\frac{dG(z)}{dz}. \tag{9}$$

Finally, using $k' = k-1$ ($k = k'+1$), we have

$$\sum_{k\geq 1} p_{k-1}z^k = \sum_{k\geq 0} p_k z^{k+1} = zG(z). \tag{10}$$

Using these results in eq. (5) gives

$$\rho\frac{dG(z)}{dz} - \rho p_1 - G(z) + p_0 - \rho z\frac{dG(z)}{dz} + zG(z) = 0 \tag{11}$$

or

$$\frac{dG(z)}{dz} - \frac{1}{\rho}G(z) = \frac{\rho p_1 - p_0}{\rho(1-z)}. \tag{12}$$

But the boundary condition for eq. (4) says that $\rho p_1 - p_0 = 0$, so

$$\frac{dG(z)}{dz} - \frac{1}{\rho}G(z) = 0. \tag{13}$$

Thus, if we can solve eq. (13) for $G(z)$ and expand the solution in a power series, we will know the solution of eq. (4).

In Section 6.1.2 we will briefly consider some of the techniques for solving differential equations. Eq. (13) can be solved by separating variables and integrating. We have

$$\frac{dG(z)}{G(z)} = \frac{dz}{\rho}, \tag{14}$$

or, after integrating,

$$\ln G(z) = \frac{z}{\rho} + C. \tag{15}$$

Exponentiating eq. (15) and using K for e^C, we get

$$G(z) = Ke^{z/\rho} \tag{16}$$

Expanding $G(z)$ in a power series gives

$$G(z) = K\sum_{k\geq 0} \frac{z^k}{k!\rho^k}. \tag{17}$$

Since by the definition of our generating function p_i is the coefficient of z^i, we have

$$p_i = \frac{K}{i!\rho^i}. \tag{18}$$

Considering eq. (18) with $i = 0$ shows that K must be p_0. Therefore, we have

$$p_i = \frac{p_0}{i!\rho^i}. \tag{19}$$

We need one more condition to determine the value of p_0. If the p_i are probabilities, then we have the condition that the sum of the p_i must be 1. In this case we get

$$1 = \sum_{i\geq 0} p_i = p_0 \sum_{i\geq 0} \frac{1}{i!\rho^i} = p_0 e^{1/\rho}. \tag{20}$$

Thus

$$p_0 = e^{-1/\rho} \tag{21}$$

and

$$p_i = \frac{1}{e^{1/\rho} i!\rho^i}. \tag{22}$$

Generalizing from the material in this section, you can see that by using generating functions you can reduce the problem of solving a general linear recurrence equation with coefficients that are polynomial functions of the index to the problem of solving a linear differential equation. The order of the differential equation is the same as the degree of the highest-degree polynomial coefficient. Moreover, any linear recurrence equation with rational coefficients can be reduced to the problem of solving an equation with polynomial coefficients by multiplying by the least common multiple of the denominators of the coefficients (except that special care is needed if any of the denominators is zero for some value of the index).

There is an extensive treatment of linear recurrence equations with polynomial coefficients in Milne-Thomson [43, Chapters 14 and 15].

EXERCISES

1. [See 21] Find the equation satisfied by the generating function for the recurrence

$$a_n = a_{n-1} + (n-1)a_{n-2}$$

with the boundary conditions $a_0 = 1$ and $a_1 = 1$.

2. A permutation is an *involution* if and only if it is its own inverse. A permutation is its own inverse if and only if its cycle representation consists of 1-cycles and 2-cycles. Show that the number of involutions of n objects is given by

$$t_n = t_{n-1} + (n-1)t_{n-2}.$$

What are the values of t_1 and t_2? Hint: To establish the recurrence equation, consider that the last element is either in a 1-cycle or a 2-cycle.

3. Find the generating function for the recurrence in the previous exercise.

6.1.1* Storage Needs

The example for the previous section comes from the following queueing problem. Assume that requests for service arrive at random, so that in a small time interval dt the probability of a request for service is $\lambda\, dt + o(dt)$. (This is the same assumption we made in Section 5.4.3.) We also assume that each unit gets service immediately (this is quite different from the assumption we made in Section 5.4.3) and that for *each* unit receiving service, the probability that the service is finished during any small time interval dt is $\mu\, dt + o(dt)$. This model of service is often called the *infinite processor model*, since each unit receives service when it requests it, no matter how many other units also want service. This model is often used for simplified descriptions of memory usage in time-shared computer systems.

If we let $p_k(t)$ be the probability that k units are being served at time t, then $p_k(t)$ obeys the equations

$$p_k(t+dt) = (1 - \lambda\, dt - k\mu\, dt)p_k(t) + ((k+1)\mu\, dt)p_{k+1}(t) + (\lambda\, dt)p_{k-1}(t) + o(dt) \tag{23}$$

where $p_{-1}(t) = 0$. Here, the first term on the right side of the equation represents the probability that k units are being served at time t, no new requests arrive, and none of the k units finishes being served in time interval dt. Comparing with the corresponding term in eq. (5.226), the factor $k\mu$ (rather than μ) is present because all k units have a chance of finishing. Similarly, the second term is the probability that there are $k+1$ units being serviced at time t and one finishes. The chance that more than one finishes is $O(dt^2)$, which is small enough to contain in the $o(dt)$ term. The third term is the probability that $k-1$ units are being serviced and an additional request arrives. Taking the limit as dt goes to zero in eq. (23) gives

$$\frac{dp_k(t)}{dt} = (k+1)\mu p_{k+1}(t) - (\lambda + k\mu)p_k(t) + \lambda p_{k-1}(t), \tag{24}$$

where $p_{-1}(t) = 0$. Using ρ for μ/λ, the steady-state solution of eq. (24) is given by

$$(k+1)\rho p_{k+1} - (1+\rho k)p_k + p_{k-1} = 0 \qquad \text{for } k \geq 0, \tag{25}$$

where $p_{-1} = 0$, which is the same as eq. (4). Therefore

$$p_i = \frac{1}{e^{1/\rho} i! \rho^i}. \tag{26}$$

Equations of the type considered in this section are covered in much more detail in Saaty [50, Section 4.3], Fisz [28, Section 8.5], and Feller [27, Section 17.7].

6.1.2* Differential Equations

Generating functions often result in the conversion of recurrence equations into related differential equations. We refer the reader to a general text, such as Ford, for a full discussion of the techniques to use to solve differential equations. Here, we will just give a few of the simpler methods.

There are many similarities between the methods used to solve recurrence equations and the methods used to solve differential equations. For example, it is straightforward to solve first order linear differential equations, while it is more difficult to solve nonlinear and higher order equations.

The general *first order linear differential equation* has the form

$$A(x)\frac{dy}{dx} + B(x)y + C(x) = 0. \tag{27}$$

By moving the $C(x)$ to the other side of the equal sign and dividing by $A(x)$, one can transform eq. (27) into

$$\frac{dy}{dx} + P(x)y = Q(x). \tag{28}$$

The general solution of this equation is

$$y = Ce^{-\int P(x)\,dx} + e^{-\int P(x)\,dx} \int e^{\int P(x)\,dx} Q(x)\,dx, \tag{29}$$

where C is an arbitrary constant.

The general n^{th} *order linear differential equation* has the form

$$y^{(n)} + p_{n-1}(x)y^{(n-1)} + \cdots + p_0(x)y = f(x), \tag{30}$$

where $y^{(i)}$ is the ith derivative of y with respect to x. If $f(x) = 0$, then the equation is *homogeneous*. If the $p_i(x)$ are constant for all i, the equation has constant coefficients. Homogeneous n^{th} order linear differential equations with constant coefficients have solutions of the same form as the corresponding recurrence equations (see Section 5.4.1), and their solutions can be found by the method of undetermined coefficients (see Section 5.4.2).

There are many excellent texts on differential equations. The authors are fond of Ford [29]. Also see Bender and Orszag [21].

EXERCISES

1. [See 21] Solve the recurrence $n(n-1)F_n - 2(n-1)F_{n-1} - 3F_{n-2} = 0$ with boundary conditions $F_0 = 2$ and $F_1 = 2$.

2. The Legendre polynomials are the solutions to the recurrence

$$(n+1)P_{n+1}(z) - (2n+1)zP_n(z) + nP_{n-1}(z) = 0,$$

with boundary conditions $P_0(z) = 1$ and $P_1(z) = z$. Find a closed form for the generating function $G(x,z) = \sum_{n\geq 0} P_n(z)x^n$.

3. The Bessel functions are the solutions of the recurrence

$$J_{n+1}(z) - \frac{2n}{z}J_n(z) + nJ_{n-1}(z) = 0,$$

with boundary conditions $J_n(0) = \delta_{n0}$. Find a closed form for the generating function $G(x,z) = \sum_{n\geq 0} J_n(z)x^n$.

4. Solve the recurrence $ncF_{n+1} + [(1+c)n + a]F_n + (b-n)F_{n-1} = 0$ in terms of a hypergeometric function.

6.2* PROPERTIES OF LINEAR RECURRENCES

This section gives some of the fundamental properties of linear finite-history recurrence equations

$$a_0(n)A_n + a_1(n)A_{n-1} + a_2(n)A_{n-2} + \cdots + a_k(n)A_{n-k} = b(n), \qquad \textbf{(31)}$$

where k is the order of the equation. The *singular points* of eq. (31) are values of n such that the first coefficient, $a_0(n)$, or the last, $a_k(n-k)$, (with a shift of k in the argument) is zero. As long as one is not at a singularity, one can use eq. (31) to compute A_n from $A_{n-1}, \ldots, A_{n-k}$, or from $A_{n+1}, \ldots, A_{n+k}$ (i.e., one can calculate the next value from the previous values when calculating in the direction of either increasing or decreasing index). The equation is *homogeneous* if $b(n) = 0$. Linear recurrence equations with order greater than 1 are much harder to solve than linear equations of order 1. The previous section contains techniques for special cases, where it is known how to solve eq. (31). Here we will consider some general *simplification techniques.*

6.2.1* Linear Combinations

If A_n is a solution of a homogeneous linear recurrence equation, then so is cA_n for any constant c. Likewise, if A_n and B_n are two solutions of a homogeneous linear recurrence equation, then $c_1A_n + c_2B_n$, where c_1 and c_2 are any constants, is also a solution. The homogeneous equation corresponding to the nonhomogeneous equation eq. (31) is obtained by replacing $b(n)$ with zero. If P_n is a solution of a nonhomogeneous equation, and A_n and B_n are solutions of the corresponding homogeneous equation, then $c_1A_n + c_2B_n + P_n$ is a solution of the nonhomogeneous equation for any constants c_1 and c_2. The function P_n is called a *particular solution.* All the results in this paragraph can be obtained by plugging $c_1A_n + c_2B_n + P_n$ into eq. (31) and simplifying by using $a_0(n)A_n + a_1(n)A_{n-1} + a_2(n)A_{n-2} + \cdots + a_k(n)A_{n-k} = 0$, $a_0(n)B_n + a_1(n)B_{n-1} + a_2(n)B_{n-2} + \cdots + a_k(n)B_{n-k} = 0$, and $a_0(n)P_n + a_1(n)P_{n-1} + a_2(n)P_{n-2} + \cdots + a_k(n)P_{n-k} = b(n)$.

The most general solution of eq. (31) has the form

$$A_n = c_1B_1(n) + c_2B_2(n) + \cdots + c_kB_k(n) + P_n, \tag{32}$$

where $B_1(n), B_2(n), \ldots, B_k(n)$ are k linearly independent solutions of the corresponding homogeneous equation and P_n is a solution of the nonhomogeneous equation.

Usually it is rather obvious whether or not a set of solutions is linearly independent. For example, polynomials of different degrees are independent. Exponentials are independent of polynomials and of exponentials with other bases. There is, however, a formal theory for testing for linear independence, which should be used when it is not clear whether or not a set of solutions is independent.

We need to define the *Casoratian* (also called the *Wronskian* by some authors), $W_n = W(B_1(n), B_2(n), \ldots, B_k(n))$, of k sequences, which is the $k \times k$ determinant

$$W_n = \det \begin{vmatrix} B_1(n) & B_2(n) & \cdots & B_k(n) \\ B_1(n+1) & B_2(n+1) & \cdots & B_k(n+1) \\ \vdots & \vdots & & \vdots \\ B_1(n+k-1) & B_2(n+k-1) & \cdots & B_k(n+k-1) \end{vmatrix}. \tag{33}$$

The functions $B_1(n), B_2(n), \ldots, B_k(n)$ are linearly dependent over the range $n_1 \le n \le n_2$ if and only if $W_n = 0$ over the same range. In the case $k = 2$ the Casoratian is just $W_n = B_1(n)B_2(n+1) - B_1(n+1)B_2(n)$.

Now, the Casoratian of a linear homogeneous difference equation obeys the linear first order difference equation

$$W_{n+1} = (-1)^k \frac{a_k(n+k)}{a_0(n+k)} W_n, \tag{34}$$

which has the solution

$$W_n = (-1)^{kn} W_0 \prod_{k \le i \le n+k-1} \frac{a_k(i)}{a_0(i)}. \tag{35}$$

Thus, when $a_k(i) \neq 0$ for $0 \leq i \leq n$, W_n is equal to zero if and only if W_0 is equal to zero. Thus when $a_k(i)$ is well behaved, you only need to evaluate W_0 to determine whether your proposed solution set is linearly independent.

To prove eq. (34), notice that

$$W_{n+1} = \det \begin{vmatrix} B_1(n+1) & B_2(n+1) & \cdots & B_k(n+1) \\ B_1(n+2) & B_2(n+2) & \cdots & B_k(n+2) \\ \vdots & \vdots & & \vdots \\ B_1(n+k) & B_2(n+k) & \cdots & B_k(n+k) \end{vmatrix}. \tag{36}$$

Now replace the last row of the determinant by the sum over i of the ith row multiplied by $a_{k-i}(n+k)/a_0(n+k)$. This does not change the value of the determinant since the last row is multiplied by 1 when it is combined into this sum. The jth element of the new last row is

$$\frac{a_{k-1}(n+k)}{a_0(n+k)} B_j(n+1) + \frac{a_{k-2}(n+k)}{a_0(n+k)} B_j(n+2) + \cdots + \frac{a_0(n+k)}{a_0(n+k)} B_j(n+k). \tag{37}$$

But the recurrence equation [eq. (31) with $b(n) = 0$ and with n replaced by $n+k$] gives

$$-\frac{a_k(n+k)}{a_0(n+k)} A_n = \frac{a_{k-1}(n+k)}{a_0(n+k)} A_{n+1} + \frac{a_{k-2}(n+k)}{a_0(n+k)} A_{n+2} + \cdots + \frac{a_0(n+k)}{a_0(n+k)} A_{n+k}. \tag{38}$$

Each B_j is a solution of this equation, so the sum in eq. (37) equals

$$\frac{-a_k(n+k)}{a_0(n+k)} B_j(n). \tag{39}$$

So, replacing the last row of the modified determinant, we get

$$\det \begin{vmatrix} B_1(n+1) & B_2(n+1) & \cdots & B_k(n+1) \\ B_1(n+2) & B_2(n+2) & \cdots & B_k(n+2) \\ \vdots & \vdots & & \vdots \\ -\frac{a_k(n+k)}{a_0(n+k)} B_1(n) & -\frac{a_k(n+k)}{a_0(n+k)} B_2(n) & \cdots & -\frac{a_k(n+k)}{a_0(n+k)} B_k(n) \end{vmatrix}. \tag{40}$$

Now move the last row to the top, and move all the other rows down one. This will multiply the value of the determinant by $(-1)^{k+1}$. Finally, you can factor out $-a_k(n+k)/a_0(n+k)$ to obtain eq. (34). (The exercises suggest that you redo this proof without the use of determinants for the case $k = 2$.)

Casoratians are considered by Bender and Orszag [21, Section 2.4] and by Milne-Thomson [43, Section 12.11].

EXERCISES

1. Show that any solution $B(n)$ of a kth order linear homogeneous recurrence equation which is equal to zero for k consecutive integers is equal to zero for all integers, provided $a_0(i) \neq 0$ for any i and provided $a_k(i) \neq 0$ for any i. Hint: Compute $B(n)$ where n is the point just after the consecutive zeros and use induction.
2. Show that, under the assumption $a_0(i) \neq 0$ and $a_k(i) \neq 0$ for all i, a kth order linear homogeneous recurrence equation can have at most k linearly independent solutions. Hint: Use the previous exercise.
3. Prove eq. (34) directly (without use of determinant notation) for the case $k = 2$.

6.2.2* Reducing the Order with a Solution

If you manage somehow to find a solution of a kth order linear homogeneous recurrence equation, then you can use that solution to obtain a new equation of degree $k-1$ that has all the remaining solutions of the original equation (the solutions that are linearly independent of the solution you found).

Let the original equation be

$$a_0(n)A(n) + a_1(n)A(n-1) + a_2(n)A(n-2) + \cdots + a_k(n)A(n-k) = 0, \quad \textbf{(41)}$$

and let $A_1(n)$ be a known solution of the equation. We will now make the substitution

$$A(n) = A_1(n)B(n), \qquad \textbf{(42)}$$

and try to solve eq. (41) for $B(n)$. The substitution gives

$$a_0(n)A_1(n)B(n)+a_1(n)A_1(n-1)B(n-1) + a_2(n)A_1(n-2)B(n-2) + \cdots \\ + a_k(n)A_1(n-k)B(n-k) = 0, \qquad (43)$$

or

$$\sum_{0\le i<k+1} a_i(n)A_1(n-i)B(n-i) = 0. \qquad (44)$$

We will apply summation by parts to this formula; to avoid confusion, rewrite the summation-by-parts formula [eq. (2.72)] and the related formula [eq. (2.74)] replacing a_i with r_i, b_i with s_i, and c_i with t_i. Apply the rewritten formula using $t_i(n) = a_i(n)A_1(n-i)$, $s_i(n) = B(n-i)$, and

$$r_i(n) = \sum_{0\le j<i} a_j(n)A_1(n-j). \qquad (45)$$

The value of $s_{i+1}(n) - s_i(n)$ is $B(n-i-1) - B(n-i)$. The result of using summation by parts on eq. (44) is

$$\sum_{0\le i<k} r_{i+1}(n)[B(n-i) - B(n-i-1)] = r_0(n)B(n) - r_{k+1}(n)B(n-k). \quad (46)$$

But $r_0(n)$ is zero because there are no terms in the sum of eq. (45) when $i = 0$; and $r_{k+1}(n)$ is zero because $A_1(n)$ is a solution of eq. (41), which has a left side equal to eq. (45) when $i = k+1$. So eq. (46) can be written as

$$\sum_{0 \le i < k} r_{i+1}(n)[B(n-i) - B(n-i-1)] = 0. \tag{47}$$

Letting $D(n) = B(n) - B(n-1)$, eq. (47) may be written as

$$\sum_{0 \le i < k} r_{i+1}(n) D(n-i) = 0, \tag{48}$$

which is a linear homogeneous equation of degree $k-1$.

This gives us the following procedure for solving eq. (41) when we know one solution, $A_1(n)$, which is not identically zero.

1. Define $r_i(n)$ by eq. (45).
2. Solve eq. (48) for $D(n)$. This is a $k-1$ order linear equation.
3. Solve the equation $B(n) - B(n-1) = D(n)$ for $B(n)$. This is a first order linear equation.
4. Obtain the solutions of the original problem by solving $A(n) = A_1(n)B(n)$ for $A(n)$. This is an algebraic equation.

Let's consider an example:

$$(n^2+n+1)A(n) = 2nA(n-1) + 4(n^2+1)A(n-2). \tag{49}$$

One solution of this equation is $A_n = 2^n$. Using eq. (45), we get

$$r_1 = (n^2+n+1)2^n \tag{50}$$

$$r_2 = (n^2+n+1)2^n - 2n2^{n-1} = (n^2+1)2^n \tag{51}$$

$$r_3 = (n^2+n+1)2^n - 2n2^{n-1} - 4(n^2+1)2^{n-2} = 0. \tag{52}$$

The equation for $D(n)$ is

$$(n^2+n+1)2^n D(n) + (n^2+1)2^n D(n-1) = 0, \tag{53}$$

and the solution is

$$D(n) = c_1 \prod_{1 \le i \le n} \frac{-(i^2+1)}{i^2+i+1}, \tag{54}$$

where c_1 is a constant equal to $D(0)$. Now, $B(n)$ is the solution of the equation $B(n) - B(n-1) = D(n)$, so

$$B(n) = c_2 + \sum_{2 \le i \le n} D(i) = c_2 + c_1 \sum_{1 \le i \le n} \prod_{1 \le j \le i} \frac{-(j^2+1)}{j^2+j+1}, \tag{55}$$

where c_1 and c_2 are constants. The solution of the original equation is

$$A(n) = c_2 2^n + c_1 2^n \sum_{1 \le i \le n} \prod_{1 \le j \le i} \frac{-(j^2+1)}{j^2+j+1}. \tag{56}$$

This answer is not very simple, and it is not in closed form, but that is because the problem does not have a simple closed-form answer.

Reduction of linear recurrence equations by use of a solution is discussed in Bender and Orszag [21, Section 2.4] and Milne-Thomson [43, Section 12.3].

EXERCISES

1. [See 21] Solve $(n-2)^2 F_n - (n-1)(n-2)F_{n-1} + F_{n-2} = 0$ using the knowledge that one solution is $F_n = n$.
2. Solve the equation $(n+4)F_{n+2} + F_{n+1} - (n+1)F_n = 0$. You may use the fact that $F_n = 1/[(n+1)(n+2)]$ is a solution.
3. Solve $F_{n+2} - \dfrac{2n+1}{n}F_{n+1} + \dfrac{n}{n+1}F_n = n(n+1)$. Hint: $F_n = n-1$ is a solution.
4. The *Euler recurrence equation* is

$$c_N n^{\overline{N}} \Delta^N F_n + c_{N-1} n^{\overline{N-1}} \Delta^{N-1} F_n + \cdots + c_1 n \Delta F_n + c_0 F_n = 0.$$

Show that this equation is satisfied by $n^{\underline{r}}$ provided r is chosen correctly. Hint: The correct way to choose r is so that

$$c_N r^{\underline{N}} + c_{N-1} r^{\underline{N-1}} + \cdots + c_1 r + c_0 = 0.$$

5. Solve $n(n+1)\Delta^2 F_n + n\Delta F_n - \frac{1}{4}F_n = 0$.
6. Solve $n(n+1)\Delta^2 F_n + \frac{1}{4}F_n = 0$.

6.2.3* Reducing the Order by Factoring

In this section we will write our linear recurrence equation as

$$\sum_{0 \le i < k} a_i(n) X(n+i) = b(n). \tag{57}$$

In operator notation eq. (57) is

$$\left(\sum_{0 \le i < k} a_i(n) \mathbf{E}^i \right) X(n) = b(n). \tag{58}$$

Now, $\sum_{0 \le i < k} a_i(n)\mathbf{E}^i$ is a polynomial of degree k in $\mathbf{E}$. Calling this polynomial $P(\mathbf{E})$, eq. (58) can be written as

$$P(\mathbf{E})X(n) = b(n). \tag{59}$$

If $P(\mathbf{E})$ factors into $P_1(\mathbf{E})P_2(\mathbf{E})$, where $P_1(\mathbf{E})$ has degree k_1 and $P_2(\mathbf{E})$ has degree k_2 (with $k = k_1 + k_2$), then you can solve eq. (58) by first solving

$$P_1(\mathbf{E})Y(n) = b(n) \tag{60}$$

and then solving

$$P_2(\mathbf{E})X(n) = Y(n). \tag{61}$$

Factoring replaces solving one equation of degree k with solving two equations (of degrees k_1 and $k - k_1$).

If the coefficients of $P(\mathbf{E})$ are constant, then $P(\mathbf{E})$ can be factored in the same way as an ordinary polynomial. Let's consider the example from Section 5.3 again. We rewrite eq. (5.105) as

$$A_{n+2} - 5A_{n+1} + 6A_n = 0 \qquad \text{for } n \geq 0, \tag{62}$$

$$A_0 = 0, \qquad A_1 = 1. \tag{63}$$

Using operators, eq. (62) can be written as

$$\mathbf{E}^2 A_n - 5\mathbf{E}A_n + 6A_n = 0 \qquad \text{for } n \geq 0, \tag{64}$$

or

$$(\mathbf{E}^2 - 5\mathbf{E} + 6)A_n = 0 \qquad \text{for } n \geq 0, \tag{65}$$

which is a second degree polynomial in $\mathbf{E}$.

When factoring expressions that contain operators, you must be careful of the effects of the operators on the coefficients of the polynomial. In this case we have $\mathbf{E}c = c\mathbf{E}$ for any constant c, so there is no problem. The polynomial can be factored as

$$\mathbf{E}^2 - 5\mathbf{E} + 6 = \mathbf{E}^2 - 3\mathbf{E} - 2\mathbf{E} + 6 = \mathbf{E}(\mathbf{E} - 3) - 2(\mathbf{E} - 3) = (\mathbf{E} - 2)(\mathbf{E} - 3). \tag{66}$$

To solve eq. (65), we first need to solve

$$(\mathbf{E} - 2)Y_n = 0, \tag{67}$$

or equivalently

$$Y_{n+1} - 2Y_n = 0. \tag{68}$$

The solution of $Y_{n+1} - 2Y_n = 0$ is $Y_n = Y_0 2^n$. Now we need to solve

$$(\mathbf{E} - 3)A_n = Y_0 2^n, \tag{69}$$

or equivalently

$$A_{n+1} - 3A_n = Y_0 2^n. \tag{70}$$

The solution is

$$A_n = [A_0 + Y_0]3^n - Y_0 2^n, \tag{71}$$

which can be written as

$$A_n = k_1 3^n + k_2 2^n. \tag{72}$$

Our boundary conditions are $A_0 = 0$ and $A_1 = 1$, so

$$0 = k_1 + k_2, \tag{73}$$

$$1 = 3k_1 + 2k_2, \tag{74}$$

giving $k_1 = 1$ and $k_2 = -1$. Thus the solution of eq. (62) and its boundary conditions is

$$A_n = 3^n - 2^n. \tag{75}$$

By factoring, we reduced the problem of solving a second order recurrence to that of solving two first order recurrences.

If the coefficients of $P(\mathbf{E})$ depend on the variable upon which $\mathbf{E}$ operates, then factoring is more complex. You must allow for the effect of the operator in the first factor upon any occurrence of the variable in the second factor. For example, consider the equation

$$A_{i+2} - (i+2)A_{i+1} + iA_i = 0 \tag{76}$$

which can be written as

$$(\mathbf{E}^2 - (i+2)\mathbf{E} + i)A_i = 0. \tag{77}$$

Since $\mathbf{E}i = (i+1)\mathbf{E}$, we can factor $\mathbf{E}^2 - (i+2)\mathbf{E} + i$ as

$$\mathbf{E}^2 - (i+2)\mathbf{E} + i = \mathbf{E}^2 - \mathbf{E} - (i+1)\mathbf{E} + i \tag{78}$$

$$= \mathbf{E}^2 - \mathbf{E} - \mathbf{E}i + i = (\mathbf{E}-1)(\mathbf{E}-i). \tag{79}$$

Notice that the "multiplication" $\mathbf{E}i = (i+1)\mathbf{E}$ is noncommutative. Care is needed to ensure that the proper adjustments are made when you interchange the order of an $\mathbf{E}$ and a factor involving the variable. We write $\mathbf{E}i = (i+1)\mathbf{E}$ rather than $\mathbf{E}i = i+1$ because we wish to apply the resulting expression to A_i. If the result of the operation of $\mathbf{E}i$ were our final answer, we would write it as $i+1$, since the trailing $\mathbf{E}$ would have no effect. When writing an operator formula that is supposed to be valid whether or not the answer is a final result, retain any trailing operators.

To solve our original problem, we first need to solve

$$(\mathbf{E}-1)Y_i = 0, \tag{80}$$

which has the constant solution

$$Y_i = Y_0. \tag{81}$$

Then we need to solve the equation

$$(\mathbf{E}-i)A_i = Y_0, \tag{82}$$

which has the solution

$$A_i = (i-1)!\left(A_1 + Y_0 \sum_{2\le j\le i} \frac{1}{j!}\right). \tag{83}$$

For large i the sum is close to $e-1$.

Reduction by factoring is discussed in Greene and Knuth [6, Section 2.1.1.2].

EXERCISES

1. A *derangement* is a permutation that leaves no elements fixed. Show that the number of derangements of n objects is given by the recurrence equation

$$D_n = (n-1)D_{n-1} + (n-1)D_{n-2},$$

with the boundary conditions $D_1 = 0$ and $D_2 = 1$. Hint: Consider a sequence of numbers in the range 1 to n where the last number is n. Then consider interchanging the last number with the ith one. When does this result in a derangement?

2. Solve the recurrence in the previous exercise. Hint: The solution is

$$D_n = n! \sum_{1 \leq i \leq n} \frac{(-1)^{i+1}}{i!}.$$

3. Show that the function D_n from the previous exercise satisfies the equation $D_n = (n!/e)(1 \pm \epsilon)$, where $\epsilon < 1/n!$.

6.2.4* Variation of Parameters

Many of the methods we have previously considered, including generating functions and reduction of order with known solutions, can often be used to solve nonhomogeneous recurrence equations. It is usually quicker to first try those methods and then try variation of parameters if the earlier methods do not quickly lead to a solution. Variation of parameters reduces the problem of solving nonhomogeneous linear recurrence equations to the evaluation of a sum, provided the general solution of the associated homogeneous equation is known.

If you have a complete independent set of solutions for a homogeneous linear equation, then you can use variation of parameters to obtain a solution of the associated nonhomogeneous equation. Since the details of this technique are somewhat involved, we will consider second order equations first. Let the original equation be

$$a_n A_n + b_n A_{n-1} + c_n A_{n-2} = d_n. \tag{84}$$

Let B_n and D_n be two linearly independent solutions of the corresponding homogeneous equation.

We will look for solutions of eq. (84) that have the form

$$A_n = x_n B_n + y_n D_n, \tag{85}$$

where x_n and y_n are unknown functions of n. Plugging eq. (85) into eq. (84) gives

$$\begin{aligned} &a_n x_n B_n + b_n x_{n-1} B_{n-1} + c_n x_{n-2} B_{n-2} \\ &\qquad + a_n y_n D_n + b_n y_{n-1} D_{n-1} + c_n y_{n-2} D_{n-2} = d_n. \end{aligned} \tag{86}$$

Since B_n is a solution of the homogeneous equation, $a_n B_n = -b_n B_{n-1} - c_n B_{n-2}$, and a similar formula holds for D_n. Using this, eq. (86) can be reduced to

$$b_n B_{n-1}[x_{n-1} - x_n] + c_n B_{n-2}[x_{n-2} - x_n] \\ + b_n D_{n-1}[y_{n-1} - y_n] + c_n D_{n-2}[y_{n-2} - y_n] = d_n. \quad (87)$$

Now, eq. (87) is one recurrence equation with two undetermined functions, x_n and y_n. We can impose one more condition relating y_n to x_n and still solve eq. (87) for x_n. We will select this additional condition to make eq. (87) easy to solve. One such condition is

$$B_{n-1}[x_{n-1} - x_n] + D_{n-1}[y_{n-1} - y_n] = 0. \quad (88)$$

Using eq. (88) to simplify eq. (87) gives

$$c_n B_{n-2}[x_{n-2} - x_n] + c_n D_{n-2}[y_{n-2} - y_n] = d_n. \quad (89)$$

We would now like to find the simultaneous solution of eq. (88) and eq. (89). First, however, we need one more simplification. Eq. (89) has $x_{n-2} - x_n$, while eq. (88) has $x_{n-1} - x_n$, and likewise for y. We want to make the two equations more alike.

Eq. (88) is required to hold for all n, so it is still true if n is replaced by $n - 1$. This replacement gives

$$B_{n-2}[x_{n-2} - x_{n-1}] + D_{n-2}[y_{n-2} - y_{n-1}] = 0, \quad (90)$$

or

$$B_{n-2}x_{n-2} + D_{n-2}y_{n-2} = B_{n-2}x_{n-1} + D_{n-2}y_{n-1}. \quad (91)$$

Using eq. (91) to eliminate x_{n-2} and y_{n-2} in eq. (89) gives

$$c_n B_{n-2}[x_{n-1} - x_n] + c_n D_{n-2}[y_{n-1} - y_n] = d_n. \quad (92)$$

If you now regard $x_{n-1} - x_n$ and $y_{n-1} - y_n$ as unknowns, then eqs. (88, 92) are a set of two equations in two unknowns. Using linear algebra, one gets

$$x_{n-1} - x_n = \frac{D_{n-1}d_n}{[B_{n-2}D_{n-1} - B_{n-1}D_{n-2}]c_n} = \frac{D_{n-1}d_n}{W_{n-2}c_n}, \quad (93)$$

$$y_{n-1} - y_n = \frac{-B_{n-1}d_n}{[B_{n-2}D_{n-1} - B_{n-1}D_{n-2}]c_n} = \frac{-B_{n-1}d_n}{W_{n-2}c_n}, \quad (94)$$

where W is the Casoratian. A solution of eqs. (93, 94) is

$$x_n = -\sum_{2 \le i \le n} \frac{D_{i-1}d_i}{W_{i-2}c_i}, \qquad y_n = \sum_{2 \le i \le n} \frac{B_{i-1}d_i}{W_{i-2}c_i}. \quad (95)$$

(The choice of the lower limit for the summations is arbitrary.) Thus a general solution of eq. (84) is

$$A_n = pB_n + qD_n - B_n \sum_{2 \le i \le n} \frac{D_{i-1}d_i}{W_{i-2}c_i} + D_n \sum_{2 \le i \le n} \frac{B_{i-1}d_i}{W_{i-2}c_i}, \quad \mathbf{(96)}$$

where p and q are constants.

EXERCISES

1. Solve $F_{n+2} + F_{n+1} + F_n = 2^n$.
2. Solve $F_{n+2} - 4F_{n+1} + 4F_n = 2^n$.
3. Solve $n^2 F_{n+2} - n(n+1)F_{n+1} + F_n = 2^n$.
4. Solve $(n+4)F_{n+2} + F_{n+1} - (n+1)F_n = 1$. The homogeneous equation has the solutions $F_n = 1/[(n+1)(n+2)]$ and $F_n = (-1)^{n+1}(2n+3)/[4(n+1)(n+2)]$.

6.2.5* Variation of Parameters: The General Case

We will now show how to generalize variation of parameters so that it applies to the general linear k^{th} order recurrence equation. (You may find it helpful to compare this section with the previous one.) The equation we want to solve is

$$\sum_{0 \le i \le k} a_i(n) A(n-i) = b(n). \tag{97}$$

Let $B_1(n), B_2(n), \ldots, B_k(n)$ be k linearly independent solutions of the corresponding homogeneous equation. We will look for solutions of eq. (97) that have the form

$$A(n) = \sum_{1 \le i \le k} x_i(n) B_i(n). \tag{98}$$

Plugging eq. (98) into eq. (97) gives

$$\sum_{0 \le i \le k} a_i(n) \sum_{1 \le j \le k} x_j(n-i) B_j(n-i) = b(n). \tag{99}$$

Since each $B_j(n)$ is a solution of the homogeneous equation,

$$a_0(n) B_j(n) = -\sum_{1 \le i \le k} a_i(n) B_j(n). \tag{100}$$

Now, eq. (100) is one recurrence equation with k undetermined functions, $x_j(n)$ for $1 \le j \le k$. We can impose $k-1$ more conditions. We will select the additional conditions to make eq. (100) easy to solve. The conditions are:

$$\sum_{1 \le j \le k} B_j(n-i)[x_j(n-i) - x_j(n)] = 0 \qquad \text{for } 1 \le i \le k-1. \tag{101}$$

Using eq. (101) to simplify eq. (100), we get

$$a_k(n) \sum_{1 \le j \le k} B_j(n-k)[x_j(n-k) - x_j(n)] = b(n). \tag{102}$$

We would like to find a simultaneous solution of eq. (101) and eq. (102). First, however, we need additional simplifications. We want to convert all the equations so that the unknowns have the form $x_j(n-1) - x_j(n)$.

Now, eq. (101) is required to hold for all n, so it is still true if n is replaced by $n-1$, giving

$$\sum_{1\le j\le k} B_j(n-i-1)[x_j(n-i-1)-x_j(n-1)]=0 \qquad \text{for } 1\le i\le k-1, \tag{103}$$

$$\sum_{1\le j\le k} B_j(n-i)[x_j(n-i)-x_j(n-1)]=0 \qquad \text{for } 2\le i\le k, \tag{104}$$

$$\sum_{1\le j\le k} B_j(n-i)x_j(n-i)=\sum_{1\le j\le k} B_j(n-i)x_j(n-1) \quad \text{for } 2\le i\le k. \tag{105}$$

Using eq. (105) to eliminate $x_j(n-i)$ for $2\le i\le k$ in eq. (101) and eq. (102) gives

$$\sum_{1\le j\le k} B_j(n-i)[x_j(n-1)-x_j(n)]=0 \qquad \text{for } 1\le i\le k-1, \tag{106}$$

$$a_k(n)\sum_{1\le j\le k} B_j(n-k)[x_j(n-1)-x_j(n)]=b(n). \tag{107}$$

Now if you regard the $x_j(n-1)-x_j(n)$ for $1\le j\le k$ as the k unknowns, eqs. (106, 107) are k equations in k unknowns. To find the general solution, solve these equations for the $x_j(n-1)-x_j(n)$. Call the solutions $X_j(n)$, so that

$$x_j(n-1)-x_j(n)=X_j(n) \tag{108}$$

$$=\frac{b(n)\det\begin{vmatrix} B_1(n-k) & \cdots & B_{j-1}(n-k) & B_{j+1}(n-k) & \cdots & B_k(n-k)\\ \vdots & & \vdots & \vdots & & \vdots\\ B_1(n+j-k-1) & \cdots & B_{j-1}(n+j-k-1) & B_{j+1}(n+j-k-1) & \cdots & B_k(n+j-k-1\\ B_1(n+j-k+1) & \cdots & B_{j-1}(n+j-k+1) & B_{j+1}(n+j-k+1) & \cdots & B_k(n+j-k+1\\ \vdots & & \vdots & \vdots & & \vdots\\ B_1(n-1) & \cdots & B_{j-1}(n-1) & B_{j+1}(n-1) & \cdots & B_k(n-1)\end{vmatrix}}{a_k(n)\det\begin{vmatrix} B_1(n-k) & B_2(n-k) & \cdots & B_k(n-k)\\ B_1(n-k+1) & B_2(n-k+1) & \cdots & B_k(n-k+1)\\ \vdots & \vdots & & \vdots\\ B_1(n-1) & B_2(n-1) & \cdots & B_k(n-1)\end{vmatrix}} \tag{109}$$

and notice that the denominator of this last expression is equal to $a_k(n)W(n-k)$. A solution for the $x_j(n)$ is

$$x_j(n)=\sum_{k\le i\le n} X_j(i). \tag{110}$$

Thus, a general solution of eq. (97) is

$$A(n)=\sum_{1\le j\le k} c_jB_j(n)+\sum_{1\le j\le k} B_j(n)\sum_{k\le i\le n} X_j(i). \tag{111}$$

A nonhomogeneous equation can often be solved with the same technique that is used to solve the corresponding homogeneous equation (for example, the technique of generating functions). For those linear equations where this is not the case, the method of variation of parameters can be used to reduce the original problem to one of solving the homogeneous equation plus the problem of summing the $X_j(i)$ in eq. (111).

Variation of parameters is discussed by Fort [30, Section 7.5] and by Bender and Orszag [21, Section 2.4].

EXERCISE

1. Solve $4F_{n+3} + 3F_{n+2} + 2F_{n+1} + F_n = n$.

6.3* GENERATING FUNCTIONS OF PROBABILITIES

Consider a process that produces a *score* that ranges over the nonnegative integers. Let p_i be the probability that the process produces score i. For example, in Section 6.1.1 we considered a process where the number of units being served could be considered as a score. For any process that generates scores, the *probability generating function*

$$G(z) = \sum_{i\geq 0} p_i z^i \tag{112}$$

has several interesting and useful properties. Evaluating $G(z)$ at $z = 1$ gives

$$G(1) = \sum_{i\geq 0} p_i = 1 \tag{113}$$

because the sum of the probabilities of all possible scores is 1.

The derivative of $G(z)$ is

$$G'(z) = \sum_{i\geq 0} ip_i z^{i-1}. \tag{114}$$

Evaluating at $z = 1$ gives

$$G'(1) = \sum_{i\geq 0} ip_i = A, \tag{115}$$

where A is the average of the scores. The second derivative of $G(z)$ is

$$G''(z) = \sum_{i\geq 0} i(i-1)p_i z^{i-2} = \sum_{i\geq 0} i^2 p_i z^{i-2} - \sum_{i\geq 0} ip_i z^{i-2}. \tag{116}$$

Since the variance V of a process is given by

$$V = \sum_{i\geq 0} i^2 p_i - A^2, \tag{117}$$

where A is the average of the process (see eq. (1.57)), we can use eq. (115) and eq. (116) to obtain

$$V = G''(1) + G'(1) - G'(1)^2. \tag{118}$$

Eq. (115) gives a simple way to compute the average score when the generating function is known. For example, eq. (16) gives the generating function for the process of Section 6.1.1. Since $G(1)$ must be 1, the K in eq. (16) is given by $1 = Ke^{1/\rho}$ or $K = e^{-1/\rho}$, so

$$G(z) = e^{(z-1)/\rho}. \tag{119}$$

Taking the derivative gives

$$G'(z) = \frac{1}{\rho}e^{(z-1)/\rho}. \tag{120}$$

Evaluating at $z = 1$ gives the average:

$$A = G'(1) = \frac{1}{\rho}. \tag{121}$$

The second derivative is

$$G''(z) = \frac{1}{\rho^2}e^{(z-1)/\rho}, \tag{122}$$

so the variance is

$$V = G''(1) + G'(1) - G'(1)^2 = \frac{1}{\rho}. \tag{123}$$

Many of the techniques for solving recurrences lead directly to a generating function for the problem. (See Section 6.1, for example.) It is not always easy or even possible to obtain explicit closed-form formulas for the probabilities, but it is usually straightforward to take the derivatives required to use eq. (115) and eq. (118).

Suppose $A(z) = \sum_{i\geq 0} a_i z^i$ and $B(z) = \sum_{i\geq 0} b_i z^i$ are generating functions (not necessarily probability generating functions) and $C(z) = A(z)B(z)$. If we write $C(z)$ in the form $C(z) = \sum_{i\geq 0} c_i z^i$, we get

$$c_0 + c_1 z + c_2 z^2 + \cdots = a_0 b_0 + (a_0 b_1 + a_1 b_0)z + (a_0 b_2 + a_1 b_1 + a_2 b_0)z^2 + \cdots. \tag{124}$$

Since eq. (124) must hold for all z, the coefficients of each power of z on the two sides of the equation must be equal. We therefore have

$$c_0 = a_0 b_0, \tag{125}$$

$$c_1 = a_0 b_1 + a_1 b_0, \tag{126}$$

$$c_2 = a_0 b_2 + a_1 b_1 + a_2 b_0, \tag{127}$$

and in general

$$c_n = \sum_{0\leq i\leq n} a_i b_{n-i}. \tag{128}$$

This sum is called the *convolution* of the two sequences.

If $A(z)$ and $B(z)$ are the probability generating functions for the processes A and B, then $C(z)$ is the probability generating function for the process C that consists of doing process A and process B and summing the two scores. To see this, notice that in process C the probability of score zero is $a_0 b_0$, because both A and B must get a score of zero for C to have a score of zero. Likewise, the probability of score 1 for C is $a_1 b_0 + a_0 b_1$, because one of A and B must get a score of 1 and the other must get a score of zero. In general the probability of score n for process C is given by

$$c_n = \sum_{0 \le i \le n} a_i b_{n-i}, \tag{129}$$

where the ith term in the sum gives the probability that process A gets score i and process B gets score $n-i$. This is the same as eq. (128). As in our other examples, this formula allows us to calculate $C(z)$ from the functions $A(z)$ and $B(z)$ without having to calculate the a's and b's explicitly.

When $C(z)$ is the product of probability generating functions the rule for differentiating products can be used to obtain

$$A_C = A_A + A_B, \tag{130}$$

$$V_C = V_A + V_B, \tag{131}$$

where A_X is the average of process X and V_X is the variance.

Suppose $A(z) = \sum_{i \ge 0} a_i z^i$ and $B(z) = \sum_{i \ge 0} b_i z^i$ are probability generating functions and $C(z) = A(B(z))$. Then $C(z)$ is also a probability generating function. Let the score for process A indicate the number of times that process B will be done. Then C is the process whose score is the total score for all the times that process B is done.

To see what is happening, let's compute the first few coefficients of $C(z)$:

$$\begin{aligned} C(z) &= a_0(b_0 + b_1 z + b_2 z^2 + \cdots)^0 \\ &\quad + a_1(b_0 + b_1 z + b_2 z^2 + \cdots)^1 \\ &\quad + a_2(b_0 + b_1 z + b_2 z^2 + \cdots)^2 \\ &\quad + \cdots \end{aligned} \tag{132}$$

$$\begin{aligned} &= [a_0 + a_1 b_0 + a_2 b_0^2 + \cdots] \\ &\quad + [a_1 b_1 + 2a_2 b_0 b_1 + 3a_3 b_0^2 b_1 + \cdots] z \\ &\quad + [a_1 b_2 + a_2(2b_0 b_2 + b_1^2) + a_3(3b_0^2 b_2 + 3b_0 b_1^2) + \cdots] z^2 \\ &\quad + \cdots \end{aligned} \tag{133}$$

The coefficient of z^0 combines the probabilities for the various ways that a total score of zero can be obtained (not running B at all, running it once and getting a score of zero, running it twice and getting a score of zero both times, etc.). In general the coefficient of z^i combines the probabilities for the various ways that a total score of i can be obtained. Each term of the coefficient has an a_j multiplying a polynomial in b. The term gives the total probability of obtaining score i when B is run j times. The polynomial in b is made up of those terms in $B(z)^j$ which give a total score of i [i.e., the coefficient of z^i in $B(z)^j$]. From our previous discussion of products of probability generating functions we know that the polynomial in b gives the probability that the score is i when process B is run j times.

The mean and variance of the process C with generating function $C(z) = A(B(z))$ can be calculated by using the chain rule for derivatives. The results are

$$A_C = A_A A_B, \tag{134}$$
$$V_C = V_A A_B^2 + V_B A_A. \tag{135}$$

Composition of generating functions is particularly important for describing processes that produce children that have the same characteristics as the parent. In this case the score is the number of children. Let $G(z) = \sum_{i\geq 0} p_i z^i$ be a probability generating function where p_i is the probability that a process produces i children. Then $G(G(z))$ is the generating function for the probability distribution of the number of grandchildren and $G(G(G(z)))$ is the generating function for the number of greatgrandchildren.

Generating functions of probabilities are discussed in Knuth [9, Section 1.2.10] and Feller [27, Chapter 11]. Several interesting applications are given in Karp and Pearl [102].

EXERCISES

1. Derive eq. (130) and eq. (131).

2. Derive eq. (134) and eq. (135).

3. Let p_{nk} be the probability that n rolls of a 6-sided die result in a total of k. Find the closed form for $G(z) = \sum_{i\geq 0} p_{nk} z^k$. What are the average and variance of the total?

4. Let p_{nk} be the probability that n rolls of an m-sided die with numbers 1 to m result in a total of k. Find the closed form for $G(z) = \sum_{i\geq 0} p_{nk} z^k$. What is the average and variance of the total.

5. Suppose an elephant has zero children with probability $\frac{1}{3}$, one child with probability $\frac{1}{3}$, and two children with probability $\frac{1}{3}$. What is the generating function for the probability distribution of children? What is the generating function for the probability distribution of children in the nth generation? What is the average number of children in the nth generation? What is the variance in the number of children in the nth generation? (Assume that no elephants mate with relatives and that there is an adequate supply of mates.)

6. In the previous problem what is the probability that there will be no children in the nth generation?

7. Show that for any probability distribution of children with the generating function $G(z)$ the probability that there will be no children after an infinite number of generations is a root of the equation $z = G(z)$.

6.3.1* Find the Maximum

In Section 3.7.1 we considered the probability p_{ni} that the Maximum Algorithm finds i preliminary maxima when looking for the maximum of n numbers. We obtained the recurrence equation

$$p_{n,i} = \frac{1}{n}p_{n-1,i-1} + \frac{n-1}{n}p_{n-1,i} \tag{136}$$

with the boundary conditions $p_{1i} = \delta_{01}$ for $i \geq 0$. If we define the set of generating functions

$$G_n(z) = \sum_{i\geq 0} p_{ni}z^i, \tag{137}$$

then, multiplying eq. (136) by z^i and summing over i, we obtain

$$G_n(z) = \frac{z}{n}G_{n-1}(z) + \frac{n-1}{n}G_{n-1}(z) \tag{138}$$

$$= \frac{z+n-1}{n}G_{n-1}(z). \tag{139}$$

This is a first order linear recurrence equation for $G_n(z)$. Since $p_{1i} = \delta_{0i}$, we have the boundary condition $G_1(z) = 1$. Before solving for $G_n(z)$, let's notice that we can use eq. (115) to quickly find the average number of preliminary maxima ($\sum_i ip_{ni}$). Taking the derivative of eq. (139) gives

$$G'_n(z) = \frac{1}{n}G_{n-1}(z) + \frac{z+n-1}{n}G'_{n-1}(z). \tag{140}$$

Setting $z = 1$ and using eq. (113) and eq. (115) gives

$$G'_n(1) = \frac{1}{n} + G'_{n-1}(1). \tag{141}$$

The initial condition is $G'_1(1) = 0$. Solving eq. (141) gives

$$G'_n(1) = \sum_{2\leq i\leq n} \frac{1}{i} = H_n - 1. \tag{142}$$

Notice how little effort was needed to find the average by using eq. (115).

If we want the values of the p_{ni}, then we can apply eq. (5.29) to eq. (139) and obtain

$$G_n(z) = \prod_{2\leq i\leq n} \frac{z+i-1}{i} = \frac{1}{n!}\prod_{2\leq i\leq n}(z+i-1) \tag{143}$$

$$= \frac{1}{n}\binom{z+n-1}{n-1}. \tag{144}$$

The easiest way to expand eq. (144) in a power series is to first modify it so that z is on the top of the binomial by itself. Applying eq. (3.69) with $-r = z+n-1$ and $k = n-1$ gives

$$G_n(z) = \frac{(-1)^{n-1}}{n}\binom{-z-1}{n-1}. \tag{145}$$

Applying eq. (3.55) gives

$$G_n(z) = \frac{(-1)^{n-1}}{z}\binom{-z}{n-1}. \tag{146}$$

Our equation is now in the right form to use eq. (3.176), giving

$$G_n(z) = \frac{(-1)^{n-1}}{n!z}\sum_i (-1)^{n-i}\left[{n \atop i}\right](-z)^i = \frac{1}{n!}\sum_i \left[{n \atop i}\right] z^{i-1}, \tag{147}$$

so

$$p_{ni} = \frac{1}{n!}\left[{n \atop i+1}\right]. \tag{148}$$

Notice that for this problem much more effort was required to obtain the p_{ni} than was required to obtain the average, $\sum_i ip_{ni}$.

If we want the variance of the number of preliminary maxima, then we can apply eq. (118) and eq. (131) to eq. (143). Eq. (143) is the product of functions of the form $(z+i-1)/i$. These functions are probability generating functions, and they have first derivatives (with respect to z) equal to $1/i$ and second derivatives equal to zero. Thus

$$V = \sum_{2\le i\le n}\left(0 + \frac{1}{i} - \frac{1}{i^2}\right) = H_n - \sum_{1\le i\le n}\frac{1}{i^2}. \tag{149}$$

The value of the second sum in eq. (149) varies between 1 and $\pi^2/6$ as n varies between 1 and infinity (see Exercise 4.54–4), so it is not very important for large n.

Knuth [9, Section 1.2.10] was a source for this section.

6.4* FIRST ORDER LINEAR RECURRENCES

There is a general method for solving many types of first order recurrence equations. For suitable definitions of closed form, this method will tell you whether or not a first order linear recurrence equation has a closed-form solution, and will give you the solution if it exists. The method also covers the summation of series because the problem of summing the series

$$\sum_{0\le i\le n} T_i \tag{150}$$

is equivalent to the problem of solving the first order linear recurrence

$$S_i = S_{i-1} + T_i. \tag{151}$$

Unfortunately this method is too complex for us to present completely. We will therefore present one part of it. This part will serve as an introduction to the entire method, and it is also useful in its own right.

We are concerned with finding rational solutions — solutions that are the ratio of two polynomials. However, we have a very general idea of what a polynomial is. We form a *tower* in the following way. We start with a set of constants and a variable x. The constants form the zeroth level of the tower. Now we can form polynomials $a_0 + a_1x + a_2x^2 + \cdots + a_nx^n$, where the a's are from the set of constants and n is an integer, and we can form rational functions $P_1(x)/P_2(x)$ where P_1 and P_2 are polynomials. This forms the first level of the tower. If we stop here, we have a problem that can be solved with the techniques of Section 5.2. This process can be continued, however. Now we take some function $f(x)$ that is not equal to any ratio of polynomials in x, such as $x!$, and add another level to the tower by forming polynomials of the form $p_0(x) + p_1(x)f(x) + p_2(x)f(x)^2 + \cdots + p_n(x)f(x)^n$, where the coefficients $p_i(x)$ are polynomials from the previous level of the tower. We also form rational functions of the form $P_1(x)/P_2(x)$ where P_1 are P_2 are polynomials from the current level of the tower. This process can be continued by using additional functions that are not expressible as rational functions at the current level of the tower. In particular, functions of the form c^x for any $c > 1$ can be added to a tower that already has x, $x!$, and powers of the form d^x for $d > 1$ where $d \neq c$.

The method does not tell you which tower to form, although clearly the tower should contain, at a minimum, all the functions used to state the problem. Once you have formed a tower, the method does tell you whether or not your problem has a solution in that tower.

We illustrate the method on the recurrence

$$S_i - i^2 S_{i-1} = (i^2 - 2i)i!. \qquad (152)$$

We will look for a solution in the tower that has the factorial function on the second level (and no additional levels). We will also look only for polynomial solutions, although the complete theory also covers rational solutions. (The rational case is similar to, but much more complex than, the polynomial case.) What we want to know, then, is whether eq. (152) has a solution of the form

$$S_i = \sum_{0 \leq j \leq n} p_j(i)(i!)^j \qquad (153)$$

for some integer n, where $p_j(i)$ is a polynomial in i. The first step is to determine an upper limit on n. We only need an upper limit, because if we have a limit that is too high, the only harm will be that we do extra (but finite) work. When our value for n is too large, the $p_j(i)$ will be zero for large j. On the other hand, if our value for n is too low, the method just does not work.

Finding a suitable value for n is difficult, but Karr [103] suggests that often 1 more than the degree of the right side is large enough, although there are cases where a higher value is needed. For eq. (153) the right side has degree 1 (in $i!$), and it turns out that $n = 2$ is suitable.

The next step in solving the problem is to use the method of undetermined coefficients (Section 5.4.2), with one difference. Our undetermined coefficients are polynomials from the previous level of the tower. So we are looking for a solution of the form

$$S_i = p_2(i)(i!)^2 + p_1(i)i! + p_0(i). \tag{154}$$

Plugging eq. (154) into eq. (152) gives

$$p_2(i)(i!)^2 + p_1(i)i! + p_0(i) - i^2\{p_2(i-1)[(i-1)!]^2 + p_1(i-1)(i-1)! + p_0(i-1)\} = (i^2 - 2i)i!. \tag{155}$$

Since the $i!$ are independent of the $p_j(i)$, eq. (155) reduces to the following three first order equations for the $p_j(i)$:

$$p_2(i)(i!)^2 = i^2 p_2(i-1)[(i-1)!]^2, \tag{156}$$
$$p_1(i)(i!) = i^2 p_1(i-1)(i-1)! + (i^2 - 2i)i!, \tag{157}$$
$$p_0(i) = i^2 p_0(i-1). \tag{158}$$

Using $i(i-1)! = i!$ and cancellation of the $i!$, these equations simplify to

$$p_2(i) = p_2(i-1), \tag{159}$$
$$p_1(i) = ip_1(i-1) + i^2 - 2i, \tag{160}$$
$$p_0(i) = i^2 p_0(i-1). \tag{161}$$

So our original problem is now reduced to determining whether this set of three equations has polynomial solutions. Again we have to bound the degrees of the solutions. Also, we are only looking for polynomial solutions.

For $p_2(i)$ we will look for solutions of degree 1. So let $p_2(i) = q_{21}i + q_{20}$ and plug into eq. (159) to get

$$q_{21}i + q_{20} = q_{21}i - q_{21} + q_{20}. \tag{162}$$

This gives us two equations for the q's:

$$q_{21} = q_{21}, \tag{163}$$
$$q_{20} = -q_{21} + q_{20}. \tag{164}$$

The first equation contains no information, but the second one gives $q_{21} = 0$, so $p_2(i) = q_{20}$.

For $p_1(i)$ we will look for solutions of degree 3. So let $p_1(i) = q_{13}i^3 + q_{12}i^2 + q_{11}i + q_{10}$ and plug into eq. (160) to get

$$q_{13}i^3 + q_{12}i^2 + q_{11}i + q_{10} = i(q_{13}(i-1)^3 + q_{12}(i-1)^2 + q_{11}(i-1) + q_{10}) + i^2 - i, \tag{165}$$

which simplifies to

$$q_{13}i^3 + q_{12}i^2 + q_{11}i + q_{10} = q_{13}i^4 - 3q_{13}i^3 + 3q_{13}i^2 - q_{13}i + q_{12}i^3 - 2q_{12}i^2 + q_{12}i + q_{11}i^2 - q_{11}i + q_{10}i + i^2 - i. \tag{166}$$

From the various powers of i we get

$$q_{13} = 0, \tag{167}$$

$$4q_{13} - q_{12} = 0, \tag{168}$$

$$-3q_{13} + 3q_{12} - q_{11} = 1, \tag{169}$$

$$q_{13} - q_{12} + 2q_{11} - q_{10} = -1, \tag{170}$$

$$q_{10} = 0. \tag{171}$$

The solution of this set of equations is $q_{13} = 0$, $q_{12} = 0$, $q_{11} = -1$, $q_{10} = 0$, so $p_1(i) = -i$.

For $p_0(i)$ we will look for solutions of degree 1. So let $p_0(i) = q_{01}i + q_{00}$ and plug into eq. (161) to get

$$q_{01}i + q_{00} = q_{01}i^3 - q_{01}i^2 + q_{00}i^2. \tag{172}$$

This gives us four equations for the q's; they all say that some q must be zero, so $p_0(i) = 0$.

Putting all the parts together, we get for our solution

$$S_i = q_{20}(i!)^2 - i\,i!, \tag{173}$$

where q_{20} is an arbitrary constant.

To summarize, we can find out whether a first order difference equation has a solution in some tower by using a generalized version of undetermined coefficients, where the coefficients come from the previous level of the tower. From the description in this book, you should often be able to use the method to find solutions of problems.

To prove that problems do *not* have solutions in a tower, however, you need to read Karr's [103] paper to obtain three additional pieces of information: (1) which kinds of towers his method applies to, (2) how to bound the degrees of the solutions you are looking for, and (3) how to handle rational functions.

CHAPTER 7

Full-History and Nonlinear Recurrences

7.1 FULL HISTORY

Recurrence equations with full history are usually more difficult to solve than those with finite history. In some cases we can eliminate the history by combining carefully chosen versions of the equations. In other cases we can use generating functions of one form or another. Finally there are methods that make systematic use of guessing.

7.1.1 Elimination of the History

Sometimes you can find a way to combine versions of a full-history recurrence equation with itself so as to convert the equation to one of finite history. The simplest example is the equation

$$T_n = \sum_{0 \le i < n} T_i \qquad \text{for } n \ge 1, \tag{1}$$

with the boundary condition $T_0 = 1$. By replacing n with $n-1$, eq. (1) can be written as

$$T_{n-1} = \sum_{0 \le i < n-1} T_i \qquad \text{for } n \ge 2. \tag{2}$$

Subtracting eq. (2) from eq. (1) gives the first order equation

$$T_n - T_{n-1} = T_{n-1} \qquad \text{for } n \ge 2, \tag{3}$$

or

$$T_n = 2T_{n-1} \qquad \text{for } n \ge 2. \tag{4}$$

Eq. (1) with $T_0 = 1$ gives $T_1 = 1$. The solution of eq. (4) with the boundary condition $T_1 = 1$ is

$$T_n = 2^{n-1} \qquad \text{for } n > 0, \tag{5}$$

so it is also the solution of eq. (1).

Elimination of the history is covered in Greene and Knuth [6, Section 2.1.2.1].

EXERCISES

1. Can you solve eq. (3) with the boundary condition $T_0 = 1$? Explain.
2. Solve the equation $t_n = n + \sum_{1 \le i \le n-1} t_i$ with a suitable boundary condition.
3. Solve the equation $t_n = n + 2\sum_{1 \le i \le n-1} t_i$ with a suitable boundary condition.

7.1.1.1 *Quicksort (Average Time)*

Quicksort (Algorithm 2.2) uses $n + 1$ comparisons of data elements to divide a file into two parts. (Only $n - 1$ comparisons are necessary to divide the file into parts, but the use of sentinels, while saving $n - 1$ comparisons of indices, does result in two extra comparisons of data items.) Once the file is divided, Quicksort calls itself recursively to sort each part. Therefore the average number of comparisons used by Quicksort obeys the recurrence

$$C_n = n + 1 + \sum_{0 \le i < n} (p_i C_i + p_{n-i-1} C_{n-i-1}), \tag{6}$$

where i represents the size of the first part of the file (and thus $n-i-1$ is the size of the second part) and p_i is the probability that the first part has i elements. The boundary conditions are $C_1 = C_0 = 0$. If the file being sorted is initially in random order (any permutation of n distinct values being equally likely), then the value of p_i is just $1/n$. In this case $\sum_{0 \le i < n} p_i C_i = \sum_{0 \le i < n} p_{n-i-1} C_{n-i-1}$, so eq. (6) simplifies to

$$C_n = n + 1 + \frac{2}{n} \sum_{0 \le i < n} C_i, \tag{7}$$

which is a linear recurrence equation with full history.

Eq. (7) has a form that suggests that you should consider the method of eliminating the history. But before applying the method it is necessary to transform the equation. Multiplying eq. (7) by n gives

$$nC_n = n^2 + n + 2 \sum_{0 \le i < n} C_i. \tag{8}$$

Now if we make a new version of eq. (8) by replacing n with $n-1$, and then subtract the new version from the old version, we obtain

$$\begin{aligned} nC_n - (n-1)C_{n-1} &= n^2 - (n-1)^2 + n - (n-1) \\ &\qquad + 2\sum_{0\le i<n} C_i - 2\sum_{0\le i<n-1} C_i \qquad (9) \\ &= 2n + 2C_{n-1}, \qquad (10) \end{aligned}$$

which eliminates the history. Rearranging eq. (10) gives

$$C_n = \frac{n+1}{n} C_{n-1} + 2, \qquad (11)$$

a first order linear equation. The solution for the boundary condition $C_1 = 0$ is

$$\begin{aligned} C_n &= \sum_{2\le i\le n} 2 \prod_{i<j\le n} \frac{j+1}{j} \qquad (12) \\ &= 2\sum_{2\le i\le n} \frac{n+1}{i+1} = 2(n+1)(H_{n+1} - \tfrac{3}{2}). \qquad (13) \end{aligned}$$

Since $H_{n+1} = \ln n + O(1)$, the average number of comparisons used by Quicksort is

$$C_n = 2(n+1)(H_{n+1} - \tfrac{3}{2}) = 2n\ln n + O(n). \qquad (14)$$

The average time for Quicksort is computed in Knuth [11, Section 5.2.2].

EXERCISE

1. Eq. (6) is as simple as it is only because the average number of comparisons is a linear function. The variance is not a linear function. Let p_{nk} be the probability that Quicksort uses k comparisons when sorting a file with n distinct items. Write a recurrence equation obeyed by p_{nk}. (The resulting equation is nonlinear and cannot be solved by the techniques in this book or by any other techniques known to the authors. It is possible, however, to use it to obtain equations satisfied by the average and variance. Section 11.6 and its subsections show how to do this for a similar problem.)

7.1.1.2 Binary Search Trees

A *binary tree* is a tree where each node has two possible children, a *left child* and *right child.* A *binary search tree* is a binary tree that is used to organize items so that they can be found quickly. Such a tree has the property that the item at the root of a subtree is larger than any element in its left subtree and smaller than any element in its right subtree. Figure 7.1 shows a binary search tree.

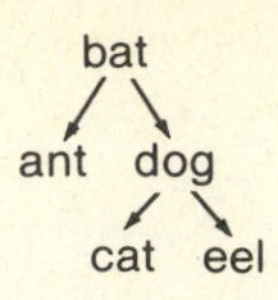

FIGURE 7.1 A binary search tree.

We will use the fields *data*, *llink*, and *rlink* to represent the binary tree. The variable *root* points to the root of the tree. The variable *avail* points to the available space list. The available space list is linked with the *llink* field. Initially *root* is *empty* and all the nodes are on the available space list.

The following algorithm inserts a node into the binary tree and sets its *data* field to *value*.

Algorithm 7.1 Insert(*root*,*value*): Input: *root* points to the root of the search tree, and *value* is the data that will be inserted into the new node. Global variable: *avail* points to the available space list. Result: A new node with a *data* field equal to *value* is added to the tree using a node with an *empty* link field. This is done in a way that produces a new search tree. The procedure calls itself recursively.

Step **1.** [Place found?] If $root = empty$, then set $root \leftarrow avail$, $avail \leftarrow llink(avail)$, $llink(root) \leftarrow empty$, $rlink(root) \leftarrow empty$, $data(root) \leftarrow value$, and return.

Step **2.** [Test] If the relation between *value* and *data*(*root*) is:

Step **3.** [Go left] $value < data(root)$, then call Insert($llink(root), value$).

Step **4.** [Go right] $value \geq data(root)$, then call Insert($rlink(root), value$).

The time needed to look up an item in a binary search tree is proportional to 1 plus the length of the path from the root to the node containing the item. The average time to look up an item is therefore closely related to the average distance from the root to a node. The sum of the distances from the root to every node in a tree is called the *internal path length* of the tree. For example, the internal path length for the tree in Figure 7.1 is 6. The internal path length divided by the number of nodes in the tree gives the average distance to the root.

Inserting an item into a binary search tree can be thought of as looking up an item that is not there. If we think of a nonexistent item at each position where the tree has a null pointer, then the sum of the distances to all these imaginary items is called the *external path length* of the tree. The external path length of the tree in Figure 7.1 is 16.

There is a simple formula relating the internal and external path lengths. For a tree with one node, the internal path length is zero and the external path length is 2. Consider what happens when a new node is added to a binary search tree. Suppose the position of the new node is a distance x from the root. Then the internal path length is increased by x. The external path length is decreased by x (since one imaginary node has been replaced by a real node) and increased by $2x+2$, for a total increase of $x+2$. This shows that in an n-node search tree the external path length is $2n$ plus the internal path length.

The number of comparisons needed to look an item up is 1 plus the number of comparisons needed to insert it, since an extra comparison is needed to be sure we have found the item.

We are interested in the average number of comparisons needed to look up an item in a binary search tree. As usual, to study average time, we need a definition of what a random tree is like. The appropriate notion of randomness for binary search trees is to consider a random sequence of insertions. That is, we assign equal probability to each permutation of the numbers from 1 to n (as in sorting, it is only the relative sizes of the numbers that matters). Different permutations of the items can lead to trees that have the same shape; we assign probabilities to the trees based on the number of permutations that produce them.

We will let S_n be the average number of comparisons in a successful search, and U_n the average number in an unsuccessful search. Let I_n be the average internal path length for a tree with n nodes, and let E_n be the average external path length for a tree with n nodes. From our remarks above, we have that

$$S_n = 1 + \frac{I_n}{n} \tag{15}$$

and

$$U_n = \frac{E_n}{n+1}. \tag{16}$$

Since $E_n = I_n + 2n$,

$$S_n = \left(1 + \frac{1}{n}\right) U_n - 1. \tag{17}$$

The node we are looking up is equally likely to have been the first, second, third, etc. node inserted in the tree. We have observed that the number of comparisons needed to look up an item is 1 plus the number needed to insert it, and that the number needed to insert it is the number needed to do an unsuccessful lookup. This leads to the following equation:

$$S_n = 1 + \frac{U_0 + U_1 + \cdots + U_{n-1}}{n}. \tag{18}$$

Using eq. (17) to eliminate S_n from eq. (18) gives

$$(n+1)U_n = 2n + U_0 + U_1 + \cdots + U_{n-1}. \tag{19}$$

Notice that eq. (19) with $n = 0$ implies that $U_0 = 0$.

The solution of eq. (19) is

$$U_n = 2H_{n+1} - 2, \tag{20}$$

where H_n is the nth harmonic number. Applying eq. (17) to this answer gives

$$S_n = 2\left(1 + \frac{1}{n}\right) H_n - 3. \tag{21}$$

The proof of these formulas is left to the exercises.

Binary search trees are discussed by Knuth [11, Section 6.2.2], Reingold and Hansen [61, Section 5.3], and Aho et al. [1, Section 4.4]. They were first developed and analyzed by Windley [139], Booth and Colin [74], and Hibbard [93].

EXERCISE

1. Solve eq. (19) by the method of eliminating history.

7.1.2 Multiplying Generating Functions

Suppose $A(z) = \sum_{i\geq 0} a_i z^i$, $B(z) = \sum_{i\geq 0} b_i z^i$, and $C(z) = \sum_{i\geq 0} c_i z^i$. In addition, suppose that $C(z) = A(z)B(z)$. Then we have

$$c_n = \sum_{0\leq i\leq n} a_i b_{n-i}. \tag{22}$$

[See eq. (6.128).] This sum is the convolution of the sequences for a and b.

Eq. (22) has many applications to the use of generating functions to solve recurrence equations. Whenever you have a recurrence equation that has a part that looks like the right side of eq. (22), you should consider using it. There are many ways this can happen. In this section we consider the case where one of the factors (say, the a_i) is equal to the unknown function in the recurrence while the other factor is a known function.

We start with eq. (1). Multiplying by z^n and summing over n gives

$$\sum_{n\geq 1} T_n z^n = \sum_{n\geq 1} z^n \sum_{0\leq i<n} T_i. \tag{23}$$

Adding T_0 to both sides gives

$$\sum_{n\geq 0} T_n z^n = T_0 + \sum_{n\geq 1} z^n \sum_{0\leq i<n} T_i. \tag{24}$$

(Notice the need to be extremely careful with the summation limits.)

If we let $a_i = T_i z^i$ and $b_{n-i} = z^{n-i}$, then the sum on the right side of eq. (24) almost has the form of eq. (22). The only difference is that the sum in eq. (24) goes to $n-1$ while the one in eq. (22) goes to n. We may use $n \geq 0$ instead of $n \geq 1$ in the sum over n in eq. (24) because there are no terms for $n = 0$ in any case. Thus, using $G(z) = \sum_{n\geq 0} T_n z^n$ and $B(z) = \sum_{n\geq 0} z^n$, from eq. (24) we get

$$G(z) = T_0 + \sum_{n\geq 0} \sum_{0\leq i<n} (T_i z^i)(z^{n-i-1})z \tag{25}$$

$$= T_0 + z\Big(\sum_{i\geq 0} T_i z^i\Big)\Big(\sum_{i\geq 0} z^i\Big) = T_0 + zG(z)\sum_{i\geq 0} z^i \tag{26}$$

$$= T_0 + \frac{zG(z)}{1-z}. \tag{27}$$

Multiplying by $1 - z$ and solving for $G(z)$ gives

$$(1-z)G(z) = (1-z)T_0 + zG(z), \tag{28}$$

$$(1-2z)G(z) = (1-z)T_0, \tag{29}$$

$$G(z) = \frac{1-z}{1-2z}T_0 = T_0 + \frac{z}{1-2z}T_0. \tag{30}$$

Expanding $G(z)$ in a power series gives

$$G(z) = T_0 + T_0 \sum_{n\geq 1} 2^{n-1} z^n, \tag{31}$$

so the solution of eq. (1) is

$$T_0 = 1, \qquad T_n = 2^{n-1} \quad \text{for } n > 0. \tag{32}$$

The technique of generating functions is a little more complicated to use on easy problems than the method of eliminating history, but with complex problems generating functions are usually easier to apply.

Multiplying generating functions is covered in Knuth [9, Section 1.2.9] and in Riordan [48, Section 2.2].

EXERCISE

1. Use generating functions to solve eq. (6).

7.1.2.1 Traversing Threaded Binary Trees*

Traversing a binary tree consists of visiting all the nodes of the tree. There are three standard ways to systematically traverse a binary tree: preorder, inorder, and postorder.

In *preorder* traversal, you first visit the root, then visit the left subtree (in preorder), and finally visit the right subtree (in preorder). Traversing the tree in Figure 7.2 in preorder visits the nodes in the order 1, 2, 3, 4, 5. In *inorder* traversal, you first visit the left subtree (in inorder), then the root, and finally the right subtree (in inorder). Traversing the tree in Figure 7.2 in inorder visits the nodes in the order 2, 1, 4, 3, 5. In *postorder* traversal, you first visit the left subtree (in postorder), then the right subtree (in postorder), and finally the

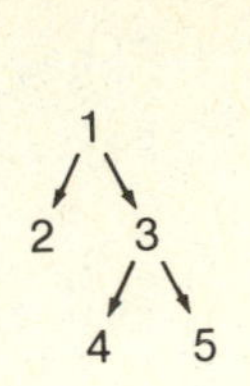

FIGURE 7.2 A binary tree.

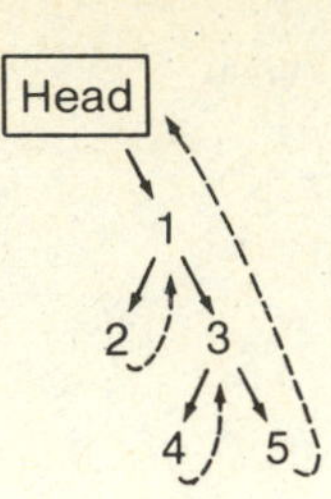

FIGURE 7.3 The tree of Figure 7.2 with right threads and a header node added.

root. Traversing the tree in Figure 7.2 in postorder visits the nodes in the order 2, 4, 5, 3, 1.

A *threaded binary tree* is a binary tree in which the child pointers that would ordinarily be null are used to point to an ancestor of a node. In particular, a null left link is replaced by a pointer to the inorder predecessor, and a null right link is replaced by a pointer to the inorder successor. These extra links are called *threads*. An extra *tag bit* is used with each link to distinguish threads from normal links. A threaded tree also has an extra node, called the *header*. (We do not count this node when counting the number of nodes in the tree.) Both links of the header point to the root of the tree. The left thread from the leftmost node in the tree points to the header (the leftmost node has no inorder predecessor), and the right thread of the rightmost node also points to the header. Figure 7.3 shows the threaded binary tree that corresponds to the binary tree in Figure 7.2.

The following algorithm does one step of an inorder traversal. From any node p, it finds the inorder successor of p. (If p is the header node, the algorithm finds the first node to be visited in an inorder traversal.)

Algorithm 7.2 Inorder Successor(p)**:** Input: A node p of a threaded binary tree (p may be the header node). Intermediate variables: q is a node, $llink_p$ and $rlink_p$ denote the left and right links of node p, $ltag_p$ and $rtag_p$ indicate whether a link is a thread or a normal link. Output: q, the inorder successor of p.

Step **1.** Set $q \leftarrow rlink_p$.

Step **2.** If $rtag_p \neq thread$, then

Step **3.** While $ltag_q \neq thread$ do $q \leftarrow llink_q$, end while.

Step **4.** End if.

If you start with p pointing to the header and call the Inorder Successor Algorithm repeatedly (setting $p \leftarrow q$ before each new call) until q points to the root, you will visit each node in inorder.

Now let's consider the average time required by this algorithm. We will solve this problem in a way that illustrates the use of products of generating functions. (Exercise 9.3.2–4 suggests a much easier way to obtain the most important parts of the final results.) Steps 1 and 2 are executed once per node. The difficult part of the analysis is to compute the average number of times that Step 3 is done. We will do this by computing the total number of times it is done for *all* nodes in *all* n-node binary trees and dividing by the total number of nodes in all n-node binary trees. This gives the correct average under the assumption that all n-node binary trees are equally likely to occur.

There are $C_n = \binom{2n}{n}/(n+1)$ binary trees with n nodes (see Section 7.2.3). (This number is the nth *Catalan number.*) The generating function for C_n is also given in Section 7.2.3.

Let K_n be the total number of times that Step 3 is executed while traversing all the n-node binary trees. Now $K_0 = 0$ because the only zero-node tree (the empty tree) does not have any nodes to visit. For larger trees we compute the answer as the sum of two parts: the number of times Step 3 is done while looking for the successor of the root plus the number of times it is done on the rest of the traversal.

The inorder successor of the root is found by following the right link for one step and then following left links until reaching a node with no left child (indicated by the left link being a thread). The number of times Step 3 is done when looking for the inorder successor of the root is equal to the number of left links that are followed. A *left shell node* is a node that can be reached from the root by following zero or more (nonthread) left links. The number of times Step 3 is done while finding the inorder successor of the root of a particular tree is equal to the number of nodes in the left shell of the right subtree of the root.

Let Q_n be the total number of left shell nodes summed over all the n-node binary trees. To compute the number of times Step 3 is done, consider the set of n-node binary trees with i nodes in the left subtree and $n-i-1$ nodes in the right subtree. Altogether the right subtrees contain Q_{n-i-1} left shell nodes. Each right subtree, and therefore each of the Q_{n-i-1} left shell nodes, occurs in combination with each of the C_i possible left subtrees. Therefore the total number of left shell nodes for the right subtrees of the roots of the n-node binary trees which have i nodes in the left subtree (and $n-i-1$ in the right subtree) is

$$C_i Q_{n-i-1}. \tag{33}$$

This is the total number of times that Step 3 is done while finding the inorder successor of the root in those binary trees with n nodes and i nodes in the left subtree. Summing eq. (33) over i gives the total number of times that Step 3 is done while finding the inorder successor of the root for all n-node binary trees.

We also need the total number of times that Step 3 is done while finding the inorder successors of the nonroot nodes. The total number of times it is done for left subtrees is the same as the total number of times it is done for right subtrees, so we will compute the number of times it is done for left subtrees and multiply by 2. If we count each left subtree of size i once, then the total number of times that Step 3 is done while searching for inorder successors of nodes in the left subtree is K_i, by the definition of K. But each left subtree of size i appears with each of the C_{n-i-1} right subtrees of size $n-i-1$, so the total number of times that Step 3 is done for nodes in the left subtree is

$$K_i C_{n-i-1}. \tag{34}$$

Combining eq. (33) and eq. (34) gives the recurrence equation for K_n:

$$K_n = \sum_{0 \le i \le n-1} (2C_{n-i-1}K_i + C_i Q_{n-i-1}). \tag{35}$$

Our boundary condition is $K_0 = 0$. To solve this recurrence, we need to know C_n (given in Section 7.2.3) and Q_n.

We obtain a recurrence equation for Q_n by considering all binary trees with n nodes which contain i nodes in the left subtree. The root of each of these trees is a left shell node. There are $C_i C_{n-i-1}$ such trees, and there are the same number of roots. The remaining left shell nodes are left shell nodes of the left subtrees of the roots. If we count each left shell node in each of the C_i left subtrees once, we get Q_i. But each left subtree occurs in combination with every right subtree of size $n-i-1$, so the total number of nonroot left shell nodes in binary trees with a total of n nodes and i nodes in the left subtree is $C_{n-i-1}Q_i$. Therefore the recurrence for Q_n is

$$Q_n = \sum_{0 \le i \le n-1} (C_{n-i-1}Q_i + C_i C_{n-i-1}). \tag{36}$$

The empty tree does not have any left shell nodes, so $Q_0 = 0$.

Using eq. (156) from Section 7.2.3, we have

$$C_n = \sum_{0 \le i \le n-1} C_i C_{n-i-1}, \tag{37}$$

so eq. (36) can be rewritten as

$$Q_n = \sum_{0 \le i \le n-1} C_{n-i-1}Q_i + C_n. \tag{38}$$

We will solve this recurrence using the generating function

$$Q(x) = \sum_{n \ge 0} Q_n x^n = \sum_{n \ge 1} Q_n x^n \tag{39}$$

(since Q_0 is zero). Also define the generating function

$$C(x) = \sum_{n \ge 0} C_n x^n = \sum_{n \ge 1} C_n x^n + 1 \tag{40}$$

(since $C_0 = 1$). From Section 7.2.3 [eq. (161)], the closed-form expression for $C(x)$ is

$$C(x) = \frac{1}{2x}\left(1 - \sqrt{1-4x}\right). \tag{41}$$

Multiplying eq. (38) by x^n and summing for $n \geq 0$ gives

$$Q(x) = \sum_{n\geq 1}\Bigl(\sum_{0\leq i\leq n-1} C_{n-i-1}Q_i + C_n\Bigr)x^n. \tag{42}$$

The sum of the C_n terms in this sum simplifies to $C(x)$, while the sum of the $C_{n-i-1}Q_i$ terms is a convolution. Writing x^n as xx^ix^{n-i-1} and using the change of variable $n' = n - i - 1$ on eq. (42) gives

$$Q(x) = x\Bigl(\sum_{i\geq 0} Q_i x^i\Bigr)\Bigl(\sum_{n\geq 0} C_n x^n\Bigr) + C(x) - 1 \tag{43}$$

$$= xQ(x)C(x) + C(x) - 1. \tag{44}$$

Solving eq. (44) for $Q(x)$ gives

$$Q(x) = \frac{C(x) - 1}{1 - xC(x)}. \tag{45}$$

To simplify this equation, we can use the fact that $C(x)$ obeys eq. (160). We can rewrite eq. (160) as

$$1 - xC(x) = \frac{1}{C(x)}, \tag{46}$$

or

$$C(x)^2 = \frac{C(x) - 1}{x}. \tag{47}$$

Using first eq. (46) and then eq. (47) on eq. (45) gives

$$Q(x) = C(x)^2 - C(x) = x^{-1}C(x) - x^{-1} - C(x). \tag{48}$$

Now we replace $C(x)$ by $\sum_{n\geq 0} C_n x^n$ and use $C_0 = 1$ to obtain

$$Q(x) = x^{-1}\sum_{n\geq 0} C_n x^n - x^{-1} - \sum_{n\geq 0} C_n x^n \tag{49}$$

$$= \sum_{n\geq 1} C_n x^{n-1} - \sum_{n\geq 0} C_n x^n = \sum_{n\geq 0}(C_{n+1} - C_n)x^n. \tag{50}$$

Since Q_n is the coefficient of x^n in $Q(x)$, we have

$$Q_n = C_{n+1} - C_n. \tag{51}$$

Now that we have the value of Q_n, we can simplify eq. (35) to

$$K_n = \sum_{0\leq i\leq n-1} [2C_{n-i-1}K_i + C_i(C_{n-i} - C_{n-i-1})]. \tag{52}$$

We can apply eq. (37) (with n replaced by $n-1$) to the last term of eq. (52) to obtain

$$K_n = \sum_{0\leq i\leq n-1} (2C_{n-i-1}K_i + C_iC_{n-i}) - C_n. \tag{53}$$

The C_iC_{n-i} term is almost in the right form to apply eq. (37), but the sum in eq. (52) goes only to $n-1$, while eq. (37) goes to n. When $n \geq 1$, we can add and subtract the missing term ($C_0C_n = C_n$) and apply eq. (37). No correction is needed when $n = 0$. Thus we have

$$\sum_{0\le i\le n-1} C_iC_{n-i} = C_{n+1} - C_n + \delta_{n0}. \tag{54}$$

Thus we can simplify eq. (53) to

$$K_n = \sum_{0\le i\le n-1} 2C_{n-i-1}K_i + C_{n+1} - 2C_n + \delta_{n0}. \tag{55}$$

This is a convolution equation.

Define the generating function

$$K(x) = \sum_{n\ge 0} K_n x^n. \tag{56}$$

Since $K_0 = 0$, the sum in eq. (56) can just as well start at $n = 1$. Multiplying eq. (55) by x^n and summing over $n \ge 1$ gives

$$K(x) = \sum_{n\ge 1}\sum_{0\le i\le n-1} 2C_{n-i-1}K_ix^n + x^{-1}C(x) - x^{-1} - 2C(x) + 1. \tag{57}$$

Using the same techniques that were used on eq. (42), eq. (57) reduces to

$$K(x) = 2x\sum_{i\ge 0} K_ix^i \sum_{n\ge 0} C_nx^n + x^{-1}C(x) - x^{-1} - 2C(x) + 1 \tag{58}$$

$$= 2xC(x)K(x) + x^{-1}C(x) - x^{-1} - 2C(x) + 1. \tag{59}$$

Solving for $K(x)$ gives

$$K(x) = \frac{x^{-1}C(x) - x^{-1} - 2C(x) + 1}{1 - 2xC(x)}. \tag{60}$$

We will now use the brute-force approach [rather than using clever identities on $C(x)$] to find the power series for $K(x)$. The brute-force approach often leads to longer (and therefore more error-prone) calculations, but it does have the advantage that no new insights are needed. Substituting the value of $C(x)$ from eq. (41) into eq. (60) gives

$$K(x) = \frac{1}{2x^2\sqrt{1-4x}} - \frac{2}{x\sqrt{1-4x}} - \frac{1}{2x^2} + \frac{1}{x} + \frac{1}{\sqrt{1-4x}}. \tag{61}$$

Now, by the binomial theorem

$$\frac{1}{\sqrt{1-4x}} = \sum_{n\ge 0}(-4)^n\binom{-\frac{1}{2}}{n}x^n, \tag{62}$$

so

$$K(x) = \frac{1}{2}\sum_{n\geq 0}(-4)^n\binom{-\frac{1}{2}}{n}x^{n-2} - 2\sum_{n\geq 0}(-4)^n\binom{-\frac{1}{2}}{n}x^{n-1} - \frac{1}{2x^2} + \frac{1}{x} + \sum_{n\geq 0}(-4)^n\binom{-\frac{1}{2}}{n}x^n. \quad (63)$$

From Exercise 3.2–1, we have that

$$\binom{-\frac{1}{2}}{n} = (-\tfrac{1}{4})^n\binom{2n}{n}. \quad (64)$$

Applying this to eq. (63) gives

$$K(x) = \frac{1}{2}\sum_{n\geq 0}\binom{2n}{n}x^{n-2} - 2\sum_{n\geq 0}\binom{2n}{n}x^{n-1} - \frac{1}{2x^2} + \frac{1}{x} + \sum_{n\geq 0}\binom{2n}{n}x^n. \quad (65)$$

The terms with negative powers of x cancel out. (If they did not, we would know we had an error because K_n is zero for $n \leq 0$.) Replacing n by $n+2$ in the first sum and by $n+1$ in the second sum gives

$$K(x) = \frac{1}{2}\sum_{n\geq 0}\binom{2n+4}{n+2}x^n - 2\sum_{n\geq 0}\binom{2n+2}{n+1}x^n + \sum_{n\geq 0}\binom{2n}{n}x^n. \quad (66)$$

Applying eq. (3.55) and eq. (3.64) gives

$$K(x) = \sum_{n\geq 0}\left[2\frac{(2n+3)(2n+1)}{(n+2)(n+1)}\binom{2n}{n} - 4\frac{2n+1}{n+1}\binom{2n}{n} + \binom{2n}{n}\right]x^n \quad (67)$$

$$= \sum_{n\geq 0}\frac{n(n-1)}{(n+1)(n+2)}\binom{2n}{n}x^n. \quad (68)$$

This gives

$$K_n = \frac{n(n-1)}{(n+1)(n+2)}\binom{2n}{n}. \quad (69)$$

Since $C_n = \binom{2n}{n}/(n+1)$, this can also be written as

$$K_n = \frac{n(n-1)}{(n+2)}C_n. \quad (70)$$

Now we are ready to complete the analysis of the Inorder Successor Algorithm. Step 3 is done K_n times. The test in Step 3 gives the result *true* $(n-1)C_n/2$ times (see Exercises 2 to 4). Thus the assignment in Step 3 is done

$$K_n - \frac{n-1}{2}C_n = \frac{(n-1)(n-2)}{2(n+2)}C_n \quad (71)$$

times. There are a total of nC_n nodes in the n-node binary trees, so the test in Step 3 is done $(n-1)/(n+2)$ times per node on the average and the assignment is done $(n-1)(n-2)/[2n(n+2)]$ times on the average. The algorithm also

n	0	1	2	3	4	5
C_n	1	1	2	5	14	42
Q_n	0	1	3	9	28	90
K_n	0	0	1	6	28	120

TABLE 7.1. The values of C_n, Q_n, and K_n for small n.

traverses one right link per node visited, so the average total number of links traversed per node visited is

$$1 + \frac{(n-1)(n-2)}{2n(n+2)} = \frac{3n^2+n+2}{2n(n+2)} = \frac{3}{2} + O\left(\frac{1}{n}\right). \tag{72}$$

It is difficult to perform derivations like the ones in this section without making errors, particularly if you do not know the answer in advance. It is therefore important to check your work as you go along. One good place for a check is at the stage where the recurrence formula is first obtained. You should compare the numbers generated by the recurrence with values you can calculate from the definitions. Eqs. (37, 38, 35) can be used to compute C_n, Q_n, and K_n for small values of n. The results in Table 7.1 were obtained this way.

To check these numbers for $n \le 2$, construct all the binary trees with zero, one, and two nodes. Count the number of trees, the number of left shell nodes, and the number of times that Step 3 of the Inorder Algorithm is done for all the trees of each size. If you have correctly constructed the trees and correctly performed the counts, your results will agree with Table 7.1.

Once you have the first few values for K_n, you can use them to check your derivations. For example, when the authors were deriving eq. (54), we at first omitted the δ_{n0} term. As a result, our initial version of eq. (60) was missing the $+1$ term on top. We noticed this problem when we were checking eq. (60). Using the values from Table 7.1, we know that the power series for $C(x)$ is

$$C(x) = 1 + x + 2x^2 + 5x^3 + \cdots, \tag{73}$$

while the power series for $K(x)$ is

$$K(x) = x^2 + 6x^3 + \cdots. \tag{74}$$

Taking the first few terms of $C(x)$ from eq. (73), you can use eq. (60) and division of polynomials to compute the first few terms for $K(x)$. They should agree with eq. (74), but if you make a mistake and omit the $+1$ term from the top of eq. (60), you get 1 for K_1 instead of zero. Once we knew we had a problem, we were able to quickly find it by checking formulas in a binary search order. (Binary search works well if you have only a few errors, but checking every formula in order works better if you have a lot of errors.)

The first time we derived eq. (70), we had an $n-1$ in the numerator instead of the correct $n+2$. Again, testing the formulas against the first few values for K_n helped us quickly isolate the mistake, which turned out to be a simple algebra error. This testing of formulas is as helpful in avoiding errors in formulas as testing programs is in avoiding errors in programs. Likewise, poring over your formulas to find errors is about as helpful as the corresponding technique for programs. Usually it is useful to both carefully check your derivations and also do independent checks for correctness of selected steps.

Since it is so easy to make minor errors in derivations (which often lead to major errors in the final results), all long derivations should be carefully tested as well as checked. If a formula can correctly calculate several known values, there is a good chance that it is correct. The longer the formula is, the more values are needed for a given level of confidence. In particular, enough values should be calculated to test all parts of the formula. Also, if the first few tests disclose errors, you should both repeat the original tests on the corrected formulas and do some new ones.

It is worth emphasising again that the average number of links traversed was obtained by considering the number of links traversed in all n-node trees and then dividing by the number of n-node trees. This is the same technique that was used in Section 4.3.1 to compute the average time for Backtracking. This approach is often quite helpful in average-time studies.

The results of this section are average-case results, so they are dependent on the model used to define the average case. In this section we have assumed that each n-node binary tree is equally likely. Many of the algorithms which build binary trees are more likely to build trees of certain shapes than they are to build trees of other shapes.

The results in this section are from Brinck and Foo [76]. In [77] Brinck analyzes the same algorithm for the distribution of trees that results from using trees to sort items. This time the result is that on the average

$$\frac{3}{2} + O\left(\frac{\ln n}{n}\right) \tag{75}$$

links are traversed.

EXERCISES

1. The analysis in this section included the time to find the inorder successor to the last node in a threaded tree, but it omitted the time to find the first node in inorder (the inorder successor of the header node). How many links are traversed if you include finding the first node? How many links per node are traversed? (Remember that for this problem, one more node is visited than for the problem worked in the text.) Hint: This problem does not require a new complex derivation. You just need to combine already computed functions together.

2. Prove that the total number of nodes with nonnull left subtrees, over all n-node binary trees, is given by the recurrence

$$L_n = \sum_{0 \le i \le n-1} 2L_i C_{n-i-1} + C_n - C_{n-1},$$

with boundary condition $L_0 = 0$.

3. Show that the function $L(x) = \sum_{n\geq 0} L_n x^n$, where L_n is defined in the previous exercise, satisfies the equation

$$L(x) = \frac{C(x) - xC(x) - 1}{1 - 2C(x)}.$$

4. Show that the quantity L_n from the previous two exercises is given by

$$L_n = \frac{n-1}{2} C_n.$$

7.1.3 Exponential Generating Functions

When using generating functions on full-history recurrence equations, it is often worthwhile to consider modifications of the definition of a generating function. Here we will consider the most common modification. The *exponential generating function* of a sequence T_n is

$$G(z) = \sum_{n\geq 0} \frac{T_n z^n}{n!}. \tag{76}$$

The exponential generating function $G(z)$ contains complete information about the sequence T_n. If you expand $G(z)$ in a power series, then you can determine T_n by taking the coefficient of z^n and multiplying by $n!$.

Exponential generating functions are often useful for solving recurrence equations that contain binomial coefficients. Often the factorials in the exponential generating functions can be combined with the factorials in the binomial coefficients in a way that results in a great simplification. Suppose $A(z) = \sum_{i\geq 0} a_i z^i/i!$, $B(z) = \sum_{i\geq 0} b_i z^i/i!$, and $C(z) = \sum_{i\geq 0} c_i z^i/i!$. In addition suppose that $C(z) = A(z)B(z)$. Then we have

$$\frac{c_0}{0!} + \frac{c_1 z}{1!} + \frac{c_2 z^2}{2!} + \cdots = \frac{a_0}{0!}\frac{b_0}{0!} + \left(\frac{a_0}{0!}\frac{b_1}{1!} + \frac{a_1}{1!}\frac{b_0}{0!}\right) z + \left(\frac{a_0}{0!}\frac{b_2}{2!} + \frac{a_1}{1!}\frac{b_1}{1!} + \frac{a_2}{2!}\frac{b_0}{0!}\right) z^2 + \cdots. \tag{77}$$

Since eq. (77) must hold for all z, the coefficients of each power of z must be equal. Therefore we have

$$\frac{c_0}{0!} = \frac{a_0}{0!}\frac{b_0}{0!}, \tag{78}$$

$$\frac{c_1}{1!} = \frac{a_0}{0!}\frac{b_1}{1!} + \frac{a_1}{1!}\frac{b_0}{0!}, \tag{79}$$

$$\frac{c_2}{2!} = \frac{a_0}{0!}\frac{b_2}{2!} + \frac{a_1}{1!}\frac{b_1}{1!} + \frac{a_2}{2!}\frac{b_0}{0!}, \tag{80}$$

or in general

$$\frac{c_n}{n!} = \sum_{0\leq i\leq n} \frac{a_i}{i!}\frac{b_{n-i}}{(n-i)!}. \tag{81}$$

Multiplying by $n!$ and combining the ratio of factorials into a binomial gives

$$c_n = \sum_{0 \leq i \leq n} \binom{n}{i} a_i b_{n-i}. \qquad \textbf{(82)}$$

Exponential generating functions should be considered whenever you have a recurrence with a right side similar to eq. (82).

Exponential generating functions are covered in Knuth [9, Section 1.2.9] and in Riordan [48, Sections 2.4 and 2.5].

EXERCISES

1. The Hermite polynomials are the solutions of the recurrence

$$H_{n+1}(z) = zH_n(z) - nH_{n-1}(z),$$

with boundary conditions $H_0(z) = 1$ and $H_1(z) = z$. Find a closed form for the generating function $f(x,z) = \sum_{n \geq 0} H_n(z) x^n / n!$.

2. Find the exponential generating function for the Bernoulli numbers. Their recurrence is given in Section 4.5.2.

7.1.3.1 A Silly Algorithm*

In this section we will consider another algorithm for solving the satisfiability problem. (See Section 4.3.1.) Our analysis will show that the algorithm is not very good. The algorithm sets the variables one at a time, very much as Backtracking does. After each variable is set, the algorithm simplifies the predicate. Like the Backtracking Algorithm, this algorithm tries additional values for each variable once a value has been tested. The only difference is that instead of testing at each stage whether the predicate cannot possibly be satisfied (i.e., has an empty clause), this algorithm tests whether the predicate has become trivial (i.e., has no clauses, so it can be satisfied with any values for the remaining variables). The idea behind the algorithm is plausible. It is only after the analysis is done that it becomes clear that this algorithm is usually very slow. (Also Exercise 7.1.4–2 shows that, with additional refinement, the idea behind the algorithm in this section can be useful.)

Algorithm 7.3 Silly: Input: A set $\{x_1, \ldots, x_n\}$ of variables, where each variable can have the values *true* and *false*, and a predicate $P(x_1, x_2, \ldots, x_n)$ that is the conjunction of clauses. Output: All sets $i, x_1, x_2, \ldots, x_i$ such that $P(x_1, x_2, \ldots, x_n)$ is *true* for all values of x_j $(i < j \leq n)$.

Step **1.** [Initialize] For $1 \le i \le n$, set $L_i \leftarrow empty$. Set $i \leftarrow 0$.

Step **2.** [Bottom?] If $i \ne n$, then go to Step 3, otherwise go to Step 6.

Step **3.** [Search deeper] Set $i \leftarrow i + 1$.

Step **4.** [First value] Set $x_i \leftarrow false$. Remove from $P(x_1, \ldots, x_n)$ all clauses that contain the literal $\neg x_i$ and put the removed clauses on the list L_i. If $P(x_1, \ldots, x_n)$ has any remaining clauses, go to Step 2.

Step **5.** [Output] Output $(i, x_1, x_2, \ldots, x_i)$.

Step **6.** [Restore] Remove the clauses from L_i and put them back into the predicate $P(x_1, \ldots, x_n)$. If x_i is *true*, then go to Step 8.

Step **7.** [Second value] Set $x_i \leftarrow true$. Remove from $P(x_1, \ldots, x_n)$ all clauses that contain the literal x_i and put the removed clauses on the list L_i. If $P(x_1, \ldots, x_n)$ has any remaining clauses, go to Step 2; otherwise go to Step 5.

Step **8.** [Backtrack] If $i = 0$, then stop. Otherwise set $i \leftarrow i - 1$ and go to Step 6.

Let $A(t, v)$ be the average time used when the algorithm is run on a random predicate with t clauses, v variables, and probability p of a literal appearing in a clause (see Section 4.3.1 for a discussion of random predicates). Now, $A(t, v)$ may also be a function of p, but our notation does not need to show the p dependence of A because only the t and v dependences need to be considered when writing our recurrence equation.

For purposes of analysis we will assume that the algorithm needs time at (for some constant a) plus the time for the subproblems that it generates. (This assumption is not likely to be strictly true, but it is a good approximation and it leads to equations that can be solved.) When a variable is set to *false*, the probability that the newly generated problem has $t-i$ clauses is $\binom{t}{i}p^i(1-p)^{t-i}$. Here p^i is the probability that the set variable occurs in i clauses, $(1-p)^{t-i}$ is the probability that the set variable does not occur in the remaining $t-i$ clauses, and $\binom{t}{i}$ is the number of ways that i of t clauses can be selected. When a variable is set to *true*, the probability that the newly generated problem has $t-i$ clauses is also $\binom{t}{i}p^i(1-p)^{t-i}$. If the newly generated problem has $t-i$ clauses, then $A(t-i, v-1)$ is (by definition) the average number of times that Step 5 will be done while solving it. Thus $A(t, v)$ obeys the recurrence

$$A(t, v) = at + 2\sum_i \binom{t}{i} p^i (1-p)^{t-i} A(t-i, v-1). \tag{83}$$

We still need some boundary conditions. The algorithm stops generating subproblems when $v = 0$, so in eq. (83) we should use $A(t, 0) = 0$.

The right side of eq. (83) is similar to eq. (82), so the use of exponential generating functions should be considered. Define

$$G_v(z) = \sum_{i \geq 0} A(i, v) \frac{z^i}{i!}. \tag{84}$$

Divide eq. (83) by $t!$, multiply by z^t, and sum over t to obtain

$$\sum_{t \geq 0} A(t, v) \frac{z^t}{t!} = \sum_{t \geq 0} at \frac{z^t}{t!} + 2 \sum_{t \geq 0} z^t \sum_i \frac{1}{i!(t-i)!} p^i (1-p)^{t-i} A(t-i, v-1) \tag{85}$$

$$= a \sum_{t \geq 0} \frac{z^t}{(t-1)!} + 2 \left(\sum_{t \geq 0} A(t, v-1) \frac{(1-p)^t z^t}{t!} \right) \left(\sum_i \frac{z^i}{i!} p^i \right). \tag{86}$$

Now $\sum_{t \geq 0} A(t, v) z^t / t! = G_v(z)$, $\sum_{t \geq 0} A(t, v)(1-p)^t z^t / t! = G_{v-1}((1-p)z)$, $\sum_{t \geq 0} z^t / (t-1)! = z e^z$, and $\sum_i z^i p^i / i! = e^{pz}$, so eq. (86) reduces to

$$G_v(z) = aze^z + 2e^{pz} G_{v-1}((1-p)z). \tag{87}$$

Since $A(t, 0) = 0$, we have $G_0(z) = 0$.

In contrast to our previous examples, this does not lead to an algebraic equation for the generating function. We do, however, get a first order linear recurrence equation of sorts. [We have both a recurrence in v and a change of argument from z to $(1-p)z$.] If we keep plugging eq. (87) into itself, we get

$$G_v(z) = aze^z + 2e^{pz} G_{v-1}((1-p)z) \tag{88}$$

$$= aze^z + 2e^{pz} \left[a(1-p) z e^{(1-p)z} + 2e^{p(1-p)z} G_{v-2}((1-p)^2 z) \right] \tag{89}$$

$$= aze^z + 2e^{pz} \left[a(1-p) z e^{(1-p)z} + 2e^{p(1-p)z} (1-p)^2 z \Big(e^{(1-p)^2 z} + 2e^{p(1-p)^2 z} G_{v-3}((1-p)^3 z) \Big) \right] \tag{90}$$

$$\vdots$$

$$= az \sum_{0 \leq i < v} 2^i (1-p)^i e^{(1-p)^i z} \prod_{0 \leq j < i} e^{p(1-p)^j z} + 2^v \prod_{0 \leq j < v} e^{p(1-p)^j z} G_0((1-p)^v z) \tag{91}$$

Since $G_0(z) = 0$, this simplifies to

$$G_v(z) = az \sum_{0 \leq i < v} 2^i (1-p)^i e^{(1-p)^i z} \prod_{0 \leq j < i} e^{p(1-p)^j z}. \tag{92}$$

The product of exponentials is equal to the exponential of a sum, so

$$\prod_{0 \leq j < i} e^{p(1-p)^j z} = \exp\left(pz \sum_{0 \leq j < i} (1-p)^j \right) = \exp\left[z(1 - (1-p)^i) \right]. \tag{93}$$

Using eq. (93) to simplify eq. (92) results in additional cancellation to give

$$G_v(z) = a \sum_{0 \le i < v} 2^i (1-p)^i z e^z = a \frac{[2(1-p)]^v - 1}{1-2p} z e^z. \tag{94}$$

Expanding $G_v(z)$ in a power series gives

$$G_v(z) = a \sum_{t \ge 0} \frac{([2(1-p)]^v - 1) z^{t+1}}{(1-2p)(t-1)!}, \tag{95}$$

so

$$A(t,v) = \frac{at\{[2(1-p)]^v - 1\}}{(1-2p)}, \tag{96}$$

because $A(t,v)$ is the coefficient of $z^t/t!$. The complete tree has $2^{v+1}-1$ nodes, so when $p < \frac{1}{2}$, this algorithm is more efficient than searching all the nodes by a factor of about $[2(1-p)]^v$ for large v. This is an exponential improvement for any fixed p. Often, however, one wishes to do satisfiability for problems where p is quite small, say, $O(1/v)$. In such cases this algorithm is not much better than the naïve algorithm that searches every node.

This analysis was first done by Purdom and Brown [88].

EXERCISES

1. If you assume the time for the Silly Algorithm is atv plus the time for the subproblems, the general outline of the analysis is the same as given above, but the details are different. What is the recurrence for this assumption, and what is the solution of the recurrence?
2. Solve the recurrence $T_n = T_{n-1}T_{n-2}$ with boundary conditions $T_1 = 1$ and $T_2 = 2$.
3. Solve the recurrence $T_n T_{n-2} = n T_{n-1}^2$.
4. Solve the recurrence $T_{n+1} = 2T_n(1-T_n)$. Hint: The equation is much simpler if the change of variable $S_n = a + T_n$ is made with the right choice for a.

7.1.4* Repertoire

The method of *repertoire* applies to linear recurrence equations, including those of full history. The general form of such equations is

$$\sum_{0 \le i \le n} a_{ni} T_i = b_n. \tag{97}$$

One of the ideas behind the method of repertoire is to start with a guess G_i. Then find the function d_n satisfying

$$\sum_{0 \le i \le n} a_{ni} G_i = d_n. \tag{98}$$

The value of d_n is computed by evaluating the sum on the left side of eq. (98) (this will be easy for some guesses and difficult for others). Suppose we have a set G_{ik} of guesses, and have computed the corresponding d_{nk}. Let $H_i = \sum_k l_k G_{ik}$ be any linear combination of the G_{ik} and let $e_n = \sum_k l_k d_{nk}$ be the corresponding linear combination of the d_{nk}. Then, since eq. (98) is a linear equation, we have

$$\sum_{0 \le i \le n} a_{ni} H_i = e_n. \tag{99}$$

This brings us to the second idea. If we have an appropriate set of guesses G_{ik}, then we can form a linear combination of the corresponding d_{nk} such that e_n is equal to b_n. In this case the corresponding linear combination of the G_{ik} is a solution of the original equation [eq. (97)]. The key to the successful use of the method is in choosing the G_{ik}: they must be summable, and it must be possible to find a linear combination of the sums that is equal to the right-hand side of the original equation.

To see how the method works, let's use it to solve the recurrence for the average time required by Quicksort. Our equation is

$$C_n = n + 1 + \frac{2}{n} \sum_{0 \le i < n} C_i. \tag{100}$$

For our first set of guesses, let's use $G_{ik} = \binom{i}{k}$. This set of guesses is rather general (due to the parameter k), and it is easy to sum. Since G_{ik} is a polynomial of degree k in i, this set of guesses will be adequate if and only if C_n is a polynomial function of n.

For this choice of guesses, the sum in eq. (100) is easy to simplify. From eq. (3.50) we get $\sum_{0 \le i < n} G_{ik} = \binom{n}{k+1}$. Thus we have

$$G_{nk} - \frac{2}{n} \sum_{0 \le i < n} G_{ik} = \binom{n}{k} - \frac{2}{n}\binom{n}{k+1} = \binom{n}{k} \frac{kn - n + 2k}{(k+1)n}. \tag{101}$$

If we examine eq. (101) for small values of k, we find

$$G_{n0} = -1 + \frac{2}{n} \sum_{0 \le i < n} G_{i0}, \tag{102}$$

$$G_{n1} = 1 + \frac{2}{n} \sum_{0 \le i < n} G_{i1}, \tag{103}$$

$$G_{n2} = \frac{(n-1)(n+4)}{6} + \frac{2}{n} \sum_{0 \le i < n} G_{i2}. \tag{104}$$

We have guesses that result in d_n being constant and in d_n being a quadratic function of n. This set does not contain any linear functions of n, and a linear function is needed to satisfy eq. (100). Therefore we need to increase our "repertoire" of functions.

Good sorting algorithms require a time that increases like $n \ln n$, so we should consider guesses that vary this way. Now $n \ln n$ itself is not a useful guess because there is no easy way to sum it. On the other hand, harmonic numbers vary like $n \ln n$ for large n and are also easy to sum. Therefore, let's now consider guesses of the form $T_{ik} = \binom{i}{k} H_i$. Using the answer to Exercise 4.5.4–3, we have that

$$\sum_{0 \le i < n} T_{ik} = \binom{n}{k+1}\left(H_n - \frac{1}{k+1}\right). \tag{105}$$

Using eq. (105) and $T_{nk} = \binom{n}{k} H_n$ we obtain the following relation:

$$T_{nk} = \binom{n-1}{k}\frac{kn - n + 2k}{(n-k)(k+1)} H_n + \binom{n}{k+1}\frac{2}{(k+1)n} + \frac{2}{n}\sum_{0 \le i < n} T_{ik}. \tag{106}$$

If we examine eq. (106) for small values of k, we find

$$T_{n0} = -H_n + 2 + \frac{2}{n}\sum_{0 \le i < n} T_{i0}, \tag{107}$$

$$T_{n1} = H_n + \frac{n-1}{2} + \frac{2}{n}\sum_{0 \le i < n} T_{i1}. \tag{108}$$

Notice that eq. (108) has a term that is linear in n.

We now have four useful equations. Eq. (108) can be used to introduce a linear term, eq. (107) can be used to cancel out the harmonic number that is in eq. (108), and eq. (102) and eq. (103) can be used to get the constants to come out right. We now need to find constants l_1, l_2, l_3, and l_4 such that the $n+1$ from eq. (100) is expressed as a linear combination of the terms from eqs. (102, 103, 107, 108). In other words, we need to find l's such that

$$n + 1 = -l_1 + l_2 + l_3(-H_n + 2) + l_4\left(H_n + \frac{n-1}{2}\right) \tag{109}$$

for all n. Since n, H_n, and 1 are linearly independent functions of n, we need to balance the coefficients of each function separately. This gives

$$1 = \tfrac{1}{2} l_4, \tag{110}$$

$$0 = -l_3 + \tfrac{1}{2} l_4, \tag{111}$$

$$1 = -l_1 + l_2 + 2l_3 - \tfrac{1}{2} l_4. \tag{112}$$

Solving for the l's gives $l_4 = 2$, $l_3 = 1$, and $l_2 = l_1$. So far we can have any value for l_1. The solution of eq. (100) is

$$C_n = l_1 + l_1 n + H_n + 2nH_n = 2nH_n + l_1 n + H_n + l_1 + 2. \tag{113}$$

We now need to eliminate the constant l_1. To do this, we use the boundary condition $C_0 = 0$. For $n = 0$, eq. (113) gives $C_0 = l_1 + 2$, so l_1 must be -2. The solution of eq. (100) with the boundary condition $C_0 = 0$ is therefore

$$C_n = 2nH_n - 2n + 2H_n, \tag{114}$$

which agrees with the answer obtained in Section 7.1.1.1.

Repertoire was developed as a formal method by Greene and Knuth [6, Section 2.1.2.2].

EXERCISES

1. A variation on Quicksort examines three elements before selecting the splitting element. The middle element of the three is selected for the splitting element. This usually results in a more even division of the file. The average number of comparisons used by this Median-of-three Quicksort obeys the recurrence

$$C_n = n + 1 + \sum_{1 \le k \le n} \frac{\binom{k-1}{1}\binom{n-k}{1}}{\binom{n}{3}} (C_{k-1} + C_{n-k}).$$

What is the solution of this recurrence? Hint: This can be solved using either generating functions or the method of repertoire.

2. The method of repertoire can often be used to obtain limits on the size of the solution of a recurrence. This is particularly straightforward to do when the recurrence has a highest order term that is a positive linear combination of the lower order terms. In Section 7.1.3.1, we considered the the recurrence

$$A(t, v) = at + 2\sum_{i} \binom{t}{i} p^i (1-p)^{t-i} A(t-i, v-1)$$

for the average running time of the Silly Algorithm. One way to make the Silly Algorithm run much faster (on the average) is to check at Steps 4 and 6 to see whether the literals x_i and $\neg x_i$ occur in the predicate. If both occur, proceed with the algorithm as stated in Section 7.1.3.1. If only $\neg x_i$ occurs, skip over Step 4. If only x_i occurs, skip over Step 6. If neither occurs, skip Step 4. This modified algorithm contains a simplified version of the *pure literal rule.* It will find solutions if and only if the problem has solutions. It does not usually find all the solutions. The average running time of this algorithm is given by

$$B(t, v) = at + 2\sum_{i \ge 1} \binom{t}{i} p^i (1-p)^{t-i} B(t-i, v-1) + (1-p)^{2t} B(t, v-1).$$

with boundary conditions $B(t, 0) = 0$ and $B(0, v) = 0$. Define $B(t)$ to be the solution of the recurrence

$$B(t) = at + 2\sum_{i \ge 1} \binom{t}{i} p^i (1-p)^{t-i} B(t-i) + (1-p)^{2t} B(t)$$

Show that $B(t) \le ct^x$ for $x > -1/\lg(1-p)$ and some constant c. Hint: This is surely true for all t below any fixed bound. Plug in the proposed limit and see what condition x must satisfy for ct^x to be a bound for large t. See Goldberg et al. [88] for the relation of $B(t)$ to the average running time.

7.2 NONLINEAR EQUATIONS

Nonlinear recurrence equations are often much more difficult to solve than linear recurrence equations. There are three basic techniques to consider. First, sometimes you can transform the nonlinear equation into an equivalent linear equation. Second, you can sometimes solve nonlinear equations with generating functions. Third, you may be lucky and be able to guess a solution of your equation. This approach is particularly promising if the values generated by your equation appear to follow a simple pattern. In the subparts of this section we will consider the first two approaches. We will mainly use the third approach in Sections 7.3 and 7.4.

7.2.1 Transformation to Linear Recurrences

Perhaps the simplest type of nonlinear recurrence equation is one that has the form

$$\sum_{0 \le i \le n} c_{in} g(T_i) = h_n, \qquad \textbf{(115)}$$

where g is some function. In this case, you just define a new variable $G_n = g(T_n)$. Now, eq. (114) is a linear equation in G, so our previous techniques can be used.

Consider the problem
$$T_n^2 - T_{n-1}^2 = 1. \qquad (116)$$

This is a linear first order equation in T_n^2. The solution is

$$T_n^2 = T_0^2 + n, \qquad (117)$$

or

$$T_n = \sqrt{T_0^2 + n}. \qquad (118)$$

There are several more complex types of equations that can be solved by a transformation of variables. Consider equations of the form

$$\prod_{0 \le i \le n} T_i^{k_{ni}} = h_n. \qquad (119)$$

If we take the logarithm of this equation, we get

$$\sum_{0 \le i \le n} k_{ni} \ln T_i = \ln h_n, \qquad (120)$$

which is a linear equation in the variable $\ln T_n$.

Consider the equation

$$T_n = T_{n-1}^2, \qquad (121)$$

with the boundary condition $T_0 = 2$. Taking logarithms, we get

$$\lg T_n = 2 \lg T_{n-1}, \qquad (122)$$

with the boundary condition $\lg T_0 = 1$. (Notice that the use of logarithms to the base 2 made the boundary condition simple). The solution is

$$\lg T_n = 2^n. \qquad (123)$$

Expressing the solution in terms of T_n rather than $\lg T_n$ gives

$$T_n = 2^{2^n}. \qquad (124)$$

Equations of the form

$$T_n T_{n+2} = a_n T_{n+1}^2 \qquad \textbf{(125)}$$

can be reduced to first order linear equations by the transformation $G_n = T_{n+1}/T_n$. (Such an equation can also be transformed to a linear second order equation by the use of logarithms.) Eq. (125) becomes

$$G_{n+1} = a_n G_n. \tag{126}$$

Transformations for nonlinear equations are considered in Greene and Knuth [6, Section 2.2.2], in Bender and Orszag [21, Section 2.5], and Benton [47, Section 6.9.3].

EXERCISE

1. [See 47] Solve $f_{n+1}f_n + af_{n+1} + bf_n + c = 0$. Hint: Consider the transformation $f_n = g_n + A$, where $A^2 + (a+b)A + c = 0$, and the transformation $h_n = 1/g_n$.

7.2.1.1 Decision Tables

A *decision table* relates a set of tests to a set of actions that are to be carried out. Each column in the table corresponds to a set of test results and leads to a particular set of actions. The entire set of columns covers every set of test results that leads to actions. Decision tables provide a very convenient way to organize the logic of some types of programs.

One problem with decision tables is that they do not automatically lead to efficient programs. Often the best approach is a branching program that does one test and then branches to one of two places depending on the outcome of the test. At each of the two places this approach is repeated for the remaining tests until all the needed tests are done. Two programs of this type that solve the same problem can differ markedly in efficiency depending on the order in which the tests are done. To get the most efficient program, you could just consider all such branching programs to find the most efficient one. In this section we will show that even for programs with a moderate number of tests there are too many such programs for the most efficient one to be found by searching over all possible programs.

The first step is to find the recurrence relation that describes the process. Suppose that the program must make a total of n tests. There are n possible choices for the first test of the program. For each choice it is still necessary to provide for the remaining $n-1$ tests. If P_n is the number of programs for n tests, then P_n obeys the recurrence

$$P_n = nP_{n-1}^2 \tag{127}$$

with the boundary condition $P_1 = 1$.

With equations of this form, it is helpful to take logarithms. Let $L_n = \ln P_n$. Then eq. (127) is equivalent to

$$L_n = 2L_{n-1} + \ln n \tag{128}$$

with the boundary condition $L_1 = 0$. This is a first order linear equation with the solution

$$L_n = \sum_{0 \le j < n} 2^j \ln(n-j). \tag{129}$$

Using $P_n = e^{L_n}$ gives

$$P_n = \prod_{0 \le j < n} (n-j)^{2^j}. \tag{130}$$

From this product it is easy to tell that $2^{2^{n-2}} < P_n < n^{2^{n-1}}$ for $n \ge 3$. Both the upper and lower limits are very large functions, but they are also quite far apart. Let's see what more we can determine by a more careful analysis.

In eq. (129), if we replace j by $n-j$, we get

$$L_n = 2^n \sum_{1 \le j \le n} 2^{-j} \ln j. \tag{131}$$

Now,

$$\sum_{1 \le j \le n} 2^{-j} \ln j = \sum_{j \ge 1} 2^{-j} \ln j - \sum_{j > n} 2^{-j} \ln j. \tag{132}$$

The value of the first sum on the right side of eq. (132) is some constant, which we will call C_0. We can write eq. (132) as

$$\sum_{1 \le j \le n} 2^{-j} \ln j = C_0 - \sum_{j > n} 2^{-j} \ln j. \tag{133}$$

The sum on the right side of eq. (133) is very small. We can obtain an upper limit on its value as follows:

$$\sum_{j > n} 2^{-j} \ln j < \int_n^\infty 2^{-x} \ln x \, dx \tag{134}$$

$$< \int_n^\infty 2^{-x} x \, dx = \left[\frac{n}{\ln 2} + \frac{1}{(\ln 2)^2} \right] e^{-n \ln 2} \tag{135}$$

$$< \frac{n+1}{\ln 2} 2^{-n}. \tag{136}$$

From eq. (136) we also conclude that the sum $\sum_{j \ge 1} 2^{-j} \ln j$ converges rapidly. We can approximate its value by summing the first 20 terms to get $C_0 \approx 0.50780$.

Plugging eq. (136) and eq. (133) into eq. (131) gives

$$\ln P_n = (C_0 - |O(n 2^{-n})|) 2^n, \tag{137}$$

or, letting $K = e^{C_0} \approx 1.6617$,

$$P_n = K^{2^n} e^{-|O(n)|}. \tag{138}$$

The error term in eq. (138) is still exponential, but it does vary much less rapidly than K^{2^n} does. Notice that the value of K^{2^n} for large n depends dramatically on the value of K, so that even a very small error in K (such as one part in

10^{-4}) has a much larger effect on the final answer for large n than the $e^{-|O(n)|}$ factor does.

Decision tables are discussed by Metzer and Barnes [115].

7.2.2* Almost Linear Recurrences

In this section we will consider recurrences of the form

$$T_{n+1} = a_n T_n + f(n, T_n), \tag{139}$$

where $a_n > 1$ and $f(n, T_n)$ is small. The function $f(n, T_n) = \ln(1 + 1/T_n)$ is an example of a possible $f(n, T_n)$. Such equations are almost linear. If $f(n, T_n)$ is small enough, then it is possible to obtain interesting asymptotic results.

Such equations often arise from taking the logarithm of a nonlinear equation. For example, in the last section we considered the equation for the number of programs for a decision table:

$$P_n = nP_{n-1}^2. \tag{140}$$

After taking logarithms, we obtained

$$L_n = 2L_{n-1} + \ln n, \tag{141}$$

which is equivalent to eq. (139) with $a_n = 2$ and $f(n, T_n) = \ln(n+1)$.

The basic idea for solving an almost linear equation is to treat $f(n, T_n)$ as known [i.e., as though $f(n, T_n)$ did not depend on T_n] and use eq. (5.29) to solve eq. (139). The only problem is that since $f(n, T_n)$ is not known, the answer contains an unknown function [i.e., the b_i in eq. (5.29) depends on $f(n, T_n)$]. If, however, $a_n > 1$ and $f(n, T_n)$ is small enough, then the sum from eq. (5.29) will converge rapidly and it can be approximated by its first few terms.

To illustrate the method, let's consider another problem related to decision tables. A decision table program tests one variable and then has two subparts to test the remaining variables. Suppose we wish to consider two decision table programs equivalent if their only difference is in the order of the subparts. The number of inequivalent programs is given by the recurrence

$$P_n = n\frac{P_{n-1}(P_{n-1}+1)}{2}, \tag{142}$$

with the boundary condition $P_1 = 1$. Taking logarithms and using L_n for $\ln P_n$ gives

$$L_n = 2L_{n-1} + \ln n - \ln 2 + \ln\left(1 + \frac{1}{P_{n-1}}\right), \tag{143}$$

with the boundary condition $L_1 = 0$. We now apply eq. (5.29) with $a_n = 2$ and $b_n = \ln n - \ln 2 + \ln(1 + 1/P_{n-1})$ to obtain

$$L_n = \sum_{2 \le i \le n} 2^{n-i} b_i = 2^n \sum_{2 \le i \le n} 2^{-i} b_i. \tag{144}$$

n	P_n	K^{2^n}
1	1	2.16
2	2	4.68
3	9	21.86
4	180	478.01
5	81450	228497.22
6	1.990×10^{10}	5.221×10^{10}

TABLE 7.2. The number of inequivalent programs, P_n, for an n-test decision table compared with K^{2^n} where $K = 1.4705$.

Now we need to study the rightmost sum in eq. (144). First notice that P_n is an increasing function of n, so $P_n \geq 1$ and $\ln(1+1/P_n) \leq \ln 2$, which means that this sum is less than the sum considered in eq. (132). Using eq. (136) and letting $C_0 = \sum_{i \geq 2} 2^{-i} b_i$, we get

$$\sum_{2 \leq i \leq n} 2^{-i} b_i = C_0 - |O(n2^{-n})|. \tag{145}$$

From the first 20 terms of the sum we find that $C_0 \approx 0.3856$. Letting $K = e^{C_0} \approx 1.4705$, we find that the solution of the original equation is

$$P_n = K^{2^n} e^{-|O(n)|}. \tag{146}$$

Table 7.2 compares the first few values computed by the recurrence eq. (142) with the values computed by eq. (146). The two formulas show about the same rate of growth, but the actual values generated are somewhat different. Had b_i not grown so rapidly, the results would have been closer. Even a small relative error in L_n can be important when computing P_n since $P_n = e^{L_n}$.

Our second example uses binary trees. A *rooted binary tree* has a special node called the *root.* We will also consider the empty tree, which has no nodes, to be a rooted tree. Each node in a rooted tree has two children. Each child is either empty or the root of a subtree. (Often one does not count the empty children, so we may say that a node with one empty child has one child, and a node with two empty children has no children.) The two children (the left and right children) are distinct. The number of such trees with height no more than n is given by

$$X_n = 1 + X_{n-1}^2 \tag{147}$$

with boundary condition $X_0 = 1$, because a tree of height no more than n consists either of a root or of a root and two subtrees of height no more than $n-1$. The first term in the recurrence allows for the root and no subtrees. The second term allows for the number of ways to have two subtrees with height no more than $n-1$.

n	X_n	K_{2^n}
0	1	1.50284
1	2	2.25852
2	5	5.10091
3	26	26.01924
4	677	677.00074
5	458330	458330.00000

TABLE 7.3. The rooted trees, X_n, for a maximum height n compared with K^{2^n} where $K = 1.502837$.

Using $L_n = \ln X_n$ and taking the logarithm of eq. (147) (after transforming the right side by factoring out X_{n-1}^2) gives

$$L_n = 2L_{n-1} + \ln\left(1 + \frac{1}{X_{n-1}^2}\right) \tag{148}$$

with boundary condition $L_0 = 0$. The solution of this equation is

$$L_n = 2^n \sum_{1 \le i \le n} 2^{-i} \ln\left(1 + \frac{1}{X_{n-1}^2}\right). \tag{149}$$

Now consider the sum in eq. (149). Since X_n is an increasing function, the logarithm is no larger than $\ln 2$. Thus

$$\sum_{1 \le i \le n} 2^{-i} \ln\left(1 + \frac{1}{X_{n-1}^2}\right) = \sum_{i \ge 1} 2^{-i} \ln\left(1 + \frac{1}{X_{n-1}^2}\right) - \sum_{i > n} 2^{-i} \ln\left(1 + \frac{1}{X_{n-1}^2}\right) \tag{150}$$

$$= C_0 - \left| O\left(\ln\left(1 + \frac{1}{X_{n-1}^2}\right)\right)\right|, \tag{151}$$

where $C_0 = \sum_{i \ge 1} 2^{-i} \ln(1 + 1/X_{n-1}^2) \approx 0.407355$.

Exponentiating eq. (149) and using eq. (151) with $K = e^{C_0} \approx 1.502837$ gives

$$X_n = K^{2^n} e^{-|O(\ln(1+1/X_{n-1}^2))|}. \tag{152}$$

Since X_n is an increasing function, this gives

$$X_n = K^{2^n}(1 + O(X_{n-1}^{-2})). \tag{153}$$

Since X_{n-1} is very close to $K^{2^{n-1}}$, $K^{2^n} X_{n-1}^{-2}$ is very close to 1. Thus we have

$$X_n = K^{2^n} + O(1). \tag{154}$$

If we take slightly more care, we can find that the constant in the big O is less than one-half, so X_n is the integer nearest to K^{2^n}. Of course we must know the value of K to high accuracy to use this last result effectively. Table 7.3 shows some of the values for X_n and for K^{2^n}. This time the values are quite close, because this time the logarithm of the original equation was quite close to linear.

Almost linear recurrences are considered by Greene and Knuth [6, Section 2.2.3]. Their presentation is based on the paper by Aho and Sloane [64], and our section is based on both sources.

EXERCISES

1. A balanced binary tree is a rooted binary tree where at every node the height of the right subtree differs by no more than 1 from the height of the left subtree. Show that the number of balanced binary trees of height n is given by the recurrence

$$T_n = T_{n-1}^2 + 2T_{n-1}T_{n-2},$$

with boundary conditions $T_0 = 1$, $T_1 = 3$.

2. Find an approximate solution of the recurrence in the previous exercise. Hint: The transform $X_n = T_n + T_{n+1}$ leads to an easier to work with recurrence.

3. Find an approximate solution of the recurrence

$$T_n = T_0 T_1 \cdots T_{n-1} + r,$$

with boundary condition $T_0 = 1$. Hint: First notice that the recurrence is equivalent to

$$T_n = (T_{n-1} - r)T_{n-1} + r,$$

with boundary condition $T_1 = r + 1$. Then make the substitution $X_n = T_n - \frac{1}{2}r$.

7.2.3 Generating Functions and Convolutions

The sum

$$\sum_{0 \le i \le n} T_i G_{n-i} \tag{155}$$

is the *convolution* of T and G. We saw some examples of convolutions in Section 7.1.2 when considering linear full-history equations. The approach recommended there was to use generating functions. If T and G are the same function, then the convolution is a nonlinear function. In this case generating functions are still often useful.

One application of a convolution type of recurrence equation is to the problem of counting how many different binary trees can be formed with n nodes. If we have n nodes, then one node must be the root. If we then use i nodes for the left subtree, we must use $n-i-1$ nodes for the right subtree. Each value of i leads to a different set of subtrees. Also, for each left subtree with i nodes you get a different overall tree for each right subtree with $n-i-1$ nodes. Thus, if we let C_n be the number of binary trees with n nodes, C_n obeys the recurrence

$$C_n = \sum_{0 \le i \le n-1} C_i C_{n-i-1} \qquad \text{for } n \ge 1. \tag{156}$$

Since there is only one tree with zero nodes (the empty tree), we have the boundary condition $C_0 = 1$.

To solve eq. (156) with generating functions, first multiply it by z^n, sum over $n \geq 1$, and add $C_0 = 1$. Then interchange the order of summation and use j for $n - i - 1$ (see Exercise 3.3–2). We get

$$\sum_{n\geq 0} C_n z^n = \sum_{n\geq 0} z^n \sum_{0\leq i\leq n-1} C_i C_{n-i-1} + 1 \tag{157}$$

$$= z\Big(\sum_{i\geq 0} C_i z^i\Big)\Big(\sum_{j\geq 0} C_j z^j\Big) + 1. \tag{158}$$

Now define the generating function $G(z) = \sum_{n\geq 0} C_n z^n$. The generating function obeys the equation

$$zG(z)^2 - G(z) + 1 = 0. \tag{159}$$

Using the formula for quadratic equations to solve for $G(z)$ gives

$$G(z) = \frac{1}{2z}\left(1 \pm \sqrt{1-4z}\right). \tag{160}$$

We also have that $G(0) = C_0 = 1$, which can be true only if we take the minus sign in eq. (160), so

$$G(z) = \frac{1}{2z}\left(1 - \sqrt{1-4z}\right). \tag{161}$$

Now we need the power series for $G(z)$. We can use the binomial theorem to expand the square root. This gives

$$G(z) = \frac{1}{2z}\left(1 - \sum_{n\geq 0}\binom{\frac{1}{2}}{n}(-4z)^n\right). \tag{162}$$

Cancelling the $n = 0$ term with the 1, multiplying the $1/(2z)$ into the sum, and replacing n by $n+1$ gives

$$G(z) = \sum_{n\geq 0}\binom{\frac{1}{2}}{n+1}(-1)^n 2^{2n+1} z^n. \tag{163}$$

Since

$$\binom{\frac{1}{2}}{n+1} = \frac{(-1)^n 2^{-2n-1}}{n+1}\binom{2n}{n}, \tag{164}$$

(see Exercise 3.2–1), eq. (163) simplifies to

$$G(z) = \sum_{n\geq 0}\frac{1}{n+1}\binom{2n}{n} z^n. \tag{165}$$

Thus there are

$$C_n = \frac{1}{n+1}\binom{2n}{n} \tag{166}$$

ways to form binary trees with n nodes.

The first few values of C_n are given in Table 7.4. Using Stirling's approximation, you can show that

$$C_n = 4^n/(n\sqrt{\pi n}) + O(4^n n^{-5/2}). \tag{167}$$

$C_0 = 1$	$C_1 = 1$	$C_2 = 2$	$C_3 = 5$
$C_4 = 14$	$C_5 = 42$	$C_6 = 132$	$C_7 = 429$
$C_8 = 1430$	$C_9 = 4862$	$C_{10} = 16796$	$C_{11} = 58786$

TABLE 7.4. The number of binary trees with n nodes.

The number C_n is the nth Catalan number.

This calculation of the number of binary trees is given in Knuth [9, Section 2.3.4.4].

EXERCISES

1. Compute the number of possible ways you can combine n associative items. For example, the addition $a+b+c$ can be done two ways. You can compute $(a+b)+c$ or you can compute $a+(b+c)$. Let W_n be the number of ways that you can combine n items. It obeys the recurrence

$$W_n = \sum_{1 \le i \le n-1} W_i W_{n-i}$$

with the boundary condition $W_1 = 1$. (Explain why this is so.) Solve this recurrence to obtain the final answer to the problem. Hint: It is often better to relate your problem to a previously solved problem rather than to solve it from scratch. Try the substitution $Y_i = W_{i+1}$.

2. [See 23] A *diagonal triangulation* of a convex n-gon is a division of the convex n-gon into $n-2$ triangles by drawing $n-3$ noncrossing lines between vertices. Let T_n be the number of diagonal triangulations of an n-gon. Show that T_n obeys the recurrence

$$T_n = T_2 T_{n-1} + T_3 T_{n-2} + \cdots + T_{n-1} T_2,$$

with the boundary condition $T_2 = 1$. Hint: For $n > 3$ some triangle must connect the first two vertices and a third vertex. This triangle will divide the n-gon into three regions.

3. Find a closed form for T_n from the previous exercise.

4. Let p be the probability that a coin lands heads up, $1-p$ the probability that it lands tails up, and ϕ_n the probability that after n flips the coin has for the first time landed heads up one more time than it has landed tails up. Show that

$$\phi_n = (1-p)(\phi_1 \phi_{n-2} + \phi_2 \phi_{n-3} + \cdots + \phi_{n-2} \phi_1) \qquad \text{for } n \ge 1,$$

$\phi_0 = 0$, and $\phi_1 = p$. Hint: For $n \ge 1$, if the first flip results in tails up, then there is a string of flips to get back up to the same number of heads as tails, followed by a string of flips that results in one more head than tail.

5. Find a closed form for the generating function $\Phi(x) = \sum_{n \ge 0} \phi_n x^n$, where ϕ_n is defined in the previous exercise. What is the probability that some number of flips results in one more head than tail? What is the average number of flips required to obtain one more head than tail?

6. Obtain a power series for ϕ_n (see the previous exercise).

7.3* INTEGER PART

Previously we studied divide and conquer recurrences that had the form

$$T_n = a_n T_{n/c} + f_n, \tag{168}$$

$$T_1 = b. \tag{169}$$

This form leads to a solution only when n is a power of c, but usually the recurrence (with slight modification) describes an algorithm that works correctly even when n is not a power of c. The recurrence that describes the performance of the algorithm for general n may have the form

$$T_n = a_n T_{\lceil n/c \rceil} + f_n, \tag{170}$$

$$T_1 = b. \tag{171}$$

or even something that appears harder to analyze, such as

$$T_n = T_{\lceil n/2 \rceil} + T_{\lfloor n/2 \rfloor} + f_n, \tag{172}$$

$$T_1 = b. \tag{173}$$

In this section we consider briefly a few techniques that can be used to solve such recurrences. We will proceed mainly by way of examples.

If a recurrence has the form of eq. (170), then you can use the ordinary technique for first order equations: repeatedly use the equation to simplify the right side. Since $\lceil \lceil n/c^i \rceil / c \rceil = \lceil n/c^{i+1} \rceil$, we obtain

$$T_n = T_1 \prod_{0 \le i \le \lceil \log_c n \rceil} a_{\lceil n/c^i \rceil} + \sum_{0 \le i \le \lceil \log_c n \rceil} b_{\lceil n/c^i \rceil} \prod_{i < j \le \lceil \log_c n \rceil} a_{\lceil n/c^j \rceil}. \tag{174}$$

Eq. (174) is comparatively easy to evaluate when n is a power of c since the ceiling function can then be dropped. For other values of n, eq. (174) is likely to be much more difficult to evaluate exactly.

If in eq. (170) you replace the ceiling function with the floor function, then eq. (174) with the same replacement gives the solution.

A simple example is given by the equation

$$T_n = 1 + T_{\lceil n/2 \rceil} \tag{175}$$

with the boundary condition $T_1 = 0$. The solution of this equation is $T_n = \lceil \lg n \rceil$.

Often you can determine the solution of an equation of this type by generating the first few values and then guessing the pattern. Of course, after you guess the pattern, you must prove that it is correct. It is usually possible to do this using proof by induction. Let's try this approach on the previous equation. Table 7.5 shows the first few values.

The values appear to follow the pattern $T_n = \lceil \lg n \rceil$. Let's now prove this is correct by induction. Our induction hypothesis is that $T_n = \lceil \lg n \rceil$ for $n \ge 1$. For $n = 1$ the hypothesis gives $T_1 = 0$, which agrees with the boundary condition. Now we assume the hypothesis is true for all $n < n_*$, where $n_* > 1$,

$T_1 = 0$	$T_2 = 1$	$T_3 = 2$	$T_4 = 2$
$T_5 = 3$	$T_6 = 3$	$T_7 = 3$	$T_8 = 3$
$T_9 = 4$	$T_{10} = 4$	$T_{11} = 4$	$T_{12} = 4$
$T_{13} = 4$	$T_{14} = 4$	$T_{15} = 4$	$T_{16} = 4$

TABLE 7.5. The solution of the equation $T_n = 1 + T_{\lceil n/2 \rceil}$.

and try to use the recurrence equation to compute T_{n_*}. Since $n_* > 1$, $\lceil n_*/2 \rceil < n_*$, and we get

$$T_{n_*} = 1 + T_{\lceil n_*/2 \rceil} = 1 + \left\lceil \lg \left\lceil \frac{n_*}{2} \right\rceil \right\rceil. \tag{176}$$

Now define k to be $\lceil \lg n_* \rceil$, so $2^{k-1} < n_* \leq 2^k$. The value of $n_*/2$ obeys the inequalities $2^{k-2} < n_*/2 \leq 2^{k-1}$, and so $\lceil n_*/2 \rceil$ obeys the inequalities $2^{k-2} < \lceil n_*/2 \rceil \leq 2^{k-1}$ when $n_* \geq 2$ (2^{k-1} is even for $n_* \geq 3$ and $n_*/2 = \lceil n_*/2 \rceil$ for $n_* = 2$). Thus we have

$$T_{n_*} = 1 + \left\lceil \lg \left\lceil \frac{n_*}{2} \right\rceil \right\rceil = 1 + k - 1 = \lceil \lg n_* \rceil. \tag{177}$$

Thus if our hypothesis is true for all n in the range $1 \leq n < n_*$, then it is also true for n_* (where $n_* > 1$). Previously we showed our hypothesis is true for $n_* = 1$. Therefore, by induction, the hypothesis is true for all $n_* \geq 1$.

For the next example, let's consider the recurrence

$$D_n = n + D_{\lfloor n/2 \rfloor} \tag{178}$$

with the boundary condition $D_0 = 0$. The first few values are given in Table 7.6. These values have a somewhat stranger behavior than the values for the last example.

Applying eq. (174) to eq. (178) gives

$$D_n = \sum_{0 \leq i \leq \lfloor \lg n \rfloor} \left\lfloor \frac{n}{2^i} \right\rfloor. \tag{179}$$

Since $\lfloor n/2^i \rfloor = 0$ for $i > \lg n$, the upper limit on the sum can be dropped. The floor function on $\lfloor n/2^i \rfloor$ does not affect the value much (by no more than 1), so let's look at the sum

$$\sum_{i \geq 0} \frac{n}{2^i} = 2n. \tag{180}$$

The right side of eq. (179) would be $2n$ if we could ignore the floor function; the actual value should be a little less.

Let's now compare the values in Table 7.6 with $2n$. Five of the values (D_1, D_2, D_4, D_8, and D_{16}) are 1 less. Six of the values (D_3, D_5, D_6, D_9, D_{10}, D_{12}) are 2 less. Four of the values (D_7, D_{11}, D_{13}, D_{14}) are 3 less. One of the values (D_{15}) is 4 less. Perhaps you are beginning to see a pattern. The items

$D_1 = 1$	$D_2 = 3$	$D_3 = 4$	$D_4 = 7$
$D_5 = 8$	$D_6 = 10$	$D_7 = 11$	$D_8 = 15$
$D_9 = 16$	$D_{10} = 18$	$D_{11} = 19$	$D_{12} = 22$
$D_{13} = 23$	$D_{14} = 25$	$D_{15} = 26$	$D_{16} = 31$

TABLE 7.6. The solution of the equation $D_n = n + D_{\lfloor n/2 \rfloor}$.

that are 1 less have indices whose binary representation has one 1-bit. The items that are 2 less have indices whose binary representation has two 1-bits, etc.

If you think of what happens when you compute the sum of $\lfloor n/2^i \rfloor$, you can see that, when bit i of n is a zero, the floor function does not have much effect, but when bit i of n is a 1, the floor function causes you to lose $\frac{1}{2}$ for the $i+1$ term, $\frac{1}{4}$ for the $i+2$ term, $\frac{1}{8}$ for the $i+3$ term, etc.

The arguments of the last few paragraphs have been typical of the type of arguments that you should first make when trying to solve a problem. They are designed to lead to the answer fairly quickly. They are too informal, however, for you to be really sure that no errors were made. (We gained most of our speed by being informal.) We now need a formal proof that the answer is correct. We can either go back over the arguments and fill in the missing details, or we can construct a proof by induction. In this case it is easier to construct a proof by induction (see Exercise 1).

Now let's consider the recurrence for the average time that the Binary Search Algorithm (Algorithm 1.6) uses to look up an item in a table. Each time we do the test for the loop, the algorithm decides whether the item is in the upper or lower half of the remaining part of the table. If the size of the remaining part is not even, then one part has just over half the items and the other part has just under half. A random item is slightly more likely to be in the larger part.

We will consider two approaches to computing the average time for general n. First, we will find a recurrence relation and determine its solution. Second, we will step back and look at the overall nature of the problem. With the proper insight there is a simple way to solve the problem.

The recurrence equation for the average number of times to do the test is

$$T_n = 1 + \frac{\lceil n/2 \rceil}{n} T_{\lceil n/2 \rceil} + \frac{\lfloor n/2 \rfloor}{n} T_{\lfloor n/2 \rfloor} \tag{181}$$

with the boundary condition $T_1 = 1$. The first few values for T_n are given in Table 7.7. If we write n in the form $n = 2^k + j$ where $0 \le j < 2^k$, then the values in Table 7.7 obey the formula

$$T_{2^k+j} = k + 1 + \frac{2j}{2^k + j} \qquad \text{for } 0 \le j < 2^k. \tag{182}$$

Since $k = \lfloor \lg n \rfloor$ and $j = n - 2^{\lfloor \lg n \rfloor}$, this can also be written as

$$T_n = \lfloor \lg n \rfloor + 3 - \frac{2^{\lfloor \lg n \rfloor + 1}}{n}. \tag{183}$$

$T_1 = 1$	$T_2 = 2$	$T_3 = 2\frac{2}{3}$	$T_4 = 3$
$T_5 = 3\frac{2}{5}$	$T_6 = 3\frac{2}{3}$	$T_7 = 3\frac{6}{7}$	$T_8 = 4$
$T_9 = 4\frac{2}{9}$	$T_{10} = 4\frac{2}{5}$	$T_{11} = 4\frac{6}{11}$	$T_{12} = 4\frac{2}{3}$
$T_{13} = 4\frac{10}{13}$	$T_{14} = 4\frac{6}{7}$	$T_{15} = 4\frac{14}{15}$	$T_{16} = 5$

TABLE 7.7. The solution of the equation
$T_n = 1 + (\lceil n/2 \rceil / n) T_{\lceil n/2 \rceil} + (\lfloor n/2 \rfloor / n) T_{\lfloor n/2 \rfloor}$.

You can prove by induction that this is the correct answer.

The other way to look at the original problem is to realize that we are looking up items that are on the leaves of a binary tree. If there are n leaves where $n = 2^k + j$ with $0 \le j < 2^k$, then some of the leaves are on level k and some are on level $k+1$. The number of times the program will do the test for the loop is equal to the level of the leaf that is being looked up plus 1. Each leaf is equally likely to be looked up. When $j = 0$ all the leaves are on level k. When $j = 1$, two leaves must be on level $k+1$: one leaf must go down to level $k+1$ because only 2^k leaves will fit on level k and one leaf must go down so that there will be a place to attach the pair of leaves that are on level $k+1$. When $j = 2$, four leaves must go on level $k+1$, etc. In general there are $2j$ leaves on level $k+1$ (and $2^k - j$ on level k), so the average number of times to do the test is

$$1 + \frac{2^k - j}{2^k + j} k + \frac{2j}{2^k + j}(k+1) = k + 1 + \frac{2j}{2^k + j}. \tag{184}$$

We have now seen several examples that show that by examining the values generated by the recurrence, doing some analysis, and developing some insight, you can often find exact solutions of simple recurrences that involve floor and ceiling functions. Often, however, it is difficult or impossible to develop useful exact solutions. In such cases you will probably want to find an approximate solution. This may be difficult or easy, depending in part on how much accuracy you desire.

We will consider just one example. In Section 5.2.3 we did some analysis of the worst-case time used by the Faster Multiplication Algorithm (Algorithm 5.2). Let's look at the analysis in a little more detail. Assume the algorithm has been modified so that when the problem does not have an even number of digits the algorithm breaks it up into subparts with $\lceil n/2 \rceil$ and $\lfloor n/2 \rfloor$ digits. To be definite let's also assume that the running time is given by the recurrence

$$T_n = n + T_{\lceil n/2 \rceil + 1} + T_{\lceil n/2 \rceil} + T_{\lfloor n/2 \rfloor} \tag{185}$$

with the boundary conditions $T_3 = 9$ and $T_2 = 4$. We are making several assumptions in writing this equation. We are assuming that the time for the algorithm is n plus the time for the subproblems that are generated. (The real time will probably have the form $t_1 n + t_2$. We are assuming that time will be measured in units where $t_1 = 1$ and that t_2 can be neglected.) We are also assuming that the algorithm will generate smaller problems until it gets down to size two or three. As pointed out in Section 5.2.3.1, this is not a good idea. We do these things to obtain a simpler analysis suitable for a textbook. Obviously if you want a complete analysis of the real problem, you must make the assumptions appropriate to the real problem. (Simplified analyses, by the way, are often quite helpful in getting at the general characteristics of a problem without getting bogged down in too many details.)

Clearly T_n is an increasing function of n. If we round up at several places in eq. (185), we obtain the equation

$$U_n = n + 3U_{n/2+1}, \tag{186}$$

provided n is even. The boundary condition is $U_3 = 9$. Clearly $T_n \le U_m$ if $n \le m$. The solution of eq. (186) is

$$U_n = 12(n-2)^{\lg 3} - 2n + 3. \tag{187}$$

Since $T_n \le U_{n+1}$, we have $T_n \le 12(n-1)^{\lg 3} - 2n + 1$.

Likewise if we round down in several places we have

$$L_n = n + 3L_{n/2}, \tag{188}$$

provided n is even. The boundary condition is $L_2 = 4$. Clearly $T_n \ge L_m$ if $n \ge m$. The solution of eq. (188) is

$$L_n = \tfrac{8}{3} n^{\lg 3} - 2n. \tag{189}$$

Since $T_n \ge L_{n-1}$, we have $T_n \ge \frac{8}{3}(n-1)^{\lg 3} - 2n + 2$. For large n these bounds imply that T_n increases like a function somewhere between $\frac{8}{3}n^{\lg 3}$ and $12n^{\lg 3}$. If you assume that it increases like $4\sqrt{2}\, n^{\lg 3}$, you won't be off by more than a factor of 2.2. This is not very accurate, but it does give you a general idea of how the time varies. It is more difficult to obtain a better approximation.

Eq. (179) is considered in Knuth [9, Section 1.2.5]. Binary search is analyzed (using a much different approach from this one) in [11, Section 6.2.1].

EXERCISES

1. Prove that the solution of eq. (178) is $D_n = 2n - d_n$ where d_n is the number of 1-bits in the binary representation of n.
2. Prove that eq. (183) is the solution of eq. (181).
3. [See 1] There is a version of the Merge Sort Algorithm that uses a number of comparisons given by

$$T_n = n - 1 + T_{\lceil n/2 \rceil} + T_{\lfloor n/2 \rfloor},$$

with the boundary condition $T_1 = 0$. Give a closed-form solution for T_n.
4. Investigate eq. (185) numerically. What do your results indicate?
5. Show that the number of times that 2 evenly divides $n!$ is given by

$$T_n = \sum_{i \ge 0} \left\lfloor \frac{n}{2^i} \right\rfloor$$

and also by the recurrence

$$T_n = \left\lfloor \frac{n}{2} \right\rfloor + T_{\lfloor n/2 \rfloor},$$

with the boundary condition $T_0 = 0$. Solve the recurrence and say what the relation is between the solution and the number of 1-bits in the binary representation of n.

7.4* MIN AND MAX

Divide and conquer algorithms often lead to recurrences containing the min function. Such recurrences usually come about as follows. You have a problem of size n, which can be broken up into smaller problems in several different ways. Each way that you can break it up leads to a recurrence. You want to break your problem up in the best way, so you obtain a recurrence that contains a min over all the ways that the problem might be broken up.

There are two basic approaches to recurrences with min and max. We call the first approach the *clever approach.* In this approach you formulate the problem in such a way that its solution is obvious. We recommend that you use this approach whenever you can. When it works, it usually leads quickly to a solution. The only trouble with the clever approach is that it is not a general method. When you are not able to think of a suitable formulation of your problem, the clever approach is of no help.

We call the second approach the *brute-force approach.* It usually leads to long calculations, so you should not try it until you have given the clever approach a try. The advantage of the brute-force approach is that it is usually clear how to proceed. The more you work on your problem, the more you learn about the nature of the solution. Often this approach eventually leads to a solution. Even when it does not, it leads to bounds on the solution. Your whole analysis is not dependent on whether or not you can come up with a clever idea.

EXERCISES

1. Minimize

$$p_1 \ln p_1 + p_2 \ln p_2 + \cdots + p_n \ln p_n$$

subject to

$$p_1 + p_2 + \cdots + p_n = 1$$

and

$$p_1 \ge 0, \quad p_2 \ge 0, \ldots, p_n \ge 0.$$

Hint: This is equivalent to finding $f_n(1)$, where

$$f_n(x) = \min_{0 \le y \le x} \{y \ln y + f_{n-1}(y - x)\}$$

and $f_0(x) = 0$. Making each p_j the same size is promising.

2. Minimize

$$x_1^2 + x_2^2 + \cdots + x_n^2$$

subject to

$$x_1 + x_2 + \cdots + x_n = 1.$$

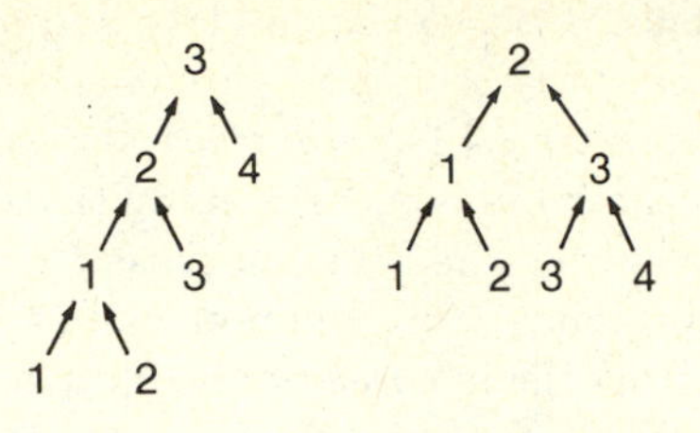

FIGURE 7.4 Two binary trees that correspond to a generalized binary search algorithm.

7.4.1* The Clever Approach

For example, let's consider generalizing the Binary Search Algorithm (Algorithm 1.6). At Step 3, instead of setting $m = \lfloor (b+t)/2 \rfloor$, let's set m to the best possible value, i.e., for each b and t have the algorithm set m to the value that leads to the smallest average number of comparisons. The average number of two-way comparisons used by this generalized binary search (to search for an item that is in the table) is given by the recurrence

$$C_n = 1 + \min_{0 \le i \le n} \left\{ \frac{i}{n} C_i + \frac{n-i}{n} C_{n-i} \right\} \quad \text{for } n > 1 \tag{190}$$

with the boundary condition $C_1 = 1$. With one comparison we can divide the file into two parts, one with i elements and the other with the rest of the elements. Since the minimum does not occur with $i = 0$ or $i = n$, we can replace eq. (190) with

$$C_n = 1 + \min_{1 \le i \le n-1} \left\{ \frac{i}{n} C_i + \frac{n-i}{n} C_{n-i} \right\} \quad \text{for } n > 1. \tag{191}$$

To solve eq. (191) using the clever approach, we must find the right way to view the problem so that the solution becomes obvious. We need an inspired idea. In this case we need to notice the connection between eq. (191) and binary trees. Consider a binary tree where each node has degree 2 or zero (see Figure 7.4.). Label the leaves of the tree from left to right with the integers from 1 to n. The labels on leaves represent table positions where items are found. Label each internal node with the largest number that is used to label any of the leaves that are its descendants.

Any rule for doing the comparisons implied by eq. (191) can be represented by some such tree. (A tree that corresponds to the set of possible comparisons for a search algorithm is called a *search tree.*) The number of comparisons needed to find an item is equal to the number of internal nodes on the path from the root to the leaf containing the item (counting the root). This number is equal to the length of the path from the root to the leaf node. The time to find all the items once is equal to n plus the sum of the lengths of all paths from leaves to the root. The average time is 1 more than the sum of the lengths divided by n.

By considering the properties of path length, we can now find the solution of our recurrence. Consider an old tree and a new tree that are the same, except that two leaves with a common parent in the old tree have a different common parent in the new tree. (The two trees correspond to two algorithms that differ in the way m is chosen in one particular situation.) This movement of two siblings creates one new leaf (the common parent in the old tree) and converts one leaf into an internal node (the common parent in the new tree). For example, in Figure 7.4, letting the left tree be the old tree and the right tree be the new tree, we see that leaf nodes 1 and 2 of the old tree have been moved to positions 3 and 4 of the new tree. Internal node 1 of the old tree has become leaf 1 of the new tree and leaf node 4 of the old tree has become internal node 3 of the new tree. No other nodes have been moved (although the other nodes have had their numbers changed by the renumbering).

What effect does moving a pair of nodes have on the path length? Let d_1 be the distance from the root to a moved node in the old tree. Removing the nodes from their place in the old tree caused a decrease of $2d_1$ in the total path length in the old tree, but this decrease is partially offset by an increase in path length of $d_1 - 1$ caused by their former parent becoming a leaf. The sum of the changes is $-d_1 - 1$. Let d_2 be the distance from the root to a moved node in the new tree. Adding the nodes to the new tree causes an increase in the total path length of $2d_2$, but this increase is partially offset by a decrease in path length of $d_2 - 1$ caused by the new parent becoming an internal node. The sum of these changes is $d_2 + 1$. The total change is thus $d_2 - d_1$. Moving a pair of nodes closer to the root therefore decreases the total path length, moving them further away increases the total path length. To make the total path length as small as possible, it is necessary to have all the leaves equidistant from the root, or as close to equidistant as possible.

This results in a tree with all the leaves on at most two levels. If the tree had leaves on three or more levels, you could move some leaves from the bottom level to be children of a leaf node on the level of leaves closest to the root and shorten the total path length. (On a tree with leaves on two levels, this transformation does not change the total path length, because the moved nodes stay the same distance from the root.)

Any positive integer n can be written in the form $n = 2^k + j$ where $k = \lfloor \lg n \rfloor$ and $j = n - 2^k$. The value of j is in the range $0 \leq j < n$. If a tree with all the leaves on one or two levels has n nodes, then it has $2j$ nodes on the bottom level and $2^k - j$ nodes on the next to bottom level. The path length to a node on the next to bottom level is k, and the path length to a node on the bottom level is $k+1$. The total path length is $kn + 2j$. The average number of comparisons is therefore

$$C_n = 1 + \frac{kn + 2j}{n} = \lfloor n \rfloor + 3 - \frac{2^{\lfloor n \rfloor + 1}}{n}. \tag{192}$$

If you solve a recurrence by the clever approach, and are not entirely sure that your answer is correct, then you should use the recurrence equation and proof by induction to prove that your answer is correct. When using the clever method, you become very hopeful that you have a correct insight, and so it becomes very easy for you to fool yourself.

EXERCISE

1. If you generalize the version of binary search that uses three-way comparisons (greater than, equal, or less than), then the number of comparisons is given by the recurrence

$$C_n = 1 + \min_{1 \le i \le n} \left\{ \frac{i-1}{n} C_{i-1} + \frac{n-i}{n} C_{n-i} \right\} .$$

with the boundary condition $C_0 = 0$. Solve this recurrence. Hint: Consider a three-way tree with the restriction that only two of the children of a node are permitted to have children of their own. (Greene and Knuth [6, Section 2.2.1] consider this problem with the boundary conditions $C_0 = 0$, $C_1 = 0$. In that case the best strategy is no longer to guess the middle element. Our boundary conditions give $C_1 = 1$.)

7.4.2* The Brute-Force Approach

To use the clever approach, you need a clever idea. When you don't have any clever ideas, then you should try the brute-force approach. You will still need some cleverness, but your work will be directed toward some definite short-range goals. Unfortunately you will probably have to do a lot of calculation.

The brute-force way to solve a recurrence with a min over positive terms is:

1. Guess where the minimum occurs.
2. Solve the recurrence using the presumed minimum.
3. Use the solution with the presumed minimum to prove that the guess about the minimum is correct.

If all three steps are completed successfully, then you have found the solution of the recurrence. If your initial guess is wrong, then at Step 3 you will not be able to show that the minimum occurs at the place you guessed. You will, however, obtain additional information about where the true minimum is located. Even if your guess is wrong, you will still have an upper limit on the correct solution. A similar method can be used for recurrences with max.

There may be several values of i that give a minimum in eq. (191). Any of them will do for solving the recurrence, although some may be easier to work with than others. Usually Steps 2 and 3 of the above procedure require considerable work, and so it is a good idea to be fairly sure of your guess before you put too much work in on using it. When guessing where the minimum occurs, particular attention should be given to obvious places such as 0, 1, $\lfloor n/2 \rfloor$, $\lceil n/2 \rceil$, $n-1$, and n.

To continue with eq. (191), an obvious guess is that the minimum occurs at $\lfloor n/2 \rfloor$. Let's first plug the guess into the recurrence; this gives us the presumed solution of the recurrence, which we will use to see whether the minimum occurs at the place we guessed. Let T_i be the solution of the recurrence under the

n	T_{n1}	T_{n2}	T_{n3}	T_{n4}	T_{n5}	T_{n6}	T_{n7}	T_{n8}	T_{n9}
1	1								
2	2								
3	$2\frac{2}{3}$	$2\frac{2}{3}$							
4	$3\frac{1}{4}$	3	$3\frac{1}{4}$						
5	$3\frac{3}{5}$	$3\frac{2}{5}$	$3\frac{2}{5}$	$3\frac{3}{5}$					
6	4	$3\frac{2}{3}$	$3\frac{2}{3}$	$3\frac{2}{3}$	4				
7	$4\frac{2}{7}$	4	$3\frac{6}{7}$	$3\frac{6}{7}$	4	$4\frac{2}{7}$			
8	$4\frac{1}{2}$	$4\frac{1}{4}$	$4\frac{1}{8}$	4	$4\frac{1}{8}$	$4\frac{1}{4}$	$4\frac{1}{2}$		
9	$4\frac{2}{3}$	$4\frac{4}{9}$	$4\frac{1}{3}$	$4\frac{2}{9}$	$4\frac{2}{9}$	$4\frac{1}{3}$	$4\frac{4}{9}$	$4\frac{2}{3}$	
10	$4\frac{9}{10}$	$4\frac{3}{5}$	$4\frac{1}{2}$	$4\frac{2}{5}$	$4\frac{2}{5}$	$4\frac{2}{5}$	$4\frac{3}{5}$	$4\frac{3}{5}$	$4\frac{9}{10}$

TABLE 7.8. Values for T_{ni}.

assumption that the minimum is at $i = \lfloor n/2 \rfloor$. That is, T_i is the solution of the recurrence

$$T_i = 1 + \frac{\lfloor i/2 \rfloor}{i} T_{\lfloor i/2 \rfloor} + \frac{\lceil i/2 \rceil}{i} T_{\lceil i/2 \rceil} \tag{193}$$

with the boundary condition $T_1 = 1$. The solution for eq. (193) was given in eq. (183); it is

$$T_i = \lfloor \lg i \rfloor + 3 - 2^{\lfloor \lg i \rfloor + 1}/i. \tag{194}$$

We now define

$$T_{ni} = 1 + \frac{i}{n} T_i + \frac{n-i}{n} T_{n-i}. \tag{195}$$

If our guess about the location of the minimum is correct, then $\min_i T_{ni}$ should occur at $i = \lfloor n/2 \rfloor$. Table 7.8 shows the first few values for T_{ni}. We can see that for as far as we have tested there is always a minimum at $i = \lfloor n/2 \rfloor$.

Now let's try to prove that there is a minimum at $i = \lfloor n/2 \rfloor$ (i.e., $T_{n,i} - T_{n,\lfloor n/2 \rfloor} \geq 0$). Rather than investigate $T_{n,i} - T_{n,\lfloor n/2 \rfloor}$ directly, we will look at $T_{n,i+1} - T_{n,i}$, because the algebra will be simpler. The difference $T_{n,i+1} - T_{n,i}$ is the discrete analog of the derivative. Using it to find the minimum is similar to using a derivative. The algebra is fairly simple, but if the function has more than one minimum, additional work is required to find the global minimum. Our table, however, suggests that we may be lucky and have only one minimum.

If there is a minimum at $i = \lfloor n/2 \rfloor$, then $T_{n,i+1} - T_{n,i}$ should be greater than or equal to zero for $i \leq (n-1)/2$ and less than or equal to zero for $i \geq (n-1)/2$. If we define $D_i = (i+1)T_{i+1} - iT_i$, then $T_{n,i+1} - T_{n,i} = (D_i - D_{n-i-1})/n$. [Use eq. (195) to replace $T_{n,i+1}$ and $T_{n,i}$ in $T_{n,i+1} - T_{n,i}$; then use the definition of D_i to replace the single index T's.] The value of D_i is

$$\begin{aligned} D_i &= (i+1)T_{i+1} - iT_i && (196) \\ &= (i+1)\lfloor \lg(i+1) \rfloor - i\lfloor \lg i \rfloor + 3(i+1) - 3i - 2^{\lfloor \lg(i+1) \rfloor + 1} + 2^{\lfloor \lg i \rfloor + 1} && (197) \\ &= i(\lfloor \lg(i+1) \rfloor - \lfloor \lg i \rfloor) + \lfloor \lg(i+1) \rfloor + 3 - 2^{\lfloor \lg(i+1) \rfloor + 1} + 2^{\lfloor \lg i \rfloor + 1}. && (198) \end{aligned}$$

To continue simplifying eq. (198) we will consider two cases. The first case is $\lfloor \lg(i+1) \rfloor = \lfloor \lg i \rfloor$. This case applies for most values of i. In this case everything cancels out except for one logarithm and the 3. We get

$$D_i = \lfloor \lg(i+1) \rfloor + 3 = \lfloor \lg i \rfloor + 3. \tag{199}$$

The other case is $\lfloor \lg(i+1) \rfloor = \lfloor \lg i \rfloor + 1$. This case applies when $i = 2^k - 1$ for some k. In this case $2^{\lfloor \lg(i+1) \rfloor} = i + 1$ and $2^{\lfloor \lg i \rfloor} = (i+1)/2$. We get

$$D_i = i + \lfloor \lg(i+1) \rfloor + 3 - 2(i+1) + i + 1 = \lfloor \lg(i+1) \rfloor + 2 \tag{200}$$
$$= \lfloor \lg i \rfloor + 3. \tag{201}$$

Notice that with the last step of this simplification we get the same answer in this case as we did in the first case. If you like to avoid unnecessary work, it is important to notice such simplifications when they are possible. Even if we had not noticed this simplification, we could continue, but we would have two cases for the value of D_i. By the time we considered the possible values for both D_i and D_{n-i-1}, we would have four combinations of cases to consider. Thus this little simplification has reduced the remaining work by a factor of 4.

The value of $T_{n,i+1} - T_{n,i}$ is given by

$$T_{n,i+1} - T_{n,i} = \frac{D_i - D_{n-i-1}}{n} = \frac{\lfloor \lg i \rfloor - \lfloor \lg(n-i-1) \rfloor}{n}. \tag{202}$$

Since $\lfloor \lg i \rfloor$ is a nondecreasing function, the right side of eq. (202) is less than or equal to zero for $i \leq n - i - 1$ [i.e., for $i \leq (n-1)/2$]. Thus $T_{n,i}$ increases (or at least does not decrease) as i goes from 1 to $\lfloor (n+1)/2 \rfloor$, while it decreases (or at least does not increase) as i goes from $\lceil (n-1)/2 \rceil$ to $n-1$. Thus there is a global minimum at $i = \lfloor n/2 \rfloor$, which is what we wanted to prove.

Recurrences with min or max usually arise when trying to construct the best possible algorithm (over some class of algorithms). The final algorithm, of course, will have to compute the appropriate index at each step. For example, the Binary Search Algorithm sets the index to the midpoint of the remaining region. If the formula for the best index is unknown or overly complex, then you cannot construct a *practical* best algorithm.

Recurrences with min and max functions are considered by Greene and Knuth [6, Section 2.2.1].

EXERCISE

1. Reingold and Tarjan [127] obtain the following recurrence when bounding the worst case performance of a matching algorithm:

$$T_n = \min_{\substack{2 \leq i \leq n-2 \\ \alpha \geq 1-\alpha-\beta > 0 \\ \beta \geq 1-\alpha-\beta > 0}} [\alpha T_i + \beta T_{n-i}],$$

with the boundary condition $T_2 = 1$. Solve this recurrence. Hint: First show that $\alpha = \frac{1}{3}$ and $\beta = \frac{1}{3}$. Then show that the recurrence reduces to

$$T_{2i} = \begin{cases} \frac{2}{3}T_i & \text{when } i \text{ is even} \\ \frac{1}{3}T_{i+1} + T_i & \text{when } i \text{ is odd} \end{cases}$$

for even indices and a similar equation for odd indices.

7.4.3 Dynamic Programming

Dynamic programming is an approach to solving certain optimization problems. It can be applied to problems which can be decomposed into subproblems that have the same form as the original problem. To have a nontrivial dynamic programming problem, there must be several different possible decompositions. Solving a dynamic programming problem involves selecting the best decomposition and solving the resulting subproblems. For dynamic programming to apply, the subproblems must be complete self-contained problems, i.e., a subproblem cannot depend on any hidden extra conditions (such as the sequence of events that led to generation of the subproblem). The solution that is best for a subproblem in isolation must be the solution that is best for any and all higher-level problems that invoke that subproblem.

A nice example of a problem of this type is the multiplication of a series of matrices of various sizes. Multiplying an $n \times p$ matrix by a $p \times m$ matrix requires npm multiplications when the traditional Matrix Multiplication Algorithm is used. Let M_p^n denote an $n \times p$ matrix. Suppose we want to multiply $M_3^5 \cdot M_1^3 \cdot M_4^1 \cdot M_6^4$. We can do this problem in five different ways. Each way gives the same result but takes a different number of operations:

Order	Operations
$((M_3^5 \cdot M_1^3) \cdot M_4^1) \cdot M_6^4$	$5 \cdot 3 \cdot 1 + 5 \cdot 1 \cdot 4 + 5 \cdot 4 \cdot 6 = 155$
$(M_3^5 \cdot (M_1^3 \cdot M_4^1)) \cdot M_6^4$	$3 \cdot 1 \cdot 4 + 5 \cdot 3 \cdot 4 + 5 \cdot 4 \cdot 6 = 192$
$(M_3^5 \cdot M_1^3) \cdot (M_4^1 \cdot M_6^4)$	$5 \cdot 3 \cdot 1 + 1 \cdot 4 \cdot 6 + 5 \cdot 1 \cdot 6 = 69$
$M_3^5 \cdot ((M_1^3 \cdot M_4^1) \cdot M_6^4)$	$3 \cdot 1 \cdot 4 + 3 \cdot 4 \cdot 6 + 5 \cdot 3 \cdot 6 = 174$
$M_3^5 \cdot (M_1^3 \cdot (M_4^1 \cdot M_6^4))$	$1 \cdot 4 \cdot 6 + 3 \cdot 1 \cdot 6 + 5 \cdot 3 \cdot 6 = 132$

The fastest way to multiply this series of matrices is the method that uses 69 operations.

To view multiplying a series of n matrices as a dynamic programming problem, notice that you need to multiply the first i matrices (in the best order for multiplying them), then you need to multiply the last $n - i$ matrices (in their best order), and finally you need to multiply the product of the first and last series of matrices. Each value for i gives a way to break the original problem into two subproblems.

Recurrence equations with min (or max) are also problems to which dynamic programming obviously applies. When computing C_n with eq. (191), for example, we use the values of C_i where $i < n$. The subproblems (computing the C_i) are problems in their own right, so we can use dynamic programming.

Two steps are needed to apply dynamic programming. First, formulate the problem so that it is expressed in the correct form (as described above). Second, solve the equations in an efficient way.

There are three approaches to consider for solving the equations that arise from dynamic programming. The first method is the *blind top-down* method. You start from the original problem and consider all subproblems that can be generated. You solve each one in turn and combine the answers together to form a solution of the overall problem. You do not remember any intermediate results for use in case the same subproblem arises again (this is the *blind* part of the method.) This method is very easy to program, and it usually has a huge running time, which can easily be exponential in the size of the problem. The same subproblem often occurs over and over again. The blind top-down method is so bad that we would not even mention it, except that it is the approach that naïve use of top-down programming naturally leads to, and so a lot of blind top-down programs are written.

The second approach is the *blind bottom-up* approach. In this approach you solve all possible subproblems of the original problem, starting with the smallest. Each time you solve a subproblem, you record the answer so that you do not have to work it again. By working the subproblems from smallest to largest, the solutions of subproblems are available during solution of larger problems. Thus each problem can be solved rapidly. This approach usually leads to polynomial-time solutions. The method is also easy to program. Time is often wasted, however, because many subproblems may be solved which are not needed for the solution of the original problem. (This is why we call it the *blind* bottom-up method.)

The third approach combines the best of the top-down and bottom-up methods. You proceed top down to find out what problems will need to be solved. As you solve each subproblem, you save the answer so that, if it occurs again, you do not have to re-solve the subproblem. In the more sophisticated variations of this method, you determine when the answer to a subproblem will be needed and save only those answers that will be needed again. This approach is often the fastest, but it does require more effort to program, particularly in its more sophisticated variations.

When solving dynamic programming equations with methods that remember the solutions of subproblems, care is needed to avoid using excessive memory.

7.4.3.1 Multiplying Series of Matrices

When you need to compute the matrix product

$$M_{c_1}^{r_1} M_{c_2}^{r_2} \cdots M_{c_n}^{r_n}, \tag{203}$$

what is the best order to do the multiplications, where matrix $M_{c_i}^{r_i}$ has r_i rows and c_i columns, and where $r_{i+1} = c_i$?

The most naïve way to figure out how best to multiply n matrices is to list each possible way and compute how long it takes. By Exercise 1 in Section 7.2.3, there are $\Theta(4^n/n^{3/2})$ ways to consider. This approach is okay for small n (say, $n \leq 4$) but it is extremely slow for large n.

The blind top-down approach is somewhat better. We compute T_{1n} as the minimum over i of T_{1i}, the number of operations needed to compute the product of the first i matrices (in the best way), plus $T_{i+1,n}$, the number of operations needed to compute the product of the last $n-i$ matrices, plus $r_1 c_i c_n$, the time to multiply the product of the first i matrices by the product of the last $n-i$ matrices. Using the blind top-down approach, the time to discover the best solution of our problem is given by the recurrence

$$t_n = n + \sum_{1 \leq i \leq n-1} (t_i + t_{n-i}), \tag{204}$$

with boundary condition $t_1 = 1$, under the assumption that we compute the answer to the two subproblems (for each value of i) and then require n units of time to select the best answer. The solution of this recurrence is $t_n = (3^n - 1)/2$. (See Exercise 7.1.1–3.) This is much better than listing all the cases [by about a factor of $(4/3)^n$], but it is still terribly slow for large n. In many cases you would be better off just to multiply the matrices in some fixed order rather than to first calculate the best order in this way.

The problem of finding the best way to multiply a series of matrices is equivalent to solving the equations

$$s_{ij} = \min_{i \leq k \leq j-1} \{r_i c_k c_j + s_{ik} + s_{k+1,j}\} \tag{205}$$

for s_{1n}. (Here s_{ij} is the number of operations needed to multiply the matrices in positions i to j using the best approach.) This equation can be solved bottom-up, saving all the intermediate results, by the following algorithm:

Algorithm 7.4 Series Matrix Multiplication: Input: The number of rows $r_1, r_2, \ldots, r_n$ and the number of columns $c_1, c_2, \ldots, c_n$ for a series of matrices, where $r_{i+1} = c_i$ for $1 \leq i \leq n-1$. Output: the matrices S, where s_{ij} is the minimum number of operations needed to compute the product of matrices i through j, and H, where the best way to compute the product of matrices i through j is to first compute the product of matrices i through h_{ij}, then $h_{ij}+1$ through j, and then multiply the results of the two subproblems.

Step **1.** For $1 \leq i < n$ set $s_{ii} \leftarrow 0$.

Step **2.** For $w \leftarrow 1$ to n do

Step **3.** For $i \leftarrow 1$ to $n-w$ do

Step **4.** Set $s_{i,i+w} \leftarrow \min_{i \leq j < i+w}\{s_{ij} + s_{j+1,i+w} + r_i c_j c_{i+w}\}$, and set $h_{i,i+w}$ to the index j that results in the minimum.

Step **5.** End for i.

Step **6.** End for w.

This algorithm calculates the minimum number of operations in time $O(n^3)$. It also stores in the H array the index of the last multiplication to do. Once you know the last multiplication to do, you can use the H array to find the next-to-last multiplication to do, and in this way quickly determine the entire sequence of multiplications that is associated with the best solution of the problem.

Although multiplying series of matrices makes a nice illustration of dynamic programming, straightforward application of dynamic programming does not lead to the best algorithm for the problem. Using more sophisticated techniques, Hu and Shing [96] show how to solve the problem in time $O(n \log n)$.

Aho et al. [1, Section 2.8] was a source for this section.

EXERCISES

1. The *string edit* problem is to find the cheapest way to modify two strings so that they are the same. The permitted operations are deletions, insertions, and substitutions. Let the two strings be $a_1 a_2 \ldots a_m$ and $b_1 b_2 \ldots b_n$, where each a_i and each b_j is a character from the character set S. If s_i and s_j are two characters of S, then the cost of deleting s_i is $D_i > 0$, the cost of inserting s_i is $I_i > 0$, and the cost of changing s_i to s_j is $C_{ij} \geq 0$. Assume that $C_{ij} = C_{ji}$ and that $C_{ij} = 0$ if and only if $i = j$. Give equations for the cost of the string edit problem such that the equations can be solved in time $O(mn)$.

2. Develop an algorithm that can solve the string edit problem in time $O(mn)$. Hint: The cheapest way of matching two strings of length i and j has a cost that is the minimum of (1) the cost of a single insertion plus the cost of matching two substrings, (2) the cost of a single deletion plus the cost of matching two substrings, (3) the cost of a doing a substitution plus the cost of matching two substrings. The substrings have the same length or are one character shorter than the original string. The details are left to you.

3. Consider the following recursive algorithm to calculate the number of multiplications needed to multiply a series of matrices.

Algorithm 7.5 Slow(i, j)**:** Input: The number of rows $r_1, r_2, \ldots, r_n$ and the number of columns $c_1, c_2, \ldots, c_n$ for a series of matrices, where $r_{i+1} = c_i$ for $1 \leq i \leq n-1$. Output: The minimum number of operations needed to compute the product of matrices i through j.

Step **1.** If $i = j$, then set $Slow(i, j) \leftarrow 0$ and return.
Step **2.** Set $min = \infty$.
Step **3.** For $k \leftarrow i$ to $j - 1$ do
Step **4.** Set $t \leftarrow r_i c_k c_j + Slow(i, k) + Slow(k + 1, j)$.
Step **5.** If $t < min$, then set $min \leftarrow t$.
Step **6.** End for.
Step **7.** Set $Slow(i, j) \leftarrow min$ and return.

What is the running time of the Slow Algorithm when Slow$(1, n)$ is called? Why is Slow such a slow algorithm?

4. Here is another recursive algorithm to calculate S. Before this algorithm is called, the array S is initialized to -1, which serves as a marker to indicate that a value has not yet been calculated. (The true value of an entry in S is never negative.)

Algorithm 7.6 Fast (i,j): Input: The number of rows r_1, r_2, ..., r_n and the number of columns c_1, c_2, ..., c_n for a series of matrices, where $r_{i+1} = c_i$ for $1 \le i \le n-1$. Output: The matrix S, where s_{ij} is the minimum number of operations needed to compute the product of matrices i through j.

Step **1.** If $i = j$, then set $s_{ij} \leftarrow 0$, $Fast(i,j) \leftarrow 0$, and return.

Step **2.** If $s_{ij} \neq -1$, then set $Fast(i,j) \leftarrow s_{ij}$ and return.

Step **3.** Set $min = \infty$.

Step **4.** For $k \leftarrow i$ to $j-1$ do

Step **5.** Set $t \leftarrow r_i c_k c_j + Fast(i,k) + Fast(k+1,j)$.

Step **6.** If $t < min$, then set $min \leftarrow t$.

Step **7.** End for.

Step **8.** Set $s_{ij} \leftarrow min$, $Fast(i,j) \leftarrow min$, and return.

What is the running time of the Fast Algorithm when Fast$(1,n)$ is called? Why is Fast so much faster than Slow?

5. Here is yet another recursive algorithm to calculate S. Before this algorithm is called, the array S is initialized to -1, which serves as a marker to indicate that it has not yet been calculated. (The true value of an entry in S is never negative.)

Algorithm 7.7 Faster (i,j): Input: The number of rows r_1, r_2, ..., r_n and the number of columns c_1, c_2, ..., c_n for a series of matrices, where $r_{i+1} = c_i$ for $1 \le i \le n-1$. Output: The matrices S, where s_{ij} is the minimum number of operations needed to compute the product of matrices i through j.

Step **1.** If $i = j$, then set $s_{ij} \leftarrow 0$, $Faster(i,j) \leftarrow 0$, and return.

Step **2.** If $s_{ij} \neq -1$, then set $Faster(i,j) \leftarrow s_{ij}$ and return.

Step **3.** Set $min = \infty$.

Step **4.** For $k \leftarrow i$ to $j-1$ do

Step **5.** Set $t \leftarrow r_i c_k c_j$. If $t < min$, then

Step **6.** Set $t \leftarrow t + Faster(i,k)$. If $t < min$, then

Step **7.** Set $t \leftarrow t + Faster(k+1,j)$. If $t < min$, then

Step **8.** Set $min \leftarrow t$.

Step **9.** End if.

Step **10.** End if.

Step **11.** End if.

Step **12.** End for.

Step **13.** Set $s_{ij} \leftarrow min$, $Faster(i,j) \leftarrow min$, and return.

This is an example of a clever top-down algorithm which saves intermediate results and which (sometimes) avoids solving all the subproblems. Why is the Faster Algorithm sometimes faster than Fast? Why is it more difficult to decide how to approach an analysis of Faster than it is to decide how to analyze Fast? Hint: Does the time for $Fast(1,n)$ depend in any significant way on anything besides the value of n? What about the time for $Faster(1,n)$?

Production	*Cost*
$S \to E$	0
$E \to E + T$	0
$E \to T$	0
$T \to T \times F$	0
$T \to TF$	1
$T \to F$	0
$F \to (E)$	0
$F \to i$	0

TABLE 7.9. A sample grammar.

7.4.3.2 Least-Cost Parsing*

To further illustrate dynamic programming, we now consider the problem of least-cost parsing. We have a *grammar* G which is a *starting symbol* S and a list of *productions* of the form $X \to X_1 X_2 \ldots X_p$. A production says that the symbol on the left side can be replaced by the string of symbols on the right side. The *length* of the production is p. A *sentential form* is a string that can be obtained from S by zero or more applications of the productions. The symbols that occur on the left side of productions are called *nonterminal* symbols. The symbols that do not occur on the left side of productions are called *terminal* symbols. A *sentence* is a sentential form that contains only terminal symbols. The *language* of a grammar is the set of all sentences that can be generated by the grammar.

Associated with each production we have a nonnegative cost. The cost of a sentence is the sum of the costs of the productions used to generate it. If a sentence can be generated more than one way, then the cost is the minimum of the costs associated with the various ways. You can think of the productions with cost zero as giving the correct way of generating strings, and the productions with positive costs as giving various erroneous ways of generating strings. The greater the cost, the worse the error.

Table 7.9 shows a sample grammar, one which generates arithmetic expressions. The production $T \to TF$ with cost 1 allows for sentences where the multiplication operator has been omitted. We want to take a grammar such as the one in Table 7.9 and a sentence such as $(i+i)(i+i)$ and determine the cost of generating the sentence.

The input to the algorithm will be the string $s_1 s_2 \ldots s_n$. The algorithm maintains an array T whose elements are costs. Entry t_{ijX} contains the lowest cost found so far for obtaining the string $s_{i+1} s_{i+2} \ldots s_j$ from X by zero or more applications of the productions. Initially all entries in T are infinity.

The algorithm is expressed in terms of the function $Match(P, i, j)$, which returns $k_0 + k_1 \cdots + k_p$, where production P is of the form $X \to X_1 X_2 \ldots X_p$ with cost k_0, $k_1 = t_{i,i_1,X_1}$, $k_2 = t_{i_1,i_2,X_2}$, $\ldots$, $k_{p-1} = t_{i_{p-2},i_{p-1},X_{p-1}}$, and $k_p = t_{i_{p-1},j,X_p}$, for some $i_1, \ldots, i_{p-1}$. The value infinity indicates that there is no such production. In other words, $Match(P, i, j)$ gives the cost of explaining all the input from position $i+1$ to position j with production P, k_0 is the direct cost of using production P, and k_l is the cost associated with matching X_l with an appropriate subsegment of the input.

The algorithm needs to eventually record into t_{0nS} the least-cost way of matching the input from position 1 to position n with symbol S. The optimized version does this in time at most $O(n^3)$ if T is filled in a bottom-up way. [The worst-case time is $\exp(O(n))$ if an algorithm which does not remember intermediate results is used.]

Algorithm 7.8 Parse: Input: A string $s_1 s_2 \ldots s_n$ and a grammar G where each production of G has an associated cost. (The grammar is used by the function *Match.*) Intermediate variable: r is *true* when a new minimum has been found. Output: The cost k of the lowest-cost derivation of the input string if the input string can be derived using the grammar. The cost is infinity if the input cannot be derived from the grammar.

Step **1.** [Initialize] For $0 \le j < i \le n$ and for all symbols A in G set $t_{ijX} \leftarrow \infty$.

Step **2.** [Each input position] For $j \leftarrow 1$ to n do:

Step **3.** [Self cost] Set $t_{j-1,j,s_j} \leftarrow 0$. (A terminal symbol can explain itself at no cost.)

Step **4.** [Each previous position] For $i \leftarrow j-2$ down to 0 do:

Step **5.** [Each production] Set $r \leftarrow$ *false*. For each production P in grammar G do:

Step **6.** [Match] Let X be the symbol on the left side of P. If $Match(P,i,j) < t_{i,j,X}$ then set $t_{i,j,X} \leftarrow Match(P,i,j)$ and $r \leftarrow$ *true*.

Step **7.** End for P.

Step **8.** [Complete?] If $r =$ *true*, then go to Step 5. (This step is needed to correctly handle productions where the right side consists of a single nonterminal.)

Step **9.** End for i.

Step **10.** End for j.

Step **11.** [Final Answer] Set $k \leftarrow t_{0nS}$.

In addition to finding the cost of the best derivation, the Parse Algorithm remembers part of the information needed to reconstruct the derivation. If you need the derivation, then you should have the parser save in t_{ijX} both the minimum cost and the production associated with the minimum cost.

Let q be the length of the longest production. The Parse Algorithm runs in time $O(n^{q+1})$ if $q \geq 1$. Match takes time $O(n^{q-1})$ for $q \geq 1$. The two loops in Parse contribute an additional factor of n^2.

There is a fairly simple way to improve the algorithm so that its speed is $O(n^3)$. To obtain the new grammar, retain all productions of length 1 and 2. Replace each production

$$X \leftarrow X_1 X_2 \ldots X_p \qquad \text{where } p \geq 3$$

with

$$X \leftarrow Z_{i,p-1} X_p$$
$$Z_{i,p-1} \leftarrow Z_{i,p-2} X_{p-1}$$
$$\ldots$$
$$Z_{i,2} \leftarrow X_1 X_2$$

where i is the number of the production and Z_{ij} is a nonterminal that is not used elsewhere in the grammar. The cost of the first production is the cost of the corresponding production in the original grammar. The productions with a Z on the left side have cost zero. The modified grammar generates the same language as the original grammar, and a sentence has the same cost with each grammar. The new grammar is close to what is known as Chomsky Normal form. (A grammar is in Chomsky Normal form if each production has a right side consisting of a single terminal symbol or of two nonterminal symbols.) The derivations are essentially the same; the modified grammar must do in a series of steps what the original grammar does in one step. The Parse Algorithm takes time $O(n^3)$ when run on the modified grammar.

Although we described how to modify the grammar to obtain an $O(n^3)$ running time, it is straightforward to make modifications to the algorithm instead of the grammar. In other words, you can produce an algorithm that will run on the original grammar in just the same way that the Parse Algorithm runs on the modified grammar.

This blind bottom-up algorithm has worst-case running time that is very good. Very few methods can parse grammars in a worst-case time better than $O(n^3)$. There are, however, ways to improve the constant that is hidden in the big O, and there are ways to make the algorithm run faster than $\Theta(n^3)$ for many grammars. [Many important grammars can be parsed in time $O(n)$.] These matters are discussed in detail in a paper by Graham et al. [91].

The parsing method described in this section is adapted from Graham et al.'s [91] description of the Cocke-Kasami-Younger algorithm [81, 104, 141]. This algorithm was independently invented three times in the late 1960s. Before that most parsing methods required exponential worst-case time. Two things make the parsing problem hard. First, a string of characters contains no markers showing how it should be broken up. There may be one way to parse the first three characters and a second way to parse the characters in positions 2 through 4. From local information it may be impossible to tell which of these overlapping ways will be part of the complete parse. Second, it may be possible to parse a fixed string of characters into two different nonterminals. Again there may be no local test to tell which nonterminal is part of the overall solution. These difficulties blinded many people to the fact that there is a simple dynamic approach to the problem. So it is in many applications of dynamic programming. Once the problem has been solved using dynamic programming, it is fairly obvious that the approach leads to a good solution. The hard part is discovering that you should use dynamic programming in the first place.

A thorough introduction to dynamic programming is given in Bellman and Dreyfus [20]. Hu [8, Chapter 3] also discusses the techniques.

EXERCISES

1. The *minimum-cost search tree* problem is as follows. There is a set of keys, $K_1 < K_2 < \cdots < K_n$, that will be put on the nodes of the search tree. If K_i is put at the root, then $K_1, \ldots, K_{i-1}$ are on the left subtree and $K_{i+1}, \ldots, K_n$ are on the right subtree. Likewise, if a subtree has $K_i, \ldots, K_k$ with K_j at the root, then $K_i, \ldots, K_{j-1}$ are on the left subtree and $K_{j+1}, \ldots, K_k$ are on the right. The probability that a search for K_i is done is given by p_i. The cost of the search for K_i is 1 more than the level of K_i in the tree. The problem is to rapidly determine how to build the optimum tree and to determine the cost of the tree. It will not do to try each possible tree, because there are $O(4^n/n^{3/2})$ trees to consider. Hint: Show that the minimum cost is given by $C_{1,n}$, where

$$C_{i,i} = 0$$
$$C_{i,j} = p_{i+1} + p_{i+2} + \cdots + p_j + \min_{i<k\leq j} \{C_{i,k-1} + C_{k,j}\}, \quad \text{for} \quad i < j.$$

Show how to solve these equations in time $O(n^3)$. (If you are clever, the equations can be solved in time $O(n^2)$; see Knuth [11, Section 6.2.2].)

7.5* ADVANCED ASYMPTOTIC METHODS

For methods of obtaining asymptotic information directly from recurrence equations see Bender and Orszag [21, Chapter 5].

There is a technique, Darboux's method, which uses complex variable theory to obtain asymptotic results about a sequence directly from the generating function for the sequence. The reader wishing to learn about this method should compare the treatment of Greene and Knuth [6, Section 4.3.1] with that of Olver [46, Section 8.9].

Bender [68, part II] also gives techniques for determining asymptotic properties of a function from its generating function.

CHAPTER 8

Multidimensional Recurrences

8.1 SYSTEMS OF EQUATIONS

Some problems produce a set of recurrence equations that contain several unknown functions. One approach in such cases is to solve one or more of the equations for one of the unknown functions in terms of the remaining unknown functions. Substituting back produces a new set of equations that contains one less unknown function and one less equation. The process is repeated until you have just one equation, which you solve directly. Generally the order of the equations increases as additional functions are solved for and eliminated.

A second approach is to define a generating function for each unknown function and use the recurrence equations to obtain a system of equations that the generating functions satisfy.

8.1.1* Transformations

A system of recurrence equations where some equations have an order higher than 1 can be converted to an equivalent system where each equation has order 1. Suppose you have the kth order set of equations

$$f_1(T_1(n), \ldots, T_1(n-k), \ldots, T_m(n), \ldots, T_m(n-k), n) = 0 \tag{1}$$

$$\vdots$$

$$f_m(T_1(n), \ldots, T_1(n-k), \ldots, T_m(n), \ldots, T_m(n-k), n) = 0. \tag{2}$$

Then let $T_i(n-j) = S_{ij}(n)$ for $1 \leq i \leq m$ and $0 \leq j \leq k$, and replace all occurrences of T with S in eqs. (1–2). This will give

$$f_1(S_{10}(n), \ldots, S_{1k}(n), \ldots, S_{m0}(n), \ldots, S_{mk}(n), n) = 0 \tag{3}$$

$$\vdots$$

$$f_m(S_{10}(n), \ldots, S_{1k}(n), \ldots, S_{m0}(n), \ldots, S_{mk}(n), n) = 0, \tag{4}$$

which is an algebraic equation relating the S's. Combining eqs. (3–4) with

$$S_{i,j}(n) = S_{i,j+1}(n-1) \tag{5}$$

for $1 \le i \le m$ and $0 \le j < k$ gives a first order set of recurrence equations that has essentially the same solution as eqs. (1–2). After solving eqs. (3–5), you can obtain the solution of eqs. (1–2) by using $T_i(n) = S_{i0}(n)$. Thus, any system of recurrences can be converted to a first order system.

Any linear system of recurrences with constant coefficients can be converted into a single linear recurrence. The basic idea is to solve one of the recurrences for one of the unknown functions, and then to use that result (with shifts of index as required) to replace all occurrences of the function in the remaining equations. After you run out of equations with a single occurrence of an unknown function, you may still have more than one recurrence left. These equations will have each unknown appearing with more than one index. You can solve one of the equations for a linear combination of an unknown function, where the unknown function occurs in the linear combination with several index values. If you take the corresponding linear combination of each remaining equation, with the corresponding index shifts, then you can eliminate that unknown function from the linear combination, to obtain a new equation. Each step reduces the number of functions by 1 and reduces the number of equations by 1.

When an equation is replaced by a linear combination, the resulting equation may have more solutions than the original equation did. Thus, when linear combinations are formed as part of the process for generating the final equation in one unknown function, the resulting solutions must be tested to see which ones obey the original equations.

For example, consider the equations

$$S_n = S_{n-1} + 2T_{n-1} - 2T_{n-2} \tag{6}$$

$$T_n = S_{n-1} + S_{n-2} - T_{n-1}. \tag{7}$$

Each equation contains two S's and two T's. From eq. (6), however, we can obtain

$$S_n - S_{n-1} = 2T_{n-1} - 2T_{n-2}. \tag{8}$$

In eq. (8) we have obtained a linear combination of S's in terms of the remaining variable. Now, if we take the corresponding linear combination of eq. (7) (i.e., we subtract eq. (7) with all occurrences of n replaced by $n-1$ from the original equation with no change in n), we obtain

$$T_n - T_{n-1} = S_{n-1} - S_{n-2} + S_{n-2} - S_{n-3} - T_{n-1} + T_{n-2}. \tag{9}$$

Now we can use eq. (8) to eliminate $S_{n-1} - S_{n-2}$ and $S_{n-2} - S_{n-3}$ from eq. (9), giving

$$T_n - T_{n-1} = 2T_{n-2} - 2T_{n-3} + 2T_{n-3} - 2T_{n-4} - T_{n-1} + T_{n-2}, \tag{10}$$

which simplifies to

$$T_n = 3T_{n-2} - 2T_{n-4}. \tag{11}$$

Any solution of eqs. (6–7) must use values of T_n that obey eq. (11).

EXERCISES

1. Convert $T_n = T_{n-1} + T_{n-2}$ into a system of first order recurrences. What is the solution of the system of recurrences?
2. Give the general solution of the recurrence $T_n = 2T_{n-1}$. Subtract the recurrence from itself with a shift of 1 to obtain $T_n = 3T_{n-1} - 2T_{n-2}$. What is the general solution of the new recurrence? What is the relation of the solutions of the new recurrence and the solutions of the original recurrence?
3. Find the general solution of eqs. (6–7).

8.1.1.1 Parallel Or*

To illustrate the substitution technique, we will consider a parallel computer with n memory cells and n processors. The processors are synchronous; they all execute their steps in lockstep. On going from step i to step $i+1$, each processor does the following: (1) it reads one cell of memory (any number of processors can read the same cell), (2) it does a calculation based on its index and on the information it has read so far, and (3) it writes a result into a memory cell (or does nothing if its program does not call for it to write on this step). Writing is subject to a major restriction: during one step, at most *one* processor can write into any particular memory cell. The problem we want to do on this parallel computer is to set memory cell 1 to the logical *or* of the bits initially in memory.

The object of this problem is to compute the logical or as fast as possible. We will use the following strange and clever algorithm.

Algorithm 8.1 Parallel Or: Computer model and input: Memory cells $M_1, M_2, \ldots, M_n$, each of which holds one bit. Let $X_1, X_2, \ldots, X_n$ be the initial values of the corresponding memory cells. Processors: $P_1, P_2, \ldots, P_n$, where P_j has the local variable Y_j. Initially $Y_j = \mathit{false}$ for $1 \le j \le n$. Output: At the end of execution M_1 has the logical *or* of the initial values of $M_1, M_2, \ldots, M_n$. **This algorithm is done synchronously by each processor.** Also, F_j is the jth Fibonacci number.

Step **1.** Set $t \leftarrow 0$.

Step **2.** While $F_{2t+1} < n$ do:

Step **3.** If $i + F_{2t} \le n$, then set $Y_i \leftarrow Y_i \vee M_{i+F_{2t}}$. ($Y_i$ now contains the logical *or* of X_i through $X_{i+F_{2t}}$).

Step **4.** If $i > F_{2t+1}$ and $Y_i = 1$, then set $M_{i-F_{2t+1}} \leftarrow 1$. ($M_i$ now contains the logical *or* of X_i through $X_{i+F_{2t+1}}$.)

Step **5.** Set $t \leftarrow t + 1$.

Step **6.** End while.

Each time through the loop each processor reads (at most) one memory cell. No two processors read from the same cell. Then each processor writes to at most one memory cell. Again no two processors write to the same cell.

With this algorithm both the processors and the memory cells accumulate information about more of the input on each step. When a processor has a zero result, it passes it on to the memory by not writing; thus the memory elements as well as the processors do logical *or*'s each time through the loop. This algorithm is therefore able to compute the logical *or* rather rapidly, probably faster than you thought was possible.

To analyze this algorithm, it will be helpful to reuse some of the letters used to state the algorithm in a new way. Let M_t be the number of bits of input that a cell knows about at the beginning of step t. That is, at step t each cell contains the logical *or* of M_t bits of memory, if you disregard the cells of high index that contain less information. (It is helpful to think of the input as containing n unknown bits plus some additional zero bits. Then you can think of the high-index processors as knowing about some of the unknown bits plus enough of the zero bits so that all processors know about the same number of bits. To get this effect in the algorithm, provide memory above cell n, initialize it to zero, and omit the test in Step 2.) Let P_t be the number of cells that each processor knows about at the beginning of step t. We count the number of bits that each processor knows about in the same way as we did for memory cells.

We have the initial conditions $P_0 = 0$ and $M_0 = 1$. During step t each processor remembers the P_t bits it already knew and also learns about M_t additional bits. So

$$P_{t+1} = P_t + M_t. \tag{12}$$

During step t the memory cells remember the M_t bits of information that they already knew about and learn about the P_{t+1} additional bits of information that some processor knows about. So

$$M_{t+1} = P_{t+1} + M_t. \tag{13}$$

We will now show how to solve this set of equations with the substitution approach described above. We first solve eq. (13) for P_{t+1}, obtaining

$$P_{t+1} = M_{t+1} - M_t. \tag{14}$$

Reducing t by 1 in eq. (14) gives

$$P_t = M_t - M_{t-1}. \tag{15}$$

Plugging eq. (14) and eq. (15) into eq. (12), we get

$$M_{t+1} - M_t = M_t - M_{t-1} + M_t, \tag{16}$$

which on rearranging gives

$$M_{t+1} - 3M_t + M_{t-1} = 0. \tag{17}$$

Increasing t by 1 gives

$$M_{t+2} - 3M_{t+1} + M_t = 0. \tag{18}$$

This is a second order equation, so we need two boundary conditions. We already have that $M_0 = 1$. From eq. (12) we can compute that $P_1 = 1$. Then we can use eq. (13) to obtain $M_1 = 2$.

Eq. (18) can be written as

$$(\mathbf{E}^2 - 3\mathbf{E} + 1)M_t = 0. \tag{19}$$

Solving the characteristic equation $x^2 - 3x + 1 = 0$ gives the roots

$$x_1 = \frac{3+\sqrt{5}}{2}, \tag{20}$$

$$x_2 = \frac{3-\sqrt{5}}{2}, \tag{21}$$

so the general solution of eq. (18) is

$$M_t = a_1\left(\frac{3+\sqrt{5}}{2}\right)^t + a_2\left(\frac{3-\sqrt{5}}{2}\right)^t. \tag{22}$$

The boundary conditions $M_0 = 1$ and $M_1 = 2$ give

$$1 = a_1 + a_2, \tag{23}$$

$$2 = \frac{3+\sqrt{5}}{2}a_1 + \frac{3-\sqrt{5}}{2}a_2, \tag{24}$$

which has the solution

$$a_1 = \frac{5+\sqrt{5}}{10}, \qquad a_2 = \frac{5-\sqrt{5}}{10}. \tag{25}$$

Therefore the particular solution of eq. (18) is

$$M_t = \frac{5+\sqrt{5}}{10}\left(\frac{3+\sqrt{5}}{2}\right)^t + \frac{5-\sqrt{5}}{10}\left(\frac{3-\sqrt{5}}{2}\right)^t. \tag{26}$$

To complete the solution, we now need the value of P_t. Plugging our value for M_t into eq. (15) gives

$$P_t = \frac{5+\sqrt{5}}{10}\left(\frac{3+\sqrt{5}}{2}\right)^t + \frac{5-\sqrt{5}}{10}\left(\frac{3-\sqrt{5}}{2}\right)^t - \frac{5+\sqrt{5}}{10}\left(\frac{3+\sqrt{5}}{2}\right)^{(t-1)} - \frac{5-\sqrt{5}}{10}\left(\frac{3-\sqrt{5}}{2}\right)^{(t-1)} \tag{27}$$

$$= \frac{5+\sqrt{5}}{10}\left(1 - \frac{2}{3+\sqrt{5}}\right)\left(\frac{3+\sqrt{5}}{2}\right)^t + \frac{5-\sqrt{5}}{10}\left(1 - \frac{2}{3-\sqrt{5}}\right)\left(\frac{3-\sqrt{5}}{2}\right)^t \tag{28}$$

$$= \frac{5+\sqrt{5}}{10}\left(\frac{1+\sqrt{5}}{3+\sqrt{5}}\right)\left(\frac{3+\sqrt{5}}{2}\right)^t + \frac{5-\sqrt{5}}{10}\left(\frac{1-\sqrt{5}}{3-\sqrt{5}}\right)\left(\frac{3-\sqrt{5}}{2}\right)^t \tag{29}$$

$$= \frac{10+6\sqrt{5}}{10(3+\sqrt{5})}\left(\frac{3+\sqrt{5}}{2}\right)^t + \frac{10-6\sqrt{5}}{10(3-\sqrt{5})}\left(\frac{3-\sqrt{5}}{2}\right)^t \tag{30}$$

$$= \frac{\sqrt{5}}{5}\left(\frac{3+\sqrt{5}}{2}\right)^t - \frac{\sqrt{5}}{5}\left(\frac{3-\sqrt{5}}{2}\right)^t \tag{31}$$

If you are given a large value M and want to know how large t must be so that $M_t \geq M$, then it is useful to rewrite eq. (26) as

$$M_t = \frac{5+\sqrt{5}}{10}\left(\frac{3+\sqrt{5}}{2}\right)^t \left[1 + O\left(\left(\frac{3-\sqrt{5}}{3+\sqrt{5}}\right)^t\right)\right]. \tag{32}$$

Taking logarithms gives

$$\ln M_t = \ln\frac{5+\sqrt{5}}{10} + t\ln\frac{3+\sqrt{5}}{2} + O\left(\left(\frac{3-\sqrt{5}}{3+\sqrt{5}}\right)^t\right). \tag{33}$$

Solving for the most important term with a t in it gives

$$t = \frac{\ln M_t}{\ln\dfrac{3+\sqrt{5}}{2}} - \frac{\ln\dfrac{5+\sqrt{5}}{10}}{\ln\dfrac{3+\sqrt{5}}{2}} + O\left(\left(\frac{3-\sqrt{5}}{3+\sqrt{5}}\right)^t\right). \tag{34}$$

Since $(3-\sqrt{5})/(3+\sqrt{5})$ is less than 1, for large M_t eq. (34) says that t is equal to $\ln M_t$ times a constant plus a constant plus a term that becomes small. Therefore we can write eq. (34) as

$$t = \frac{\ln M_t}{\ln\dfrac{3+\sqrt{5}}{2}} - \frac{\ln\dfrac{5+\sqrt{5}}{10}}{\ln\dfrac{3+\sqrt{5}}{2}} + O\left(M_t^{\ln(3-\sqrt{5})/\ln(3+\sqrt{5})-1}\right) \tag{35}$$

$$= \frac{\ln M_t}{\ln\dfrac{3+\sqrt{5}}{2}} - \frac{\ln\dfrac{5+\sqrt{5}}{10}}{\ln\dfrac{3+\sqrt{5}}{2}} + O\left(M_t^{-1.163}\right) \approx 1.039\ln M_t - 0.335. \tag{36}$$

The Parallel Or Algorithm was invented by Cook and Dwork [83].

EXERCISES

1. Justify in detail the derivation of the big O term in eq. (35).

2. In a lower-bound analysis Cook and Dwork [83] obtain the recurrence

$$P_{t+1} = P_t + M_t$$
$$M_{t+1} = 3P_t + 4M_t$$

with boundary conditions $P_0 = 0$ and $M_0 = 1$. What is the solution of their recurrence?

8.1.2* Discrete Linear Systems

One common source of systems of recurrence equations is *discrete linear systems.* For such systems there are a number of variables, $S_1(t)$, $S_2(t)$, ..., $S_n(t)$ that describe the state of the system at time t. The value of each variable at time $t+1$ is a linear function of the values of the variables at time t. Thus the system is described by the set of equations

$$S_1(t+1) = a_{11}(t)S_1(t) + a_{12}(t)S_2(t) + \cdots + a_{1n}(t)S_n(t) \tag{37}$$
$$S_2(t+1) = a_{21}(t)S_1(t) + a_{22}(t)S_2(t) + \cdots + a_{2n}(t)S_n(t) \tag{38}$$
$$\vdots$$
$$S_n(t+1) = a_{n1}(t)S_1(t) + a_{n2}(t)S_2(t) + \cdots + a_{nn}(t)S_n(t). \tag{39}$$

When $n = 2$, the method of solving the second equation for S_1 and then eliminating S_1 from the first equation works well. If $n > 2$, however, that method does not work well. The first step goes okay, but the second step is not so simple. For $n > 2$ you should consider using generating functions.

If the $a_{ij}(t)$ are constant, then applying generating functions to eqs. (37–39) will lead to a set of linear equations for the generating functions. Define $G_i(z) = \sum_{t\geq 0} S_i(t)z^t$, multiply each equation [eqs. (37–39)] by z^t, and sum over t. This gives

$$\sum_{t\geq 0} S_1(t+1)z^t = a_{11}\sum_{t\geq 0} S_1(t)z^t + a_{12}\sum_{t\geq 0} S_2(t)z^t + \cdots + a_{1n}\sum_{t\geq 0} S_n(t)z^t, \tag{40}$$
$$\sum_{t\geq 0} S_2(t+1)z^t = a_{21}\sum_{t\geq 0} S_1(t)z^t + a_{22}\sum_{t\geq 0} S_2(t)z^t + \cdots + a_{2n}\sum_{t\geq 0} S_n(t)z^t, \tag{41}$$
$$\vdots$$
$$\sum_{t\geq 0} S_n(t+1)z^t = a_{n1}\sum_{t\geq 0} S_1(t)z^t + a_{n2}\sum_{t\geq 0} S_2(t)z^t + \cdots + a_{nn}\sum_{t\geq 0} S_n(t)z^t. \tag{42}$$

The right sides are now in the form of generating functions. For each left side make the replacement

$$\sum_{t\geq 0} S_i(t+1)z^t = \frac{1}{z}\sum_{t\geq 0} S_i(t+1)z^{t+1} = \frac{1}{z}\sum_{t\geq 1} S_i(t)z^t \tag{43}$$
$$= \frac{1}{z}\sum_{t\geq 0} S_i(t)z^t - \frac{1}{z}S_i(0) = \frac{G_i(z) - S_i(0)}{z} \tag{44}$$

Making these substitutions gives

$$\frac{G_1(z) - S_1(0)}{z} = a_{11}G_1(z) + a_{12}G_2(z) + \cdots + a_{1n}G_n(z), \tag{45}$$

$$\frac{G_2(z) - S_2(0)}{z} = a_{21}G_1(z) + a_{22}G_2(z) + \cdots + a_{2n}G_n(z), \tag{46}$$

$$\vdots$$

$$\frac{G_n(z) - S_n(0)}{z} = a_{n1}G_1(z) + a_{n2}G_2(z) + \cdots + a_{nn}G_n(z). \tag{47}$$

Rearranging the equations gives

$$-\frac{S_1(0)}{z} = \left(a_{11} - \frac{1}{z}\right)G_1(z) + a_{12}G_2(z) + \cdots + a_{1n}G_n(z), \tag{48}$$

$$-\frac{S_2(0)}{z} = a_{21}G_1(z) + \left(a_{22} - \frac{1}{z}\right)G_2(z) + \cdots + a_{2n}G_n(z), \tag{49}$$

$$\vdots$$

$$-\frac{S_n(0)}{z} = a_{n1}G_1(z) + a_{n2}G_2(z) + \cdots + \left(a_{nn} - \frac{1}{z}\right)G_n(z). \tag{50}$$

These equations can now be solved for the generating functions. The final solutions will be in the form of a ratio of polynomials in z. The generating functions can be expanded in power series to obtain the $S_i(t)$. This is the same type of results as those obtained in Section 5.4 for single linear recurrences with constant coefficients. In general, each S_i is given by the sum over the roots of the characteristic equation (for the entire problem) of exponentials times polynomials, where the base of each exponential is given by the reciprocal of the corresponding root, and where the degree of the polynomial is given by 1 less than the multiplicity of the corresponding root.

8.1.3* Characteristic Equations

Although generating functions give a straightforward way to solve eqs. (37–39), they lead to a lot of algebraic manipulation that can be a source of error. Computationally, the easiest way to solve eqs. (37–39) when the $a_{ij}(t)$ are constants is to find the roots of the characteristic equation and then use the method of undetermined coefficients (see Sections 5.4.1 and 5.4.2).

To obtain the characteristic equation, we start by looking for solutions of eqs. (37–39) of the form $S_i(t) = c_i\lambda^t$. Plugging this form into the equations gives

$$c_1\lambda^{t+1} = a_{11}c_1\lambda^t + a_{12}c_2\lambda^t + \cdots + a_{1n}c_n\lambda^t, \tag{51}$$

$$c_2\lambda^{t+1} = a_{21}c_1\lambda^t + a_{22}c_2\lambda^t + \cdots + a_{2n}c_n\lambda^t, \tag{52}$$

$$\vdots$$

$$c_n\lambda^{t+1} = a_{n1}c_1\lambda^t + a_{n2}c_2\lambda^t + \cdots + a_{nn}c_n\lambda^t. \tag{53}$$

If we divide out λ^t and move everything to the right side, we get

$$0 = (a_{11} - \lambda)c_1 + a_{12}c_2 + \cdots + a_{1n}c_n, \tag{54}$$

$$0 = a_{21}c_1 + (a_{22} - \lambda)c_2 + \cdots + a_{2n}c_n, \tag{55}$$

$$\vdots$$

$$0 = a_{n1}c_1 + a_{n2}c_2 + \cdots + (a_{nn} - \lambda)c_n. \tag{56}$$

This equation can be written in matrix form as

$$0 = (A - \lambda I)C, \tag{57}$$

where A is the matrix of the a_{ij}, I is the identity matrix, and C is the vector of c_i's.

Eq. (57) has the solution $C = 0$ for all values of λ. This solution, however, is not very interesting. You cannot fit many boundary values with it. To have a nontrivial solution (one with $C \neq 0$), it is necessary and sufficient for

$$\det |A - \lambda I| = 0, \tag{58}$$

which is the case only for certain values of λ. If you multiply out the determinant in eq. (58), you obtain an nth-degree polynomial equation for λ. This *characteristic equation* has n roots, some of which may be repeated. Once you have the roots of the characteristic equation, you can apply the method of undetermined coefficients (Section 5.4.2) to the boundary values to finish solving the problem.

The theory of systems of recurrence equations has many similarities to the theory of systems of differential equations, which is covered by Goldberg and Schwartz [32]. In particular, Section 2.3 of [32] covers linear first order equations with constant coefficients.

8.1.3.1* *Tape Sorting: Four Tapes*

In Section 5.3.3 we considered the problem of sorting with three tape drives. We will now consider the problem of sorting with four tape drives. We will make the same assumptions as last time concerning the properties of the tape drives and the time required for various operations. Remember, the results apply only for the particular assumptions being made, although it is easy to modify the analysis to accommodate other assumptions.

In Section 5.3.3.1 we analyzed Simple Merge Sorting for any number of tape drives. The time required for four drives is

$$T_{\text{Simple Merge}} = L(2\log_3 N - 1) \approx 1.262L \lg N. \tag{59}$$

For a perfect Polyphase Merge, the number of blocks on each tape obeys a set of recurrences which we can deriv‹ by considering the operation of the algorithm. Let $x_j(l)$ be the number of blocks on the tape that has the jth most blocks at the beginning of the lth from last phase. Thus, $x_1(l)$ is the number of blocks on the tape that has the most blocks at the beginning of the lth from

Phase	*Tape 1*		*Tape 2*		*Tape 3*		*Tape 4*	
	Blocks	*Length*	*Blocks*	*Length*	*Blocks*	*Length*	*Blocks*	*Length*
7	37	1	24	1	Output		44	1
6	13	1	Output		24	3	20	1
5	Output		13	5	11	3	7	1
4	7	9	6	5	4	3	Output	
3	3	9	2	5	Output		4	17
2	1	9	Output		2	31	2	17
1	Output		1	57	1	31	1	17
0	1	105						

TABLE 8.1. The arrangement of blocks on tapes for Polyphase Merge with four tape units and seven phases. Block sizes are given for the case where the initial block size is 1.

last phase, and $x_{k-1}(l)$ is the number of blocks on the input tape that has the fewest blocks. The output tape has zero blocks at the beginning of the phase.

During phase l from the end, $x_{k-1}(l)$ blocks from each input tape are merged and written to the output tape. This tape will have the most blocks of any tape on the next phase (phase $l-1$ from the end), so $x_1(l-1) = x_{k-1}(l)$. The tape that had $x_{k-1}(l)$ blocks will be empty; it is the new output tape. The rest of the tapes have $x_{k-1}(l)$ fewer blocks than they had at the beginning of phase l. The tape that had $x_1(l)$ blocks at the beginning of phase l will have the most blocks remaining, so $x_2(l-1) = x_1(l) - x_{k-1}(l)$. In general for $2 \le j \le k-1$, $x_j(l-1) = x_{j-1}(l) - x_{k-1}(l)$. We thus obtain the following equations:

$$x_1(l-1) = x_{k-1}(l), \tag{60}$$

$$x_j(l-1) = x_{j-1}(l) - x_{k-1}(l) \qquad \text{for } 2 \le j \le k-1. \tag{61}$$

Replacing $x_{k-1}(l)$ in eq. (61) with $x_1(l-1)$, followed by rearranging eqs. (60, 61) so that the $x_j(l)$ terms are on the left side of the equations, gives:

$$x_j(l) = x_{j+1}(l-1) + x_1(l-1) \qquad \text{for } 1 \le j \le k-2, \tag{62}$$

$$x_{k-1}(l) = x_1(l-1). \tag{63}$$

The characteristic equation for eqs. (62, 63) has the form $\det|A - \lambda I| = 0$, where I is the identity matrix and A is the matrix with $a_{j1} = 1$ for $1 \le j \le k-1$, $a_{j,j+1} = 1$ for $1 \le j \le k-2$, and $a_{lj} = 0$ otherwise. Let's consider the case $k = 4$ in detail. The characteristic equation is

$$\det \begin{vmatrix} 1-\lambda & 1 & 0 \\ 1 & -\lambda & 1 \\ 1 & 0 & -\lambda \end{vmatrix} = 0. \tag{64}$$

Multiplying out the determinant in eq. (64) gives

$$\lambda^3 - \lambda^2 - \lambda - 1 = 0. \tag{65}$$

Calling the roots of this equation λ_1, λ_2, and λ_3, the form of the solution of eqs. (62, 63) is

$$x_j(l) = c_{j1}\lambda_1^l + c_{j2}\lambda_2^l + c_{j3}\lambda_3^l, \tag{66}$$

where the c_{jp} are constants that can be determined by fitting the boundary conditions.

Eq. (65) is a cubic equation. Using the formula for the roots of a cubic equation (see [16, Section 3.8.2]), we get the roots

$$\lambda_1 \approx 1.839, \tag{67}$$

$$\lambda_2 \approx -0.419 + 0.606i, \tag{68}$$

$$\lambda_3 \approx -0.419 - 0.606i, \tag{69}$$

where i is the square root of -1. The magnitude of λ_2 (and of λ_3) is approximately 0.737. (We will use the convention that λ_j is the root with the jth largest absolute value.)

Using $x_j(l) = c_{j1}\lambda_1^l + c_{j2}\lambda_2^l + c_{j3}\lambda_3^l$ for $j = 1$ and $0 \le l \le 2$ gives

$$0 = c_{11} + c_{12} + c_{13}, \tag{70}$$

$$1 = c_{11}\lambda_1 + c_{12}\lambda_2 + c_{13}\lambda_3, \tag{71}$$

$$2 = c_{11}\lambda_1^2 + c_{12}\lambda_2^2 + c_{13}\lambda_3^2, \tag{72}$$

Solving for c_{11}, we obtain

$$c_{11} = \frac{2 - \lambda_2 - \lambda_3}{(\lambda_1 - \lambda_2)(\lambda_1 - \lambda_3)}. \tag{73}$$

We can express the value of c_{11} in terms of λ_1 by using some of the properties of the roots of equations. If the roots of the equation

$$x^n + a_{n-1}x^{n-1} + a_{n-2}x^{n-2} + \cdots + a_0 = 0 \tag{74}$$

are x_1, x_2, ..., x_n, then a_{n-1} is equal to the negative of the sum of the roots, a_{n-2} is the sum of the products of pairs of roots, etc. For our problem we have

$$1 = \lambda_1 + \lambda_2 + \lambda_3 \tag{75}$$

$$-1 = \lambda_1\lambda_2 + \lambda_1\lambda_3 + \lambda_2\lambda_3. \tag{76}$$

Using these results to eliminate λ_2 and λ_3 from eq. (73) gives

$$c_{11} = \frac{\lambda_1 + 1}{(3\lambda_1 + 1)(\lambda_1 - 1)}. \tag{77}$$

We may now write eq. (66) as

$$x_j(l) = c_{j1}\lambda_1^l + O(|\lambda_2|^l) = c_{j1}\lambda_1^l + O(0.738^l). \tag{78}$$

[We have 0.738^l rather than 0.737^l to be sure the error term is large enough. Remember the big O term gives an upper bound on the error; $O(0.737^l)$ is not quite large enough to be an upper bound.]

Returning to the problem for general k, let λ_1, λ_2, ... be the roots for the associated characteristic equation in order of decreasing size (see Exercise

8.1.3.1-1). We simplify the notation by using c_j for c_{j1} and λ for λ_1. A generalization of the above presentation shows that the following analog of eq. (78) holds:

$$x_j(l) = c_j\lambda^l + O(|\lambda_2|^l). \tag{79}$$

Plugging into eq. (61) and taking the limit as l goes to infinity gives

$$c_j = \lambda c_{j-1} - c_1 \qquad \text{for } 2 \le j \le k-1. \tag{80}$$

Since $\lambda > \lambda_2$, the solution of eq. (80) is

$$c_j = \left(\lambda^{j-1} - \sum_{0 \le l \le j-2} \lambda^l\right)c_1 = \frac{\lambda^j - 2\lambda^{j-1} + 1}{\lambda - 1}c_1. \tag{81}$$

[For $k = 4$ we have $c_2 = (\lambda - 1)c_1$ and $c_3 = (\lambda^2 - \lambda - 1)c_1$.]

From eq. (60) we get that

$$c_{k-1} = c_1/\lambda. \tag{82}$$

[Since λ is a root of the characteristic equation, eq. (81) with $j = k - 1$ is equivalent to eq. (82).]

On phase 1 (phase n from the end), the number of blocks on the input tapes must be the same as the starting number of blocks, so we have

$$N = x_1(n) + x_2(n) + \cdots + x_{k-1}(n) \tag{83}$$

For $k = 4$ this reduces to

$$N = [1 + (\lambda - 1) + (\lambda^2 - \lambda - 1)]c_1\lambda^n + O(0.738^n) \tag{84}$$

$$= \frac{(\lambda+1)^2}{3\lambda+1}\lambda^n + O(0.738^n). \tag{85}$$

Solving eq. (85) for λ^n gives

$$\lambda^n = \frac{(3\lambda+1)N}{(\lambda+1)^2} + O(0.738^n) \tag{86}$$

and solving for n gives

$$n \lg \lambda = \lg N + O(1) \tag{87}$$

or

$$n = \frac{\lg N}{\lg \lambda} + O(1) \approx 1.137 \lg N. \tag{88}$$

Consideration of the case with general k also leads to

$$n = \frac{\lg N}{\lg \lambda} + O(1). \tag{89}$$

The size of the output block produced on phase l from the end is

$$x_{k-1}(l) = c_{k-1}\lambda^l + O(\lambda_2^l), \tag{90}$$

which for $k = 4$ gives

$$x_3(l) = \frac{(\lambda+1)\lambda^{l-1}}{(3\lambda+1)(\lambda-1)} + O(0.738^l) \approx 0.283\lambda^l. \tag{91}$$

Now let's consider the size of the blocks that result from Polyphase Merge. On the nth phase from the end the blocks are all of size L/N. On each successive phase, the largest block size is the sum of the block sizes from the previous phase. The block with smallest size disappears. All the remaining blocks carry over to the next phase. In other words, if $y_j(l)$ is the size of the jth largest block on phase l from the end, then

$$y_1(l-1) = \sum_{1 \le m \le k-1} y_m(l), \tag{92}$$

$$y_j(l-1) = y_{j-1}(l) \qquad \text{for } 2 \le j \le k-1, \tag{93}$$

with the boundary conditions $y_j(n) = L/N$ for $1 \le j \le k-1$.

Eqs. (92, 93) are simple as they stand, but they do not quite have the form used in Section 8.1.2 for developing the characteristic equation given in Section 8.1.3. In eqs. (92, 93) the left side has index 1 lower than the right side rather than 1 higher. If we modify the derivation given in Section 8.1.3 to fit this case, we obtain the characteristic equation

$$\det|A - \lambda^{-1}I| = 0, \tag{94}$$

where I is is the identity matrix and A is the matrix with $a_{1j} = 1$ for $1 \le j \le k-1$, $a_{j,j-1} = 1$ for $2 \le j \le k-1$, and $a_{lj} = 0$ otherwise. For $k = 4$ we have

$$\det \begin{vmatrix} 1-\lambda^{-1} & 1 & 1 \\ 1 & -\lambda^{-1} & 0 \\ 0 & 1 & -\lambda^{-1} \end{vmatrix} = 0. \tag{95}$$

This characteristic equation (considered as a function of $1/\lambda$) is the determinant of the transpose of the matrix for eqs. (62, 63). Therefore this equation has the same solution (for $1/\lambda$) as the earlier characteristic equation has (for λ). This is not entirely a coincidence. For each phase, the collection of input tapes must contain the same total number of items. This can be the case only if the two characteristic equations have the same roots (for λ in one case and $1/\lambda$ in the other).

Let
$$y_j(l) = d_{j1}\lambda_1^{n-l} + d_{j2}\lambda_2^{n-l} + \cdots \tag{96}$$

where the λ's are the roots for the characteristic equation. [They are given by eqs. (67–69) for the $k = 4$ case.] [The constants in eq. (96) are written as $d_{ij}\lambda_i^n$ to allow easy use of the boundary conditions at $l = n$.] Now $y_1(n) = L/N$, $y_1(n-1) = (k-1)L/N$, etc., so eq. (96) can be used to calculate the values of the d's. Since we will calculate d_1 (the only d that we need) by a different method, the important thing to notice is that for a fixed value for k and for the ratio L/N, d_1 does have a fixed value (independent of n). Since the d's are independent of n, we have

$$y_j(l) = d_{j1}\lambda_1^{n-l} + O(|\lambda_2|^{n-l}). \tag{97}$$

Using λ for λ_1 and d_j for d_{j1}, we have

$$\lambda d_j = d_{j-1}, \tag{98}$$

so

$$d_j = \lambda^{-j} d_1 \tag{99}$$

and

$$\sum_{1 \le j \le k-1} d_j = \frac{1-\lambda^{-k}}{1-\lambda^{-1}} d_1 . \tag{100}$$

The number of items on the output tape of the phase 1 from the end is

$$L = \sum_{1 \le j \le k-1} y_j(n) = \sum_{1 \le j \le k-1} d_j \lambda^{n-1} + O(|\lambda_2|^n), \tag{101}$$

so

$$d_1 = \frac{\lambda^{-n+1}(1-\lambda^{-1})L}{1-\lambda^{-k}} + O\left(\left|\frac{\lambda_2}{\lambda}\right|^n\right). \tag{102}$$

(For $k = 4, |\lambda_2|/\lambda_1 = 0.401$).

The number of items per block output on phase l from the end is

$$\sum_{1 \le l \le k-1} y_j(l) = L\lambda^{-l+1} + O(|\lambda_2|^{n-l}\lambda^{-n}). \tag{103}$$

The total time for Polyphase Merge is the number of blocks on the output tape times the length of a block summed over all the phases. For $k = 4$ we have

$$T_{\text{Polyphase Merge}} = \sum_{1 \le j \le n} \frac{L(\lambda+1)}{(3\lambda+1)(\lambda-1)} + O(0.401^l) \tag{104}$$

$$= \frac{nL(\lambda+1)}{(3\lambda+1)(\lambda-1)} + O(1) \tag{105}$$

$$= \frac{\lambda+1}{(3\lambda+1)(\lambda-1)\lg\lambda} L \lg N + O(1) \approx 0.590 L \lg N . \tag{106}$$

Comparing eq. (106) with eq. (59), we see that with four tapes Polyphase Merge is a little more than twice as fast as Simple Merge (under the timing assumptions that we are making).

Since Polyphase Merge uses all the tapes except one as input tapes and does a $k-1$ way Merge, you might think that Polyphase Merge is the best way (or close to it) to do tape merges. (There are, of course, a few practical problems such as rewind time, but there are modifications to Polyphase Merge to handle rewinding efficiently.) Although Polyphase Merge does quite well with a small number of tape units, it has a problem that becomes more important as the number of tape drives increases. The tapes being merged have different size blocks. The tapes with small block sizes are not well utilized when they are being merged with tapes with large block sizes. The more tape units there are, the worse this problem is. Therefore, when there are a large number of tape drives, other clever methods of merging are even better than Polyphase Merge. See Knuth [11, Section 5.4] for the details.

EXERCISES

1. Show that in general the characteristic equation for Polyphase Merge [eqs. (62, 63)] is

$$\lambda^{k-1} - \sum_{0 \le l \le k-2} \lambda^l = 0.$$

Hint: Add $1/\lambda$ times the last row of the determinant to the next-to-last row. (This does not change the value of the determinant.) Next add $1/\lambda$ times the new next to last row to the previous row. Continue like this until you have a new first row. The determinant will now have zeros everywhere above the main diagonal, so its value is the product of the entries on the diagonal.

2. Compare the approximations in this section with the numbers in Table 8.1. How well do the approximations, which were designed for a large number of passes, work for seven passes?

3. Redo the analysis in this section for $k = 5$. (The instructor may wish to limit this problem to some portion of the analysis.)

8.1.4* Steady-State Solutions

Many problems involve systems that can be in one of n states. Let $P_i(t)$ be the probability that the system is in state i at time t. Also suppose the next state of the system is a random function of the current state, where p_{ij} is the probability that the system will go to state j at time $t+1$ if it is in state i at time t. Then the probability that the system will be in any state at time $t+1$ is given by the solution of the following set of equations:

$$P_1(t+1) = p_{11}P_1(t) + p_{12}P_2(t) + \cdots + p_{1n}P_n(t), \tag{107}$$
$$P_2(t+1) = p_{21}P_1(t) + p_{22}P_2(t) + \cdots + p_{2n}P_n(t), \tag{108}$$
$$\vdots$$
$$P_n(t+1) = p_{n1}P_1(t) + p_{n2}P_2(t) + \cdots + p_{nn}P_n(t), \tag{109}$$

subject to the appropriate boundary conditions. If the system is in state 1 at time zero, for example, then the boundary conditions are $P_1(0) = 1$, $P_2(0) = 0$, $P_3(0) = 0, \ldots, P_n(0) = 0$. Systems that obey eqs. (107–109) are called *Markov chains.*

This set of equations can be solved by the techniques of the previous section. Often, however, we need less information, and we would like to do less work. It may be that all we need to know is the limit of the $P_i(t)$ as t goes to infinity. We will make several assumptions. The first assumption is that $\sum_{1 \le i \le n} p_{ij} = 1$. If this sum is greater than 1, then the p_{ij} clearly are not probabilities. If this sum is less than 1, then something other than going to the next state is possible. We will also assume that $p_{ij} \ge 0$ for all i and j. Next we will assume that, from the starting state, you can get to any state by going through some series of states such that $p_{ij} > 0$ if i and j are consecutive states on the path. Such a path is called a *path of positive probability.* If there is no path of positive probability from the starting state to some other state, then the state in question cannot ever be reached from the starting state. In such cases we can eliminate the unreachable states and apply the method to the reduced problem. We also assume that there is a chain of positive probability connecting every pair of states . (The techniques can be extended to handle problems where this is not the case.) Finally, we assume that there are no chains where every probability on the chain is 1. (Such chains are easy to analyze; it is just that our present technique does not apply.)

Under the above assumptions our system will tend to a steady-state where, for each i, $\lim_{t\to\infty}(P_i(t+1)-P_i(t))=0$. We can find the steady-state distribution by setting $P_i(t+1)=P_i(t)$ in eqs. (107–109), dropping the t dependence, and solving for the P_i. Doing this we get

$$0=(p_{11}-1)P_1+p_{12}P_2+\cdots+p_{1n}P_n, \tag{110}$$

$$0=p_{21}P_1+(p_{22}-1)P_2+\cdots+p_{2n}P_n, \tag{111}$$

$$\vdots$$

$$0=p_{n1}P_1+p_{n2}P_2+\cdots+(p_{nn}-1)P_n. \tag{112}$$

The determinant of the coefficients is zero for this set of equations because each row of coefficients sums to 1, and therefore the rows are linear combinations of each other. The set of equations thus has solutions where the P_i are not all zero.

The equations, since they are homogeneous, only give you the ratios of the P_i. The final condition that you use to get their actual value is that the sum of all the P_i must be 1.

This type of analysis is often done by people who do not check to see that all the assumptions needed are actually satisfied. When the assumptions are not satisfied, the equations often let you know that things went wrong by having solutions where the P_i do not all have the same sign. Since the P_i are probabilities, this is a clear signal of trouble. Another way that the equations can let you know that something went wrong is by giving an answer where not all the P_i are connected by ratios.

Feller [27, Chapter 15] has a much more complete introduction to Markov chains. Also see Fisz [28, Chapter 7] and Knuth [9, exercise 2.3.4.2-26].

EXERCISES

1. Suppose you have n autos and one mechanic, where the probability that a working auto breaks between time t and time $t+dt$ is $\lambda\,dt+o(dt)$. Whenever there are any broken autos, the mechanic fixes them one at a time. The probability that the mechanic fixes a broken auto between time t and $t+dt$ is $\mu\,dt+o(dt)$. Show that the probability that k autos work at time t obeys the equations

$$\frac{dP_0(t)}{dt}=-n\lambda P_0(t)+\mu P_1(t), \tag{113}$$

$$\frac{dP_i(t)}{dt}=-[(n-i)\lambda+\mu]P_i(t)+(n-i+1)\lambda P_{i-1}(t)+\mu P_{n+1}(t), \tag{114}$$

$$\frac{dP_n(t)}{dt}=-\mu P_i(t)+\lambda P_{i-1}(t). \tag{115}$$

2. Find the steady-state solution of the previous exercise.
3. Modify Exercise 1 to have r mechanics, so that up to r autos can be repaired at a time.
4. Find the steady-state solution of the previous exercise.

8.2 RECURRENCES IN SEVERAL VARIABLES

Several times previously we have considered recurrences in two variables. For example, in Section 6.3.1 we studied the equation

$$p_{n,i} = \frac{1}{n} p_{n-1,i-1} + \frac{n-1}{n} p_{n-1,i}. \tag{116}$$

In Section 7.1.3.1 we studied the equation

$$A(t,v) = at + 2\sum_i \binom{t}{i} p^i (1-p)^{t-i} A(t, v-1). \tag{117}$$

Recurrence equations in several variables are often much more difficult to solve than equations in one variable. There are five main techniques to consider for such equations.

The first technique is indexed generating functions. Thus in Section 6.3.1 we defined the generating function

$$G_n(z) = \sum_{i \geq 0} p_{ni} z^i, \tag{118}$$

while in Section 7.1.3.1 we defined the generating function

$$G_v(z) = \sum_{i \geq 0} A(i,v) \frac{z^i}{i!}. \tag{119}$$

Indexed generating functions are particularly promising when the recurrence equation that needs to be solved is first order in one variable. (In that case the variable in which the equation is first order should be tried as the index for the generating function.)

The second technique for solving recurrences in several variables is multidimensional generating functions. For eq. (116), for example, we can define the generating function

$$G_n(x,y) = \sum_{\substack{i \geq 0 \\ n \geq 0}} p_{ni} x^n y^i. \tag{120}$$

Multiplying eq. (116) by $nx^n y^i$ and summing over i and n gives

$$\sum_{\substack{i \geq 0 \\ n \geq 0}} n p_{ni} x^n y^i = \sum_{\substack{i \geq 0 \\ n \geq 0}} p_{n-1,i-1} x^n y^i + \sum_{\substack{i \geq 0 \\ n \geq 0}} (n-1) p_{n-1,i} x^n y^i \tag{121}$$

or

$$x \frac{\partial G(x,y)}{\partial x} = xyG(x,y) + x^2 \frac{\partial G(x,y)}{\partial x} \tag{122}$$

(using the fact that $p_{-i,i} = 0$), which reduces to

$$(1-x)\frac{\partial G(x,y)}{\partial x} = yG(x,y). \tag{123}$$

This partial differential equation, along with the appropriate boundary conditions, can be solved to give $G(x,y)$. We will not do that here, since the approach of indexed generating functions works so well for this problem.

The third technique is operator methods. Eq. (116) can be written as

$$(n+1)p_{n+1,i+1} - p_{n,i} + np_{n,i+1} = 0, \tag{124}$$

or as

$$[(n+1)\mathbf{E}_n\mathbf{E}_i - 1 + n\mathbf{E}_i]p_{ni} = 0, \tag{125}$$

where $\mathbf{E}_n$ is the operator that replaces n with $n+1$ and $\mathbf{E}_i$ is the operator that replaces i with $i+1$. In some cases, after writing a recurrence in operator form, you will be able to apply the techniques of Section 6.2.3.

The fourth technique is transformation of your equation to an equation with a known solution. These transformations can also be combined with the three previous techniques. They will be considered in more detail in the following sections.

The fifth technique is guessing the solution and then using proof by induction. Some rather complex equations happen to have relatively simple solutions. Using the recurrence to generate the answer for small indices, followed by guessing at the general form of the solution and proof by induction, is often the best way to solve such equations. This technique was used in Sections 1.12.1.1 and 1.12.1.2.

The simplest multidimensional recurrences have three terms. For example, we have seen the recurrence

$$F_{ni} = F_{n-1,i} + F_{n-1,i-1}, \tag{126}$$

which is the same as eq. (3.65), so this equation is satisfied by

$$F_{ni} = \binom{n}{i}. \tag{127}$$

This is a *particular solution*, which satisfies the boundary conditions

$$F_{0i} = \delta_{i0}. \tag{128}$$

The function

$$F_{ni} = \binom{n}{i-j} \tag{129}$$

is a solution that satisfies eq. (126) and the boundary conditions

$$F_{0i} = \delta_{ij}. \tag{130}$$

The *general solution* of eq. (130) is the function

$$F_{ni} = \sum_j a_j \binom{n}{i-j} \tag{131}$$

where the a_j are an infinite set of arbitrary constants. (The general solution of a multidimensional recurrence equation involves an infinite set of parameters.)

Recurrences with two indices are considered by Liu [41, Section 3.5] and briefly by Knuth [9, Section 1.2.10].

8.2.1* Linear Transformation of Indices

Transformation of the indices on a multidimensional recurrence can have a major effect on its appearance. We will illustrate this with the general linear three-term recurrence, which has the form

$$x(n,i)H_{n,i} = y(n,i)H_{n-a,i-b} + z(n,i)H_{n-c,i-d}. \tag{132}$$

Any such equation can be transformed into the form

$$x'(n,i)F_{n,i} = y'(n,i)F_{n-1,i} + z'(n,i)F_{n-1,i-1}, \tag{133}$$

unless the points (n,i), $(n-a,i-b)$, and $(n-c,i-d)$ lie on a straight line. [If the three points lie on a straight line, then eq. (132) is a one-dimensional equation in disguise.]

To do this transformation, we will begin by introducing the function F_{ni} defined by

$$F_{pn+qi,rn+si} = H_{ni}, \tag{134}$$

where p, q, r, and s were chosen so that F_{ni} satisfies an equation with the form of eq. (133). Written with the variable F_{ni}, eq. (132) is

$$\begin{aligned} x(n,i)F_{pn+qi,rn+si} &= y(n,i)F_{pn+qi-ap-bq,rn+si-ar-bs} \\ &\quad + z(n,i)F_{pn+qi-cp-dq,rn+si-cr-ds}. \end{aligned} \tag{135}$$

We call this step "transforming the indices" because, while there is a close correspondence between the functions F and H, they depend on their indices in a different way.

We want the difference of the first index of the term on the left side and the first index of the first term on the right side to be 1 [just as it is in eq. (133)]:

$$pn + qi - (pn + qi - ap - bq) = 1. \tag{136}$$

The difference of the second indices of these two terms should be zero:

$$rn + si - (rn + si - ar - bs) = 0. \tag{137}$$

Likewise, the difference of the first index of the term on the right side and the first index of the second term on the left side should be 1:

$$pn + qi - (pn + qi - cp - dq) = 1, \tag{138}$$

and the difference of the second indices of these terms should be 1:

$$rn + si - (rn + si - cr - ds) = 1. \tag{139}$$

Simplifying these four equations gives

$$ap + bq = 1, \qquad ar + bs = 0, \tag{140}$$

$$cp + dq = 1, \qquad cr + ds = 1. \tag{141}$$

The solution of these equations is

$$p = \frac{-b+d}{ad-bc}, \qquad q = \frac{a-c}{ad-bc}, \tag{142}$$

$$r = \frac{-b}{ad-bc}, \qquad s = \frac{a}{ad-bc}. \tag{143}$$

The denominator, $ad - bc$, is zero if and only if the three points are in a straight line.

Using the values given in eqs. (142, 143), eq. (135) simplifies to

$$x(n,i)F_{pn+qi,rn+si} = y(n,i)F_{pn+qi-1,rn+si} + z(n,i)F_{pn+qi-1,rn+si-1}. \tag{144}$$

Now that the three terms have indices with the right relationships, we do a transformation of variables so that the indices on the term on the left side are (n,i). The required transformation is

$$n' = pn + qi \tag{145}$$

$$i' = rn + si. \tag{146}$$

Since we wish to replace n's and i's with n''s and i''s, we must solve eqs. (145, 146) for n and i in terms of n' and i'. The solution is

$$n = \frac{sn' - qi'}{ps - qr}, \qquad i = \frac{-rn' + pi'}{ps - qr}. \tag{147}$$

The results of applying the transformation of variables to eq. (144) (and dropping the primes) is

$$x\left(\frac{sn-qi}{ps-qr}, \frac{-rn+pi}{ps-qr}\right)F_{ni} = y\left(\frac{sn-qi}{ps-qr}, \frac{-rn+pi}{ps-qr}\right)F_{n-1,i} + z\left(\frac{sn-qi}{ps-qr}, \frac{-rn+pi}{ps-qr}\right)F_{n-1,i-1}. \tag{148}$$

Eq. (148) has the desired form [eq. (133)] if we use the definitions

$$x'(n,i) = x\left(\frac{sn-qi}{ps-qr}, \frac{-rn+pi}{ps-qr}\right), \tag{149}$$

$$y'(n,i) = y\left(\frac{sn-qi}{ps-qr}, \frac{-rn+pi}{ps-qr}\right), \tag{150}$$

$$z'(n,i) = z\left(\frac{sn-qi}{ps-qr}, \frac{-rn+pi}{ps-qr}\right). \tag{151}$$

The next step is to solve eq. (148) for F_{ni}. To obtain the desired particular solution, we need to transform the boundary conditions for eq. (132) to match eq. (148). For some problems we are given the boundary conditions. In these cases we just need to use eq. (134) to obtain the boundary conditions to use with

F. In other cases, we are not given any particular boundary conditions; instead part of the problem is to find suitable boundary conditions. In the second case we have two choices: (1) we can work out some boundary conditions for H and proceed as in the first case, or (2) we can determine a boundary contour that makes eq. (148) easy to solve and use the inverse of eq. (134) to determine the corresponding contour for H. (Our example of this method will illustrate this second choice.) After we solve eq. (148) for F_{ni}, we use eq. (134) to transform the solution of eq. (148) into the solution of eq. (132). All occurrences of n in the solution are replaced with $pn+qi$, and all occurrences of i are replaced with $rn+si$.

Let's now see how transformation of indices works on an example. We will compute how many strings of 0's and 1's of length n have exactly i pairs of 1's and no pairs of 0's. (Three adjacent 1's count as two pairs, four as three pairs, etc.) Let a_{ni} be the number of such sequences, let b_{ni} be the number of such sequences where the first symbol is a 0, and let c_{ni} be the number of such sequences where the first symbol is a 1. Clearly,

$$a_{ni} = b_{ni} + c_{ni}, \tag{152}$$

because each string must start with a 0 or a 1. Also

$$b_{ni} = c_{n-1,i}, \tag{153}$$

because a string that starts with a 0 is converted into one that starts with a 1 when the initial zero is removed, and the number of pairs of 1's is not changed. Finally

$$c_{ni} = b_{n-1,i} + c_{n-1,i-1}, \tag{154}$$

because removing the initial symbol from a string that starts with a 1 gives either a string which starts with a 0, has one less symbol, and the same number of pairs; or a string which starts with a 1, has one less symbol, and one less pair.

Combining eq. (153) and eq. (154) gives

$$c_{ni} = c_{n-2,i} + c_{n-1,i-1}, \tag{155}$$

which has the same form as eq. (132) with $a = 2$, $b = 0$, $c = 1$, and $d = 1$. Also x, y, and z are all constant, so the transformations will have no effect on them. Eqs. (142, 143) give

$$p = \tfrac{1}{2}, \quad q = \tfrac{1}{2}, \quad r = 0, \quad s = 1. \tag{156}$$

Using $F_{n/2+i/2,i} = c_{ni}$ in eq. (155) gives

$$F_{n/2+i/2,i} = F_{n/2+i/2-1,i} + F_{n/2+i/2-1,i-1}. \tag{157}$$

Replacing $\frac{1}{2}n + \frac{1}{2}i$ with n gives

$$F_{ni} = F_{n-1,i} + F_{n-1,i-1}, \tag{158}$$

so F is a linear combination of binomial coefficients. To find out which linear combination, we must consider the boundary values for the problem.

The nicest boundary conditions for eq. (158) are the values of F_{0i}, so we will see if we can easily compute these values. If we use $F_{n/2+i/2,i} = c_{ni}$, then we must set $\frac{1}{2}n + \frac{1}{2}i = 0$ to find out which c_{ni} corresponds to F_{0i}. We get

$$F_{0i} = c_{-i,i}. \tag{159}$$

But $c_{00} = 1$ while $c_{-i,i} = 0$ for $i \neq 0$, so

$$F_{0i} = \delta_{i0}. \tag{160}$$

The solution of eq. (158) with this boundary condition is

$$F_{ni} = \binom{n}{i} \tag{161}$$

(see eq. (3.65)). Using $F_{n/2+i/2,i} = c_{ni}$, we get

$$c_{ni} = \binom{\frac{n+i}{2}}{i} \tag{162}$$

as the solution of the original recurrence.

We are not done with the original recurrence, however! Eq. (161) is valid only for integer n and integer i. Now, $F_{n/2+i/2,i}$ has an integer first index only if $n+i$ is even. Therefore we have solved the original recurrence only for the case where $n+i$ is even. The integer points in the space for c_{ni} are mapped onto the half-integer points for n in the space for F_{ni}. These mappings to noninteger points occur whenever the determinant of the coefficients associated with the transformation of indices is less than 1. (The inverse mapping from F_{ni} space to c_{ni} space will map integer points onto noninteger points if the determinant is larger than 1.)

Notice that eq. (154) is an equation relating three terms where the first index plus the second index always has the same parity. In other words, if $n+i$ is even, then the sum of the indices on the left-side term is even, and the same is true for each of the terms on the right side. If $n+i$ is odd, then all three terms have indices with an odd sum. Thus the original recurrence gives no connection between the points where $n+i$ is even and the points where $n+i$ is odd. By forcing eq. (154) into the form of eq. (133), we forced consideration of noninteger indices for the transformed equation.

To finish the problem, we need a set of boundary conditions for the case where $n+i$ is odd. Previously we considered the case where $\frac{1}{2}n + \frac{1}{2}i = 0$. Now let's consider the nearest case where $n+i$ will be odd, namely, $\frac{1}{2}n + \frac{1}{2}i = \frac{1}{2}$. We get

$$F_{1/2,i} = c_{1-i,i}. \tag{163}$$

Now $c_{10} = 1$ while $c_{1-i,i} = 0$ for $i \neq 0$, so

$$F_{1/2,i} = \delta_{i0}. \tag{164}$$

The solution of eq. (158) with this boundary condition is

$$F_{ni} = \binom{n-\frac{1}{2}}{i}. \tag{165}$$

Using $F_{n/2+i/2,i} = c_{ni}$, we get

$$c_{ni} = \binom{\frac{n+i-1}{2}}{i} \tag{166}$$

as the solution of the original recurrence when $n+i$ is odd.

We can write eq. (162) and eq. (166) as a single equation:

$$c_{ni} = \binom{\lfloor \frac{n+i}{2} \rfloor}{i}. \tag{167}$$

To finish solving the original problem, we use eq. (153) to obtain

$$b_{ni} = \binom{\lfloor \frac{n+i-1}{2} \rfloor}{i}, \tag{168}$$

and eq. (152) to obtain

$$a_{ni} = \binom{\lfloor \frac{n+i}{2} \rfloor}{i} + \binom{\lfloor \frac{n+i-1}{2} \rfloor}{i}, \tag{169}$$

The example of strings is adapted from Liu [41, section 3.5]. The first two exercises are adapted from Cohen [23, Chapter 3], who attributes them to Irving Kaplansky.

EXERCISES

1. Let $f(n,k)$ be the number of k-element subsets that can be selected from the set $\{1, 2, \ldots, n\}$ and that do not contain two consecutive integers. Show that

$$f(n,k) = f(n-2, k-1) + f(n-1, k).$$

Hint: If n is selected, then $n-1$ cannot be, so the remaining numbers come from $\{1, 2, \ldots, n-2\}$, while if n is not selected, the remaining numbers come from the set $\{1, 2, \ldots, n-1\}$.

2. Let $f(n,k)$ be defined as in the previous exercise. Find a closed form for it.

3. What is the general solution of the equation

$$F_{ni} = nF_{n-1,i} + F_{n-1,i-1}?$$

Hint: This recurrence is similar to eq. (3.185).

4. Solve the recurrence of the previous exercise subject to the boundary conditions $F_{0i} = \delta_{i0}$.

8.2.2* Separation of Variables

In this section we continue our study of transformations that can be used to simplify multidimensional recurrence equations. We will continue to concentrate on recurrences with three terms. In Chapter 3 we had three three-term recurrences:

$$F_{ni} = F_{n-1,i} + F_{n-1,i-1}, \tag{170}$$

$$F_{ni} = (n-1)F_{n-1,i} + F_{n-1,i-1}, \tag{171}$$

$$F_{ni} = iF_{n-1,i} + F_{n-1,i-1} \tag{172}$$

[see eqs. (3.65, 3.185, 3.189)]. The general solution of eq. (170) is a linear combination of binomial coefficients, the general solution of eq. (171) is a linear combination of Stirling numbers of the first kind, and the general solution of eq. (172) is a linear combination of Stirling numbers of the second kind.

In the previous section we saw that any linear three-term recurrence equation where the three terms do not lie on a straight line can be transformed so that the indices have same the pattern as eqs. (170–172). Many equations, however, do not have coefficients that match any of these three patterns. In this section we consider transformations that can be used to simplify the coefficients.

Suppose F_{ni} is a solution of the recurrence

$$x(n,i)F_{n,i} = y(n,i)F_{n-1,i} + z(n,i)F_{n-1,i-1}. \tag{173}$$

Define H_{ni} as

$$f(n)g(i)H_{ni} = F_{ni}, \tag{174}$$

where f and g are arbitrary functions. Then H satisfies the equation

$$\begin{aligned} x(n,i)f(n)g(i)H_{n,i} = y(n,i)f(n-1)g(i)H_{n-1,i} \\ + z(n,i)f(n-1)g(i-1)H_{n-1,i-1}. \end{aligned} \tag{175}$$

Dividing by $f(n-1)$ and $g(i)$ gives

$$x(n,i)\frac{f(n)}{f(n-1)}H_{n,i} = y(n,i)H_{n-1,i} + z(n,i)\frac{g(i-1)}{g(i)}H_{n-1,i-1}. \tag{176}$$

Notice that since f is an arbitrary function of n, $f(n)/f(n-1)$ is also an arbitrary function of n. Similar remarks apply to $g(i-1)/g(i)$.

We will use eq. (176) as follows. When we have a three-term recurrence equation, we will first use the technique of the previous section to get the indices into standard form. (There are six different ways that transformation of indices can be applied, depending on which of the terms in the original equation we transform into which term in the resultant equation.) Then we will see if our equations can be made to match one of eqs. (170–172). If we match one of the equations directly, then we use the technique of the previous section. If the match is exact for the first term on the right, off by only a function of n on the left term, and off by only a function of i on the second term on the right, then we define $f(n)$ and $g(i)$ to cancel out the factors that prevent a perfect match and apply eq. (174) in the reverse direction to obtain the solution of our original problem. The example in the following section will bring out the details of the method.

These transformations can also be combined with generating function techniques. Indexed one-dimensional generating functions are particularly useful when one of the indices does not appear as a coefficient in the recurrence equation. Transformations can often be used to remove an unwanted index from a coefficient.

We will now mention several additional three-term recurrences that do not reduce to the ones given at the beginning of this section.

A *run of increasing numbers* is a sequence of increasing numbers which is not contained in a longer sequence of increasing numbers. The number of permutations of n numbers which contain k runs of increasing numbers is given by the *Euler number* $\left\langle {n \atop k} \right\rangle$, which obeys the recurrence

$$\left\langle {n \atop k} \right\rangle = k\left\langle {n-1 \atop k} \right\rangle + (n-k+1)\left\langle {n-1 \atop k-1} \right\rangle. \tag{177}$$

The Euler numbers are discussed at length in Knuth [11, section 5.1.3].

The *Lah numbers* are given by the recurrence

$$L_{n,k} = (n+k-1)L_{n-1,k} + L_{n-1,k-1}, \tag{178}$$

which is satisfied by

$$L_{nk} = \frac{n!}{k!}\binom{n-1}{k-1}. \tag{179}$$

This recurrence was studied by Lah [109]. See Riordan [48, pp. 43–44] to see how to use exponential generating functions on this recurrence.

Riordan [48, p. 85] also discusses a recurrence equivalent to

$$F_{n,i} = F_{n-1,i} + (n+i-1)F_{n-1,i-1}, \tag{180}$$

which has the solution

$$F_{n,i} = \frac{(n+i)!}{i!(n-i)!2^i}. \tag{181}$$

The term *separation of variables* usually refers to techniques that find solutions of equations where the solution has the form $f(n)g(i)$. Here, we are using the term in a more general sense.

Monier [117] discusses how to apply techniques similar to the ones we have to the problem of solving multidimensional divide and conquer recurrences.

EXERCISE

1. Show that the recurrence

$$F_{i_1,i_2,\ldots,i_n} = F_{i_1-1,i_2,\ldots,i_n} + F_{i_1,i_2-1,\ldots,i_n} + \cdots + F_{i_1,i_2,\ldots,i_n-1}$$

is satisfied by

$$F_{i_1,i_2,\ldots,i_n} = \binom{i_1+i_2+\cdots+i_n}{i_1,i_2,\ldots,i_n}.$$

8.2.2.1 Binary Search Tree

Consider the following algorithm, which builds a binary search tree. Each new item is stored in the tree. The first item is stored at the root. After that any item larger than the root is stored in the right subtree, while any item smaller than the root is stored in the left subtree. The same principle applies recursively to the subtrees. This algorithm usually results in most items being stored near the root.

Algorithm 8.2 Binary Search Tree: Input: A query q and a binary search tree with the head "Head". Each node in the tree has a *Key* field that can store a query, an *llink* field pointing to the subtree for smaller queries, and an *rlink* field pointing to the subtree for larger queries. Convention: We assume that new nodes come with empty *llink* and *rlink* fields. (If not, the fields should be set to empty in Step 7.) Output: The position i of q in the tree. The item q is inserted into the tree if it is not already there.

Step **1.** [Initialize] Set $p \leftarrow$ Head, and $i \leftarrow$ *llink*(Head).

Step **2.** [Test] If i is *empty*, set i to point to a new node, set *llink*(Head) $\leftarrow i$, and go to Step 7.

Step **3.** [Compare] Set $p \leftarrow i$. If the relation between q and *Key*(p) is:

Step **4.** [Found] $q =$ *Key*(p), then stop, the position for q has been found.

Step **5.** [Left] $q <$ *Key*(p), then set $i \leftarrow$ *llink*(p). If i is not *empty*, then go to Step 3. Otherwise set i to point to a new node, set *llink*(p) $\leftarrow i$, and go to Step 7.

Step **6.** [Right] $q >$ *Key*(p), then set $i \leftarrow$ *rlink*(p). If i is not empty, then go to Step 3. Otherwise set i to point to a new node, set *rlink*(p) $\leftarrow i$, and go to Step 7.

Step **7.** [Insert] Set *Key*(i) $\leftarrow q$.

We wish to consider the height of the last node inserted into an n-node binary search tree. The height of the node is the number of nodes that come before it on the path to the root. Thus for a one-node tree the height of the last node is always zero, and for a two-node tree the height of the last node is always 1. For a three-node tree, the height of the last node is either 1 or 2, depending on whether the last and next-to-last queries both have the same relation to the first query or a different relation (i.e., one is smaller and one is larger). The maximum height for the last node is $n-1$, and the minimum height is 1 when $n > 1$.

Now let's consider the average height. We will assume that the input is a random permutation of the integers from 1 to n. Define H_{ni} to be the number of permutations of n items that result in the nth item having height i. The first step of the analysis is to find a recurrence obeyed by H. To develop the recurrence, focus on the effect of the next-to-last item on the height of the last item. When the first $n-1$ items have been inserted into the tree, there are n places where the nth item can go. (It can go before the first, after the first, after the second, etc.) For each place, there are the same number of permutations that result in the item going to that place. [For each place there are $(n-1)!$ associated permutations.] If the nth item is either 1 less than or 1 greater than the next-to-last item, it will be inserted as a child of that item. In this case the last item will have height 1 greater than it would have had if the next-to-last item had been omitted. In all other cases, the last item has the same height whether or not the next-to-last item is included. This gives the following recurrence:

$$H_{ni} = (n-2)H_{n-1,i} + 2H_{n-1,i-1}. \tag{182}$$

This recurrence has almost the same form as eq. (171). Let

$$g(i)H_{ni} = F_{ni}, \tag{183}$$

where $g(i)$ is a solution of the equation

$$\frac{g(i-1)}{g(i)} = 2. \tag{184}$$

To be definite, let $g(i) = 2^{-i}$. Then F_{ni} obeys the equation

$$F_{ni} = (n-2)F_{n-1,i} + F_{n-1,i-1}, \tag{185}$$

so F is a linear combination of Stirling numbers of the first kind.

When $n = 1$, H_{ni} is zero except for $i = 1$, and $H_{10} = 1$, so F obeys the boundary conditions

$$F_{1i} = \delta_{i0}. \tag{186}$$

Thus we have

$$F_{ni} = \begin{bmatrix} n-1 \\ i \end{bmatrix}, \tag{187}$$

and since $g(i)H_{ni} = F_{ni}$,

$$H_{ni} = 2^i \begin{bmatrix} n-1 \\ i \end{bmatrix}. \tag{188}$$

Dividing by the number of permutations of n objects, we get that the average height of the last node added to the search tree is

$$\frac{2^i}{n!} \begin{bmatrix} n-1 \\ i \end{bmatrix}. \tag{189}$$

The Binary Tree Search Algorithm is covered in Knuth [11, Section 6.2.2]. The average height of the last node was first obtained by Lynch [112].

EXERCISES

1. Compute the average height of a node in a random n-node binary search tree.

2. What is the general solution of the equation

$$F_{ni} = g(n-i)F_{n-1,i} + F_{n-1,i-1},$$

where $g(x)$ is a given function?

3. The *Gaussian binomial coefficients* are defined by the equation

$$\binom{n}{k}_q = \frac{q^n - 1}{q - 1} \cdot \frac{q^{n-1} - 1}{q^2 - 1} \cdots \frac{q^{n-k-1} - 1}{q^k - 1},$$

where $\binom{n}{0}_q = 1$. Show that they obey the recurrence of the previous exercise and find the appropriate $g(n)$.

4. Use the result of the previous problem to prove that the Gaussian binomial coefficients are polynomials in q.

5. Show that Guassian binomial coefficients obey the recurrence

$$\binom{n}{k}_q = \binom{n-1}{k}_q + q^{n-k}\binom{n-1}{k-1}_q.$$

6. Show that Guassian binomial coefficients obey the recurrence

$$\binom{n}{k}_q = q^k\binom{n-1}{k}_q + \binom{n-1}{k-1}_q.$$

7. Show that $(1+x)(1+qx)\cdots(1+q^{n-1}) = \sum_{0\le i\le n} \binom{n}{i}_q q^{i(i-1)/2}x^i$.

8. Show that $\lim_{q\to 1} \binom{n}{i}_q = \binom{n}{i}$.

9. [Rényi, see [80]] The number of n-node labelled trees with exactly i vertices of degree 1 obeys the recurrence

$$\frac{i}{n}T_{ni} = (n-i)T_{n-1,i-1} + iT_{n-1,i}.$$

Give a closed-form expression for T_{ni}. Hint: Consider the change of variable $n' = n - i$ and the appropriate change of index so that the equation has the standard form for three-term recurrences.

10. The number of k-cubes in an n-cube obeys the recurrence

$$N_{kn} = 2N_{k,n-1} + N_{k-1,n-1},$$

where N_{0n} is the number of vertices of an n-cube. What is the value of N_{kn}?

8.2.3* Simplifying Multiple-Variable Recurrences

One of the more general forms for a recurrence equation is

$$F(x_1,\ldots,x_n) = \begin{cases} \text{if } p(x_1,\ldots,x_n) \text{ is } \textit{true}, \text{ then } a(x_1,\ldots,x_n) \\ \text{otherwise } b(x_1,\ldots,x_n,F(C_1(x_1,\ldots,x_n)),\ldots,F(C_m(x_1,\ldots,x_n))), \end{cases} \tag{190}$$

where a, b, and C_1, $\ldots$, C_m are given functions of $x_1,\ldots,x_n$. The C_i are vector-valued functions; they return a vector of n values. There are many interesting cases where this general form can be greatly simplified.

If $m = 1$, then a change of variables can be used to reduce the problem to a problem with one index. Using C for C_1, eq. (190) becomes

$$F(x_1,\ldots,x_n) = \begin{cases} \text{if } p(x_1,\ldots,x_n) \text{ is } \textit{true}, \text{ then } a(x_1,\ldots,x_n) \\ \text{otherwise } b(x_1,\ldots,x_n,F(C(x_1,\ldots,x_n))). \end{cases} \tag{191}$$

Define the following functions:

$$p'(i) = p(C^{[i]}(x_1,\ldots,x_n)), \tag{192}$$

$$a'(i) = a(C^{[i]}(x_1,\ldots,x_n)), \tag{193}$$

$$F'(i) = F(C^{[i]}(x_1,\ldots,x_n)), \tag{194}$$

and

$$b'(i,y) = b(C^{[i]}(x_1,\ldots,x_n),y). \tag{195}$$

In terms of the primed variables, eq. (191) becomes

$$F'(i) = b'(i-1,F'(i-1)) \tag{196}$$

with the boundary conditions $F'(i) = a'(i)$ when $p'(i)$ is *true*. This is a first order equation with the single index i.

To illustrate the method, let's see how it applies to eq. (7.87):

$$G_v(z) = aze^z + 2e^{pz}G_{v-1}((1-p)z) \tag{197}$$

with $G_0(z) = 0$. The two variables are v and z. To put it in the form of eq. (191), we let $x_1 = v$ and $x_2 = z$, so $F(x_1,x_2) = G_v(z)$. Then

$$b(x_1,x_2) = ax_2e^{x_2} + 2e^{px_2}F(x_1-1,(1-p)x_2), \tag{198}$$

$$C(x_1,x_2) = (x_1-1,(1-p)x_2), \tag{199}$$

$$p(x_1,x_2) = (x_1 = 0), \tag{200}$$

$$a(x_1,x_2) = 0. \tag{201}$$

Applying C i times gives

$$C^{[i]}(x_1,x_2) = (x_1-i,(1-p)^ix_2). \tag{202}$$

Our primed functions are thus

$$p'(i) = (x_1 - i = 0), \tag{203}$$

$$a'(i) = 0, \tag{204}$$

$$F'(i) = F(x_1 - i, (1-p)x_2), \tag{205}$$

$$b'(i, y) = a(1-p)^i x_2 e^{(1-p)^i x_2} + 2e^{p(1-p)^i x_2} y. \tag{206}$$

and our recurrence becomes

$$F'(i) = a(1-p)^i x_2 e^{(1-p)^i x_2} + 2e^{p(1-p)^i x_2} F'(i-1), \tag{207}$$

which is a first order linear equation in $F'(i)$. Eq. (207) can be solved using eq. (5.29). After replacing $F'(i)$ with G, this gives eq. (7.91).

We now consider the case where $m = 2$. The generalizations needed for larger values of m are obvious. We use $C(x_1, \ldots, x_n)$ for $C_1(x_1, \ldots, x_n)$ and $D(x_1, \ldots, x_n)$ for $C_2(x_1, \ldots, x_n)$. Thus eq. (190) becomes

$$\begin{aligned} &F(x_1, \ldots, x_n) \\ &\quad = \begin{cases} \text{if } p(x_1, \ldots, x_n) \text{ is } \textit{true}, \text{ then } a(x_1, \ldots, x_n), \\ \text{otherwise } b(x_1, \ldots, x_n, F(C(x_1, \ldots, x_n)), F(D(x_1, \ldots, x_n))). \end{cases} \end{aligned} \tag{208}$$

There are several interesting special cases to consider.

One of the simplest of the special cases is the one where we can find a function $H(x_1, \ldots, x_n)$ such that for some positive integers r and s, where the greatest common divisor of r and s is 1, $H^{[r]}(x_1, \ldots, x_n) = C(x_1, \ldots, x_n)$ and $H^{[s]}(x_1, \ldots, x_n) = D(x_1, \ldots, x_n)$. In this case we can reduce eq. (208) to a recurrence in a single index. The approach is quite similar to the technique of secondary recurrences that was developed in Section 5.2.1.

For example, let $n = 1$ and use x for x_1. Let $p(x)$ be *true* when $x \leq 1$ and *false* otherwise. Let $a(0) = 0$ and $a(1) = 1$. Let $C(x) = x-1$ and $D(x) = x-2$. Finally let $b(x, F(x-1), F(x-2))$ be the function $F(x-1) + F(x-2)$. Our function has the form of eq. (208). We can use $H(x) = x-1$, so $C(x) = H^{[1]}(x)$ and $D(x) = H^{[2]}(x)$. Our recurrence is the Fibonacci recurrence $F(x) = F(x-1) + F(x-2)$ with the boundary conditions $F(0) = 0$ and $F(1) = 1$.

In general, we can simplify equations where

$$H^{[r]}(x_1, \ldots, x_n) = C(x_1, \ldots, x_n) \tag{209}$$

and

$$H^{[s]}(x_1, \ldots, x_n) = D(x_1, \ldots, x_n) \tag{210}$$

by defining some new functions as follows. Let

$$p'(i) = p(H^{[i]}(x_1, \ldots, x_n)), \tag{211}$$

$$a'(i) = a(H^{[i]}(x_1, \ldots, x_n)), \tag{212}$$

$$F'(i) = F(H^{[i]}(x_1, \ldots, x_n)), \tag{213}$$

and

$$b'(i, y, z) = b(H^{[i]}(x_1, \ldots, x_n), y, z). \tag{214}$$

In terms of the primed variables, eq. (208) becomes

$$F'(i) = b'(i, F'(i-r), F'(i-s)) \tag{215}$$

with the boundary conditions $F'(i) = a'(i)$ when $p'(i)$ is *true*. This is an equation in the single index i.

The next case of interest is the case where

$$D(C(x_1, \ldots, x_n)) = C(D(x_1, \ldots, x_n)) \tag{216}$$

and

$$C^{[r]}(x_1, \ldots, x_n) = D^{[s]}(x_1, \ldots, x_n) \tag{217}$$

for some positive integers r and s, where the greatest common divisor of r and s is 1. Again we can reduce eq. (208) to an equation with one index.

Since r and s are relatively prime, any integer n can be written uniquely in the form $ir + js$ where $0 \leq j < r$. [The Euclidean Algorithm (Algorithm 1.5) guarantees that any integer can be written in the form $ir + js$, where i and j are integers.] Since $ir + js = (i+s)r + (j-r)s$, we can restrict j to the range $0 \leq j < r$. We now show that the representation is unique. If $n = ir + js = i'r + j's$, then $j' = j + kr$ for some integer k and $i' = i - ks$. If $j = j'$, then $i = i'$, while if $j \neq j'$ and $0 \leq j < r$, then j' is not in the range $0 \leq j' < r$. Thus the representation is unique.

The following definitions are needed:

$$p'(ir + js) = p(C^{[j]}(D^{[i]}(x_1, \ldots, x_n))), \tag{218}$$

$$a'(ir + js) = a(C^{[j]}(D^{[i]}(x_1, \ldots, x_n))), \tag{219}$$

$$F'(ir + js) = F(C^{[j]}(D^{[i]}(x_1, \ldots, x_n))), \tag{220}$$

and

$$b'(ir + js, y, z) = b(C^{[j]}(D^{[i]}(x_1, \ldots, x_n)), y, z). \tag{221}$$

We assume that $0 \leq j < r$. In terms of the primed variables, eq. (208) becomes

$$F'(i) = b'(i, F'(i-r), F'(i-s)) \tag{222}$$

with the boundary conditions $F'(i) = a'(i)$ when $p'(i)$ is *true*. This is again an equation in the single index i.

The final case that we will consider is the case where

$$D(C(x_1, \ldots, x_n)) = C(D(x_1, \ldots, x_n)). \tag{223}$$

The commutative property lets us reduce eq. (208) to a recurrence in two indices. We need the definitions

$$p'(i, j) = p(C^{[j]}(D^{[i]}(x_1, \ldots, x_n))), \tag{224}$$

$$a'(i, j) = a(C^{[j]}(D^{[i]}(x_1, \ldots, x_n))), \tag{225}$$

$$F'(i, j) = F(C^{[j]}(D^{[i]}(x_1, \ldots, x_n))), \tag{226}$$

and

$$b'(i, j, y, z) = b(C^{[j]}(D^{[i]}(x_1, \ldots, x_n)), y, z). \tag{227}$$

In terms of the primed variables, eq. (208) becomes

$$F'(i,j) = b'(i, F'(i,j-1), F'(i-1,j)) \tag{228}$$

with the boundary conditions $F'(i,j) = a'(i,j)$ when $p'(i,j)$ is *true*. This is an equation in the pair of indices i and j.

The material in this section is adapted from Cohen [80], where essentially the same approach is presented as a technique for the automatic simplification of LISP programs. Cohen also considers the case where

$$C^{[r]}(x_1, \ldots, x_n) = D^{[s]}(x_1, \ldots, x_n). \tag{229}$$

This case leads to only a slight simplification.

EXERCISE

1. Solve the recurrence $(m+1)F_{m+1,n} = nF_{m,n-1}$.

CHAPTER 9

Global Techniques

Up to this point it has usually been rather simple to obtain an exact answer for the number of times each step in an algorithm is done. The complexity in the analyses has come mainly from the mathematics required to simplify the exact answers. In this chapter we will consider the information contained in the overall structure of an algorithm, and we will consider some algorithms where the principal difficulty in analysis comes from determining *any* suitable expression for how often the steps are done. We will also consider an algorithm that can solve a whole set of problems simultaneously much more rapidly than the problems could be worked individually.

9.1 FLOW GRAPHS

Most of the algorithms that we have analyzed so far have been rather small. It has been fairly easy to determine where in the algorithm it is most important to apply analysis. The analysis itself has required most of the effort. A large algorithm, on the other hand, has many parts, most of which are usually simple to analyze, and it is important to have a systematic way to organize the preliminary investigation of the algorithm. This section shows how to efficiently extract the information that is contained in the shape of the flowchart for the algorithm.

First we need a number of definitions from graph theory. A *graph* is a set of vertices, V, and a set of edges, E. Each *edge* is an *unordered* pair of vertices. Graphs are often represented by a diagram in which the vertices are represented by dots, circles, or squares, and the edges are represented by lines that connect pairs of vertices. Figure 9.1 shows a sample graph with four vertices: 1, 2, 3, 4. The graph in Figure 9.1 has edges $\{1,2\}$, $\{2,3\}$, $\{2,4\}$, and $\{3,4\}$. An edge *connects* a pair of vertices; the edge is *incident* on the two vertices it connects. Thus, in Figure 9.1, edge $\{1,2\}$ connects vertices 1 and 2; it is incident on vertex 1 and on vertex 2.

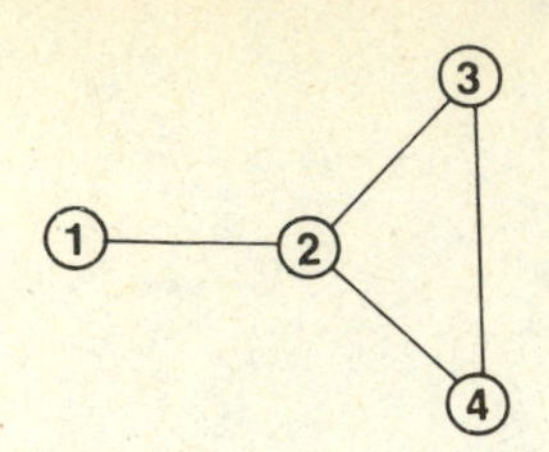

FIGURE 9.1 A graph with four nodes and four edges.

Graphs are usually used to indicate connections between objects. Vertices are used to represent the objects and edges are used to indicate which objects have a direct connection. It is appropriate to use a graph when the connections have no associated direction. If there is a connection in a graph from object A to object B, then there is also a connection from B to A. It can be helpful to think of the vertices as cities and the edges as two-way roads that connect the cites. The typical road map can be regarded as a large graph, if you disregard the distances between cities and other extraneous information.

A *directed graph* is a graph where each edge is an *ordered* pair and therefore has a direction. Figure 9.2 shows a directed graph. An edge connects the first vertex of the pair (the *initial vertex*) to the second vertex of the pair (the *terminal vertex*). In Figure 9.2 the edge $[1,2]$ has initial vertex 1 and terminal vertex 2. The edge does not provide a connection in the reverse direction. Thus in Figure 9.2 there is no connection from vertex 2 to vertex 1. When there is a connection in the reverse direction, it is indicated by a second edge. Thus, if we wanted to modify the graph in Figure 9.2 so that there was a connection from vertex 2 to vertex 1, we would have to add the edge $[2,1]$. To show such an edge on the figure, we would draw another line between vertices 1 and 2. The new line would have an arrowhead at vertex 1.

The *degree* of a vertex is the number of edges incident on the vertex. (This is the number of edges that contain the vertex.) Vertex 1 has degree one. Vertex 2 has degree three. Vertices 3 and 4 have degree two. Vertices are also called *nodes* and edges are also called *arcs*. A graph is called an *undirected graph* when necessary to avoid confusion with directed graphs (see below).

An edge goes *out* of its initial vertex and *into* its terminal vertex. The *out-degree* of a vertex is the number of edges going out of the vertex, and the *in-degree* of a vertex is the number of edges coming into the vertex. In Figure 9.2 vertex 1 has in-degree zero and out-degree one. Vertex 2 has in-degree two and out-degree one. Vertices 3 and 4 have in-degree one and out-degree one. If you think of the vertices in a directed graph as cities, then the edges correspond to one-way roads between the cities.

A *subgraph* is a subset of the vertices and a subset of the edges of a graph where the edges in the subset connect vertices in the subgraph. Thus, a subgraph is any portion of a graph that still forms a graph. For example, in Figure 9.1 the set of vertices $\{1,2,4\}$ along with the set of edges $\{\{1,2\}\}$ forms a subgraph, but the same set of vertices along with the set of edges $\{\{1,2\},\{2,3\}\}$ does not

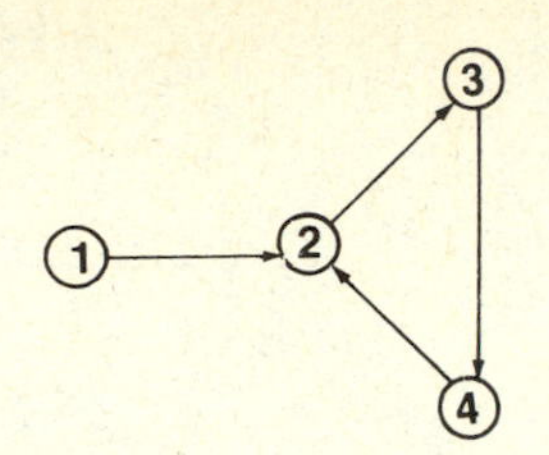

FIGURE 9.2 A directed graph with four nodes and four edges.

because edge $\{2,3\}$ contains a vertex that is not in the proposed subgraph.

A *loop* is an edge from a node to itself. A graph (or directed graph) may or may not be permitted to have loops, according to what is convenient. A *multigraph* is a graph where more than one edge may connect a pair of vertices. A *directed multigraph* is a directed graph that can have more than one edge in the same direction between a pair of vertices.

For every directed multigraph there is a corresponding multigraph obtained by ignoring the direction of the edges. For every directed graph there is a corresponding graph obtained by ignoring the direction of the edges and by removing any multiple edges that are obtained by ignoring direction.

A pair of vertices is *adjacent* if there is an edge between them. A *path* is a list of vertices $(v_1, v_2, \ldots, v_n)$ such that (v_i, v_{i+1}) is an edge for $1 \leq i < n$. The edges of a path are the edges that connect each vertex to the next vertex on the path. (A path can also be specified by listing its edges in order.) For example, the vertices $(1,2,3)$ specify a path in Figure 9.1. The same path is specified by the edges $(\{1,2\},\{2,3\})$. In a multigraph it is necessary to specify which edges make up a path. A *simple path* is a path in which no vertex appears more than once. A *directed path* is a path in a directed graph such that each vertex has an edge going to the next vertex on the path. In other words, in a directed path each edge goes the same way that the path does. For example, in Figure 9.2 $([2,3],[3,4])$ is a directed path. In directed graphs, we will also be interested in *undirected* paths, where the edges form a path if their direction is disregarded. In Figure 9.2 $([3,4],[2,3])$ is an undirected path, but not a directed one.

A *cycle* is a path where the last vertex on the path is adjacent to the first vertex on the path. The edges of a cycle are the edges that connect each vertex to the next vertex on the cycle. The edge from the last vertex to the first vertex is included in the cycle. For example, in Figure 9.1 the vertices $(2,3,4)$ specify a cycle. The same cycle is specified by the edges $(\{2,3\},\{3,4\},\{2,4\})$. A *simple cycle* is a cycle with no repeated vertices. The cycle $(2,3,4)$ is a simple cycle. A *directed cycle* is a cycle in a directed graph where each edge goes the same way that the cycle goes. In Figure 9.2, the edges $([2,3],[3,4],[4,2])$ form a directed cycle. An undirected cycle is a cycle in a directed graph where the direction of the edges is disregarded.

A *component* of a graph is a subgraph that contains all the vertices that can be reached by paths from any vertex in the component. The component also contains all the edges between vertices in the component. All vertices in

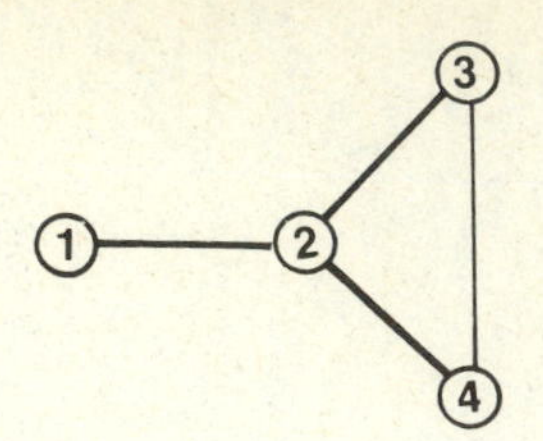

FIGURE 9.3 A spanning tree for the graph from Figure 9.1. The spanning tree consists of the three dark edges. The remaining edge of the original graph is shown as a light line.

a component are connected by paths. No vertices in different components are connected by paths. A *connected graph* is a graph with one component. The graph in Figure 9.1 is connected.

A *tree* is a connected graph that contains no cycles. In a tree there is exactly one simple path between any pair of vertices. A *rooted tree* is a tree with a designated root vertex. The subgraph of the graph in Figure 9.1 which consists of vertices $(1, 2, 3, 4)$ and edges $(\{1, 2\}, \{2, 3\}, \{2, 4\})$ is a tree.

A *spanning tree* of a graph (or multigraph) is any subgraph which is a tree and which contains all the vertices of the graph. Only connected graphs have spanning trees. The most straightforward way to form a spanning tree for a connected graph is to consider the edges of the graph one at a time and retain those edges which do not form a cycle (using only the edge in question and those edges previously retained). Figure 9.3 shows a spanning tree for the graph in Figure 9.1. The edges of the spanning tree are marked with heavy lines while the remaining edge of the graph from Figure 9.1 is shown as a light line.

Algorithm 9.1 Spanning Tree: Input: A connected graph (or multigraph) with vertex set V and edge set E. Output: A set S of edges in a spanning tree for the graph.

Step **1.** [Initialize] Set each vertex equivalent to itself. Set S to *empty.*

Step **2.** [Loop] For each edge $e = (a, b)$ do

Step **3.** [Test] If a is equivalent to b, then reject edge e. Otherwise set a equivalent to b and put e into S.

Step **4.** End for.

One important use of directed graphs is to represent the control flow of an algorithm. A *flow graph* is a directed graph with a special *source* node and one or more *sink* nodes. The flow graph for any reasonable program has a directed path from the source to any node and continuing to a sink. Each node of a flow graph represents a *block* of code: a sequence of statements that is always executed as a unit. All the statements in a block are executed the same number of times. In a flow graph two nodes A and B are connected by an edge from

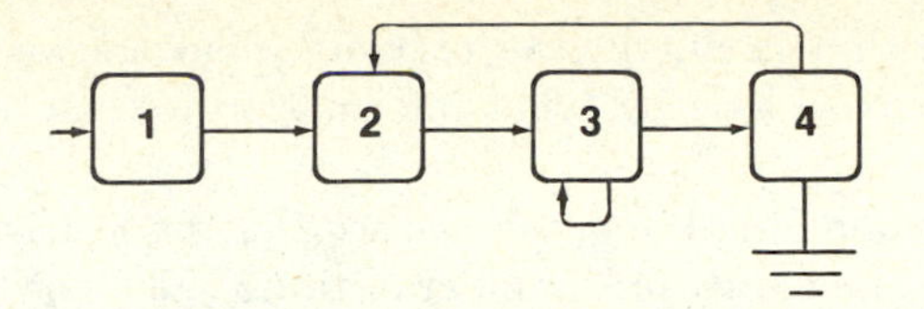

FIGURE 9.4 The flow graph for Algorithm 9.2.

node A to node B if it is possible for the flow of control to go from the last statement of block A directly to the first statement of block B. Figure 9.4 shows the flow graph for the version of the Matrix Addition Algorithm given below. (In contrast to Algorithm 1.7, great care was taken in Algorithm 9.2 to reduce the number of branches in the obvious assembly language implementation of the algorithm.) The source node of the flow graph is indicated by the short incoming arrow on node 1. The sink (node 4) is marked with an electrical ground sign.

Algorithm 9.2 Matrix Addition: Input: Two $n \times m$ matrices A and B with elements a_{ij} and b_{ij} for $1 \leq i \leq n$, $1 \leq j \leq m$. Output: The $n \times m$ matrix C with elements c_{ij} for $1 \leq i \leq n$, $1 \leq j \leq m$ where $C = A + B$.

Step **1.** [Initialize] Set $i \leftarrow 1$.

Step **2.** [Outer loop] Set $j \leftarrow 1$.

Step **3.** [Inner loop] Set $c_{ij} \leftarrow a_{ij} + b_{ij}$. Set $j \leftarrow j + 1$. If $j \leq m$, then repeat this step.

Step **4.** [End of outer loop] Set $i \leftarrow i + 1$. If $i \leq n$, then go to Step 2. Otherwise stop.

Blocks that end in **if** statements and blocks that are the exit block of a loop have out-degree 2. Blocks that end with the selecting part of a **case** statement or that end with a computed **go to** can have a large out-degree. Blocks that end with a simple statement have out-degree 1. (Flow graphs do not show the flow of control between procedures and functions; a separate flow graph is used for each routine.) Blocks that begin loops and most blocks whose first statement has a label have in-degree larger than 1. The block where the algorithm begins execution is marked as the source node. The block or blocks where the program can stop or return to the calling routine are marked as sink nodes. Normally you want each block in a flow graph to contain as many statements as possible so that there will be as few blocks as possible. Starting from any correct flow graph, you can obtain a minimum node flow graph by merging each pair of nodes, A and B, such that there is an edge from A to B, A has out-degree 1, and B has in-degree 1. The flow graph in Figure 9.4 has the minimum number of blocks.

To do a complete analysis of an algorithm, we need to compute how often each block of code is executed. Thus, to analyze Algorithm 9.2, we need to know how often each node in Figure 9.4 is gone through as we follow a path corresponding to an execution of the algorithm. Some information about this can be obtained directly from the flow graph. The rest requires us to examine the algorithm itself. In the previous chapters we studied how to do detailed analyses. Now let's see what information we can obtain directly from the flow graph.

By examining the graph, we see that node 1 is done exactly once. Nodes 2 and 4 are done the same number of times. The flow graph gives us no additional useful information. We must examine the algorithm if we want to know how often nodes 2 and 3 are done. By examining the flow graph, we have already determined how often Step 1 is done, and we know we can obtain the answer for Step 4 from the answer for Step 2. Thus the preliminary analysis saves us from having to study in detail half the steps in this algorithm.

Now consider the 40-node flow graph in Figure 9.5. This is the flow graph for the algorithm of Purdom and Moore [125] that finds immediate predominators in a flow graph. A *predominator* of a node A in a flow graph is any node B such that on any path from the source to node A it is necessary to go through node B. The source is a predominator of every node. Some nodes may have additional predominators. Among the predominators of node A there is a single node that is the last predominator on each path from the source to node A. This node is the *immediate predominator*. Some algorithms used in optimizing compilers need to compute the immediate predominator of a node.

From the flow graph in Figure 9.5 it is again easy to determine some things about the algorithm. For example, nodes 1, 7, 16, and 20 are each done once. Nodes 25, 35, and 40 are each done the same number of times. It is not easy, however, to obtain *all* the useful information from this flow graph unless you have a systematic procedure for analyzing it.

The next two sections develop such a procedure. First we convert the flow graph to a modified *flow multigraph* in such a way that the nodes in the original graph correspond to edges in the modified graph. Next we use one of Kirchhoff's laws (what comes in equals what goes out) to determine relations among the number of times each edge in the modified graph (and thus each node in the original graph) is executed. A spanning tree is used to control the application of Kirchhoff's law, so that we do not get bogged down with redundant information.

In this section we have considered graph theory briefly to prepare for the study of flow graphs and graph algorithms. There is an extensive theory of graphs. Some typical graph problems are: find the shortest path through a graph (see Section 1.6.4), find the largest flow through a graph (where each edge has a flow capacity), and find a matching for all nodes in a graph such that pairs of matched nodes are connected by arcs. These and many other such problems are considered in the excellent book by Lawler [12].

EXERCISES

1. For the graph in Figure 9.1, list all the subgraphs that are trees. How many of the subgraphs contain four nodes? How many of the subgraphs are spanning trees?

2. Draw a flow graph for the Split Algorithm (Algorithm 2.3).

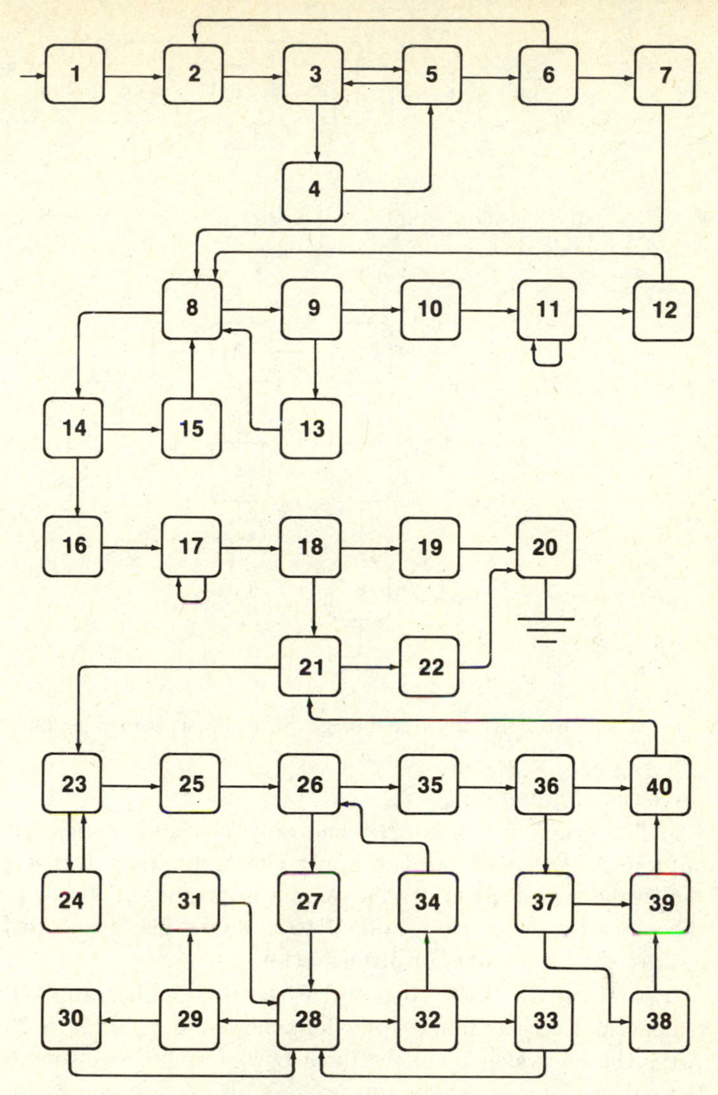

FIGURE 9.5 The flow graph for an immediate predominator algorithm.

9.1.1 Transforming Nodes into Edges

The first major step in analyzing a flow graph is to construct a multigraph in such a way that each edge of the multigraph corresponds to a node in the original graph. We require that each sink node in the original graph have no outgoing edges, and that there be only one sink node. Graphs that do not obey this condition are transformed as follows. If the graph has more than one sink node, or if the graph has a sink node with an outgoing edge, then we add a new sink node. We add an edge from each of the original sink nodes to the new sink node. Then we make all the original sink nodes into nonsink nodes. After analyzing the transformed graph, we can easily transform our answer so that it applies to

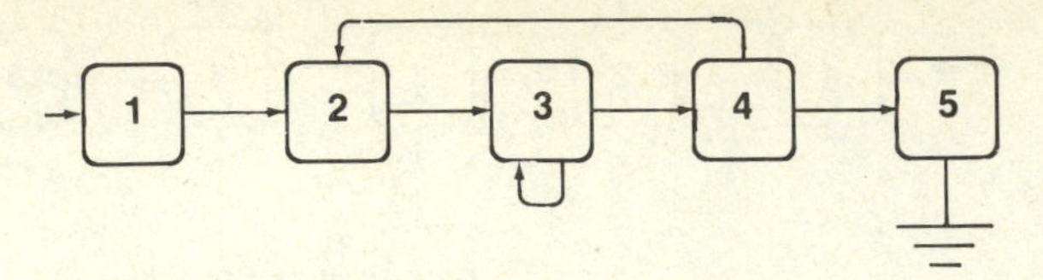

FIGURE 9.6 The flow graph for Algorithm 9.2 after conversion to a graph with a single exit node that has no outgoing edge.

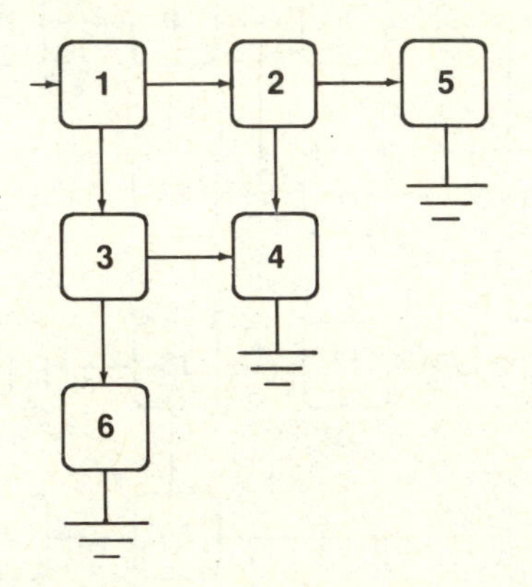

FIGURE 9.7 A flow graph to illustrate equivalence classes.

the original graph.

The graph in Figure 9.4 has only one sink node, but that node has an outgoing edge, so we need to apply the preliminary transformation. Figure 9.6 shows the transformed graph. A new sink node, node 5, has been added, and there is a new edge from node 4 to node 5. The graph in Figure 9.5 does not require the preliminary transformation.

Now we are ready to produce the flow multigraph. The idea behind the transformation is to make equivalent those nodes that have the same predecessor. After the equivalence classes have been formed, equivalent nodes are merged. **Warning:** This is a two-step process. First find the equivalence classes, then merge. Do not merge as you go. The reason that we have to be careful not to merge as we go is that the relation R, where aRb if and only if a and b have the same parent, is not an equivalence relation. For example, in Figure 9.7 nodes 4 and 5 have the same parent, and nodes 4 and 6 have the same parent. But nodes 5 and 6 do not have the same parent. The equivalence relation for this algorithm is the *minimum extension* of R. That is, the algorithm uses a relation R' defined so that $aR'b$ is *true* whenever aRb is, and $aR'b$ is also *true* at the minimum number of additional places necessary to make R' an equivalence relation.

Algorithm 9.3 Edge Flow Graph: Input: A flow graph with node set N, edge set E, a source node, and a sink node with no outgoing edges. Output: A flow multigraph whose edges correspond to the nodes in the original graph.

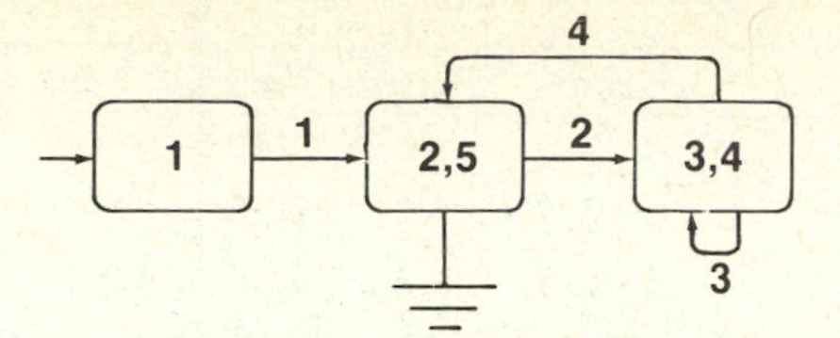

FIGURE 9.8 The edge flow multigraph for the flow graph given in Figure 9.6.

Step **1.** [Initialize] Set each node in N equivalent to itself.

Step **2.** [Loop] For each node n in N do:

Step **3.** [Find equivalences] Let M be the set of nodes that can be reached by following an outgoing edge from n. (Node m is in M if and only if E contains the edge $[n, m]$.) Set all the nodes in M equivalent to each other.

Step **4.** End for.

Step **5.** [Merge] Form a multigraph by having a node for each equivalence class from the original graph. The node for class A is connected to the node for class B by an edge with label c if and only if class A contains node c and class B contains some node d such that the original graph has an edge from node c to node d. Mark as a source node the node in the new multigraph which corresponds to the class that contains the source node of the original graph. Mark as a sink node the node in the new multigraph which corresponds to the class that contains the sink node of the original graph.

An equivalence algorithm like the one that is given in Section 9.2.3 should be used in Step 3 to form the equivalence classes. Remember that at Step 3 we are just building the equivalence classes. We do not change the original graph. The resulting multigraph has exactly one edge for each label.

If we apply the Edge Flow Graph Algorithm to the graph in Figure 9.6, the equivalence classes are (1), $(3, 4)$, and $(2, 5)$. Nodes 3 and 4 are equivalent because node 3 has outgoing edges to them. Nodes 2 and 5 are equivalent because of node 4. The resulting multigraph is shown in Figure 9.8.

Applying the Edge Flow Graph Algorithm to the graph in Figure 9.5 gives the equivalence classes shown in Table 9.1. Figure 9.9 shows the resulting edge flow graph.

The technique of converting flow graphs into edge flow multigraphs was developed by Stevenson and Knuth [132].

EXERCISE

1. Draw an edge flow graph for the Split Algorithm (Algorithm 2.3).

Parent	*Class*	*Parent*	*Class*	*Parent*	*Class*	*Parent*	*Class*
3	4,5	11	11,12	21	22,23	29	30,31
6	2,7	14	15,16	23	24,25	32	33,34
8	9,14	17	17,18	26	27,35	36	37,40
9	10,13	18	19,21	28	29,32	37	38,39

TABLE 9.1. The nontrivial equivalence classes for the flow graph in Figure 9.5. Two nodes are equivalent if there exists a node with arcs to both of them. The equivalence classes with just one node are not given.

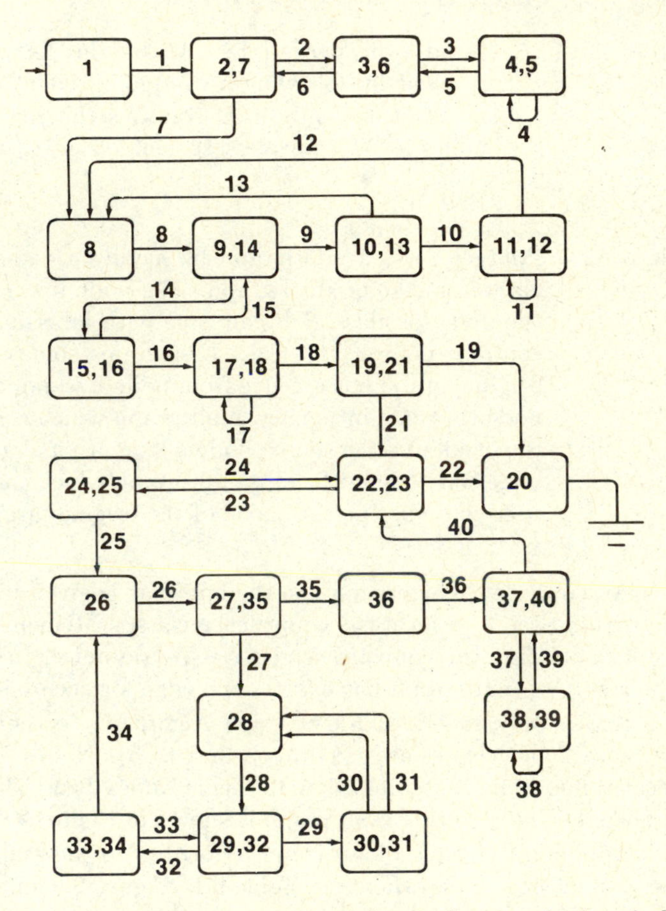

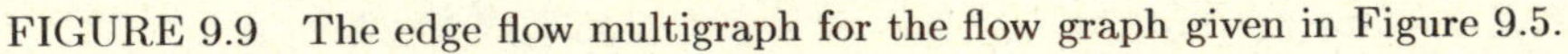
FIGURE 9.9 The edge flow multigraph for the flow graph given in Figure 9.5.

9.1.2 Kirchhoff's Law

Kirchhoff developed two laws for analyzing electrical circuits. One of his laws also applies to analyzing flow graphs. This law can be stated informally as "what goes in equals what comes out". The number of times that the flow of control goes into a node equals the number of times it comes out, except at source and sink nodes. The law even applies at source and sink nodes if we imagine an arc from the sink node to the source node which is traversed one time (we call this arc *edge zero*).

Let E_i, for $i \geq 1$, be the number of times edge i in the edge flow multigraph is traversed; E_0 is 1, the number of times the algorithm is run. Applying Kirchhoff's law directly to node 1 in Figure 9.8 gives $E_1 = E_0$. Applying it to node $(2,5)$ gives $E_1 + E_4 = E_2 + E_0$ and applying it to node $(3,4)$ gives $E_2 + E_3 = E_3 + E_4$. This set of equations is equivalent to

$$E_1 = E_0, \tag{1}$$

$$E_2 = E_4. \tag{2}$$

Also, $E_0 = 1$.

The trouble with direct application of Kirchhoff's law is that the set of equations generated can be quite large and redundant. For small graphs, such as the one in Figure 9.8, this is no problem, but it is a problem for large graphs, such as the one in Figure 9.9.

The key to obtaining nonredundant equations is to notice that any path through a flow graph, from source to sink and back, can be represented by a set of cycles. For example, in Figure 9.8 the path with edges (e_1, e_2, e_4, e_0) can be represented using the cycles (e_0, e_1), (e_4, e_2), (e_3) as "one time through (e_0, e_1) plus one time through (e_4, e_2) plus zero times through (e_3)". A set of cycles is a *fundamental set of cycles* if it is adequate to represent any path in a nonredundant way. For the graph in Figure 9.8, the set of cycles (e_0, e_1), (e_4, e_2), (e_3) is a fundamental set. So is the set (e_0, e_1), (e_4, e_3, e_2), (e_3). [For example, path (e_1, e_2, e_4, e_0) is represented as (e_0, e_1) one time plus (e_4, e_3, e_2) one time plus (e_3) minus one times.] The edges of a cycle do not all have to point the same way. We do, however, need to keep track of the way the edges go. We will therefore assign a direction to each cycle. Those edges that go the same way as the cycle are *positive edges*. The ones that go against the direction of the cycle are *negative edges* and are preceded with a minus sign. Thus, $(e_{22}, -e_{19}, e_{21})$ is a cycle in Figure 9.9.

If you take a spanning tree for a graph (or multigraph) and add one edge to the spanning tree, you form a cycle. The set of all cycles that can be formed in this way from a single spanning tree is a fundamental set of cycles. Every connected graph has a fundamental set of cycles (the set will be the empty set if the graph has no cycles); most connected graphs have many different sets of fundamental cycles.

Figure 9.10 shows a spanning tree for the flow multigraph of Figure 9.8. The heavy lines mark the edges that form the spanning tree. The set of fundamental cycles that corresponds to this tree is (e_0, e_1), (e_4, e_2), (e_3). By convention, we write fundamental cycles with the nonspanning-tree edge written first and with a positive sign.

Figure 9.11 shows the spanning tree for the flow multigraph of Figure 9.9. The set of fundamental cycles that is generated is shown in Table 9.2. There is a one-to-one correspondence between the nonspanning-tree edges and the fundamental cycles. The number of times a cycle is traversed is equal to the number

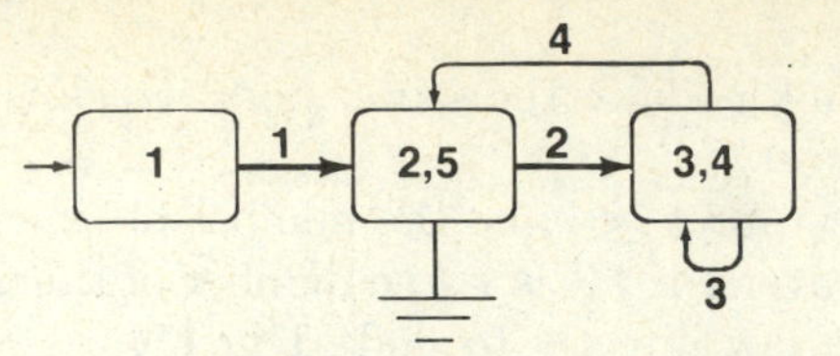

FIGURE 9.10 A spanning tree for the edge flow multigraph of Figure 9.8.

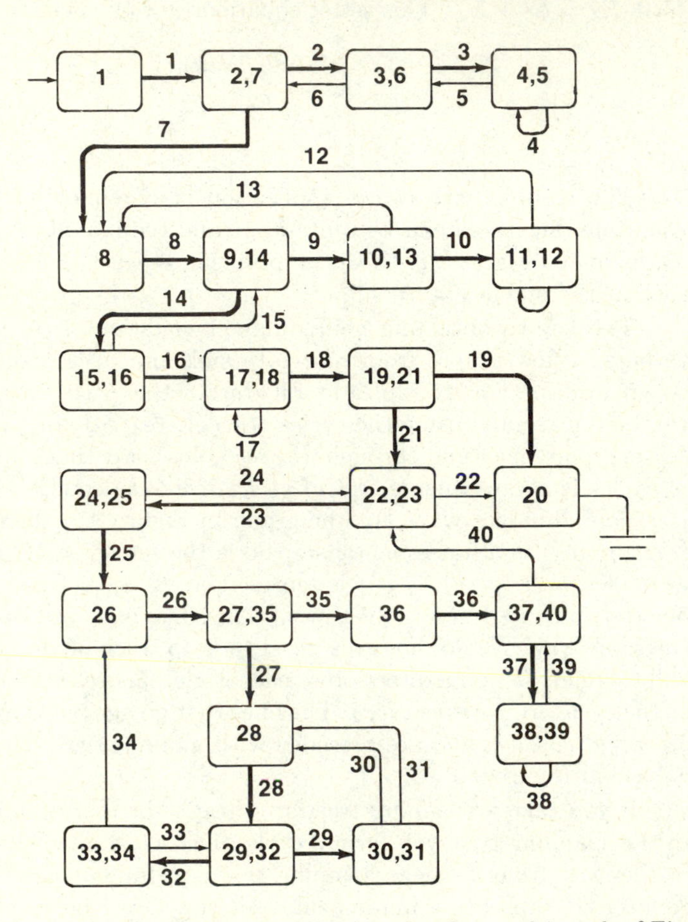

FIGURE 9.11 A spanning tree for the edge flow multigraph of Figure 9.9.

of times the nonspanning-tree edge is traversed. We will use the nonspanning tree edge of a cycle as the *name* of the cycle.

To obtain a complete independent set of equations relating the number of times each edge is traversed, we first divide the edges into two sets. The *dependent set* consists of those edges that are in the spanning tree. The *independent set* consists of those edges that are not in the spanning tree. For each *dependent* edge, e_i, we form an equation by writing "$E_i =$" followed by the sum of some E_j's where each E_j may occur positively or negatively. The variable E_j occurs in the equation for E_i if and only if edge e_i occurs in the cycle with

e_0 :	$e_1 + e_7 + e_8 + e_{14} + e_{16} + e_{18} + e_{19}$
e_4 :	
e_5 :	e_3
e_6 :	e_2
e_{11} :	
e_{12} :	$e_8 + e_9 + e_{10}$
e_{13} :	$e_8 + e_9$
e_{15} :	e_{14}
e_{17} :	
e_{22} :	$-e_{19} + e_{21}$
e_{24} :	e_{23}
e_{30} :	$e_{28} + e_{29}$
e_{31} :	$e_{28} + e_{29}$
e_{33} :	e_{32}
e_{34} :	$e_{26} + e_{27} + e_{28} + e_{32}$
e_{38} :	
e_{39} :	e_{37}
e_{40} :	$e_{21} + e_{23} + e_{25} + e_{26} + e_{35} + e_{36}$

TABLE 9.2. A set of fundamental cycles for the flow multigraph from Figure 9.11. Each cycle is written with the nonspanning tree edge first and positive. It is also separated from the other edges of the cycle with a colon.

name e_j. (Pay careful attention to the indices in this description; it is easy to get things backwards.) The variable E_j will occur positively if e_i occurs in the cycle positively, and E_j will occur negatively if e_i occurs negatively. Thus, from Figure 9.10 we get the equation $E_1 = E_0$ because edge e_1 occurs in the fundamental cycle (e_0, e_1), and it occurs in no other fundamental cycles. We also get $E_2 = E_4$. This is the complete set of independent equations for the flow multigraph given in Figure 9.10. We also know that $E_0 = 1$, so using this we get our final result: $E_1 = 1$ and $E_2 = E_4$. To find values for E_3 and E_4, we must actually analyze the algorithm. There is little more that the flow graph can tell us.

We get many more equations for the flow multigraph in Figure 9.11. There are a total of 22 dependent edges. The 22 equations are given in Table 9.3. We obtain the equation $E_1 = 1$ because edge e_1 occurs only in the cycle for edge e_0, and $E_0 = 1$. We obtain $E_8 = 1 + E_{12} + E_{13}$ because edge e_8 occurs in the cycles for edges e_0, e_{12}, and e_{13}. Equation $E_{19} = 1 - E_{22}$ holds because edge e_{19} occurs in the cycles for edges e_0 and e_{22}. It occurs negatively in the cycle for e_{22}, so E_{22} is preceded by a minus sign.

This method of forming the equations works because each of the independent variables is associated with a single cycle. The value of the independent variable says how often the associated cycle is traversed. Once we know the values of all the independent variables, we know how often each cycle is traversed. We can

$E_1 = 1$	$E_{21} = E_{40}$
$E_2 = E_6$	$E_{23} = E_{24} + E_{40}$
$E_3 = E_5$	$E_{25} = E_{40}$
$E_7 = 1$	$E_{26} = E_{34} + E_{40}$
$E_8 = 1 + E_{12} + E_{13}$	$E_{27} = E_{34}$
$E_9 = E_{12} + E_{13}$	$E_{28} = E_{30} + E_{31} + E_{34}$
$E_{10} = E_{12}$	$E_{29} = E_{30} + E_{31}$
$E_{14} = 1 + E_{15}$	$E_{32} = E_{33} + E_{34}$
$E_{16} = 1$	$E_{35} = E_{40}$
$E_{18} = 1$	$E_{36} = E_{40}$
$E_{19} = 1 - E_{22}$	$E_{37} = E_{39}$
$E_{20} = 1$	

TABLE 9.3. An independent set of equations relating the flows through the edges of the graph in Figure 9.9 (and therefore the nodes in Figure 9.5).

then deduce the values of the dependent variables by counting how often the cycles that contain them are traversed. For the cycles that are traversed in the direction that is the same as the direction associated with the dependent variable, the value must be increased by the number of times the cycle is traversed, while for cycles that are traversed in a direction contrary to the direction associated with the dependent variable, the value must be decreased.

Applying this analysis technique to the flow multigraph of Figure 9.11 gives equations for 23 dependent variables. One still must analyze the algorithm to determine the values for the 17 independent variables. But analyzing the flow graph quickly takes care of 23 variables. Since the general analysis of an algorithm is usually much more complex than analyzing the flow graph, this preliminary analysis results in a big savings.

The use of spanning trees to obtain nonredundant flow equations was developed by Knuth [9, Section 2.3.4.1].

EXERCISES

1. Do a flow analysis of the Backtracking Algorithm (Algorithm 4.1).
2. Do a flow analysis of the Split Algorithm (Algorithm 2.3).
3. Do a flow analysis of the Parse Algorithm (Algorithm 7.8).

9.2 CREDIT AND AMORTIZED COST

Some algorithms, when first considered, appear to take a long time, but when examined more carefully they turn out to be more promising. There may be a reason why the worst case cannot occur very frequently: many good cases *must* occur along with each bad case. For such algorithms we would like to do an analysis that shows that the good cases must outweigh the bad cases. We wish to *amortize* the cost associated with the bad cases over the accompanying good cases. This can be done with *credit analysis.*

In a credit analysis, along with keeping track of the time, we also keep track of a *potential.* We can define the potential as we wish; the *amortized time* equals the actual time plus the potential. For the method to be helpful, we have to find a way to define the potential so that the total amortized time increases slowly with the problem size and so that there is a bound on the total potential. If this is done properly, the cases that take a lot of time will also *reduce* the potential so that the amortized time will be reasonably small. The good cases will take a small amount of time and also build up a small amount of potential. Thus, for the good cases the amortized time will still be small, although not as small as the actual time. If the good cases can build up enough potential to cover the bad cases, then this will show that the bad cases are infrequent and that the long-term average performance of the algorithm is good. The process of building up potential is a way of spreading the cost of the bad cases over the good cases.

This type of analysis is, of course, not useful on every algorithm. It can be made to work only on those algorithms where there is a favorable balance between the occurrence of good and bad cases. What the approach does do is give a good way to demonstrate the balance when it exists.

Credit algorithms often have the nice property that they never go into debt. We will see several such algorithms below.

9.2.1 Pattern Matching on Strings

To introduce the technique of credit analysis, we will consider some algorithms for pattern matching on strings. The naïve algorithm, which we consider first, can take a long time for each character of text that it matches. It usually doesn't. The algorithm has a good average time, but its worst-case time is quite bad. The second algorithm we consider can also take a long time on any one character of text that it is matching. Its worst-case time for matching a single character is just as bad as the worst-case time of the naïve algorithm. A credit analysis, however, shows that whenever the algorithm uses a long time on one or more characters, it must also process some other characters quickly. Therefore it can process any long string of characters quickly, even in the worst case.

The task of finding a match for a relatively short string of characters (the *pattern*) in a long string (the *text*) is basic to text editing. The obvious straightforward algorithm for this problem is:

Algorithm 9.4 Simple Pattern Match: Input: A text string T which is an array of characters t_i for $1 \leq i \leq n$, and a pattern P of characters p_i for $1 \leq i \leq m$. Output: A variable k whose value is the index of the last character of the leftmost substring of T that matches P if such an index exists, and otherwise $n+1$. [That is, k is the smallest number such that $T(k+i-m) = P(i)$ for $1 \leq i \leq m$ when there is such a k.]

Step **1.** [Initialize] Set $k \leftarrow 1$ and $j \leftarrow 1$.

Step **2.** [Loop] While $k \leq n$ and $j \leq m$ do:

Step **3.** [Test] If $p_j \neq t_k$, then set $k \leftarrow k - j + 2$ and $j \leftarrow 1$. Otherwise set $j \leftarrow j + 1$ and $k \leftarrow k + 1$.

Step **4.** End while.

Step **5.** [Done] If $j > m$, then P matches starting at position $k - m$; otherwise, P does not match.

Although this algorithm is straightforward, it has a poor worst-case performance. Let a^n denote a string of n a's. Suppose the text is $a^n b$ and the pattern is $a^m b$. The algorithm will laboriously match the first m a's and fail to match the b for every value of k from 1 to $n - m$; it finally matches on the $(n - m + 1)^{\text{st}}$ try. Since $m + 1$ comparisons are done each time, the total running time is $\Theta(nm)$. Of course, this worst case is unlikely to occur very often in editing normal text, but it does illustrate the weak point of the straightforward algorithm.

The key to improving the worst-case running time of the algorithm is to avoid repeatedly matching the pattern against the same text characters. This is possible because, once the text has been matched against a portion of the pattern, we know what that text looks like—it is exactly like the portion of the pattern that was matched. Therefore, analysis of the *pattern*, done before the matching against the text is begun, will permit us to avoid redundant comparisons.

Consider some examples. If every character in the pattern is different, and five characters of the pattern have been matched before we fail on the sixth, then there can be no match starting at any of the previously matched positions in the text, since none of the characters at those positions equals the first character of the pattern. In this case we can increase k by 1, reset j to 1, and continue the algorithm. On the other hand, suppose $T = a^{10}b$ and $P = a^5 b$. At the sixth comparison a mismatch occurs, with $k = 6$ and $j = 6$. But since the first five characters of T are now known by the algorithm to be a's, we can keep $k = 6$, set $j \leftarrow 5$, and continue. Each remaining a in the text will be compared twice (once against b and once against a), and so the running time will be linear.

The following observation leads to an algorithm that is faster for some cases. If the pattern is repetitive in the sense that there is a part in the middle that looks like the beginning, and if we happen to fail just after matching that part, we may have a match for the whole string starting where the repetitive part of the pattern matched. In all other cases we can just reset j to the beginning of the pattern and resume testing at k.

Now let's consider how to analyze the pattern to find the repetitive parts. This is easiest to understand in two steps. We first define an array F that tells us, for each position j in the pattern, how long a string of characters, ending at position j, is the same as some initial sequence of the pattern. That is, f_j is the largest value of i such that $p_1 = p_{j-i+1}$, $p_2 = p_{j-i+2}$, $\ldots$, $p_{i-1} = p_{j-1}$.

Notice that for $i = 1$, no characters are matched, so $f_j \geq 1$ for $j > 1$. We define $f_1 = 0$. As an example, consider the following string:

p_j	I	'	m		a	n	g	r	y		a	n	d		I	'	m		a	n	n	o	y	e	d
j	1	2	3	4	5	6	7	8	9	10	11	12	13	14	15	16	17	18	19	20	21	22	23	24	25
f_j	0	1	1	1	1	1	1	1	1	1	1	1	1	1	1	2	3	4	5	6	7	1	1	1	1

We could base an algorithm on f_j. If a mismatch occurred on character 19, for example, we could resume matching with the text pointer at the same place and the pattern pointer at character 5. Another refinement, however, will allow us to do better. Since character 19 and character 5 are both a's, we know that a match that fails at character 19 cannot succeed at 5; we can, in fact, start over at the beginning of the pattern, knowing that no intermediate match is possible.

To use this observation, we define an array, *Next*, where $next_j$ is equal to the largest i such that $i < j$, $p_1 = p_{j-i+1}$, $p_2 = p_{j-i+2}$, $\ldots$, $p_{i-1} = p_{j-1}$, and $p_i \neq p_j$. If there is no such i, then $next_j = 0$. Notice that $next_j$ is always less than j. In our example, we have

p_j	I	'	m		a	n	g	r	y		a	n	d		I	'	m		a	n	n	o	y	e	d
j	1	2	3	4	5	6	7	8	9	10	11	12	13	14	15	16	17	18	19	20	21	22	23	24	25
$next_j$	0	1	1	1	1	1	1	1	1	1	1	1	1	1	0	1	1	1	1	1	7	1	1	1	1

The zero at position 15 means that, if a mismatch occurs there, we can start again at the beginning of the pattern and at the next character in the text, since the current text character cannot match the "I" at the beginning of the pattern. The following algorithm uses the *Next* array to do pattern matching quickly.

Algorithm 9.5 Fast Pattern Match: Input: A text string T which is a an array of characters t_i for $1 \leq i \leq n$ and a pattern P of characters p_i for $1 \leq i \leq m$. Output: A variable k whose value is the index of the last character of the leftmost substring of T that matches P if such an index exists, and otherwise $n + 1$. [That is, k is the smallest number such that $T(k + i - m) = P(i)$ for $1 \leq i \leq m$ when there is such a k.]

Step **1.** [Initialize] Set $k \leftarrow 1$ and $j \leftarrow 1$.

Step **2.** [Loop] While $k \leq n$ and $j \leq m$ do:

Step **3.** [Test] If $j > 0$ and $p_j \neq t_k$, then set $j \leftarrow next_j$ and repeat this step.

Step **4.** [Move pointers] Set $j \leftarrow j + 1$ and $k \leftarrow k + 1$.

Step **5.** End while.

Step **6.** [Done] If $j > m$, then P matches starting at position $k - m$; otherwise, P does not match.

To be complete, we must also have an algorithm to compute the *Next* array. Notice that the following algorithm is very similar to the previous algorithm for doing the pattern match.

Algorithm 9.6 Pattern Preprocessing: Input: A pattern P of characters p_i for $1 \leq i \leq m$. Output: The array *Next*, where $next_j$ equals the i such that $i < j$, $p_1 = p_{j-i+1}, \ldots, p_{i-1} = p_{j-1}$, and $p_i \neq p_j$. If there is no such i, then $next_j = 0$. This is the value for *Next* needed by the Fast Pattern Match Algorithm.

Step **1.** [Initialize] Set $k \leftarrow 1$, $j \leftarrow 0$, and $next_1 \leftarrow 0$.

Step **2.** [Loop] While $k \leq m$ do:

Step **3.** [Test] If $j > 0$ and $p_j \neq p_k$, then set $j \leftarrow next_j$ and repeat this step.

Step **4.** [Move pointers] Set $j \leftarrow j + 1$ and $k \leftarrow k + 1$.

Step **5.** [Set *Next*] If $p_k = p_j$, then set $next_k \leftarrow next_j$; otherwise, set $next_k \leftarrow j$

Step **6.** End while.

You should execute these algorithms on some sample data.

Now let's consider the running time of the Fast Pattern Match Algorithm. The algorithm has two nested loops. The outer loop is Steps 2 through 6; the inner loop is Step 3. Most of our previous algorithms with two nested loops have run in quadratic time. It is easy to see that the Fast Pattern Match Algorithm runs in time $O(mn)$. The outer loop is done once for each character in the text. In Step 3, each time j is set to $next_j$ its value is decreased by at least 1. So the inner loop is done at most m times for each iteration of the outer loop. This gives a total running time of $O(mn)$.

The actual worst-case running time, however, is much better than this rough upper-bound analysis would indicate. The time is only $\Theta(n)$ [or $\Theta(n+m)$ if the preprocessing time is counted, but $\Theta(m+n) = \Theta(n)$ for $m \leq n$]. The key to proving this is to show that the part of Step 3 after the test is executed at most n times altogether. Before reading further, examine the algorithm and see if you can discover why.

The key observation is that variables k and j are coupled in Steps 1 and 4. They are initialized to 1 in Step 1, and j is never incremented unless k is incremented also. When j is set to $next_j$ in Step 3, its value is decreased, but it can never become negative, so it cannot be decreased more than it has been increased. The total amount of increase is at most n, the same as the maximum increase in k. Therefore, the complete inner loop is executed at most n times.

Now let's consider the Fast Pattern Match Algorithm from the credit perspective. We will choose our time unit to equal the time required by the nontest part of the loop in Step 3, and we define the potential to equal the value of the variable j in the algorithm. With these choices for the potential function and the time unit, the loop in Step 3 takes zero or less units of amortized time since the amortized time is equal to the actual time (one unit per time around the loop) plus the increase in the potential (minus 1 or less each time around the

Name	*Credit*
Ann	10
Ben	50
Jim	1
Sue	15
Tom	0

TABLE 9.4. The credit customers of the Corner Ice Cream Shop.

loop). Steps 1 and 6 use some bounded amount of amortized time. The amount of amortized time used by Step 2, by the nonloop part of Step 3, and by Steps 4 and 5 is bounded by a constant times the change in the variable k. The total change in k is n. Thus the total amortized time for the algorithm is a linear function of n. The potential for this algorithm is always between zero and m; it never becomes negative. Thus the total time for the algorithm is also a linear function of n.

It is possible to obtain a more precise result with a credit analysis by including more detail in the analysis.

The Fast Pattern Match Algorithm was discovered and analyzed by Knuth et al. [108].

9.2.2 Move to Front

In this section we will use an amortized cost analysis to show that the Move to Front Algorithm is a very good algorithm for maintaining data on a linear list. We should point out, though, that when you have a large amount of data you normally do not use a linear list, because the time to access records is too large. You use trees or hash tables. The main advantage of a linear list is that it is simple.

Table 9.4 shows a list of the five credit customers for the Corner Ice Cream Shop. We will assume that the cost (computer time) to access a record has the form $t_0 + t_1 i$, where i is the position of the record on the list. We want to keep the list ordered in a way that will result in a small total access time.

The Move to Front Algorithm always moves the record it accesses to the front of the list, without changing the relative positions of the remaining items. The cost of this move is small or comparable to the cost of accessing the record, depending on what method is used to implement the list. We will assume the time required to rearrange the list is included in the formula for access time.

The move to front strategy is obviously good when the requests for each record tend to come in bunches. Even if the requests do not come in bunches, the Move to Front Algorithm will keep frequently accessed records near the front of the list. Such considerations suggest that the Move to Front Algorithm may be a good general-purpose algorithm.

If we try to prove that the Move to Front Algorithm is better than other algorithms, we run into two problems. First, for a short sequence of accesses of items near the end of the list, Move to Front will be slow. Almost any algorithm with the good luck of starting with a list that has the accessed items near the front will be faster. We can avoid this problem if we require that both algorithms start with the same list. The second problem is that there are some unusual (and unrealistic) algorithms that do an extremely good job of maintaining a linear list. The flavor of these algorithms is: "Find the item being accessed, then (at no cost) move the item that will be accessed next to the front of the list". Such an algorithm has two problems: the cost measure is unrealistic, and the algorithm knows what requests it will get before it gets them. In doing our mathematical comparison of Move to Front with other possible algorithms, we must be careful to formulate the problem in a way that excludes such unrealistic algorithms—otherwise the comparison will not be very useful.

The set of algorithms that we will compare against Move to Front is the set of algorithms of the following form:

1. Look up the record.

2. Move the record toward the front of the list zero or more positions (leaving the relative order of the remaining items unchanged).

We call this set of algorithms the *move forward* algorithms. We will assume that the cost of each algorithm is given by $t_0+t_1 i$, where t_0 and t_1 are constants (i.e., the same for every algorithm) and i is the position of the item being looked up. This set of algorithms contains many of the reasonable algorithms for maintaining a linear list. It also contains many unreasonable algorithms. These unreasonable algorithms are not going to be a problem, however, because none of them run fast. Once you have a good result, you know that the set of algorithms in your comparison set does not contain any high-performance unreasonable algorithms.

The assumption that all the algorithms have the same t_0 and t_1 is a little unrealistic. For the most part the assumption favors the algorithms that are more complex than Move to Front. The assumption about the formula for time does give the Move to Front Algorithm an advantage over the algorithm that does not change the list at all. We can expect that the value of t_1 for that algorithm is about one-half the value appropriate for Move to Front. These refinements can be allowed for in the analysis, but we choose to keep the analysis simple.

Let A be an arbitrary algorithm from the set of move forward algorithms, and let M be the Move to Front Algorithm. For the next several pages we will compare the cost (running time) for algorithm A with that for algorithm M. The time for A to process a sequence S of n requests is given by

$$t_0 + t_1 f(A, S) \tag{3}$$

for some function f that depends on the algorithm A and the sequence S. One result of our analysis will be that

$$f(M, S) \leq 2f(A, S) + \binom{m}{2} - n \qquad \text{for all } S \text{ and } A, \tag{4}$$

Algorithm	A		M		
	Order(A)	Name	Order(A)	Name	Inversions
List	1	Jim	5	Tom	4
	2	Sue	3	Ben	2
	3	Ben	4	Ann	2
	4	Ann	1	Jim	0
	5	Tom	2	Sue	0
Total					8

TABLE 9.5. Possible lists for algorithms A and M along with the number of inversions of one list with respect to the other. The order(A) column gives the position of the item on the list for algorithm A. The inversions column gives, for each item on the list for M, the number of items which follow it on the list for M but precede it on the list for A. The total of this column is the number of inversions.

where m is the number of items on the list. Thus for *any* long sequence of requests M does not take more than about twice the time required by A. (It may be much better, particularly on some sequences of requests.) For most individual sequences of requests it is possible to find some specialized algorithm that is better than M. Thus, although Move to Front is not the very best algorithm for most individual sequences, it is almost the best for every long sequence of requests.

To continue the analysis, we need to develop the appropriate potential function. Since the differences in the performances of the various move forward algorithms depend on the differences in the order of the lists they maintain, the potential function should depend on the differences in the order of the list maintained by A and the list maintained by M. If an item is moved a long way on one list and not very far on the other list, the change in the potential function should probably be large. It turns out that the number of *inversions* of one list compared to the other is an appropriate potential function. Let i be the position of an item on the list of algorithm A and let k_i be the position of the same item on the list for algorithm M. Define the inversion function to be

$$I_{ij} = \begin{cases} 0 & \text{if } k_i \le k_j, \\ 1 & \text{if } k_i > k_j. \end{cases} \tag{5}$$

The *number of inversions* (of one list with respect to the other) is defined to be

$$\sum_i \sum_{j>i} I_{ij}. \tag{6}$$

That is, the number of inversions is the number of pairs of elements that have one relative order on one list and the reverse relative order on the other list. Table 9.5 shows a possible list for each algorithm. The inversions column gives $\sum_{j>i} I_{ij}$ for each i and the total number of inversions.

Now let's compare the amortized times for accessing the same item by each algorithm. Let i be the position of the item on A's list, and let k be the position of the same item on M's list. Let x be the number of items that come after the

Algorithm	A		M		
	Order(A)	Name	Order(A)	Name	Inversions
List	1	Jim	4	Ann	3
	2	Sue	5	Tom	3
	3	Ben	3	Ben	2
	4	Ann	1	Jim	0
	5	Tom	2	Sue	0
Total					8

TABLE 9.6. The lists for algorithms A and M after algorithm M accesses *Ann* and moves the record to the front of the list.

accessed item on A's list but before the accessed item on M's list. For example, if the accessed item is Ann, then $i = 4$, $k = 3$, and $x = 1$ (Tom). Now let's consider the change in the number of inversions when the accessed item is moved only by algorithm M. Table 9.6 shows the list that results in the example. The move to front creates $k - x - 1$ inversions and destroys x inversions for a total *increase* of $k - 2x - 1$. In the example, there is one new inversion created (Ann moves in front of Ben) and one destroyed (Ann moves in front of Tom) so the net change is zero.

Suppose the accessed item is also moved forward j positions on A's list. Then the number of inversions is *decreased* by j. One way to see this is to perform the following three-step process. First, assign the accessed item position number $i - \frac{1}{2}$. (So the position of Ann in Table 6 becomes $3\frac{1}{2}$.) This does not change the number of inversions. The accessed item's number in its new position is $i - j - \frac{1}{2}$. (For Ann, $2\frac{1}{2}$.) Now all the items have the correct relative position numbers; it is just that one noninteger is used and one integer is omitted. Second, count the number of inversions. This can be done using the funny position numbers, because they have the correct relative order. Third, renumber the records so that they have integer position numbers, without changing the relative positions. This does not change the number of inversions, because the relative positions did not change. Using this method of counting inversions, it is clear that the number of inversions decreases by 1 for each position that algorithm A moves the accessed item forward. Since the move of the record for Ann decreased its position number by 1 (from $3\frac{1}{2}$ to $2\frac{1}{2}$), the entry in the first row of the inversion column in the table is reduced by 1.

Now it is time to compute the amortized time of the operation. For Algorithm A we let the amortized time just equal the cost of accessing the ith item, which we take to be i [later we will adjust for the constants t_0 and t_1 in eq. (3)]. For algorithm M we let the amortized time equal the cost of accessing the item (k) plus the increase in the potential. The increase in the potential is no more than $k - 2x - 1$, the upper limit on the change in the number of inversions. Defining C_M to be the amortized time for algorithm M, we have

$$C_M \leq 2(k - x) - 1. \tag{7}$$

Algorithm	A		M		
	Order(A)	Name	Order(A)	Name	Inversions
List	1	Jim	$2\frac{1}{2}$	Ann	2
	2	Sue	5	Tom	3
	$2\frac{1}{2}$	Ann	3	Ben	2
	3	Ben	1	Jim	0
	5	Tom	2	Sue	0
Total					8

TABLE 9.7. The lists for algorithms A and M after algorithm M accesses *Ann* and moves the record to the front of the list and algorithm A accesses it and moves it forward one position. A special way of numbering positions is used. See the text for details.

Notice that $k-x \le i$ because $i-1$ is the number of items preceding the selected item on A's list, while $k-x-1$ is the number of items that precede it on both lists. Using C_A for the cost of algorithm A, we have $C_A = i$, so

$$C_M = 2(k-x) - 1 \le 2i - 1 = 2C_A - 1 \tag{8}$$

for any single access.

Summing eq. (8) for n accesses gives

$$C'_M \le 2C'_A - n, \tag{9}$$

where C'_M is the total amortized time for algorithm M, and C'_A is the total amortized time for algorithm A. Now $f(A,S) = C'_A$ and $f(M,S)$ equals C'_M minus the change in potential produced by sequence S. But the potential is always between zero and $\binom{m}{2}$, so

$$f(M,S) \le 2f(A,S) + \binom{m}{2} - n. \tag{10}$$

If we require that the two algorithms start out with the same list, then the initial potential is zero, so the change in potential is nonnegative. In this case we can drop the $\binom{m}{2}$ term in eq. (10).

To compare the running time of the two algorithms, we should add t_0/t_1 to each side of eq. (10) and use $t_0/t_1 + f(A,S) = t_A/t_1$, to obtain

$$t_M \le 2t_A + \binom{m}{2} - n\left(1 + \frac{t_0}{t_1}\right). \tag{11}$$

Thus the time for algorithm M is no more than twice the time for algorithm A if the sequence of requests is long enough $[n \ge \binom{m}{2}/(1 + t_0/t_1)]$. Also, the time for algorithm M is no more than twice the time for algorithm A if the two algorithms start with the same lists (regardless of how long the sequence of requests is).

There are two places in the above derivation where ideas were introduced without much justification. The first was where the class of algorithms being considered was restricted to move forward algorithms. The second was where the number of inversions was chosen as the potential function. These represent places where there was a strong interaction between the development of the result and the assumptions going into the proof of the result. With simple proofs you usually decide what you want to prove, decide what additional assumptions (if any) you need to make, and then prove your result. With more complex problems these steps are usually mixed together in obtaining the first proofs of the results.

The following steps indicate one way you might go about using amortized cost to prove that an algorithm is good.

1. Select an interesting problem set where there is a sequence of data items to process. (Of course, you can also use amortized cost to analyze a single algorithm, as we did with the Fast Pattern Match Algorithm.)

2. Develop some good algorithms for the problem set.

3. Select a reasonable algorithm that appears to be good for every large problem in the problem set. If there is no good algorithm, go back to Step 2 (or give up). If the selected algorithm is quick for each individual data item in every sequence, forget about amortized cost; just prove that the algorithm is good. Assume from here on that an amortized time analysis is called for; i.e., the algorithm is good for long sequences, but not for each individual data item.

4. Consider the set of all algorithms for the problem set. See if there are any algorithms that are much more efficient than the one you are analyzing. If there are some and they are reasonable, analyze them instead. If there are some, but they are unreasonable, restrict the set of algorithms that you are considering.

5. Consider how some of the algorithms are more efficient than the selected algorithm for some items in some sequences, figure out how this happens, and try to devise a potential function so that the amortized time for the selected algorithm is good in every situation. If you succeed, go to Step 6. If you do not succeed, you may want to further restrict the class of algorithms you are considering. If you do this, however, you may exclude the algorithms that everyone else thinks are most interesting, so great care is needed if you decide to restrict the class of algorithms being considered.

6. Construct a proof that the selected algorithm is good. If trouble arises, go back to Steps 4 or 5 to see if you can overcome the difficulty.

This is the basic approach that led to restricting the previous analysis to move forward algorithms and to the use of the number of inversions as the potential function.

This section is based on a paper by of Sleator and Tarjan [130].

EXERCISES

1. Prove that for any two lists A and B the number of inversions of A with respect to B ($\sum_i \sum_{j>i} I_{ij}$) is equal to the number of inversions of B with respect to A ($\sum_i \sum_{j>i} I_{ji}$).
2. Extend the analysis of the Move to Front Algorithm by allowing for insertions in and deletions from the list.
3. Show that the Move to Front Algorithm is still good when compared to any algorithm that looks up an item on a list and moves that item to any place on the list (while maintaining the relative order of the remaining items). Pay careful attention to the assumptions that you must make to obtain your result. To obtain an interesting result, you will need to make assumptions that are similar to the ones made in the text, but some additional assumptions will also be needed.
4. The inorder tree traversal algorithm is:

   Algorithm 9.7 Inorder(r)**:** Input: The root of a binary tree, r. Result: All the nodes of the tree are visited in inorder.

   Step **1.** If $llink(r) \neq empty$, then call Inorder$(llink(r))$.

   Step **2.** Visit r.

   Step **3.** If $rlink(r) \neq empty$, then call Inorder$(rlink(r))$.

   Show that in the worst case the number of steps between one visit and the next is linear in the tree size. Then give a credit analysis showing that the time to visit all the nodes is linear in the tree size.
5. The Addition Algorithm (Algorithm 1.4) takes time $O(\log_b n)$ to add 1 to the number $n = b^k - 1$. Show that the time to start from zero and get to any positive number n by successively adding ones is $O(n)$.

9.2.3 Set Union Algorithms

Directed graphs are a good data structure for representing relations (see Section 1.11). If the relation $\rightarrow$ is over a set S, then there is a node in the corresponding graph for each element in the set. There is an edge in the graph from node x to node y if and only if $x \rightarrow y$. An undirected graph can be used to represent a symmetric relation.

One example of this use of graphs is in algorithms for computing the symmetric closure of a relation. Such algorithms construct the equivalence classes associated with the symmetric closure of the original relation and answer questions about the equivalence of two elements. In Section 9.1.2 we saw an algorithm that uses equivalence that is related to the automation of analysis of algorithms.

Algorithms of this type need a data structure for equivalences and routines to support two operations. The first operation, called Is(x, y), returns *true* if the pair (x, y) is in the relation that has been constructed so far, and returns *false* otherwise. The second operation, Set(x, y), forms the set union of the sets which contain x and y. Initially each element is in a set by itself. The data structure that is usually used for this problem is a directed forest, where each nonroot node has a pointer to its parent, and where the parent of a node is some

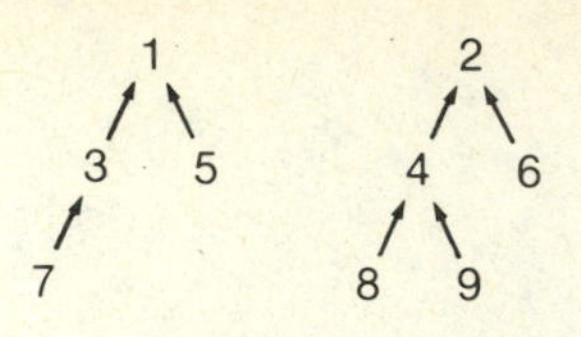

FIGURE 9.12 A representation of the sets $\{1,3,5,7\}$ and $\{2,4,6,8,9\}$ using directed trees.

node in the same set. The nodes that are in the same set are in the same tree. The element at the root of the tree serves as a representative of the set.

Figure 9.12 illustrates one forest that could be used to represent the sets $\{1,3,5,7\}$ and $\{2,4,6,8,9\}$.

With this data structure the operations Is and Set can be implemented as follows, using the auxiliary routines Find and Union:

Algorithm 9.8 Is(x, y): Input: Elements x and y. Output: *true* if x and y are in the same set and *false* otherwise.

Step **1.** Set $t \leftarrow \text{Find}(x)$ and $u \leftarrow \text{Find}(y)$.

Step **2.** If $t = u$, then return *true*, otherwise return *false.*

Algorithm 9.9 Set(x, y): Input: Elements x and y. Output: This routine modifies the data structure for the sets so that the sets for x and y are combined.

Step **1.** Set $t \leftarrow \text{Find}(x)$ and $u \leftarrow \text{Find}(y)$.

Step **2.** If $t \neq u$, then call $\text{Union}(t, u)$.

Algorithm 9.10 Find(x): Input: An element x. Output: t, the element at the root of the tree that x is in.

Step **1.** Set $t \leftarrow x$.

Step **2.** If *parent*(t) is not *empty,* then set $t \leftarrow$ *parent*(t) and repeat this step.

Step **3.** Return t.

Algorithm 9.11 Union(t, u): Input: The roots of two set trees. Output: The data structure is modified so that u points to t.

Step **1.** Set *parent*$(u) \leftarrow t$.

The Find Algorithm will run faster when the trees are not very tall. The Union Algorithm, however, sometimes builds very tall trees. This combination of algorithms can be made to run faster by modifying the Union Algorithm so that it keeps track of how tall each tree is and always connects the short tree to the tall tree. The rank of each tree is the length of the longest path from a leaf node to the root node. (For some applications the rank is actually an upper limit on the length of the longest path.) The modified Union Algorithm is:

Algorithm 9.12 Union(t, u) with Ranks: Input: The roots of two set trees. Output: The data structure is modified so that the root of the shorter tree points to the taller tree.

Step **1.** If the relation between $rank(t)$ and $rank(u)$ is:

Step **2.** $rank(t) < rank(u)$, then interchange t and u (i.e., set $a \leftarrow t$, $t \leftarrow u$, and $u \leftarrow a$, where a is a temporary variable). Go to Step 4. (Tree u was taller, so interchange the roles of the two trees.)

Step **3.** $rank(t) = rank(u)$, then set $rank(t) \leftarrow rank(t) + 1$. (The two trees have the same height, so the new tree will be one unit taller.)

Step **4.** Set $parent(u) \leftarrow t$. (Join tree u onto t.)

In this algorithm either Step 2, Step 3, or neither is executed, depending on the result of comparing the ranks of the two trees. Then Step 4 is done.

A slight variation on Union with Ranks is Union with Weights, where the number of nodes rather than the height is used to control the combining of trees.

Algorithm 9.13 Union(t, u) with Weights: Input: The roots of two set trees. Output: The data structure is modified so that the root of the smaller tree points to the larger tree.

Step **1.** If $weight(t) < weight(u)$, then interchange t and u, i.e., set $a \leftarrow t$, $t \leftarrow u$, and $u \leftarrow a$, where a is a temporary variable.

Step **2.** Set $parent(u) \leftarrow t$ and $weight(t) \leftarrow weight(t) + weight(u)$.

The naïve algorithm, which uses the unmodified Union and Find Algorithms, has a bad worst-case running time because the Union Algorithm can build tall trees and because once a tall tree is built it remains tall. The worst-case time is $O(mn)$, where m is the number of elements in the set and n is the number of calls to to Union and Find.

When the Union with Ranks Algorithm is used with the Find Algorithm, the worst-case running time is $O(n \ln m)$, because the Union part of the algorithm prevents tall trees from being built. This a much better performance than that of the naïve algorithm. One problem still remains, however. Once a moderately tall tree is built, it stays tall. A tree never loses any height. The Find with

Collapsing Algorithm overcomes this problem. Every time a Find operation is done, all the nodes on the path to the root (except the root) are connected directly to the root. The more Finds are done, the shorter the trees become. If enough different Finds are done, every node is attached directly to the root, and any future Finds can be done very quickly.

Algorithm 9.14 Find with Collapsing(x): Input: An element x. Output: t, the element at the root of the tree that x is in. Side effect: Every element on the path from x to t (except t) is attached directly to t.

Step **1.** Set $t \leftarrow x$.

Step **2.** If $parent(t)$ is not *empty*, then set $t \leftarrow parent(t)$ and repeat this step.

Step **3.** Set $y \leftarrow x$.

Step **4.** If $y \neq t$, then set $z \leftarrow y$, $y \leftarrow parent(x)$, $parent(x) \leftarrow t$, $x \leftarrow z$, and repeat this step.

Step **5.** Return t.

The algorithm that uses Union with Ranks and Find with Collapsing is quite fast, so fast that some people thought that they had proved that it ran in worst-case time $O(n)$. The algorithm is not quite that fast, but it comes very close.

Simple algorithms for the set union problem are given in Knuth [9, Section 2.3.3], Aho et al. [1, Section 4.6], Baase [2, Section 6.5], Reingold and Hansen [61, Section 5.1], Sedgewick [62, Chapter 30], Standish [63, Section 3.7.2.1], and Horowitz and Sahni [60, Section 4.6]. The Union with Ranks Algorithm was developed by Tarjan [15, Section 2.1].

EXERCISES

1. What is the worst-case time of the Union, Find pair of algorithms? That is, if you start with n elements, each in a class by itself, and if you do some combination of n Unions and Finds, how much time can be used up? For your worst-case analysis assume that Unions can be done without a preceding Find on the roots of the trees. Hint: Each Union takes constant time, so you cannot use up much time doing Unions. The time for a Find is proportional to the length of the path in the tree that is traversed. To use up a lot of time, one needs to build up a long path and then do a lot of Finds on it. Therefore, show that after k Unions the longest possible path is of length k and it is possible to have a path of length k. Assume that the time for a Union is t_1 and that the time for a Find is $t_2 + lt_3$, where l is the length of the path. If you do k Unions to build a path of length k followed by $n-k$ Finds on that path, then the total time will be $T = kt_1 + (n-k)(t_2 + kt_3)$. Set k (as a function of n) to maximize T. Show that no other strategy gives as large a time.

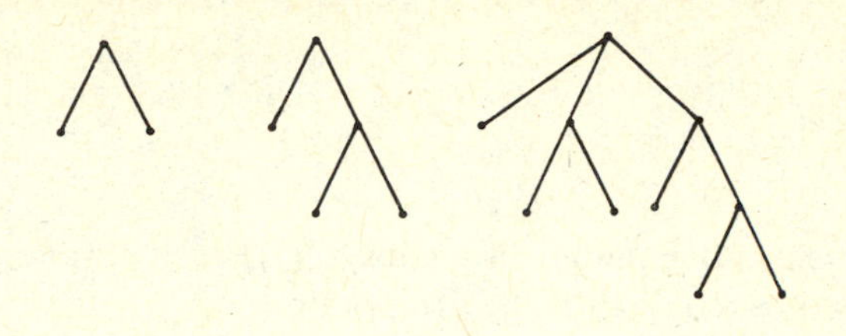

FIGURE 9.13 The tallest trees that can be built with zero, one, three, and seven Unions.

9.2.3.1 Set Union with Ranks

Let's consider the worst-case time used by Union with Ranks combined with (noncollapsing) Find. The basic idea for obtaining a large time is to first build a tall tree and then do some Finds on it. With one Union, you can build a tree of height 1 (See Figure 9.13). With two Unions, you cannot build a tree of height 2, because the light tree must be attached to the root of the heavier tree. You can, however, build two trees of height 1. With three Unions, you can use the first two Unions to build two trees of height 1 and then use the last Union to build a tree of height 2 (See Figure 9.13).

With n Unions, you can build a tree with $n+1$ nodes. Let H_n be the height of the tallest tree that you can build with n calls to Union with Ranks. The H_n obey the recurrence

$$H_n = 1 + \max_{0 \le i \le \lfloor n/2 \rfloor} \{H_i\} \tag{12}$$

with the boundary condition $H_0 = 0$. (A tree formed by a Union is at most one taller than the height of the taller of the two trees that are combined.) Since doing Unions does not decrease the height of trees, H_i is a nondecreasing function of i. Thus eq. (12) can be simplified to

$$H_n = 1 + H_{\lfloor n/2 \rfloor}. \tag{13}$$

The solution is

$$H_n = \lfloor \lg(n+1) \rfloor. \tag{14}$$

If we use i of n operations for Unions and $n-i$ for Finds, the time spent (assuming one unit of time for Union and 1 plus the path length for Finds) is no more than

$$(\lfloor \lg(i+1) \rfloor + 1)(n-i) + i. \tag{15}$$

We can find a good approximation to the maximum by ignoring the integer part operation in eq. (15), taking the derivative with respect to i, setting the derivative to zero, and solving for i.

If all we want to show is that the worst-case time is $\Theta(n \ln n)$, there is an easier way. Set $i = n/2$ to get a lower bound of

$$\max_{0 \le i \le n} \{(\lfloor \lg(i+1) \rfloor + 1)(n-i) + i\} \ge \left(\left\lfloor \lg\left(\frac{n}{2}+1\right)\right\rfloor\right) + 1\right)\left(\frac{n}{2}\right) + \frac{n}{2} = \Omega(n \ln n). \quad (16)$$

For an upper bound use $\max_i\{A(i)B(i)\} \le [\max_i\{A(i)\}][\max_i\{B(i)\}]$ for any functions $A(i)$ and $B(i)$ to obtain

$$\max_{0 \le i \le n} \{(\lfloor \lg(i+1) \rfloor + 1)(n-i) + i\} \le (\lfloor \lg(n+1) \rfloor + 1)(n) + n = O(n \ln n). \quad (17)$$

Thus the maximum time for n operations is

$$\Theta(n \ln n). \quad (18)$$

EXERCISES

1. What is the worst-case time of the Union, Find pair of algorithms when the weighted version of the Union Algorithm is used? Assume that n is large.

2. The worst-case time of the Union, Find pair of algorithms when the ranked and weighted versions of Union are used are the same to within a constant factor. Decide which version is faster for some implementation of each algorithm. Your two implementations should be in the same language, and they should also be similar in most other respects.

9.2.3.2* *Set Union with Ranks and Collapsing*

We now find an upper bound on the time used by n calls to Union with Ranks and m calls to Find with Collapsing. Consider an arbitrary sequence of Unions and Finds. Let F_t be the set forest that is present after the tth operation is done, and let F' be the set forest that would be built by doing all the Unions in the sequence and none of the Finds. Although this forest is not actually built by the algorithms we are studying, it is useful for defining several terms. Define the *rank* of node v to be the length of the longest path in F' from v to one of its descendants. So $rank(v)$ is the number of generations of descendents that v has. [If v is a leaf, then $rank(v) = 0$.] If $parent(v)$ is defined to be the parent of node v, then in any tree $rank(parent(v)) > rank(v)$. Define $d(v)$ to be the number of descendents in F' (including v itself) of node v. For any tree built by the Union with Ranks Algorithm,

$$d(v) \ge 2^{rank(v)}. \quad (19)$$

Let A_i be an arbitrary increasing sequence of numbers. We divide the nodes into groups according to their rank. Group S_i consists of those nodes v with ranks such that

$$A_i \le rank(v) < A_{i+1}. \quad (20)$$

For any forest built by the Union with Ranks Algorithm, each node with rank k has at least 2^k descendents. Therefore there are at most $n/2^k$ nodes with rank k. The number of nodes in group i is no more than

$$\sum_{A_i \le k < A_{i+1}} \frac{n}{2^k} \le \frac{2n}{2^{A_i}}. \tag{21}$$

Now let's consider the cost of n Unions and m Finds. The Unions have cost n, assuming unit cost for each Union. We will also assume unit cost for each node on the search path for a Find. A more complex method is needed, however, to get a good upper bound on the cost for Finds. We will divide the cost of Finds into two parts. The first part counts 1 for each node v on the search path of the Find, provided that v is the root, v's parent is the root, or v is a node that is in a different group from its parent. The second part counts all the remaining nodes on the path: those that are not roots nor children of the root and which are in the same group as their parent. In both cases when computing the cost of the tth operation, we are concerned with the parents in the tree F_{t-1} that is present just before the operation is done. The first part of the cost is like an amortized time. The second part is like the negative of a potential.

A suitable upper bound for the amount of the cost in the first part is $m(2+z)$, where z is the number of groups. That is, z is the smallest number such that $n \le A_z$.

Finding an upper bound for the second part of the cost is harder. First we associate each unit of the cost with the corresponding vertex. There is a limit on how much cost each vertex can have. The parent of v changes each time the find path contains a node v that contributes to the second part of the cost. Each time the parent of v changes, the new parent has a higher rank than the old parent. (Remember that ranks are computed from the forest F', which does not change as Finds are done.) Therefore, the maximum number of times that the parent of v can change, while remaining in the same group as v, is no more than the number of nodes in v's group. This is the maximum cost that can be associated with a node. If v is in group i, A_{i+1} will be a suitable upper limit on the cost associated with node v.

An upper limit on the maximum cost associated with the nodes in group i is given by multiplying the maximum cost in group i by the maximum number of nodes in group i [eq. (21)]. The total cost in the second part is no more than

$$\sum_{0 \le i \le z} \frac{2n}{2^{A_i}} A_{i+1}. \tag{22}$$

Combining the limits on the two parts of the cost, we have for any increasing sequence A_i that the total cost for Finds is no more than

$$2n + nz + \sum_{0 \le i \le z} \frac{2n}{2^{A_i}} A_{i+1}, \tag{23}$$

where z is the smallest number such that $n \le A_z$.

n	A_n
0	1
1	2
2	4
3	16
4	65536
5	2^{65536}

TABLE 9.8. The function $A_0 = 1$, $A_{i+1} = 2^{A_i}$.

The best upper bounds are the smallest ones. How should the sequence A_z be chosen? We want A_z to increase rather rapidly so that z will be small, since the answer is proportional to z. On the other hand, if A_i increases too rapidly the factor $A_{i+1}/2^{A_i}$ will be large. A very good bound results if A_i is chosen to be the solution of the equation

$$A_{i+1} = 2^{A_i}, \tag{24}$$

with the boundary condition $A_0 = 0$. With this choice for A_i the factor $A_{i+1}/2^{A_i}$ is 1, and the total cost is bounded by

$$2n + 3zn. \tag{25}$$

This lower bound is very close to linear. Table 9.8 gives the first few values. Since $A_5 = 2^{65536}$, z is less than or equal to 5 for any reasonable size problem.

There are even better upper bounds for this algorithm. Tarjan, using an approach which is a refinement of the one given above, showed that you could replace z in an equation similar to eq. (25) with the smallest z such that $\lg n \le A(z, 4\lfloor m/n \rfloor)$, where $A(i, x)$ is the solution of the recurrence

$$A(0, x) = 2x, \tag{26}$$
$$A(i, 0) = 0 \quad \text{for } i \ge 1, \tag{27}$$
$$A(i, 1) = 2 \quad \text{for } i \ge 1, \tag{28}$$
$$A(i, x) = A(i-1, A(i, x-1)) \quad \text{for } i \ge 1, \quad x \ge 2. \tag{29}$$

This function $A(i, x)$ grows much more rapidly than the function A_i used above. It is a version of Ackermann's function, and it grows more rapidly than any primitive recursive function. (A *primitive recursive function* is a function that can be defined by a recurrence equation that does not use nested application of the function being defined [see 3, Section 3.1 and 13, Section 5.3]; almost all common functions are primitive recursive.) The first few values for $A(i, x)$ are given in Table 9.9. This analysis leads to the conclusion that $z \le 3$ for all reasonable size problems.

Tarjan's analysis makes use of a series of potential functions. Each cost is divided into a number of categories, and a bound is found on the amount of cost

i	$A(i,x)$
0	$2x$
1	2^x
2	2^{A_i}

TABLE 9.9. The first few values for the function $A(i,x)$. See Table 9.8 for A_i.

in each category. The reader is referred to the original paper [134] for the details or to [15, Chapter 2]. Aho et al. [1, Section 4.7] is a source for this section.

With these upper bounds so close to linear you might begin to believe that the lower bound is linear. That is not the case. Tarjan [134] also found a lower bound on the worst-case time that has the same form as his upper bound. See Tarjan and van Leeuwen [135] for additional worst-case analyses of set union algorithms.

9.3† THE FOURIER TRANSFORM

The fast Fourier transform is one of the more important fast algorithms discovered in recent years. Whereas many asymptotically fast algorithms are good only for problems too large for present computers, the fast Fourier transform is in common use. It makes possible many calculations that otherwise would not be feasible. We will first consider some properties of the Fourier transform before showing how to compute it rapidly.

9.3.1† The Fourier Transform and Convolutions

Let $X = \{x_0, x_1, \ldots, x_{n-1}\}$ be a vector with n components. The Fourier transform of X, written $\mathbf{F}(X)$, is the n-component vector Y whose components are given by

$$y_j = \sum_{0 \le k < n} x_k \omega^{jk} \qquad \text{for } 0 \le j < n, \tag{30}$$

where ω is a principal n^{th} root of unity (see Section 3.4.1).

The inverse of the Fourier transform, written $\mathbf{F}^{-1}(Y)$, is given by

$$x_j = \frac{1}{n} \sum_{0 \le k < n} y_k \omega^{-jk} \qquad \text{for } 0 \le j < n. \tag{31}$$

Since ω^{-1} is a principal n^{th} root of unity if and only if ω is, eq. (31) is just another Fourier transform except for the division by the factor of n. Notice that since eq. (31) gives us a way to compute X from its Fourier transform, the Fourier transform contains complete information about the vector X. If you think of the components of X as being the coefficients of a polynomial, then computing the Fourier transform $\mathbf{F}(X)$ is equivalent to evaluating the polynomial $\sum_{0 \le j < n} x_j y^j$ at the points $y_k = \omega^k$ for $0 \le k < n$.

One of the main reasons for interest in the Fourier transform comes from the relationship between it and convolutions. Convolutions are used in the mathematical analysis of systems where the current behavior of the system is a linear function of its past behavior. This property holds for many systems. Convolutions are used in studying electrical circuits, removing blurring from pictures, analyzing economic data, analyzing radar echos, exploring for oil, and other many applications. Convolutions are also computed when polynomials or multidigit integers are multiplied.

The *cyclic convolution*, C, of the n-component vectors A and B where $A = \{a_0, a_1, \ldots, a_{n-1}\}$ and $B = \{b_0, b_1, \ldots, b_{n-1}\}$, denoted by $C = A \circledast B$, is the n-component vector given by

$$c_m = \sum_{0 \le k < n} a_k b_{(m-k) \bmod n} \qquad \text{for } 0 \le m < n. \tag{32}$$

The *convolution*, C, of A and B, written $C = A * B$, is the $2n-1$ component vector given by

$$c_m = \sum_{0 \le k \le m} a_k b_{m-k} \qquad \text{for } 0 \le m \le 2n-2, \tag{33}$$

where a_j and b_j are interpreted as zero for $j \ge n$ and for $j < 0$. [Alternately, we can replace the limits on the sum in eq. (33) with $\max\{0, m-n\} \le k \le \min\{m, n\}$ and avoid any special interpretation.]

The cyclic convolution can be used to compute the ordinary convolution if the vectors are first padded with sufficient zeros. Thus, using the notation $A = \{a_0, a_1, \ldots, a_{n-1}\}$, $A' = \{a_0, a_1, \ldots, a_{n-1}, 0, 0, \ldots, 0\}$, $B = \{b_0, b_1, \ldots, b_{n-1}\}$, and $B' = \{b_0, b_1, \ldots, b_{n-1}, 0, 0, \ldots, 0\}$, where A' and B' are $2n-1$ component vectors whose the last $n-1$ components are zero, we have

$$A * B = A' \circledast B'. \tag{34}$$

Similar results hold when larger amounts of padding are used.

We will use dot for component-by-component product (*dot product*) of vectors; i.e., $C = A \cdot B$ means

$$c_j = a_j b_j \qquad \text{for all } j. \tag{35}$$

The Fourier transform of a convolution is given by

$$\mathbf{F}(A \circledast B) = (\mathbf{F}A) \cdot (\mathbf{F}B), \tag{36}$$

or

$$A \circledast B = \mathbf{F}^{-1}[(\mathbf{F}A) \cdot (\mathbf{F}B)]. \tag{37}$$

Thus convolutions are very simple in the Fourier transform space; they can be done by component-by-component multiplication. Computing convolutions directly by eq. (33) takes $O(n^2)$ operations. Dot product takes $O(n)$ operations, so eq. (34) gives a quick way to compute convolutions, provided we have a *quick* way to compute Fourier transforms.

The ability to do convolutions rapidly leads immediately to several other important fast algorithms. Computing the coefficients of the product of two polynomials is the same as computing the convolution of the coefficients of the original polynomials. That is, if $P(x) = p_0 + p_1 x + \cdots + p_n x^n$, $Q(x) = q_0 + q_1 x + \cdots + q_n x^n$, and $R(x) = r_0 + r_1 x + \cdots + r_{2n} x^{2n}$, where $R(x) = P(x)Q(x)$, then

$$r_m = \sum_{0 \le k \le m} p_k q_{m-k} \qquad \text{for } 0 \le m \le 2n, \tag{38}$$

which is a convolution. Likewise, the product of two multidigit (multiword) numbers can be computed by first computing the convolution of the two series of digits (words) and then adjusting the answer to allow for carries from one digit (word) to the next. That is, integer multiplication is just like polynomial multiplication except that in integer multiplication, when one digit (word) of the answer becomes larger than the base, you reduce that digit by a multiple of the base and adjust the next digit (word) of the answer.

Convolutions are discussed by Aho et al. [1, Section 7.1] and Knuth [10, Section 4.3.3]. The use of convolutions to compute products of integers is given in [1, Section 7.5] and [10, Section 4.3.3].

EXERCISES

1. Show that eq. (31) does indeed give the inverse of the Fourier transform. That is, show that $X = \mathbf{F}^{-1}\mathbf{F}(X)$ for any vector X when $\mathbf{F}$ is defined by eq. (30) and $\mathbf{F}^{-1}$ is defined by eq. (31). Hint: Compute $\mathbf{F}^{-1}\mathbf{F}(X)$ using eq. (30), eq. (31), and Exercise 3.4.1–2.
2. Prove eq. (36). Hint: Write the definition of $(\mathbf{F}A) \cdot (\mathbf{F}B)$, multiply and collect terms so that there is just one ω in the formula, change variables so that the power of ω is in the form ik, separate the cases $k < n$ and $k \ge n$, and use $\omega^k = \omega^{k+n}$.
3. Give a fast algorithm based on convolutions for multiplying multidigit numbers.

9.3.2† The Fast Fourier Transform

We continue to use ω for a principal n^{th} root of unity. If n is even, then $\omega^{n/2} = -1$ and

$$\omega^{n/2+j} = -\omega^j. \tag{39}$$

To compute a single y_j using eq. (30) requires n multiplications and $n-1$ additions. It is easy to prove that any algorithm for computing a single y_j must use at least $n-1$ operations, because y_j depends on all n of the x_i. To compute all n of the y_j with eq. (30) takes n^2 multiplications and $n^2 - n$ additions. You might think it would be easy to obtain a matching lower bound, but it is not. If you try to apply the simple lower-bound argument used for a single element, you get that the y_j's depend on all n of the x_i's, so at least $n-1$ operations must be used. With a little more work, you can prove that n operations are needed. This is a long way from n^2. One reason you cannot do better is that n^2 operations are *not* needed. We will show this by giving an algorithm for the case where n is a power of 2 which computes the Fourier transform using $\Theta(n \ln n)$ operations.

The basic idea behind the algorithm is that the computation of a polynomial $p(x)$ can be done with even and odd parts as follows. Let n be even and

$$p(x) = \sum_{0 \le i < n} p_i x^i, \tag{40}$$

$$p_{\text{even}}(x) = \sum_{0 \le i < n/2} p_{2i} x^i, \tag{41}$$

$$p_{\text{odd}}(x) = \sum_{0 \le i < n/2} p_{2i+1} x^i. \tag{42}$$

Then

$$p(x) = p_{\text{even}}(x^2) + x p_{\text{odd}}(x^2), \tag{43}$$

and

$$p(-x) = p_{\text{even}}(x^2) - x p_{\text{odd}}(x^2). \tag{44}$$

These last two equations mean that we can evaluate an $(n-1)^{\text{st}}$ degree polynomial at two points (x and $-x$) by evaluating each of two polynomials of degree $n/2 - 1$ at one point. From the analysis of divide and conquer algorithms [eqs. (5.82–5.84)], it is clear that this is the start of a promising approach. If we need to evaluate a polynomial at both x and $-x$, we get two points by evaluating two polynomials of half the degree at one point. When x is an nth root of unity and n is a power of 2, this idea can be applied recursively.

This brings us to the recursive algorithm for the fast Fourier transform. Later we will consider a nonrecursive algorithm that has less overhead, but the recursive algorithm has the basic idea for the fast Fourier transform without the bookkeeping detail that is needed for the nonrecursive algorithm, and it is therefore easier to understand.

Algorithm 9.15 **RFFT**$(n, x_0, x_1, \ldots, x_{n-1}, m, y_0, y_1, \ldots, y_{n-1})$**:** Input: An integer $n \ge 2$ that is a power of 2, a vector $X = \{x_0, x_1, \ldots, x_{n-1}\}$ of points to be transformed, and a variable m, which is 1 when the user calls the algorithm. Output: A vector $Y = \{y_0, y_1, \ldots, y_{n-1}\}$ which is the Fourier transform of X. Assume: ω, a principal n^{th} root of unity, is available to the program.

Step **1.** If the value of n is:

Step **2.** $n = 2$, then set $y_0 \leftarrow x_0$, $y_1 \leftarrow x_1$.

Step **3.** $n \neq 2$, then call $\text{RFFT}(n/2, x_0, x_2, \ldots, x_{n/2-2}, 2m, y_0, y_1, \ldots, y_{n/2-1})$. (In this step the lower half of Y is being used for temporary storage.)

Step **4.** Call $\text{RFFT}(n/2, x_1, x_3, \ldots, x_{n/2-1}, 2m, y_{n/2}, \ldots, y_{n-1})$. (In this step the upper half of Y is being used for temporary storage.)

Step **5.** End if.

Step **6.** For $0 \leq i \leq n/2 - 1$ do the rest of this step. Set $even \leftarrow y_i$, $odd \leftarrow \omega^{mi} y_{n/2+i}$, $y_i \leftarrow even + odd$, and $y_{n/2+i} \leftarrow even - odd$.

Steps 1 and 2 stop the recursion when the base case ($n = 2$) is reached. Steps 3 and 4 calculate the even and odd parts of the transform, using the lower and upper halves of Y for temporary storage. Step 6 computes the final answer, using eqs. (39, 44).

The time for Algorithm RFFT is given by a recurrence of the form

$$T_n = 2T_{n/2} + an + b, \tag{45}$$

where the $2T_{n/2}$ comes from the two recursive calls in Steps 3 and 4, the an comes from the loop in Step 6 and the preparation of the recursive calls in Steps 3 and 4, and the b comes from the constant overhead in each step. For $n = 2^k$, it takes $k - 1$ recurrences on eq. (45) to get to the boundary case of T_2. The total time is

$$T_n = 2^{k-1}T_2 + a(k-1)n + b \sum_{0 \leq i < k-1} 2^i \tag{46}$$

$$= an \lg n + \left(\frac{b}{2} + \frac{T_2}{2} - a\right) n - b. \tag{47}$$

The fast Fourier transform, therefore, can be done in time $O(n \lg n)$.

To make practical use of the fast Fourier transform for large n, it is important to keep the running time small. Eq. (47) makes it clear that the most important place for savings (while maintaining the same basic approach) is the constant a. Any efforts at clever coding should be directed primarily toward reducing the size of a. Part of the contribution to a comes from Step 6. Surely you should code Step 6 carefully. The rest of the contribution comes from Steps 3 and 4. Here even bigger savings are possible if you take care. Steps 3 and 4 take a long time because setting up the procedure calls requires moving large amounts of data. By clever design, this data movement can be avoided. The idea is to have the data where it is needed so that it does not need to be moved.

To see what data movement is done, let's look more carefully at just how the algorithm combines the data when $n = 8$. The first call on RFFT divides the original problem into two subparts, the even indices and the odd indices. This leads to two recursive calls to RFFT, one for the each of the two subparts, which in turn produce four calls to handle the quarters of the original problem.

Level	*Calculation*
2	$y_0 \leftarrow y_0' + y_1' = x_j + x_{j+4}$
	$y_1 \leftarrow y_0' - y_1' = x_j - x_{j+4}$
1	$y_0 \leftarrow y_0' + y_2' = x_j + x_{j+4} + x_{j+2} + x_{j+6}$
	$y_2 \leftarrow y_0' - y_2' = x_j + x_{j+4} - x_{j+2} - x_{j+6}$
	$y_1 \leftarrow y_1' + \omega^2 y_3' = x_j - x_{j+4} + \omega^2 x_{j+2} - \omega^2 x_{j+6}$
	$y_3 \leftarrow y_1' - \omega^2 y_3' = x_j - x_{j+4} + \omega^2 x_{j+2} - \omega^2 x_{j+6}$
0	$y_0 \leftarrow y_0' + y_4' = x_0 + x_4 + x_2 + x_6$
	$+x_1 + x_5 + x_3 + x_7$
	$y_4 \leftarrow y_0' - y_4' = x_0 + x_4 + x_2 + x_6$
	$-x_1 - x_5 - x_3 - x_7$
	$y_1 \leftarrow y_1' + \omega y_5' = x_0 - x_4 + \omega^2 x_2 - \omega^2 x_6$
	$+\omega x_1 - \omega x_5 + \omega^3 x_3 - \omega^3 x_7$
	$y_5 \leftarrow y_1' - \omega y_5' = x_0 - x_4 + \omega^2 x_2 - \omega^2 x_6$
	$-\omega x_1 + \omega x_5 - \omega^3 x_3 + \omega^3 x_7$
	$y_2 \leftarrow y_2' + \omega^2 y_6' = x_0 + x_4 - x_2 - x_6$
	$+\omega^2 x_1 + \omega^2 x_5 - \omega^2 x_3 - \omega^2 x_7$
	$y_6 \leftarrow y_2' - \omega^2 y_6' = x_0 + x_4 - x_2 - x_6$
	$-\omega^2 x_1 - \omega^2 x_5 + \omega^2 x_3 + \omega^2 x_7$
	$y_3 \leftarrow y_3' + \omega^3 y_7' = x_0 - x_4 + \omega^2 x_2 - \omega^2 x_6$
	$+\omega^3 x_1 - \omega^3 x_5 + \omega^5 x_3 - \omega^5 x_7$
	$y_7 \leftarrow y_3' - \omega^3 y_7' = x_0 - x_4 + \omega^2 x_2 - \omega^2 x_6$
	$-\omega^3 x_1 + \omega^3 x_5 - \omega^5 x_3 + \omega^5 x_7$

TABLE 9.10. The calculations done by the RFFT Algorithm when $n = 8$. Simplifications of the form $-\omega^i = \omega^{i+4}$ have not been done so that it will be easier to compare this table with the steps in RFFT. Primes are used to indicate the original values of the y's during a level of recursion; unprimed y's contain the newly calculated values. Level 2 is done for $0 \leq j < 4$ and level 1 for $0 \leq j < 2$.

The first quarter is for those indices that are equal to zero modulo 4; the ith quarter is for those indices that are equal to $i - 1$ modulo 4.

The quarters are done directly: the even $(x_j + (-1)^{4/8} x_{j+4})$ and odd $(x_j - (-1)^{4/8} x_{j+4})$ parts are computed for $0 \leq j < 4$. Table 9.10 shows the complete set of calculations. At each level, previously computed values are multiplied by the appropriate power of ω and added in pairs.

The most obvious way to speed up the RFFT Algorithm is to replace the recursive calls with a loop and to store the intermediate results back into the array used for the input data, so that data movement is avoided. (We could also avoid data movement by using a doubly indexed array, where one index kept track of the level of the recursion, but that approach uses about $\ln n$ times as much memory.) The recursive algorithm combines y's with two indices in two different ways and then does not make any additional use of the of the original values of the y's; this is what allows us to store intermediate results on top of the original data.

These ideas lead to the following algorithm. It is rather easy to see which values need to be combined together each time through the outer loop. As you can see from Table 9.10, on the first iteration each data item needs to be combined with the item halfway up the table from it. On the second iteration data items are combined with items a quarter of the way up. On the jth iteration they are combined with items 2^{-j} of the way up. In order to store the results back on top of the data, it is necessary to store the results for both the upper and lower halves at once during the first time through the loop, and special control is needed to prevent the program from processing the upper part of the table twice. During the second time through the loop the same problem arises with quarters of the data, etc.

The pattern of the powers of the ω's is more complicated. The first time through the loop uses the two square roots of 1 (the plus and minus 1 coefficients). The second time uses the four fourth roots of 1, etc. The exact pattern of usage, however, is a little complex. It depends on reversing the lower-order bits of the index. (Refer to the algorithm for the details.) The final step of the algorithm takes care of the fact that the first two steps compute the Fourier transform correctly but leave the answers in the wrong places. (The indices have their bits reversed.)

Algorithm 9.16 FFT: Input: An integer $n \geq 2$ that equals 2^k for some integer k (i.e., $k = \lg n$) and a vector $X = \{x_0, x_1, \ldots, x_{n-1}\}$ of points to be transformed. The input data is destroyed by the algorithm. Output: The vector $X = \{x_0, x_1, \ldots, x_{n-1}\}$ which is the Fourier transform of the input. Assume: ω, a principal n^{th} root of unity, is available to the program. The binary representation of the loop index i is $i_{k-1}i_{k-2}\cdots i_0$. In all cases where the program uses the binary representation of a number, k bits are given. Thus the variable q used in Step 2 has a leading zero, followed by the j lower-order bits of the i, followed by $k-j-1$ trailing zeros.

Step **1.** For $0 \leq j < k$ do:

Step **2.** For $0 \leq i < n$ do the rest of this step. If $i_j = 0$, then set $q \leftarrow 0i_{j-1}i_{j-2}\ldots i_0 00\ldots 0$, $odd \leftarrow \omega^q x_{i+2^{k-j-1}}$, $x_{i+2^{k-j-1}} \leftarrow x_i - odd$, and $x_i \leftarrow x_i + odd$.

Step **3.** End for j.

Step **4.** For $0 \leq i < n$, if $i < i_0 i_1 \ldots i_{k-1}$ then interchange $x_i \leftrightarrow x_{i_0 i_1 \ldots i_{k-1}}$.

The above algorithm tests the jth bit of i in Step 2 to avoid doing upper and lower parts twice. Both Steps 2 and 4 need to examine index i with some or all of its bits reversed. If care is not taken, these bit computations can use excessive time. For large n, one can maintain both the value of i and of i with its bits reversed. It is straightforward to add 1 to i each time around the loop. With a little care, you can also update the reversed version of i quickly (small average time). (See Exercise 1 for details.) For small values of n, table look-up is the best way to compute the reversal of i. Just precompute a table where the ith entry contains the reversal of i.

Notice that the various computations in the inner loop are independent of each other. This means that on a parallel computer with n processors, the FFT Algorithm can be done in time $\lg n$. (The computations in Step 4 are also independent.) Some signal processing applications need rapid continuous Fourier transform calculation. The FFT Algorithm suggests how to build a pipelined computer with $n \lg n$ processors which computes a Fourier transform at every step. The input goes into the first level of n processors, where pairs of inputs are combined. As soon as the combined values are passed on to the next level, the first level is ready for the next set of inputs. After $\lg n$ levels of processing, the answer comes out. Such a computer can produce a Fourier transform at each step, although the answer comes out $\lg n$ time units after the corresponding input goes in.

The RFFT Algorithm is adapted from Baase [2, Section 5.4]. The FFT Algorithm is adapted from Aho et al. [1, Section 7.2]. See Elliott and Rao [26] for an extensive treatment of the fast Fourier transform. The first fast Fourier algorithm was described by Good [89].

EXERCISES

1. Give an algorithm that takes as input i and i with its bits reversed and which computes the value of $i+1$ and of $i+1$ with its bits reversed. Your algorithm should take total time $O(n)$ if it is called n times, once for each value of i in the range $0 \le i < n$, where $n = 2^k$. Assume the input is in arrays of digits, and that the answer is to be stored into the same arrays that contain the input.
2. Compute the running time for the FFT Algorithm.
3. Estimate the ratio of speeds for Algorithm RFFT and FFT. Do the same for storage. Give careful consideration to the storage that RFFT uses for parameters and intermediate results that are saved during recursions.

9.3.3* The All-Integer Fourier Transform

To complete our discussion of the Fourier transform, we need to consider ω, the nth root of unity, and how it is computed. Normally, we consider arithmetic for the Fourier transform to be done with complex numbers. Section 3.4 has all the information needed to compute ω. The obvious choice is

$$\omega = e^{2\pi i/n} = \cos(2\pi/n) + i \sin(2\pi/n), \tag{48}$$

where $i = \sqrt{-1}$ and the angles are measured in radians. Your local computer will probably have routines for computing the sine and cosine function with great accuracy. If not, you can refer to a standard reference such as [16, Section 4.3].

But you cannot use eq. (48) to compute ω exactly — it is not a rational number for most values of n.

There are many applications of the Fourier transform where we want exact answers. If we multiply two polynomials with integer coefficients, we want an answer that is a polynomial with integer coefficients. If we multiply two integers, we want the answer to be an integer. One approach is to use the value of ω given by eq. (48) (computed to adequate accuracy), and to round off the final answer so that we get the exact answer. Great care is needed in selecting how much accuracy is appropriate. Too little accuracy and the final answer will be wrong; too much accuracy and it will take excessive time to compute the final answer.

If we need the Fourier transform itself in an application where ω must be $e^{2\pi i/n}$, then we must make do with eq. (48) and its limitations. Often, however, we want the Fourier transform as an intermediate step in computing a convolution. Then we can use any definition for ω, provided the ω we choose has all the properties needed for eq. (37) to work.

Suppose we try to do all of our computations modulo m for some large integer m. We will want to choose m so large that the answer to our problem is equal to the answer modulo m. For example, if we want to compute the convolution of two sequences of n nonnegative numbers, where M is an upper bound on the size of each number, then we want $m > nM^2$. On the other hand, we don't want m too much bigger than this, because the larger m is, the longer the calculation will take (particularly if m is so large that it fills several computer words). In addition we need m to be relatively prime to n, because eq. (37) has a division by n. If m and n are relatively prime, then n has an inverse (modulo m) and the division can be replaced by multiplication by the inverse. Since n will be a power of 2 for the version of the fast Fourier transform that we consider, n is relatively prime to m if and only if m is odd.

We will need ω to be a principal n^{th} root of unity modulo m; i.e., $\omega^n \equiv 1 \pmod{m}$, and $\omega^k \not\equiv 1 \pmod{m}$ for $0 < k < n$. We will come back to the question of finding an nth root of unity modulo m in a moment. First, however, let's consider what other properties we need. We will assume that ω is a principal n^{th} root of unity modulo m for the rest of this section.

We need

$$\sum_{0 \le j < n} \omega^{jn} \equiv n \pmod{m}, \qquad (49)$$

but since ω is an nth root of unity this is immediately true, because $\omega^{jn} \equiv (\omega^n)^j \equiv 1^j \equiv 1$ and $\sum_{0 \le j < n} 1 \equiv n$.

Next we need

$$\sum_{0 \le j < n} \omega^{jk} \equiv 0 \pmod{m} \qquad \text{for } 0 < k < n. \qquad (50)$$

This is more difficult to arrange. We know from eq. (2.61) that this sum is equal to $(\omega^{kn} - 1)/(\omega^k - 1)$ and ω^{kn} is congruent to 1, so eq. (50) is true when $\omega^k - 1$

is relatively prime to m. We will use a somewhat different result, however. For any a and any positive integer k, we have

$$\sum_{0\leq j<n} a^j = \prod_{0\leq j<k} \left(1+a^{2^j}\right) \qquad \text{provided } n=2^k. \tag{51}$$

(See Exercise 1.) Below we will use eq. (51) to prove that eq. (50) is true for any value of ω such that $\omega^{2^j} \equiv 0$ for some j in the range $0 \leq j < k$.

Finally we need a way to compute the inverse to ω. However, $\omega\omega^{n-1} = \omega^n \equiv 1 \pmod{n}$, so ω^{n-1} is the inverse we need.

Thus, all the properties we need except for eq. (50) come from ω being a principal n^{th} of unity modulo m. We still need a way to find such an nth root. We want to be able to do this for any n that is a power of 2, but we don't need to do it for arbitrary m, just as long as the suitable values of m are not too far apart. (Remember, we want m big enough that the final answer to our convolution problem will be less than m, we don't want m to be too large or our algorithms will be slow, and we want m to be odd, but we don't have any additional requirements on m.)

It turns out that we can choose any integer greater than zero for ω provided we choose $\omega^{n/2}+1$ for m. If $\omega = 2$ gives a big enough value for m, then we use $\omega = 2$; otherwise we use a larger value of ω. Calculations modulo 2^j+1 are particularly easy to do on a binary computer (see below), so we will usually choose ω to be the smallest power of 2 that results in a large enough value of m.

To prove that our choice for m is okay, let's first show that $\omega^n \equiv 1 \pmod{m}$. Since $m > 1$, $\omega^{n/2} \equiv -1 \pmod{m}$, so $\omega^n = \omega^{n/2}\omega^{n/2} \equiv (-1)(-1) = 1$. That part is easy.

Now let's show that for the chosen value of ω

$$\sum_{0\leq j<n} \omega^{jk} \equiv 0 \pmod{m} \qquad \text{for } 0<k<n. \tag{52}$$

By eq. (51), this will be true provided $1+\omega^{k2^j} \equiv 0 \pmod{m}$ for some j in the range $0 \leq j < \lg n$. Let $k = k'2^s$, where k' is odd. Since $0 < k < n$, $0 \leq s < \lg n$. Choose $j = \lg n - s - 1$. This value of j is somewhere in the range $0 \leq j < \lg n$. For this value of j, $\omega^{k2^j} = \omega^{k'2^{\lg n-1}} = (\omega^{n/2})^{k'} = (m-1)^{k'}$. But $m-1 \equiv -1 \pmod{m}$ and k' is odd, so $\omega^{k2^j} \equiv -1 \pmod{m}$ for $j = \lg n - s - 1$ and $1+\omega^{k2^j} \equiv 0 \pmod{m}$.

Thus we see that for any integer ω we can compute convolutions with the fast Fourier transform provided $m = \omega^{n/2}+1$ and n is a power of 2. On binary computers, computations modulo m are particularly easy to do when ω is a power of 2, because for $x = \sum_{0\leq j<k} x_j\omega^{jn/2}$ with $0 \leq x_j < \omega^{n/2}$ (i.e., x has digits x_j when written in base $\omega^{n/2}$) we have

$$x \equiv \sum_{0\leq j<k} (-1)^j x_j \pmod{m} \tag{53}$$

(see Exercise 2). Thus if ω is a power of 2, we can compute x modulo m by adding and subtracting the bits of x in groups of size $n/2$.

We have gone through a lot of math to justify using the fast Fourier transform with arithmetic modulo an integer to compute convolutions rapidly, but the method itself is actually quite simple. We will summarize it here.

1. Find an upper and lower bound for the answer to your problem. Let M be the length of the range of possible answers. Choose n to be a power of 2 such that the length of the vectors being convoluted is greater than $n/4$, but no more than $n/2$. (This is done to allow for the padding that is needed to compute an ordinary convolution with Fourier transforms.)
2. Choose ω to be the smallest power of 2 such that $\omega^{n/2} + 1 > M$. Let $m = \omega^{n/2} + 1$.
3. Compute $\omega^{-1} = \omega^{n-1} \bmod m$. (You need ω^{-1} for the inverse Fourier transform.)
4. Compute the convolution with eq. (37), using the FFT algorithm for all Fourier transforms and all inverse transforms, and using arithmetic modulo m.

Aho et al. [1, Section 7.3] is a source for this section.

EXERCISES

1. [See 1] Prove eq. (51). Hint: First prove that

$$\sum_{0 \le i < n} x^i = (1 + x) \sum_{0 \le i < n/2} (x^2)^i$$

and then use induction.

2. [See 1] Prove eq. (53). Hint: First prove that $\omega^p \equiv -1 \pmod{m}$.
3. Use the all-integer Fourier transform to compute the convolution of $\{1, 2, 3, 4\}$ with $\{5, 6, 7, 8\}$. (This problem is designed to give you practice using the technique; it is not quite big enough to justify using such a sophisticated algorithm.)
4. Find integers ω, m, and n such that $\omega^n \equiv 1 \pmod{m}$, $\omega^i \not\equiv 1 \pmod{m}$, and $\sum_{0 \le i < n} \omega^i \not\equiv 0 \pmod{m}$. Hint: Try $m = 15$ and $n = 2$.

9.4* MODULAR ARITHMETIC

As we saw in the previous section, strange approaches to arithmetic are sometimes worthwhile. There are two problems with the traditional algorithms for arithmetic. On serial computers, multiplication is somewhat slow; the traditional algorithm requires time n^2 to multiply n-digit numbers. There are faster algorithms, but they are useful only for rather large numbers. On parallel computers, addition is a problem. It takes time n (worst-case) to add n-digit numbers due to the time required to propagate carries. Multiplication takes time n for the same reason.

There is a way around both of these problems — modular arithmetic. In modular arithmetic, each number is represented by its value modulo a preselected series of numbers. In the previous section we discussed doing arithmetic modulo m. That approach is straightforward as long as m fits in one computer word. When we are dealing with a very large range of numbers, where the numbers themselves may require several computer words, then modular arithmetic may be helpful. The modular arithmetic method of representing numbers is unique for a bounded range of numbers — any number that is not too large has one and only one representation. Addition, multiplication, and subtraction of n-bit numbers can be done in time $O(n)$ on serial computers. The same operations can be done in close to constant time on parallel computers with n processors.

If modular arithmetic is so great, why isn't it used all the time? Well, there are two problems. First, division by many large numbers cannot be done with modular methods. (Often, however, a problem can be reformulated to avoid divisions.) Second, comparing the size of numbers is quite slow. This operation of comparing the size of numbers, which we take for granted with traditional arithmetic, takes a long time for modular arithmetic. Therefore modular arithmetic is only used in those applications where one needs to do a large number of adds, subtracts, and multiplies for each division or comparison.

9.4.1* Representation of Numbers

To do modular arithmetic, start with a list of moduli m_1, m_2, ..., m_k. A number n is represented by the values

$$n_1 = n \bmod m_1, \tag{54}$$

$$n_2 = n \bmod m_2, \tag{55}$$

$$\vdots$$

$$n_k = n \bmod m_k. \tag{56}$$

We write the number n as $(n_1, n_2, \ldots, n_k)$ whenever we want to emphasize its modular representation. We also refer to the n_i as *digits* of n.

The most important requirement on the moduli is that they must be relatively prime, because we make extensive use of Theorem 1–4. With relatively prime moduli, every integer n in the range $0 \le n < m_1 m_2 \cdots m_k$ has a unique representation. (It is straightforward to handle a range starting with any lower limit, as long as the total length is no more than $m_1 m_2 \cdots m_k$.)

Certain special forms of the m_i make it possible to do divisions more quickly. For computers or circuits that encourage binary values, the preferred moduli have the forms 2^i, $2^i - 1$, and $2^i + 1$. However, 2^i is not relatively prime to 2^j, so at most one modulus can have the form 2^i.

The final consideration for selecting the moduli is to use moduli of an appropriate size. The appropriate size varies depending on whether you are going to do modular arithmetic on an existing computer or you are designing a computer that will use modular arithmetic in its circuits. If you are designing a computer, you probably want the moduli to be small. In general the larger the moduli are, the more circuitry your computer will use and the slower it will be. If you are using a computer with a large word size, you will want your moduli to be slightly smaller than the word size of the computer. By keeping the moduli smaller than the word size, you can do all your computations with single-word arithmetic. If the moduli are much smaller than the word size, you waste memory (and speed, because you will need more moduli). If you are using a computer with a small word size (say, eight bits), then you may have to use moduli that are larger than one word. In this case you will want to use moduli that take up no more words than necessary.

For some applications it is desirable to have large prime moduli. This rules out using 2^i as a modulus, but there are a number of primes of the form $2^i - 1$ and a few of the form $2^i + 1$ (such numbers are called Mersenne and Fermat primes, respectively).

9.4.2* Doing Arithmetic

The method of doing arithmetic is based on Theorem 1.4. Let $\circ$ be any of the operations $+$, $-$, or $*$. Then to compute $c = a \circ b$, you compute

$$c_1 = a_1 \circ b_1 \bmod m_1, \tag{57}$$

$$c_2 = a_2 \circ b_2 \bmod m_2, \tag{58}$$

$$\vdots$$

$$c_k = a_k \circ b_k \bmod m_k. \tag{59}$$

Using eqs. (57–59) the answer can be computing with k independent computations. Assuming constant time for each operation, time $O(k)$ is needed on a serial computer and time $O(1)$ is needed on a parallel computer with k processors (or parallel circuits). For serial computers, the time for multiplication is much better than the time for the traditional approach. For parallel computers, the time is much better for all three operations.

The assumption of fixed time per operation is appropriate when you have a fixed circuit that you use for various computations. When you are designing circuits, you also need to consider how fast the circuit will be as a function of the size of problem that it is designed to handle. In that case the times of the previous paragraph need to be multiplied by a factor of $\ln m$, where $m = \max\{m_1, \ldots, m_k\}$.

With modular arithmetic you can also do division by those numbers that are relatively prime to the moduli, provided you know that the final answer of your modular computation is an integer. If all your moduli are large prime numbers, then you can do division by any small integer (as long as you know your final answer is an integer).

9.4.3* Conversion to Radix Representation

To use the answers obtained from a modular calculation, it is usually necessary to convert them back to a radix representation. The best way known to compare modular numbers also requires that you do part of the work of converting the numbers to a radix representation.

Remember that any number relatively prime to m_i has an integer inverse modulo m_i, i.e., for any x relatively prime to m_i there exists a number x^{-1} such that $xx^{-1} \equiv 1 \pmod{m_i}$. All the moduli in a modular arithmetic system are relatively prime to each other, so they all have inverses. Since the moduli are fixed, these inverses can be precomputed.

To convert a modular number with (modular) digits $(n_1, n_2, \ldots, n_k)$ to a radix number, we need to do two subcalculations. The first set of calculations is

$$r_1 \leftarrow n_1 \bmod m_1, \tag{60}$$

$$r_2 \leftarrow (n_2 - r_1)m_1^{-1} \bmod m_2, \tag{61}$$

$$\vdots$$

$$r_k \leftarrow (\cdots((n_k - r_1)m_1^{-1} - r_2)m_2^{-1} - \cdots - r_{k-1})m_{k-1}^{-1} \bmod m_k. \tag{62}$$

The ith line is done using $\pmod{m_i}$ arithmetic.

Notice that the quantity m_j^{-1} is in general a different integer in each line; in the line for $\pmod{m_i}$, m_j^{-1} is the inverse of $m_j \pmod{m_i}$, which is usually not the same as the inverse $m_j \pmod{m_{i'}}$. If the m^{-1}'s are precomputed, a table with $k(k-1)/2$ entries is needed. Computing the r's requires $k(k-1)/2$ multiplications and the same number of subtractions. Each multiplication and subtraction involves single precision numbers.

The second step is to compute

$$r = (\cdots((r_k m_{k-1} + r_{k-1})m_{k-2}) + \cdots + r_2)m_1 + r_1. \tag{63}$$

This computation is done with radix arithmetic. Multiprecision numbers are used. Therefore, although only $O(n)$ arithmetic operations are done, time $O(n^3)$ is needed if the multiplications are done with the classical multiplication algorithm.

Notice that the r's computed in eqs. (60–62) are the digits of the answer in a mixed radix number system. That is, r_1 is the least significant digit for a radix of m_1, r_2 is the next least significant digit, where the radix for the second position is m_2, etc. This means we can compare the sizes of two numbers using just the first part of the conversion process. In other words, if we want to compare numbers with modular representations $(a_1, a_2, \ldots, a_k)$ and $(b_1, b_2, \ldots, b_k)$, then we first convert them to $(r_1, r_2, \ldots, r_k)$ and $(s_1, s_2, \ldots, s_k)$. If $r_k > s_k$; or $r_k = s_k$ and $r_{k-1} > s_{k-1}$; or $r_k = s_k$, $r_{k-1} = s_{k-1}$, and $r_{k-2} > s_{k-2}$; etc., then the first number is larger than the second one. Thus we can compare the size of numbers in time $O(n^2)$.

EXERCISE

1. Assuming each of the moduli takes almost one computer word and that multiplication of n word numbers takes time cn^2, compute how long it takes to do the calculation in eq. (63). Your answer should be accurate to $O(n^2)$.

9.4.4* Applications

Probably the most important problem where modular arithmetic is obviously useful is matrix multiplication. Matrix multiplication with the traditional algorithm takes n^3 multiplications and $n^2(n-1)$ additions. You do not need any divisions or comparisons. This makes it a good problem for modular methods.

Solving linear equations is an important computation that can be done rapidly with modular arithmetic. How to do it is not, however, quite so obvious as how to multiply matrices. The normal methods of solving linear equations use divisions. Worse yet, the solutions of equations with integer coefficients are usually noninteger.

The first step in solving linear equations with modular methods is to convert the equations into a form that has integer solutions. This needs to be done in such a way that the solutions of the modified problem can easily be converted into solutions of the original problem. Let the original problem be

$$\begin{aligned} a_{11}x_1 + a_{12}x_2 + \cdots + a_{1n}x_n &= b_1, \\ a_{21}x_1 + a_{22}x_2 + \cdots + a_{2n}x_n &= b_2, \\ &\vdots \\ a_{n1}x_1 + a_{n2}x_2 + \cdots + a_{nn}x_n &= b_n. \end{aligned} \tag{64}$$

We assume that the a's and b's are integers. (If they are rational, they can be converted to integers by multiplying each equation by the appropriate number.) We convert this problem to

$$\begin{aligned} a_{11}y_1 + a_{12}y_2 + \cdots + a_{1n}y_n - b_1y_0 &= 0, \\ a_{21}y_1 + a_{22}y_2 + \cdots + a_{2n}y_n - b_2y_0 &= 0, \\ &\vdots \\ a_{n1}y_1 + a_{n2}y_2 + \cdots + a_{nn}y_n - b_ny_0 &= 0. \end{aligned} \tag{65}$$

Eqs. (65) are a set of homogeneous equations (the right sides are zero). If the set has any solutions, then any multiple of a solution is also a solution. Therefore, if the coefficients are integers, then eqs. (65) have integer solutions if they have any solutions at all. If eqs. (65) have the solution

$$y_0,\ y_1,\ y_2,\ \ldots,\ y_n \tag{66}$$

where $y_0 \neq 0$, then eqs. (64) have the solution:

$$x_1 = \frac{y_1}{y_0},\quad x_2 = \frac{y_2}{y_0},\quad \ldots,\quad x_n = \frac{y_n}{y_0}. \tag{67}$$

Thus, if you can solve eqs. (65) quickly, then you can solve eqs. (64) quickly; all you have to do is to perform n divisions at the end of the computation.

To solve eqs. (65) with modular methods, you start with a list of primes m_1, m_2, ..., m_k. Next, you solve eqs. (65) modulo each prime until you have enough information to construct the solution of the original set of equations. Every time you solve eqs. (65) modulo a new prime, you use eq. (62) to convert the solution of mixed radix form. Eventually, after solving eqs. (65) modulo enough primes, you obtain an answer where all the components of the solution have a leading digit [in the mixed radix representation that eq. (62) provides] that is zero. This is probably the solution of eqs. (65), so you finish converting the solution of radix form and plug it into eqs. (65). At this stage you will occasionally find that the result does not satisfy eqs. (65). [In such cases you had the misfortune that some of your primes divided evenly some of the subdeterminants of the coefficients in eqs. (65). If all your primes are large, this is very unlikely.] If the result does not check, then continue solving eqs. (65) modulo additional primes. Usually you will find the correct solution in one more step. You are sure to find the solution in a finite number of steps.

This method of solving equations is explained in more detail by Borosh and Fraenkel [75]. They used the method to find the set of exact solutions of a set of 111 linear equations in 120 unknowns. They were able to obtain the exact answers more rapidly than they could obtain approximate answers using traditional approaches. A more advanced treatment of the method is given by McClellan [114].

For further treatment of modular arithmetic, see Knuth [10, Section 4.3.2] and Aho et al. [1, Section 8.4].

Some applications of modular arithmetic require a thorough understanding of linear algebra. See Strang [56] and Noble [45].

EXERCISES

1. Solve the equations

$$\begin{aligned} 100w + 7x + 2y + 4z &= 5, \\ 4w + 100x + 4y + 4z &= 1, \\ 9w + 9x + 100y + 6z &= 4, \\ 7w + 6x + 2y + 100z &= 6 \end{aligned}$$

using modular arithmetic. For your list of primes use 2, 3, 5, 7, 11, ...

CHAPTER 10

Lower Bounds and NP Completeness

So far in this book we have concentrated on computing the running times for algorithms. The underlying problem that the algorithm solves was a secondary consideration. But in practice the situation is usually reversed: we begin with a problem we are interested in solving and then try to find an efficient algorithm to solve it.

Some problems, such as sorting, are so common that a great deal of effort has been expended on devising algorithms for them. In this book we have looked at many algorithms for sorting, among them Heapsort, Quicksort, Merge Sort, and Insertion Sort. The best of these algorithms take time $O(n \log n)$. In a case like this you may wonder whether it is possible to obtain an even better algorithm (linear, say, in the length of the input). A useful result in answering such questions is one that places a lower limit on how good the best algorithm for a specific problem can be. Such a limit is called a *lower bound*; it is a property of the problem itself and not of any individual algorithm.

The running time of the best algorithm known for a problem gives an upper bound on the time needed to solve the problem. Once a lower bound for a problem has been obtained, we can try to find an algorithm that runs within the time limit given by the lower bound. If such an algorithm can be found, the lower bound is *tight*, and the algorithm is *optimal*. We know that we have the best possible algorithm, and we need to do no further work on developing better algorithms. When no algorithm is known that runs within the lower bound, then it may be possible to find a better algorithm, or the lower bound may need improvement.

Lower-bound results, which give a limit on the efficiency of all algorithms for solving a problem, are often very difficult to obtain. The concept of all conceivable algorithms is too vague and nebulous to be of much use in a proof;

lower-bound results have to be stated within a framework or computational model that specifies what types of operations an algorithm is permitted to use. If the model is too restrictive, the results are not interesting; if the model is too general and contains operations that are not very useful for solving the problem, the proofs may become needlessly complex. General models lead to smaller lower bounds: if model A is more general than (contains all the operations of) model B, then the lower bound for model A is less than or equal to the lower bound for model B.

Choosing an appropriate model can be quite tricky. For example, multiplication of $n \times n$ matrices takes time n^3 using only row and column operations (that is, operations that do the same thing to all the elements of a row or column). Many people believed that this meant that the ordinary matrix multiplication algorithm, which takes time n^3, was optimal. However, Strassen's [133] algorithm, which is not restricted to row and column operations, runs in time $n^{\lg 7} \approx n^{2.81}$ (see Section 5.2.5). This shows that the row and column model used in the earlier lower-bound computation was inadequate and misleading. Yet, it was actually quite plausible that the best algorithm for matrix multiplication would use only row and column operations; the existence of Strassen's algorithm is very surprising. Great care is needed in choosing the model of computation for a lower-bounds investigation.

We need to make a distinction between a problem as a generic type or set of problems (as in "the sorting problem") and a problem as a concrete instance (sort this particular list of ten elements). We use the terms *problem set* and *problem instance*, respectively, to distinguish these two meanings of "problem"; we use the word "problem" for either one when the meaning is clear from context.

It is important to realize that lower-bound results apply only to algorithms that are capable of solving *all* the instances of a problem, including the hardest. If you only need an algorithm to solve some subset of the problem instances, you can often find a fast algorithm: one that runs faster than the lower bound for the entire set. In fact, if the subset is finite, you can always solve the problem in constant time by building a table of answers into your program and using table look-up. For a reasonable size set of problems that needs to be solved repeatedly this is often a good strategy. Even for an infinite subset of the instances of a problem, there are often special algorithms that are much faster than the best possible algorithm for the general problem. For example, there is a linear time algorithm to test planar graphs for isomorphism [95], but no polynomial time algorithm has been found for the general graph isomorphism problem.

While it is common practice to call any algorithm that runs in time asymptotically equal to the lower bound *optimal*, running times of such algorithms may differ by large constant factors. In some cases an algorithm that is not asymptotically optimal may be better for real problems than the algorithm with the better asymptotic properties. For example, the break-even point between Strassen's $\Theta(n^{2.81})$ algorithm and the traditional $\Theta(n^3)$ algorithm is about size $n = 30$ for common implementations [131]. The break-even point for the $\Theta(n^{2.496})$ matrix multiplication algorithm of Coppersmith and Winograd [84] is ridiculously large; the algorithm is of theoretical interest only. The simplex algorithm for linear programming, which has an exponential worst-case time, runs in linear time on almost all practical problems, while the currently known algorithms with polynomial time bounds are usually much slower than the simplex method.

When you have a good practical algorithm that is also optimal, you know you have a good algorithm. This is the main reason for interest in lower bounds.

10.1 INPUT-OUTPUT LOWER BOUNDS

The simplest type of lower bound, one that is easily obtained for most problems, is based on the fact that for most problem sets any algorithm must read all of its input. Any algorithm definitely has to write all of its output. To show that the input must be read, the usual technique is to show that changing any part of the input has the potential to change the answer. In such cases the sum of the amount of input and of the amount of output gives a lower bound on the time for any algorithm that solves the problem.

Although an input-output lower bound may be crude, it is often the best lower bound available. For example, the $\Omega(n^2)$ lower bound for matrix multiplication can be obtained by input-output considerations, and it is (to within a constant factor) the best bound known. [The best known algorithm for matrix multiplication uses time $O(n^{2.49})$, which is much worse than the lower bound.] There are, however, some simple algorithms where the input-output lower bound is *tight*: it is the same as the time required by the best algorithm (to within a constant factor). One such algorithm is the Integer Addition Algorithm (Algorithm 1.4). As we saw in Section 1.4.1, Integer Addition takes time $\Theta(n)$. Since the answer has n digits, any algorithm for addition must take time $\Omega(n)$. Thus, Algorithm 1.4 is optimal.

The difficulty with input-output techniques is that you normally cannot obtain nonlinear lower bounds. Sometimes, however, nonlinear lower bounds have been obtained by showing that an algorithm must store a lot of intermediate answers and then look them up again.

EXERCISE

1. Show that Algorithm 1.7, Matrix Addition, is optimal.

10.2 INFORMATION THEORETIC LOWER BOUNDS

The lower-bound technique used in this section involves studying the number of tests and decisions an algorithm must make, based on the number of possible input cases that must be separated. An algorithm must take, in the worst case, at least as many steps as the number of tests it needs (in the worst case) to come up with a solution. If we can categorize the input into classes, each of which produces a different answer, then the time required is at least as large as the number of steps needed to determine which class the input belongs in.

Let n be the number of classes of input. Suppose the program does a test and performs a two-way branch based on the result of the test. (The two-way branch is the most powerful branch instruction used in many programs.) After the branch the input is known to lie in one of two subsets. Let n_1 be the number of classes in the first subset and n_2 be the number in the second. Then $n = n_1 + n_2$, and the larger of n_1 and n_2 contains at least $\lceil n/2 \rceil$ classes. The same idea works after each decision. If you consider all the cases after k decisions, the set with the most classes has at least $\lceil n/2^k \rceil$ classes. Since the final answer must correspond to a set that contains just one class of the input, the worst-case time of any program where binary branching is the most powerful branching instruction must be at least $\Omega(\lg n)$.

The same principle applies when more powerful branching operations are available. Some computers can do a three-way branch. To distinguish n classes with three-way branches requires time $\Omega(\log_3 n)$. For a computer with k-way branches, the time required is $\Omega(\log_k n)$. (An indexed jump can be thought of as a k-way branch.) All computers have a limit on the number of ways a single instruction can branch (usually the memory size is the limit for the most powerful branching instruction). Since $\Omega(\log_k n) = \Omega(\lg n)$ (because $\log_k n = (\ln n)/\ln k$), any *fixed* branching factor gives only a constant factor of improvement over two-way branching.

10.2.1 Sorting

Consider a sorting algorithm that sorts records, where each record has a distinct *key* (the records are to be sorted according to their keys). For purposes of sorting, the actual values of the keys are unimportant; all that matters is their relative order. Thus, if there are n records, we can think of the keys as the integers from 1 to n, where the value 1 is associated with the smallest key, 2 with the next smallest, and so on. On input, the records and their associated keys are arranged in some arbitrary order; on output, they will be ordered so that the keys occur in the order $1, 2, \ldots, n$. For each ordering of the numbers from 1 to n, a sorting program must rearrange the associated records differently. Thus there are $n!$ classes of input in all. Assuming that two keys can be compared in unit time regardless of their values, a program that uses binary branching as its most powerful branching instruction must, in the worst case, do $\lg n! = \Theta(n \lg n)$ operations.

Lower bounds for sorting are discussed in Knuth [11, Section 5.3] and in Baase [2, Section 2.4].

EXERCISES

1. Show that in the worst case 7 binary comparisons are necessary to sort 5 items.
2. Repeated binary insertion requires 8 comparisons to sort 5 items. Most other general sorting methods also require at least 8 comparisons. Develop an algorithm that is special for sorting 5 items which can do the sorting with only 7 comparisons.

10.2.2 Merging

Now consider the problem of merging two sorted lists, one of which contains m elements while the other contains n elements. (We assume that the elements are all distinct.) The number of ways that m items may be inserted among n items is $\binom{m+n}{m}$. Since it is possible to arrange the two lists so that the merged list corresponds to an arbitrary way of inserting m items into a list of n items, an algorithm that uses binary branching as its most powerful branching instruction must take at least $\lceil \lg \binom{m+n}{m} \rceil$ steps.

Let's investigate this lower bound in more detail for the case $m = n$. A lower limit on the number of binary comparisons needed to merge two lists of size m is given by

$$\left\lceil \lg \binom{2m}{m} \right\rceil = 2m - \tfrac{1}{2} \lg m + O(1), \tag{1}$$

where Stirling's approximation to the factorial has been used to produce the value on the right side.

An upper bound on the number of binary comparisons needed to merge two lists of size m is $2m$. With a little thought you can produce an algorithm that uses $2m - 1$ binary comparisons (see Exercise 1).

EXERCISES

1. Show that $2m - 1$ is an upper bound on the number of comparisons used for merging two lists of size m. Hint: Develop an algorithm and analyze it.
2. Consider the problem of determining for an item q and a set of n distinct elements which, if any, of the numbers in the set is equal to q. Show that if this problem is solved with binary comparisons, then at least $\lceil \lg(n+1) \rceil$ comparisons must be used.
3. Consider the problem of determining, for each query in a set of k distinct queries and a set of n distinct answers, which answer, if any, matches the query. Show that if this problem is solved with binary comparisons, then at least $\lceil \lg(n+1)^{\underline{k}} \rceil$ comparisons must be used.

10.2.3 Adversary Lower Bounds

In the previous section, the upper bound we obtained for the problem of merging two sorted lists was somewhat larger than the lower bound. This leads us to wonder which one is closest to the true optimum value. We would like, if possible, to develop an even better merging algorithm than the one for Exercise 10.2.2–1. In this case, however, it is the lower bound that is off, as we will find by use of a more powerful lower-bound technique. Any efforts to develop an algorithm that uses fewer comparisons are, therefore, wasted because *there is no such algorithm* (as long as a binary branch is the most powerful branching instruction used).

To get a better lower bound, imagine that you have developed a clever algorithm, and now you are testing it to see how long it runs. Instead of using a random input, imagine that you go to a diabolically clever adversary who selects which branch to take at every decision point in the algorithm. The adversary always selects the branches in a way that corresponds to some input, while at the same time trying to make the algorithm produce an incorrect answer. The adversary knows all the tricks of your algorithm, so if you have any shortcuts that result in the algorithm being invalid for some input, it will find an input sequence that causes an error. Such an adversary is sometimes called an *oracle* because of its great wisdom concerning your program.

The adversary argument is used to prove a lower bound that applies to all algorithms that solve the problem. It will have some strategy for deciding which branch to select at each decision point. You, the analyzer who is trying to produce the lower-bound result, know the strategy that the adversary will use (in fact you invent the adversary). From this knowledge you can determine various combinations of tests that any correct algorithm must make. If you invent an adversary with a strategy that is not very powerful, you do not get an impressive lower bound. If you invent an adversary that is too powerful, you may not be able to decide what tests the algorithm must have. Then you don't get any result. Your first problem is to invent an adversary that is powerful enough to obtain a useful result, but simple enough to analyze. After this the rest of your problem is straightforward analysis.

It is time for an example. We will use an adversary argument to show that any algorithm that uses a binary branch as its most powerful branching instruction must make $2m - 1$ comparisons to merge two lists of size m correctly. Let the input to the merge algorithm be the lists $x_1, x_2, \ldots, x_m$ and $y_1, y_2, \ldots, y_m$. The strategy for the adversary is to select branches so that they correspond to what the algorithm does when the input obeys the condition $x_1 < y_1 < x_2 < y_2 < \cdots < x_m < y_m$. (This is a particularly difficult sequence to merge correctly.) Any algorithm that does not compare x_1 to y_1, y_1 to x_2, x_2 to y_2, and so on will make a mistake for some input. If, for example, it does not compare y_1 with x_2 then it will not be able to distinguish between the input that the adversary is using for selecting branches and the input $x_1 < x_2 < y_1 < y_2 < x_3 < y_3 < \cdots < x_m < y_m$. Thus at least $2m - 1$ tests are required in the worst case.

Notice that we have said nothing about the *order* in which the tests are done; the algorithm can be extremely clever about that, but in the worst case it must do at least $2m - 1$ tests or it will fail on some input. Similar arguments can be used to improve the lower bound when the two lists are not initially the same length.

This section is based on Knuth [11, Section 5.3.2]. See Atallah and Kosaraju [67] for a simple adversary proof that sorting requires $O(n \ln n)$ time.

EXERCISES

• **1.** Show that at least $n - 1$ binary comparisons are required to determine the largest of n distinct items. Which algorithm in this book solves this problem with $n - 1$ comparisons of data items?

2. Show that you cannot use comparisons to determine the second best of n items without also determining the best. Hint: If your algorithm has two items that might be the best, how can an adversary cause trouble for its determination of the second best item?

3. Show that at least $n + \lfloor \lg n \rfloor - 2$ comparisons are required to determine the first and second largest of n items. Hint: Let a_1 be the number of elements that were second best on at least one comparison, and let a_2 be the number of elements that were second best on at least two comparisons. From the previous two exercises, $a_1 \geq n - 1$. Let p be the number of elements that were compared with the largest element. The second largest element must be one of the elements compared with the largest element, because if the algorithm does not compare the second largest element with a larger element, an adversary could set up a problem where the element that the algorithm thought was second largest was actually first largest. The second largest element must be compared with the other $p - 1$ elements that were smaller than the first largest (otherwise one of them might actually be second largest). As a final step you must show that $p \geq \lfloor \lg n \rfloor$. To do this, it is helpful to consider an adversary that arranges the data so that at any comparison between previously unbeaten elements, the element with the fewest previous comparisons will be the larger. If this is done, then on the ith comparison for an element, at most $2^i - 1$ other elements will be known to be smaller.

10.3 PROBLEMS THAT REQUIRE SIMILAR TIMES

One way to study the time needed to solve a set of problems is to show that the problems are equivalent to another set of problems, perhaps one that has already been carefully studied. Suppose you have a way to quickly transform each instance of a problem in set A into an instance of a problem in set B. Suppose you also have a method to transform the solution of the resulting problem in set B back into the solution of the original problem in set A. Then you know that you can solve the problems in set A almost as rapidly as the corresponding problems in set B. You simply transform a problem in set A into a problem in set B, use the algorithm for solving problems in set B, and transform the answer of the problem in set B so as to obtain the solution of the original problem. This means that the problems in set A are no harder to solve than those in set B. In particular, except for the overhead of the transformation, the time needed by the best algorithm for solving problems in set B is at least as large as the time for best algorithm for solving problems in set A. (Here we use "time" to mean worst-case time.) Suppose you can also find rapid transformations going the other way, so that you can solve problems in set B by transforming them into equivalent problems in set A. Then, neglecting the overhead of the transformations, the worst-case time of the best algorithms for each set of problems is the same. Likewise, any algorithm that has a good worst-case time for one problem set leads to an algorithm that has a good worst-case time for the other problem set. The ideas and insights gained from studying either problem set can be applied to both problem sets.

How small does the time for the transformations need to be? That depends on the degree of similarity of worst-case running times you wish to demonstrate for the two problem sets. Suppose that you have a series of problem sets of type A, $S_A(i)$ for $i \geq 1$, and a series of problem sets of type B, $S_B(i)$ for $i \geq 1$. [The parameter i is a measure of the difficulty of the problems in each set. If A is matrix multiplication, for example, $S_A(i)$ might be all multiplications of $i \times i$ matrices.] Suppose your transformation converts problems in $S_A(i)$ into problems in $S_B(i)$ (for each i). Let $T_B(i)$ be the worst-case time for some algorithm to solve problems in set $S_B(i)$. Finally, suppose you wish to show that $T_A(i)$, the worst-case time for solving problems in $S_A(i)$, is $O(T_B(i))$. How fast do the transformations from problems in $S_A(i)$ to corresponding problems in $S_B(i)$, and the transformation of solutions of problems in $S_B(i)$ to the solutions of corresponding problems in $S_A(i)$, need to be? Let the time for the two transformations be $t_1(i)$ and $t_2(i)$. Then

$$T_A(i) \leq t_1(i) + T_B(i) + t_2(i), \tag{2}$$

so we have $T_A(i) = O(T_B(i))$ when $t_1 = O(T_B(i))$ and $t_2(i) = O(T_B(i))$. Suppose we don't even know $T_B(i)$, but we would still like to show that $T_A(i) = O(T_B(i))$. This can be done if we have a lower bound $L_B(i)$ such that $T_B(i) \geq L_B(i)$, $t_1(i) = O(L_B(i))$, and $t_2(i) = O(L_B(i))$.

10.3.1 Boolean and Integer Matrix Multiplication

Let's illustrate how we can use this approach by comparing the *Boolean matrix multiplication* problem with the *integer matrix multiplication* problem and the *transitive closure* problem.

Boolean matrix multiplication is simply matrix multiplication modified so that the elements being multiplied are logical quantities (*true* and *false*). If A, B, and C are $n \times n$ numerical matrices with $C = AB$, then the elements of C are given by the formula

$$c_{ij} = a_{i1}b_{1j} + \cdots + a_{in}b_{nj}. \tag{3}$$

The formula is the same for Boolean matrix multiplication, except that each multiplication in eq. (3) is replaced by a logical *and* operation and each addition is replaced by a logical *or* operation. Thus in Boolean matrix multiplication the elements of the answer are

$$c_{ij} = (a_{i1} \wedge b_{1j}) \vee \cdots \vee (a_{in} \wedge b_{nj}). \tag{4}$$

It is clear that Boolean matrix multiplication can be done in time n^3 by direct use of eq. (4).

It is also easy to see that you can do Boolean matrix multiplication as fast as you can do integer matrix multiplication. To compute the product of Boolean matrices A and B, first transform A into A' by replacing each *false* in A with zero and each *true* with 1. Transform B into B' by the same method. Compute the integer matrix product $C' = A'B'$. Finally, transform C' into the answer C by replacing each nonzero element of C' with *true* and each zero element with *false*. This algorithm works because the elements of c'_{ij} count how many of the pairs $a_{ik} \wedge b_{kj}$ are *true*, while c_{ij} just tells whether or not any of the pairs are *true*. The transformation to integer multiplication takes time $O(n^2)$, and the transformation of the answer back to Boolean matrix multiplication takes time $O(n^2)$. Integer matrix multiplication takes time $\Omega(n^2)$ because every element of A and B must be examined to compute the matrix product. Thus Boolean matrix multiplication can be done in time that is big O of the time for integer matrix multiplication. This means that the fast matrix multiplication algorithms of Strassen and others can also be used for Boolean matrix multiplication.

10.3.2 Boolean Matrix Multiplication and Transitive Closure

We defined the transitive closure of a relation in Section 1.11. Superficially, there is no apparent connection between the problem of Boolean matrix multiplication and the problem of finding the transitive closure of a relation. Indeed, it is much more difficult for people to think of a fast algorithm for transitive closure. We will now show, however, that to within constant factors the two problems require the same amount of time.

First we show that there is a straightforward connection between the two problems. Consider a directed graph G on n nodes representing a relation, and an $n \times n$ Boolean matrix A that has a *true* in position i, j if and only if the graph has an arc from node i to node j. (Such a matrix is called the *adjacency matrix* of the graph.) Matrix A has *true* in position i, j if and only if there is a path of length 1 from node i to node j. Now consider A^2. Let's use $[A^2]_{ij}$ for the i, j element of A^2. Since

$$[A^2]_{ij} = (a_{i1} \wedge a_{1j}) \vee (a_{i2} \wedge a_{2j}) \vee \cdots \vee (a_{in} \wedge a_{nj}), \tag{5}$$

$[A^2]_{ij}$ is *true* if and only if there is a path of length 2 from node i to node j. Since

$$[A^3]_{ij} = ([A^2]_{i1} \wedge a_{1j}) \vee ([A^2]_{i2} \wedge a_{2j}) \vee \cdots \vee ([A^2]_{in} \wedge a_{nj}) \tag{6}$$

$[A^3]_{ij}$ is *true* if and only if there is a path of length 3 from node i to node j. In general $[A^k]_{ij}$ is *true* if and only if there is a path of length k from node i to node j. Let G^+ be the graph representing the transitive closure of the relation represented by G. Then G^+ has an *arc* from node i to node j if and only if G has a *path* of length 1 or more from node i to node j. Let A^+ be the adjacency matrix for G^+. If there is a path from node i to node j, then there is a path of length 1, or a path of length 2, or a path of length 3, etc. Therefore

$$A^+ = A \vee A^2 \vee A^3 \vee \cdots. \tag{7}$$

That is, the transitive closure is equivalent to the logical *or* of this infinite sequence of Boolean matrix products. Next notice that if there is a path from node i to node j, then there is a path whose length is no more than n. Therefore

$$A^+ = A \vee A^2 \vee A^3 \vee \cdots \vee A^n. \tag{8}$$

Eq. (8) gives us a way to compute transitive closure by first computing n matrix powers. Since the n matrix powers can be computed with n matrix products, since the logical *or*'s can be done in time n^3, and since Boolean matrix multiplication must take at least time $\Omega(n^2)$, we now know that the time for transitive closure, $T(n)$, is no more than

$$T(n) = O(nM(n)), \tag{9}$$

where $M(n)$ is the worst-case time to compute the Boolean matrix product of two $n \times n$ Boolean matrices using the best possible algorithm.

We will now, however, give a much better result. We will show that $T(n) = \Theta(M(n))$ by giving two algorithms. The first algorithm computes Boolean matrix product using a transitive closure algorithm as a subroutine. An analysis of this algorithm gives $M(n) = O(T(n))$. Next we will give a recursive algorithm that computes transitive closure by using Boolean matrix multiplication of various sized matrices and transitive closure of 1-node graphs to compute the transitive closure for a large graph (the transitive closure of a 1-node graph is the graph itself). An analysis of this algorithm will give $T(n) = O(M(n))$. Together these two results show that $T(n) = \Theta(M(n))$.

To compute AB where A and B are $n \times n$ Boolean matrices, construct the $3n \times 3n$ matrix

$$C = \begin{pmatrix} 0_n & A & 0_n \\ 0_n & 0_n & B \\ 0_n & 0_n & 0_n \end{pmatrix}, \tag{10}$$

where 0_n is an $n \times n$ matrix of zeros. Now,

$$C^2 = \begin{pmatrix} 0_n & 0_n & AB \\ 0_n & 0_n & 0_n \\ 0_n & 0_n & 0_n \end{pmatrix}, \tag{11}$$

and all higher powers of C are zero. Thus

$$C^+ = C + C^2 = \begin{pmatrix} 0_n & A & AB \\ 0_n & 0_n & B \\ 0_n & 0_n & 0_n \end{pmatrix}. \tag{12}$$

The product AB can be read off from the upper right corner of C^+. Thus we get one $n \times n$ matrix product by doing some overhead operations (copying data from one place to another) plus one $3n \times 3n$ transitive closure computation. Since the overhead takes time $O(n^2)$, we have

$$M(n) \leq T(3n) + O(n^2). \tag{13}$$

In eq. (13) since $T(n) = \Omega(n^2)$ (we must read most of the input in order to compute the transitive closure), we can eliminate the $O(n^2)$ term, and since $T(n) = O(n^4)$ (and is thus polynomial), we can remove the factor of 3. Thus we can write

$$M(n) = O(T(n)). \tag{14}$$

Now we will show how to use matrix multiplication to compute transitive closure. This algorithm is much more complicated than the previous one. Let A

be an $n \times n$ adjacency matrix for a graph G, where $n = 2^k$ (see the exercises for other values of n). Partition A into four $n/2 \times n/2$ submatrices:

$$A = \begin{pmatrix} A_{11} & A_{12} \\ A_{21} & A_{22} \end{pmatrix}. \tag{15}$$

Let V_1 be nodes 1 to $n/2$ of G and V_2 be nodes $(n/2+1)$ to n. Submatrix A_{11} represents the arcs between nodes in V_1, A_{22} represents arcs between nodes in V_2, A_{12} represents arcs from nodes in V_1 to nodes in V_2, and A_{21} represents arcs from nodes in V_2 to nodes in V_1.

Now consider A^+. Decompose it in the same way as A:

$$A^+ = \begin{pmatrix} [A^+]_{11} & [A^+]_{12} \\ [A^+]_{21} & [A^+]_{22} \end{pmatrix}. \tag{16}$$

Submatrix $[A^+]_{ij}$ represents all the paths that begin in V_i and end in V_j. A path that starts and ends in V_1 can be decomposed into segments that contain either an arc in V_1 or an arc from V_1 to V_2 followed by zero or more arcs within V_2 followed by an arc from V_2 to V_1. The first type of segment is represented by A_{11}. The second type of segment is represented by $A_{12}(I + (A_{22})^+)A_{21}$. [Here $(A_{22})^+$ is the transitive closure of the submatrix A_{22}, which is not the same as $[A^+]_{22}$.] Now $[A^+]_{11}$ represents all paths that have one or more such segments, so

$$[A^+]_{11} = (A_{11} + A_{12}(I + (A_{22})^+)A_{21})^+. \tag{17}$$

Eq. (17) gives us a way to compute one quadrant of the transitive closure of an $n \times n$ matrix by doing two $n/2 \times n/2$ transitive closures, two matrix multiplications, and two matrix additions. The other four quadrants can be computed with the same two transitive closures, some additional matrix multiplications, and some additional matrix additions. In particular,

$$[A^+]_{12} = (I + [A^+]_{11})A_{12}(I + (A_{22})^+), \tag{18}$$

$$[A^+]_{21} = (I + (A_{22})^+)A_{21}(I + [A^+]_{11}), \tag{19}$$

$$[A^+]_{22} = (A_{22})^+ + (I + (A_{22})^+)A_{21}(I + [A^+]_{11})A_{12}(I + (A_{22})^+). \tag{20}$$

These are not necessarily the most natural expressions that we could write, but they do use only two transitive closures of size $n/2 \times n/2$ to compute the transitive closure for all four quadrants. As we know from Section 5.2.2, reducing the number of $n/2 \times n/2$ transitive closure computations required is of paramount importance.

The following procedure uses eqs. (17–20) to efficiently compute all four quadrants.

Algorithm 10.1 Recursive Transitive Closure: Input: An $n \times n$ Boolean array A with quadrants A_{11}, A_{12}, A_{21}, and A_{22} where A_{11} is $\lfloor n/2 \rfloor$ by $\lfloor n/2 \rfloor$, A_{12} is $\lfloor n/2 \rfloor$ by $\lceil n/2 \rceil$, A_{21} is $\lceil n/2 \rceil$ by $\lfloor n/2 \rfloor$, and A_{22} is $\lceil n/2 \rceil$ by $\lceil n/2 \rceil$. Intermediate storage: Arrays T_1, T_2, T_3, and T_4. Output: Arrays $[A^+]_{11}$, $[A^+]_{12}$, $[A^+]_{21}$, and $[A^+]_{22}$, which are the four quadrants of A^+, the transitive closure of A.

Step **1.** $T_1 \leftarrow (A_{22})^+$. (This step is to be done by a recursive call to the algorithm unless A_{22} has size 1×1, in which case it is to be done directly).

Step **2.** $T_2 \leftarrow I + T_1$.

Step **3.** $T_3 \leftarrow A_{12}T_2$.

Step **4.** $[A^+]_{11} \leftarrow (A_{11} + T_3 A_{21})^+$. (This step is to be done by a recursive call to the algorithm unless A_{11} has size 1×1, in which case it is to be done directly).

Step **5.** $T_4 \leftarrow I + [A^+]_{11}$.

Step **6.** $[A^+]_{12} \leftarrow T_4 T_3$.

Step **7.** $T_5 \leftarrow T_2 A_{21}$.

Step **8.** $[A^+]_{21} \leftarrow T_5 T_4$.

Step **9.** $[A^+]_{22} \leftarrow T_1 + [A^+]_{21} T_3$.

When $n/2$ is an integer, the Recursive Transitive Closure Algorithm has two recursive calls for problems of size $n/2$, six $n/2 \times n/2$ matrix multiplications, and four $n/2 \times n/2$ matrix additions. Since the matrix additions and the overhead for the recursive calls are $O(n^2)$, this gives

$$T(n) \leq 2T(n/2) + 6M(n/2) + O(n^2). \tag{21}$$

Let's pick our unit of time so that $T(1) = 1$. With this boundary condition the solution of eq. (21) (when n is a power of 2, say, $n = 2^k$) is

$$T(2^k) \leq 2^k + 3 \sum_{1 \leq i \leq k} 2^i M\left(\frac{2^k}{2^i}\right) + O((2^k)^2). \tag{22}$$

Now $M(2^k) = \Omega((2^k)^2)$, so

$$\sum_{1 \leq i \leq k} 2^i M\left(\frac{2^k}{2^i}\right) \leq M(2^k) \sum_{1 \leq i \leq k} \frac{1}{2^i} < M(2^k). \tag{23}$$

(See eq. (5.38).) Thus we have

$$T(2^k) \leq 2^k + 3M(2^k) + O(2^{2k}) \tag{24}$$

or, since 2^k is small compared to $2^{2k} = (2^k)^2$,

$$T(n) \leq 3M(n) + O(n^2). \tag{25}$$

Eq. (14) and eq. (25), combined with the fact that $M(n) = \Omega(n^2)$, give

$$T(n) = \Theta(M(n)). \tag{26}$$

[Notice that we needed $M(n) = \Omega(n^2)$ here, and $T(n) = \Omega(n^2)$ to obtain eq. (26), because the transformations take time $\Theta(n^2)$.]

Since eq. (26) was proved by giving an explicit algorithm to compute transitive closure using matrix multiplication, and an explicit algorithm to compute matrix multiplication using transitive closure, whenever an improved matrix multiplication algorithm is found (as long as the improvement is by more than a small factor), it leads immediately to an improved transitive closure algorithm. (Improvement here refers to improvement in the worst-case time required by the algorithm.) Boolean matrix multiplication and transitive closure are computationally equivalent—they require the same worst-case time to within a constant factor.

This section is based on Aho et al. [1, Section 5.7 to 5.9].

EXERCISES

1. From the discussion in the text we have that $T(n) = \Theta(M(n)) + O(n^2)$. What can you say about the limit: $\lim_{n\to\infty} T(n)/M(n)$? Give upper and lower bounds on the value of this limit.

2. The recursive procedure for computing transitive closure reduces the problem of computing an $n \times n$ transitive closure to the problem of computing two $n/2 \times n/2$ transitive closures plus some matrix multiplications and some matrix additions. What two $n/2 \times n/2$ transitive closures are computed?

3. Suppose the best transitive closure algorithm we could find had used three transitive closures of size $n/2$ (instead of two), six multiplications, and three additions. Would this algorithm have been good enough to lead to eq. (26)?

4. Use the fact that any $n \times n$ matrix A can be embedded in a $2^k \times 2^k$ matrix A', i.e.,

$$A' = \begin{pmatrix} A & 0_{mn} \\ 0_{nm} & 0_{mm} \end{pmatrix},$$

where 0_{mn} is an m by n matrix of zeros and $m = 2^k - n$, to show that

$$T(n) = \Theta(M(n))$$

even when n does not have the form 2^k.

5. When you have an algorithm that requires that a matrix have a size that is of a special form, such as 2^k, embedding the original matrix into another that has the required special size usually leads to easy proofs of results about the running time of the method on matrices of arbitrary size, but it does not usually lead to the most efficient algorithm. Analyze the time required by the Recursive Transitive Closure Algorithm to solve a problem of size n. Compare this time with the time required to solve a problem of size n that is embedded in a problem of size 2^k for the smallest k such that $n \le 2^k$.

6. Why does eq. (21) have a less than or equal sign rather than an equal sign?

7. One of the more straightforward ways (when it applies) to show that two problems require the same time is to show that they have the same answer. The *maximum flow problem* has as input a source node, a destination node, and a graph where each edge has an associated capacity. The problem is to find the maximum flow from the source to the destination. Except for the source (where all the flow is away from the node) and the sink (where all the flow is toward the node), the flow into a node must equal the flow out of the node. The flow through each edge must be no more than the capacity of the edge. The *minimum cut problem* has as input two special nodes and a graph where each edge has a cost. The problem is find the minimum total cost for a set of edges that can be cut to produce a graph where there is no path from one of the special nodes to the other one. (In both problems the cost, capacities, and flows are all nonnegative.) Show that these two problems have the same answer if the appropriate matching is made between parts of the two problems. Hint: It is helpful to show that the minimum cut cannot be less than the maximum flow, and that the maximum flow cannot be less than the minimum cut.

10.4 NP-COMPLETE PROBLEMS

We now turn to a weaker type of computational equivalence. Consider two sets of problems, where the problems of size n in the first set require worst-case time $T_1(n)$ to solve and problems of size n in the second set require worst-case time $T_2(n)$ to solve. The running times $T_1(n)$ and $T_2(n)$ are *polynomially equivalent* if $T_1(n) \leq p(T_2(n))$ and $T_2(n) \leq q(T_1(n))$, where p and q are polynomials. Two problems that have worst-case times that are polynomially equivalent will be called *polynomially equivalent problems.* This is a very weak form of equivalence: problems that take time n^{27}, for example, are polynomially equivalent to problems that take linear time. Polynomial equivalence is important because there is a large class of polynomially equivalent problems, many of great practical importance, for which no fast algorithms are known. This section is about this class of problems—the NP-complete problems.

The set of *NP problems* is the set of problems for which solutions can be checked quickly (in polynomial time). Consider the following hypothetical situation. Your company has some difficult problems that it wants to solve. They have decided to hire some people who are good at guessing. Every time the bosses want a problem solved, they will give the problem to one of the guessers and have that person guess what the answer is. The bosses, however, are not fools. They know that guessers sometimes guess wrong. Therefore they ask you to write a fast program to check each guess to see whether it is correct. (This is actually a good way to solve problems if the guesses are usually correct and if your program is fast.) If, for the class of problems your company is interested in, your program always runs in time $p(n)$ on correct guesses, where p is some polynomial and n is a measure of the size of the problem, then your company's class of problems is in the set NP.

An interesting question in the theory of computer science is to what extent the use of guessing (in the above sense) leads to faster programs. In other words, do there exist problems where it is significantly quicker to check whether a guess is correct than it is to solve the problem from scratch? Our common experience suggests that this is the case. Think of how long it takes to solve a jigsaw puzzle. Yet one can tell rather quickly whether a jigsaw puzzle has been solved. Indeed all good puzzles are difficult to solve but have easily checked solutions. One particular variation of this question, "Does guessing help?", is the following: "Can the problems in NP be solved in polynomial time without using guessing?"

Well, no one knows. In what follows we will show that there are problems in NP that are at least as hard as any other problem in NP. These problems are known as the *NP-complete problems*. The solution of any NP-complete problem set requires a time that is polynomially related to the time required to solve any other NP-complete problem set. Some of the problems in this set are very famous; in some cases mathematicians have been trying for hundreds of years to find rapid solution methods, so far without success. The best-known methods for NP-complete problems require worst-case time $O(\exp(n^{1/k}))$, where k is a positive integer. On the other hand, no one has ever shown that more than polynomial time is required for this set of problems. The foremost open problem of theoretical computer science is to determine whether NP-complete problems can be solved in polynomial time without using guessing. The set of problems that can be solved in polynomial time (without using guessing) is called P, so this question is often phrased, "Does P = NP?"

When discussing the computational equivalence of problems as similar as matrix multiplication and transitive closure, it was clear how to make a reasonable choice for which elementary operations to use to express the complexity of the algorithms. For problems from a large number of vastly different domains, it is best to work with elementary operations which are simple and general enough that they constitute a completely general model of computation. The model should also be reasonable in the sense that an elementary operation can be carried out in a bounded amount of time on a real computer. For example, an operation that computed the sum of an exponential number of terms in one step would not be reasonable. The model we will use is a Turing machine. Turing machines are not much fun to write programs for, but they are very useful for proving general results about programs for real computers. Both of these conditions arise because Turing machines are very simple.

A Turing machine is an abstract model of a generalized computer. The computer has a control unit, a read-write head, and an infinite storage tape. The head reads and writes symbols on the tape. The control unit has a finite memory (represented by the state the control unit is in). Based on its current state and the symbol just read, the control unit decides what to write on the tape, which direction to move the head (one square), and what state to go to. To program the machine, we specify the following:

1. A finite alphabet Γ of tape symbols.
2. A finite set Q of states with three distinguished elements: the starting state q_0, the accepting state q_Y, and the rejecting state q_N.
3. A state transition function $\delta : (Q - \{q_Y, q_N\}) \times \Gamma \to Q \times \Gamma \times \{\text{left}, \text{right}\}$. The function δ is a function of the current state and the current symbol

under the read head. The output of the function is three items: the next state the machine will be in, the symbol that the machine will write on the tape, and the direction the head will move. This function is analogous to the program in an ordinary computer. Turing machines are like special purpose computers with the program wired in.

The Turing machine works as follows. The input tape is put into the machine, with the read head on the first square. The tape contains a problem instance to be solved. The machine starts in state q_0. It then does the following:

1. If the current state, q, is q_Y or q_N, then the computation has ended. The result is "yes" if the state is q_Y and "no" if the state is q_N.

2. Read the input symbol s on the current tape square. Let the transition for state q and input s be $\delta(q, s) = (q', s', \Delta)$. Erase symbol s from the tape and write symbol s' in its place. Move the head one square in the direction specified by Δ. Place the control unit in state q'. Go back to Step 1.

The *configuration* or *instantaneous description* of a machine is a list of all the information needed to predict its future behavior. In the case of a Turing machine, a configuration consists of the current state, the current contents of the tape, and the current position of the read head. If you are given a configuration, then you can use the above rules to generate all the future configurations.

We say that a Turing machine M *accepts* a string $x \in \Gamma^*$ (Γ^* is the set of strings of length zero or more made up of symbols drawn from Γ) if and only if M halts in state q_Y when it is run on a tape that contains x. (When the machine does not accept x, it may stop in state q_N or it may run forever.) The *language L_M recognized by M* is the set of strings that M accepts. We say that M *solves* a problem set Π if M halts in state q_Y for all problem instances in Π. When Π is given in English or some other informal language, we assume that we are really using the informal language to describe some encoding Π' of Π into a formal language, and that the machine will run on Π'.

There are many equivalent formulations of Turing machines, which differ from this one in minor details. What is important is that the Turing machine is a completely general model of computation and that the elementary operations on a standard computer are equivalent to a small number of Turing machine operations. Here "small" means that the number of operations is at most a polynomial function in the length of the input written on the tape. You can show this for any particular computer by programming a Turing machine to be an interpreter for that computer. Likewise, any general-purpose computer can be used to simulate a Turing machine and to carry out each Turing machine operation in a small number of computer operations.

The *length* of a problem instance is the number of squares of input required to express the problem instance (for a Turing machine that has been defined in a particular way). The *time* used by a Turing machine to solve a problem instance is the number of times that Step 1 in the above description of the operation of a Turing machine is carried out while the machine solves the problem. Thus the time is the same as the number of configurations that the machine goes through while solving the problem. A problem set is in the set P if there exists a Turing machine that solves all problem instances in a time that is a polynomial function of their length. If we know a problem is in set P, then we know we can program a computer to solve the problem in polynomial time.

To show that a problem set is in P, devise an algorithm that solves all instances of the problem set and prove that the algorithm runs in polynomial time for all problem instances. In this section we are interested in the set of NP-complete problems, which we will define formally below. No one has found a polynomial time algorithm for any NP-complete problem set, but no one has proved that there is no such algorithm. Thus no one knows whether the problems in the NP-complete set are also in P. What *has* been done is to show that the computing time for all problems in the set are polynomially equivalent to each other and to the hardest problem set that can be solved in polynomial time when guessing is allowed.

We now need a formal model of computations that use guessing. This model is a *nondeterministic Turing machine* (the Turing machines described above are also known as *deterministic Turing machines*). All realistic machines are, of course, like deterministic Turing machines in that they do not use guessing.

A nondeterministic Turing machine is a deterministic Turing machine with an extra *guessing module* added. The guessing module is equipped with a write-only head. When the nondeterministic Turing machine is started, the guessing module writes its guess at the answer (plus perhaps some additional information that is helpful in proving the guess to be correct) on the tape next to the input. It then turns itself off and starts the deterministic part of the program, which operates just like an ordinary deterministic Turing machine. This part simply checks that the answer written by the guessing module is indeed the solution of the problem.

You might be surprised that the guessing module doesn't even bother to look at the input. We want a model of the most powerful kind of guessing you can imagine: guessing that is so good that it is always right and does not have to worry about such mundane details as examining the input. Remember, we are constructing a model to show certain problems are hard. It is okay to make our machines very powerful, just as long as they are at least as powerful as real machines, and as long as we can obtain some useful results in the end. We have here a very powerful model of good guessing.

The guessing module can write any string of symbols from Γ on the input tape. The string can be of whatever length is necessary. A nondeterministic algorithm "solves" a problem Π if, given a problem instance I on the input tape, (1) when the guessing module guesses the correct answer to I, the machine will halt in state q_Y, and (2) when the guessing module guesses an incorrect answer, the machine will not halt in state q_Y. (It can halt in state q_N or it can run forever.) In other words, a nondeterministic Turing machine solves a problem if the deterministic part can verify, for each problem instance, whether a proposed correct solution of the problem instance is indeed correct (and will not say that any incorrect answers are correct). If a problem instance has a solution, then there must be some string that the guessing module can write that will cause the machine to stop in the accepting state. If a problem instance has no solution, then there must be no string that the guessing module can write that will cause the machine to stop in the accepting state.

In measuring the time for a nondeterministic Turing machine to finish a computation, we omit the time required for the guessing module. (Including the time for the guessing module would not change things very much if we assume that it spends most of its time writing down its guess. An efficient guessing module would only write on the squares that will later be read, so the nondeterministic machine could get its job done in about the same amount of time that the nonguessing part of the calculation takes). A nondeterministic Turing machine can have an infinite number of possible computations for a given problem instance, one for each string from Γ^* that the guessing module might produce. The time required for it to solve a problem instance I is defined to be the *minimum* number of times a new configuration is entered by the checking stage while solving I. That is, it is the number of steps the machine uses to check the easiest to check correct solution of the given problem instance. If the problem has just one solution, then the time is the number of steps needed to check it. By defining the time in this way, we capture the notion of a guesser that is always right and that always picks the best answer (the guess that allows the deterministic part solve the problem in the fastest way).

The time required for problem instances that have no solution is of no concern when defining the time required by a nondeterministic Turing machine.

A nondeterministic Turing machine runs in polynomial time if there is a polynomial p such that the accepting computation for each problem instance of size n is bounded by $p(n)$ for all n. The class NP is the class of problem sets for which there exists a polynomial time nondeterministic Turing machine acceptor.

Notice that the question of whether $\mathrm{P} = \mathrm{NP}$ does not imply anything about the degree of the polynomial time bound on any problem set. A problem set that takes time $O(n^3)$ on a nondeterministic machine might, for example, have a deterministic time of $O(n^{27})$, whether or not $\mathrm{P} = \mathrm{NP}$.

Although we do not know whether $\mathrm{P} = \mathrm{NP}$, it is easy to see that $\mathrm{P} \subset \mathrm{NP}$. If M is a deterministic Turing machine that solves a problem set Π in polynomial time, we can build a nondeterministic machine M' by attaching a guessing module to M. Machine M' ignores the guess and otherwise acts just like M.

Given a nondeterministic machine M' that solves a problem set in a time bounded by a polynomial $p(n)$, we can construct a deterministic Turing machine M for the problem that runs in exponential time. Let k be the number of symbols in the alphabet Γ. Since M' can read at most $p(n)$ tape symbols from any tape in $p(n)$ steps, we can assume that the guessed tape is no longer than $p(n)$. An exhaustive list of its possible guesses will have length at most $k^{p(n)}$. The machine M operates by systematically generating each guess and using the method of the deterministic part of M' to check the guess for correctness. To keep it from running forever on an incorrect guess, it goes on to the next guess whenever it has spent more than $p(n)$ steps on the current guess.

This horribly inefficient deterministic machine actually corresponds rather closely to the methods used in practice for solving the hardest problems in NP.

While backtracking and other techniques are used to avoid trying every possible guess, the basic approach consists of searching through a space of possible answers until an actual answer is found. Such searches take exponential time.

What are the hardest problem sets in NP like? A proposed solution of each instance can be checked in polynomial time, but, using the best-known algorithms, exponential time is needed to actually find the solutions of some of the problem instances. Cook, in a fundamental paper, showed that the satisfiability problem is essentially as hard as any other problem set in NP. He did this by showing that the computation performed by a nondeterministic Turing machine M on each input can be encoded as a satisfiability problem, and that size of the resulting satisfiability problem is polynomially related to the time required by M to solve the problem. Furthermore, this satisfiability problem can be generated in polynomial time. Thus if satisfiability is in P, so is every other problem in the class NP! A problem set is NP-complete if it is in NP and every problem set in NP can be transformed into it in polynomial time. Thus, satisfiability is NP-complete.

Recall from Section 4.3.1 the definition of the satisfiability problem. Briefly, for a problem of size n we have n Boolean variables, $\{x_1, \ldots, x_n\}$, and a set of clauses, where each clause is the disjunction (logical *or*) of some set of literals (a literal is a variable or its negation.) A predicate is the conjunction (logical *and*) of the clauses; it is satisfiable if there is some assignment of truth values to the variables that makes the predicate true.

It is easy to see that satisfiability is in NP. To check whether a predicate is satisfied by a given truth assignment takes a time that is no more than quadratic in the length of the predicate. Obtaining a satisfying truth assignment, however, is not so easy. The backtracking method, discussed in Section 4.3.1, can be refined by adding additional tricks, but the best algorithms still have exponential worst-case times. In fact, we will now show that any problem in NP can be reduced, in polynomial time, to the satisfiability problem. This result is known as Cook's Theorem, and it established satisfiability as the first of the NP-complete problems.

To show that satisfiability is NP-complete, we must show how to transform any problem in NP into a satisfiability problem in polynomial time. What we will do is to transform a description of the nondeterministic Turing machine computation that solves the problem into a conjunctive normal form formula. The resulting formula will have pieces that describe the input tape (including both the problem description and the output of the guessing module), the rules that the machine follows, and the configurations that the Turing machine can reach in polynomial time. The pieces will be combined in such a way that the resulting problem is satisfiable if and only if the original problem could be solved in polynomial time. The beauty of this method is that it allows us to specify a transformation that works for any problem in NP, since by definition any problem in NP has a corresponding nondeterministic polynomial time Turing machine computation.

Suppose M is an arbitrary nondeterministic Turing machine. Let's consider those computations whose running time is bounded by the polynomial $p(n)$, where n is the length of the input. Call the tape square on which the computation starts square 0; number the squares to the right of it 1, 2, and so on; and number those to the left -1, -2, ... The only squares that the computation can look at are those that can be reached from square 0 within $p(n)$ computation steps: squares $-p(n)$ through $p(n)$.

We will start the explanation of the construction by defining the variables that will appear in the satisfiability problem. There are three sets of variables. The first set is the set of Q_{ik} for $0 \le i \le p(n)$, $0 \le k < |Q|$, where $|Q|$ is the number of states in the Turing machine. The satisfiability problem we are constructing will be designed so that it can be satisfied with Q_{ik} *true* if and only if the machine has an accepting computation such that the machine is in state k at time i.

The second set of variables is the set of H_{ij} for $0 \le i \le p(n)$, $-p(n) \le j \le p(n)$. The satisfiability problem is constructed so that it is satisfied with H_{ij} *true* if and only if the machine has an accepting computation such that the head is scanning square j at time i.

The third set is S_{ijk} for $0 \le i \le p(n)$, $-p(n) \le j \le p(n)$, $1 \le k \le |\Gamma|$, where $|\Gamma|$ is number of symbols in the alphabet that can be written on the tape. The satisfiability problem is designed so that it is satisfied with S_{ijk} *true* if and only if the machine has an accepting computation such that at time i square j contains symbol k.

These sets of variables are adequate to describe any configuration of the Turing machine (as well as many things that are not configurations). As long as we are able to construct a predicate that forces the variables to have the properties given above, the predicate will be satisfiable if and only if the Turing machine has a computation that accepts the input.

In constructing the predicate, we will have two types of rules. One type will make sure that the configurations described by the variables are compatible with the input and with the rules of operation of the Turing machine. The rules of the other type will ensure that the variables can be set only in ways that describe configurations. For example, there is no configuration with both Q_{ik_1} and Q_{ik_2} *true* (for $k_1 \neq k_2$), so we will have a portion of the predicate that ensures that, for all i and for all $k_1 \neq k_2$, the predicate cannot be satisfied if $Q_{ik_1} = Q_{ik_2} = \mathit{true}$.

The predicate, as you might imagine, is rather complex. It contains seven groups of clauses. Each group is designed to enforce a particular type of restriction on the computation.

Group 1 contains two subgroups. The first subgroup, $Q_{i,0} \vee Q_{i,1} \vee \cdots \vee Q_{i,|Q|-1}$ for $0 \le i \le p(n)$, says that at each time i the machine is in some state. The second subgroup, $\neg Q_{ij} \vee \neg Q_{ij'}$ for $0 \le i \le p(n)$, $0 \le j < j' < |Q|$, says that at each time i the machine is in at most one state. The logical *and* of all the clauses in group 1 says that the machine is in exactly one state at each time.

The second and third groups are similar to the first. Group 2 says that at each step the machine is scanning exactly one square. Group 3 says that at each step each tape square contains exactly one symbol. The specifications of these two groups are left as exercises.

Group 4 ensures that we have the correct initial conditions. The machine must start in state q_0, the problem must be written on the input tape, and the machine must be scanning square zero. Let the input be the string $x = s_1 s_2 \ldots s_n$, where $s_i \in \Gamma$. The one-literal clauses S_{0js_j} for $1 \leq j \leq n$ ensure that the input is written on the tape. Without loss of generality we can assume that the guesser has written on the remaining squares and that we do not have to specify what is initially on them. The clause Q_{00} says that we are in state q_0 at time 0. The clause H_{00} says that we are scanning tape square zero at time 0.

Group 5 consists of one clause that says we have reached the accepting state by time $p(n)$. If the accepting state, q_Y, is state 1, then this clause is $Q_{1,1} \vee Q_{2,1} \vee \cdots \vee Q_{p(n),1}$.

Group 6 ensures that our set of clauses acts like Turing machine M. (All the clauses so far mainly took care of various bookkeeping details.) This group contains clauses that relate what happens at time i to what happens at time $i+1$. They ensure that the transitions described by the logical variables are in accordance with the transition function of M. Thus this group plays a most important part in the construction.

The clauses for Group 6 are based on the logical identity $(A \wedge B \supset C) \equiv \neg A \vee \neg B \vee C$. We would like to make statements of the form "A and B imply C", but we are forced to do it with clauses, so we say "not A or not B or C", which is logically equivalent. In particular, we want to say that, if at time i we are in state k scanning symbol l on square j, then at time $i+1$ we have moved left or right, changed our state, and changed the tape in accordance with the definition of our machine M. To say that at step i we are in state k scanning square j and reading symbol l, we write $Q_{ik} \wedge H_{ij} \wedge S_{ijl}$. To say that from this configuration we must go to state k', we can write $Q_{ik} \wedge H_{ij} \wedge S_{ijl} \supset Q_{i+1,k'}$, or in clause form $\neg Q_{ik} \vee \neg H_{ij} \vee \neg S_{ijl} \vee Q_{i+1,k'}$. To say that we move one square to the left, we write $\neg Q_{ik} \vee \neg H_{ij} \vee \neg S_{ijl} \vee H_{i+1,j-1}$. To say we move one square to the right, we write $\neg Q_{ik} \vee \neg H_{ij} \vee \neg S_{ijl} \vee H_{i+1,j+1}$ instead. Finally, to say that symbol l' is written on square j, we say $\neg Q_{ik} \vee \neg H_{ij} \vee \neg S_{ijl} \vee S_{i+1,j,l'}$.

To build our clauses for Group 6, we examine the transition function δ. Let $\delta(k, l) = (k', l', d)$, where d is the direction left or right. Then for each i $[0 \leq i < p(n)]$, the three clauses to give the next state, the direction, and the new tape symbol are constructed as described in the previous paragraph.

Group 7 takes care of one more bookkeeping detail: all of the tape except one square must be the same at time $i+1$ as it was at time i. Groups 6 and 7 are left as exercises.

Any legal computation of M that halts within time $p(n)$ corresponds to an assignment of values to the variables that satisfies all the clauses except Group 4. (This is clear if you think about the meaning of each group of clauses.) If there is a computation of M that accepts x, then the corresponding assignment of values to the variables must satisfy all the clauses, including Group 4, that we have constructed. Conversely, if there is a satisfying truth assignment for the variables, it corresponds to a legal accepting computation of M with input x.

It remains to show that the time for constructing the satisfiability problem and the size of the satisfiability problem are polynomially related to n. First let's consider the size of the clauses that are generated. Group 1 contains two subgroups: the first has $p(n) + 1$ clauses, where each clause has $|Q|$ literals. The second subgroup has $(p(n) + 1)|Q|(|Q| - 1)/2$ clauses, where each clause has two literals. The exact size of Groups 2 and 3 is left as an exercise. [The best construction leads to $O(p(n)^2)$ clauses of length $O(p(n))$.] Group 4 has $2p(n) + 3$ clauses of length 1. Group 5 has one clause of length $p(n) + 1$. Group 6 has $6p(n)(p(n) + 1)|Q||\Gamma|$ clauses of length no more than 4. Group 7 can be done with $O(p(n)^2)$ clauses of length 2. The total size of the predicate is $O(p(n)^3)$. The time to construct the transformed predicate is $O(L(n)^2)$, where $L(n)$ is the length of the predicate.

Suppose we have found a satisfying truth assignment for the predicate corresponding to giving input X to machine M. If M leaves the answer to the problem on the tape at the end of the computation, then the answer can be obtained from the values of the variables $S_{p(n)jk}$, which indicate what symbols are on the tape at time $p(n)$. The time needed to obtain this answer is linear in the number of such variables, $2|\Gamma|p(n)$. Satisfiability is therefore NP-complete.

This section is based on Garey and Johnson [4, Chapters 1 and 2]. Cook's theorem originally appeared in [82].

EXERCISES

1. Give the clauses that say that at each time i the Turing machine is scanning exactly one tape square (Group 2).
2. How many clauses did your answer to the previous problem have? How long were the clauses?
3. Give the clauses that say that at each time i each square of the tape has exactly one symbol (Group 3).
4. How many clauses did your answer to the previous problem have? How long were the clauses?
5. How many clauses of each type are there in Group 6? Expjlain your answer.
6. Give the clauses that say that all squares except the one being scanned at time i have the same symbol at times i and $i + 1$.
7. How many clauses did your answer to the previous problem have? How long were the clauses?

10.4.1 3-Satisfiability

Now that we have shown that satisfiability is NP-complete, we have a second way that we can show problems to be NP-complete. To show that a problem set Π in NP is NP-complete, we can either transform any nondeterministic Turing machine calculation into Π, or we can transform any satisfiability problem into Π. For most problem sets, it is much easier to transform satisfiability into the problem set than it is to transform nondeterministic Turing machine calculations into the set. To illustrate, let's consider 3-satisfiability. This problem is a restricted version of satisfiability where each clause contains exactly three literals.

Once we have shown that this problem is NP-complete, it will be even easier to show that other problems are NP-complete, because we will have a third technique. We will be able to show a problem set is NP-complete by transforming 3-satisfiability problems.

Proofs that a problem is NP-complete often follow a standard pattern. We first show that the problem is in NP. Then, we show how to transform some NP-complete problem to it in polynomial time. We will call this NP-complete problem the helper problem. Since the transformation requires only polynomial time, the size of the transformed problem must be polynomially related to the size of the original helper problem. Finally, we show how to obtain a solution of the original helper problem from a solution of the transformed problem in polynomial time. When chosing which helper problem to use, it is usually best to look for the most restricted problem which is known to be NP-complete and which is similar to our problem. The more restricted a helper problem is, the less effort will be needed to transform every instance of it into an instance of our problem.

It is clear that 3-satisfiability is in NP, since it is just a subset of satisfiability. Thus 3-satisfiability can be no harder than satisfiability. Now let's show that 3-satisfiability is as hard (to within polynomial transformations) as satisfiability. Let $P = c_1 \wedge c_2 \wedge \cdots \wedge c_m$ be a predicate with clauses c_1, c_2, ..., c_m over the set of variables $X = \{x_1, x_2, \ldots, x_n\}$. We wish to construct a predicate P' that is an instance of 3-satisfiability, where P' is satisfiable if and only if P is, and to do so in such a way that the transformation can be done in polynomial time.

We do this as follows. For each clause in P, we construct an equivalent set of clauses in P'. The nature of the set of clauses will depend on the length of the clause in P. Suppose clause c_j of P is $l_1 \vee l_2 \vee \ldots \vee l_k$, where the l_i are literals. If $k = 1$, then c_j consists of the single literal l_1. In this case define two new variables, y_{j1} and y_{j2}. The set of clauses in P' corresponding to c_j is $\{l_1 \vee y_{j1} \vee y_{j2},\ l_1 \vee y_{j1} \vee \neg y_{j2},\ l_1 \vee \neg y_{j1} \vee y_{j2},\ l_1 \vee \neg y_{j1} \vee \neg y_{j2}\}$. If l_1 is *true*, then all these clauses are *true*, no matter what values y_{j1} and y_{j2} have. If l_1 is *false*, then one of the clauses is *false*, regardless of the values of y_{j1} and y_{j2}. Thus this set of four clauses is satisfiable if and only if l_1 is *true*, i.e., if and only if c_j is satisfiable.

If $k = 2$, our clause c_j is $l_1 \vee l_2$. Define a new variable y_{j1}. Our set of clauses is $\{l_1 \vee l_2 \vee y_{j1},\ l_1 \vee l_2 \vee \neg y_{j1}\}$. It is easy to see that these two clauses are simultaneously satisfiable if and only if our original clause is satisfiable.

If $k = 3$, our clause c_j already has length 3, so we use it in P' unchanged.

The final case is $k > 3$. Define the new variables y_{ji} for $1 \leq i \leq k-3$. The clauses for P' are: $\{l_1 \vee l_2 \vee y_{j1},\ \neg y_{j1} \vee l_3 \vee y_{j2},\ \neg y_{j2} \vee l_4 \vee y_{j3},\ \ldots,\ \neg y_{j,k-4} \vee l_{k-2} \vee y_{j,k-3},\ \neg y_{j,k-3} \vee l_{k-1} \vee l_k\}$. For the original clause to be *true*, at least one of its literals must be *true*. We need to show that all the clauses of the above set are simultaneously satisfiable (for some set of values for the y_{ji}) if at least one of the l_i is *true*, and that at least one of the clauses is not *true* (for any set of values for the y_{ji}) when none of the l_i is *true*.

Suppose l_1 is *true*. Then the first clause is *true* regardless of the values for any other variables. So we can let y_{j1} be *false*. This makes the second clause *true*. Now let y_{j2} be *false*. This makes the third clause *true*. In fact, making all the remaining y_{ji} *false* makes all the remaining clauses *true*. We use the same set of values for the y_{ji} if l_2 is *true*. If l_{k-1} or l_k is *true*, then all the clauses can be made *true* by setting all the y_{ji} (for $1 \le i \le k$) to *true*.

Suppose l_4 is *true*. This makes the third clause *true*. To make the rest of the clauses *true*, set y_{j1} and y_{j2} to *true* and $y_{j3}, \ldots, y_{j,k-3}$ to *false*. In a similar way, if l_p is *true* for $2 < p < k-1$, then set $y_{j1}, \ldots, y_{j,p-2}$ to *true* and set $y_{j,p-1}, \ldots, y_{j,k-3}$ to *false*.

We have shown that if at least one l_i is *true*, we can find values for the y_{ji} such that all the clauses in the set are *true* simultaneously. What if all the l_i are *false*? Then to make the first clause *true*, y_{j1} must be *true*. But that means that y_{j2} must be *true*, and so on to $y_{j,k-3}$. But if $y_{j,k-3}$ is *true*, the last clause is *false*. Therefore there is no assignment of truth values to the y_{ji} that can make all the clauses *true* simultaneously.

To summarize, then, we have shown how, given P, to construct a predicate P' with three literals per clause that is satisfiable if and only if P is satisfiable. Since the number of distinct literals in a clause can be no more than $2n$, our transformation converts a clause into at most $\max\{2n-3, 4\}$ clauses. The process of deriving the clauses is quite simple. The time required for the transformation is linear in the size of the clause being transformed, so the transformation can be done in polynomial time. From a satisfying truth assignment for P', it is easy to obtain a satisfying truth assignment for P. Thus 3-satisfiability is NP-complete.

This section is based on Garey and Johnson [4, Section 3.1.1]. The original work was done by Cook [82].

EXERCISES

1. [See 126] Show that the problem of deciding satisfiability of conjunctions of binary relations that depend on three-valued variables is NP-complete. That is, deciding satisfiability of problems of the form

$$\bigwedge_i R_i(x_{a_i}, x_{b_i})$$

is NP-complete, where R_i is a relation, x_j is a variable that can have at most three values, and a_i and b_i are mappings of indices which show which pair of variables occur in each term. Hint: Show that any 3-satisfiability problem can be expressed in this form. In particular, let $Q_i(a, b)$ be the function that is *true* if $i \ne b$ or if $a = \mathit{true}$, and that is false otherwise (if $i = b$ and $a = \mathit{false}$, where $1 \le i \le 3$, $1 \le b \le 3$, and $a \in \{\mathit{false}, \mathit{true}\}$). Also consider $\neg Q_i(a, b)$. Any 3-satisfiable formula that is satisfiable has at least one literal that is *true*. Give directions for converting any 3-satisfiability problem into an equivalent problem expressed with the Q's. Use the second variable of Q to indicate which literal in the 3-satisfiability problem is *true*. You must also show that this problem is in the set NP, but this is simple, since it is easy to check if any particular assignment satisfies this relational satisfiability problem.

2. Show that 2-satisfiability takes polynomial time by giving a polynomial time algorithm. Hint: Any variable that appears only in positive literals (literals without a not sign) can be set to *true* and any variable that appears only in negative literals (literals with a not sign) can be set to *false*. (Such literals are called *pure literals.*) As these assignments are made the formula can be simplified. Show that resulting formula will be satisfiable if and only if the original formula is. After these assignments are done, the formula will contain no pure literals. Each time a variable is set one or more additional variables will have their values forced. Give the details of the algorithm and analyze its running time.

10.4.2 Vertex Cover

Now let's turn to some quite different problems. The *vertex cover* problem, as its name suggests, comes from graph theory. Given a graph G with node set V and edge set E, and a positive integer $K \leq |V|$, is there a subset $V' \leq V$ such that $|V'| \leq K$ and, for each edge $(u, v) \in E$, at least one of u and v belongs to V'? That is, can we find a set V' of K or fewer nodes such that every edge in the graph contains at least one element from V'? Such a set V' is a vertex cover of order K.

The initial step of showing that the vertex cover problem is in NP is straightforward. Given a purported vertex cover V', we can check in time proportional to $|E|$ that every edge has at least one endpoint in V'; and we can check in time $|V'|$ whether or not $|V'| \leq K$. To specify our problem, we give the nodes and edges, so the time for checking the proposed solution is bounded by a polynomial function of the length of the input.

To show that the vertex cover problem set is NP-hard (i.e., at least as hard as any problem in NP), we transform 3-satisfiability into the vertex cover problem. Suppose $P = c_1 \wedge c_2 \wedge \cdots \wedge c_m$ is a predicate over the set of variables $X = \{x_1, \ldots, x_n\}$, and that each clause c_i has three literals. We must show how to transform the problem of finding a truth assignment for $x_1, \ldots, x_n$ that satisfies P into a vertex cover problem.

For each variable x_i in X, create a pair of nodes (we will name them x_i and $\neg x_i$) with an edge connecting them. For each clause c_i in P, construct three nodes (we call them c_{i1}, c_{i2}, and c_{i3}). Make a triangular subgraph by connecting each of these three nodes to the other two; this adds edges (c_{i1}, c_{i2}), (c_{i1}, c_{i3}), and (c_{i2}, c_{i3}) to the graph. Now suppose $c_i = l_{i1} \vee l_{i2} \vee l_{i3}$. The graph contains a node corresponding to each of the literals l_{i1}, l_{i2}, and l_{i3}. Connect c_{i1} to node l_{i1}, c_{i2} to node l_{i2}, and c_{i3} to node l_{i3}. Let $K = 2m + n$. Then the vertex cover problem for this value of K and the graph we have just constructed has a solution if and only if the predicate P is satisfiable.

To prove this, let's first see how a vertex cover V' corresponds to a satisfying truth assignment. There are n edges of the form $(x_i, \neg x_i)$. At least one node of each of these edges must be in V'. We have m triangles with nodes of the form $\{c_{i1}, c_{i2}, c_{i3}\}$. In order to cover the edges of the triangles, V' must contain at least two nodes from each triangle. But now we have accounted for $2m + n$

nodes. This means that V' cannot contain any other nodes and is made up of exactly one from each set $\{x_i, \neg x_i\}$ and exactly two from each set $\{c_{i1}, c_{i2}, c_{i3}\}$.

Now suppose we set the values of the variables in such a way that the literals corresponding to nodes in V' are *true*. Consider a triangle (c_{i1}, c_{i2}, c_{i3}). Two of its nodes are in V'; the third is not. Without loss of generality, we may assume that c_{i3} is the node not in V'. An edge joins c_{i3} to the node for the third literal in clause c_i, l_{i3}. Either c_{i3} or the node for l_{i3} must be in V'. But we have assumed that c_{i3} is not in V'; therefore l_{i3} must be in V'. Since we assigned values to the variables in such a way that l_{i3} would be *true*, clause c_i is satisfied. In the same way, all the clauses of P are satisfied by the truth assignment induced by V'.

Conversely, suppose we are given a truth assignment to the variables in X that satisfies P. To construct a vertex cover, first add to V' the n nodes corresponding to the *true* literals under this assignment. Consider a triangle (c_{i1}, c_{i2}, c_{i3}). Since P is satisfied, at least one of these nodes is connected by an edge to a literal that is already in V'. Add the *other* two nodes in the set to V'. This ensures that the edges of the triangle, and the edges connecting the two selected nodes to their literals, have a node in the cover. The cover contains exactly $K = 2m + n$ nodes.

A little thought shows that all the transformations used in the proof are easily done in polynomial time. Therefore, the vertex cover problem is NP-complete.

This section is based on Garey and Johnson [4, Section 3.1.3]. Also see Karp [101].

EXERCISES

1. The *clique* problem is to determine whether a graph contains a clique of size at least k. A *clique* is a graph with an edge between each pair of nodes. Show that the clique problem is NP-complete. Hint: Show that the vertex cover problem and the clique problem are just two versions of the same problem. Define the complement G^c of a graph G to be the graph with an edge between nodes a and b if and only if G does not have an edge between nodes a and b. Show that for a graph G with the set of nodes V and vertex cover V', the graph G^c has a clique with nodes $V - V'$.

2. An *independent set* for a graph is a set of nodes such that no nodes in the set are connected by an edge. Show that determining whether a graph has an independent set of size at least k is NP-complete. Hint: This is another version of the clique problem.

10.4.3 Hamiltonian Circuit

For our final example, we turn to the Hamiltonian circuit problem. A *Hamiltonian circuit* is a simple closed path through all the nodes of the graph. In other words, for a graph G with node set $V = \{v_1, v_2, \ldots, v_n\}$ and edge set

$E = \{e_1, e_2, \ldots, e_m\}$, a Hamiltonian circuit is an ordering P of the nodes, $v_{p(1)}, v_{p(2)}, \ldots, v_{p(n)}$, such that there is an edge between $v_{p(i)}$ and $v_{p(i+1)}$ for $1 \leq i < n$, and also between $v_p(1)$ and $v_p(n)$. Following a Hamiltonian circuit around a graph, you pass through each node once and end where you started. Not every graph has a Hamiltonian circuit.

It is easy to check whether a given ordering is a Hamiltonian circuit; all that is necessary is to see whether the specified edges are in E. Therefore the Hamiltonian circuit problem is in NP. We will show that it is NP-complete by reducing the vertex cover problem to it. The reduction uses a complicated clever construction to produce a Hamiltonian circuit problem G' that has a solution if and only if the original graph $G = (V, E)$ has a vertex cover of some given degree K. The graph G' that is constructed is a "gadget" to help us in the proof.

To begin the construction, create K nodes, a_1, a_2, ..., a_K. For each edge $e_i = (v_j, v_k)$ in E, create a subgraph with nodes $b_{i,j,1}$, $b_{i,j,2}$, ..., $b_{i,j,6}$ and $b_{i,k,1}$, $b_{i,k,2}$, ..., $b_{i,k,6}$. Connect the nodes with the following 14 edges: $(b_{i,j,m}, b_{i,j,m+1})$ for $1 \leq m < 6$, $(b_{i,k,m}, b_{i,k,m+1})$ for $1 \leq m < 6$, and cross edges $(b_{i,j,1}, b_{i,k,3})$, $(b_{i,j,3}, b_{i,k,1})$, $(b_{i,j,4}, b_{i,k,6})$, and $(b_{i,j,6}, b_{i,k,4})$. The component for the edge (v_j, v_k) thus has two parts, one associated with the node v_j and one associated with the node v_k, and four edges connecting them. It is shown in Figure 10.3. Only the four corner nodes may be connected to the rest of the graph. The remainder of the construction shows how the connections to the rest of the graph are made.

This rather peculiar subgraph is chosen for its interesting properties vis-à-vis Hamiltonian circuits. Since only the four nodes $b_{i,j,1}$, $b_{i,j,6}$, $b_{i,k,1}$, and $b_{i,k,6}$ can connect a component to the outside world, any Hamiltonian path has either one or two segments passing through the component.

Suppose first that the component is connected to the rest of the graph through all four nodes: $b_{i,j,1}$, $b_{i,j,6}$, $b_{i,k,1}$, and $b_{i,k,6}$, and that we are interested in a Hamiltonian circuit that enters this component (twice) through two of these four nodes and exits through the other two. Studying the possible paths through the component shows that it is possible to send two disjoint paths through the component if the circuit enters at $b_{i,j,1}$ goes through all the "j" nodes in ascending order, and exits through $b_{i,j,6}$, and if the second part of the circuit enters through $b_{i,k,1}$, goes through the "k" nodes in ascending order, and exits through $b_{i,k,6}$. It is also possible to send two disjoint paths through the component by reversing either or both of these two paths. There are, however, no other ways to send two disjoint paths through the component and pass through all the nodes.

If we want this component to be connected to the outside world only through $b_{i,j,1}$ and $b_{i,j,6}$, then there is again exactly one way to complete a Hamiltonian circuit that enters at $b_{i,j,1}$ and exits at $b_{i,j,6}$. It uses the ordering $b_{i,j,1}$, $b_{i,j,2}$, $b_{i,j,3}$, $b_{i,k,1}$, $b_{i,k,2}$, $b_{i,k,3}$, $b_{i,k,4}$, $b_{i,k,5}$, $b_{i,k,6}$, $b_{i,j,4}$, $b_{i,j,5}$, $b_{i,j,6}$. The reverse ordering gives the only way to go through all the nodes of the component entering at $b_{i,j,6}$ and exiting at $b_{i,j,1}$. The corresponding orderings, with the j's replaced by k's and vice versa, is used to connect the component through $b_{i,k,1}$ and $b_{i,k,6}$.

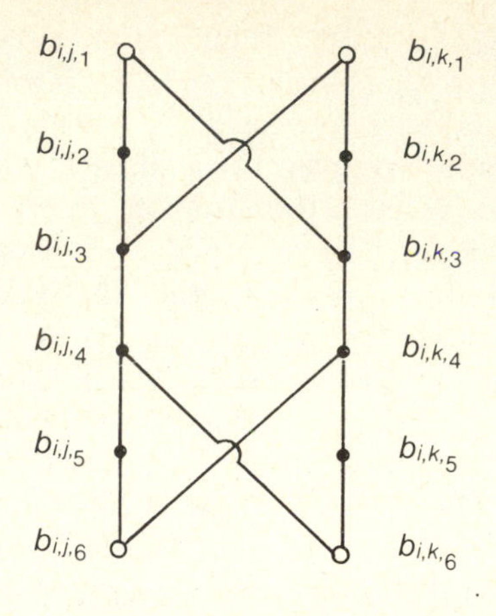

FIGURE 10.1 A component of G' corresponding to the edge $e_i = (v_j, v_k)$ of G. The open circles indicate the nodes of the component which can be connected to the rest of the graph.

We have therefore carefully constructed the component so that if we have one or two disjoint simple paths that go through all the nodes of the component and that start and end at some combination of the corner nodes, then any path that starts at $b_{i,j,1}$ must end at $b_{i,j,6}$ (and vice versa). Likewise, any path that starts at $b_{i,k,1}$ must end at $b_{i,k,6}$ (and vice versa).

To complete the construction of G', we need to connect the K nodes a_1, a_2, ..., a_K and the m subgraphs that correspond to the edges in G. The first step is to connect the edge components together. Suppose e_{j_1}, e_{j_2}, ..., e_{j_p} are the edges of G that have v_j as one of their nodes. Then add edges to G' connecting $b_{j_1,j,6}$ to $b_{j_2,j,1}$, $b_{j_2,j,6}$ to $b_{j_3,j,1}$, ..., and $b_{j_{p-1},j,6}$ to $b_{j_p,j,1}$. This creates a path through G' connecting all the b type nodes that have j as their middle index. We create such a path for every node in the original graph.

Now join each of the a nodes, $a_1, a_2, \ldots, a_K$, to each node of the form $b_{j_1,j,1}$ (the beginning of the v_j path) and to each node of the form $b_{j_p,p,6}$ (the end of the v_j path). This completes the construction of G'. We claim that G' has a Hamiltonian circuit if and only if G has a vertex cover of degree K.

To see this, notice that each node $a_1, a_2, \ldots, a_K$ can occur in the Hamiltonian circuit only once. After entering a_i, the Hamiltonian circuit must leave along an edge e that connects it to one of the edge components. This same edge e is the beginning or end of the path associated with a node v_j. The crucial observation is that, in order to get a Hamiltonian circuit of G', the nodes traversed between a_i and the next a node on the path must contain the complete path associated with some v_j. The only way this could fail to be true would be if we came into a component through a node associated with v_j an exited through a node associated with v_k. But we have already seen that it is impossible to do this with paths that cover all the nodes. Thus to have a Hamiltonian circuit, it is necessary for each path to enter and leave each component through nodes associated with the same node of G.

Once we have completed the entire path associated with v_j, the next edge in the Hamiltonian circuit must be to an a node, since there are no edges directly connecting nodes with different middle indices in different components. Thus any Hamiltonian circuit through G' begins at some a node, follows a path associated with some node of G, returns to different a node, continues with a path associated with another node of G, and so on. A total of K paths through the components are followed; this corresponds to choosing the K nodes of G for a vertex cover of degree K.

Now consider the components. If both nodes of an edge of G are in the vertex cover, then the Hamiltonian circuit of G' enters the component for that edge twice, once for the part of the component associated with the first node and once for the part associated with the second node. If only one of the nodes of an edge is in the cover, then there is just one path through the component and it enters and leaves through the corners of the component associated with the node in the cover.

Thus if G' has a Hamiltonian path, we can obtain a vertex cover for G. The vertex cover contains those nodes whose associated paths the Hamiltonian circuit traverses. This vertex cover contains K nodes. Likewise, if we have a node for G, then we can construct a Hamiltonian path for G' by traversing the corresponding paths.

This section is based on Garey and Johnson [4, Section 3.1.3]. The work was originally done by Tarjan [See 101].

EXERCISE

1. Complete the proof that the Hamiltonian circuit is NP-complete by showing that the problem G' is polynomial in the size of G and that the transformations take polynomial time.

10.5 APPROXIMATION AND HEURISTICS

When your problem set is NP-complete, you do not need to give up all hope. There are still two approaches you should consider for developing useful algorithms. A problem set can be NP-complete because of a small proportion of difficult problems. In such cases you should look for algorithms that usually run rapidly or that have a small average time. A *heuristic* is a principle that leads to rapid solution of some problems in a set. You should consider using heuristics when you desire to solve many of the problems in a difficult set rapidly.

Many NP-complete sets are related to optimization problems. An *optimization problem* set is a set of problems where each problem may have many solutions and where each solution has a value. It is necessary to find the solution with the best value. Corresponding to any optimization problem, there is "yes-no" problem of determining whether the optimization problem has a solution with a value better than a given value.

A problem set is *NP-hard* if it is at least as hard as an NP-complete set. (This is shown by showing that if you could solve all the problems in the NP-hard set in polynomial time then you could also solve all the problems in some NP-complete set in polynomial time.) If the "yes-no" problem set corresponding to an optimization set is NP-complete, then the optimization problem set is NP-hard. Most such optimization problems have not been shown to be NP-complete because no one has found a way to check in polynomial time that a proposed solution is correct. In other words, it is an open question whether or not most optimization problems are in the set NP.

An *approximation algorithm* for an optimization problem is an algorithm that always finds a value that is close to the optimum value. A rapid approximation algorithm is useful when it is not essential to find the very best solution. Some NP-hard optimization problems have known approximation algorithms that are quite good, while others do not.

One optimization problem for which a good approximation algorithm is known is the *bin packing* problem: given a set of n rational weights w_1, w_2, $\ldots$, w_n, with $0 \le w_i \le 1$ for all i, determine how to pack the weights into the minimum number of bins when the total weight in each bin can be no more than 1. (An equivalent problem is to saw a specified set of pieces out of a minimal number of 8-foot-long boards.) The bin packing problem has several generalizations (to two dimensions, for example), and it has many practical applications. The related NP-complete problem is to determine, for a given k, whether there is a way to pack the n weights into k or fewer bins.

There are a number of simple approximate methods for bin packing. In the First Fit Algorithm, each weight is placed in the first bin that has room for it. This results in a new bin being started only when all of the bins which already have items have no room for the current item. It is easy to get a rough upper bound on the effectiveness of this method. There is at most one nonempty bin that is less than half full. Using $FF(I)$ for the amount of space used by the First Fit method for problem instance I, we have

$$FF(I) < \left\lceil 2 \sum_{1 \le i \le n} w_i \right\rceil. \tag{27}$$

It is not possible to obtain a much better simple bound, because if the weights are all equal to $\frac{1}{2} + \epsilon$ for $\epsilon < 1/(2n)$, then the bound is exact. Using $OPT(I)$ for the amount of space used by the optimum packing, we have

$$OPT(I) \ge \left\lceil \sum_{1 \le i \le n} w_i \right\rceil, \tag{28}$$

so

$$FF(I) < 2OPT(I). \tag{29}$$

The conditions under which eq. (27) are a tight bound is not, however, the conditions under which eq. (28) is a tight bound.

The more careful analysis of Johnson et al. [99] leads to the result

$$FF(I) \le \tfrac{17}{10} OPT(I) + 2. \tag{30}$$

Thus, for larger problems the First Fit solution never differs from the optimal solution by more than about 70 percent.

You can do better than the First Fit Algorithm by saving the small weights to last. This prevents the filling up of bins with small weights and saves the small weights for filling up the places left by the large weights. The Decreasing First Fit Algorithm first sorts the weights and then uses the First Fit Algorithm, placing the largest weight first. Johnson [99] showed that

$$FFD(I) = \tfrac{11}{9}OPT(I) + 4, \tag{31}$$

where $FFD(I)$ is the space used by the Decreasing First Fit Algorithm. Thus, for large problems it is never more than about 22 percent worse than optimal.

The typical behavior of the Decreasing First Fit Algorithm is much better than the worst-case analysis suggests. Bentley et al. [72] measured the performance of the algorithm using n random weights from a uniform random distribution. After examining the data, Bentley conjectured that the amount of empty space for decreasing first fit packing averaged about $0.3n^{1/2}$. Lueker [111] later proved this result.

For example, the measurements showed that for packings with 128,000 weights an average of about 64,000 bins were used, with a total empty space in the used bins of about 100 units. The worst-case bound [eq. (31)] is 14,200. It is prohibitively expensive to find the best solution of such bin packing problems using an exponential-time searching method. The Decreasing First Fit method uses $n \ln n$ time (the time for sorting plus the time for binary search) to find an answer that is usually within a very small percentage of optimal.

These results show the usefulness of approximation algorithms and heuristics. Fast approximation algorithms or heuristics can be found for many hard problems. If one is willing to give up finding the very best solution or give up solving every single problem rapidly, then such methods are very useful.

A further discussion of bin packing and other approximation algorithms is contained in Garey and Johnson [4, Section 6.1].

10.6* UNDECIDABILITY

In this chapter we have investigated the notion of optimality in terms of lower bounds. Knowing that an algorithm is optimal is helpful because it prevents us from wasting time searching for a better algorithm. In this section we look at problem sets that are *undecidable*: problem sets for which *no* algorithm exists. There is often an obvious algorithm for most problems in the set; the difficulty is that the algorithm does not terminate for some problems. Thus you can think of undecidable problem sets as being those where the worst-case lower bound is infinite. If you know you have an undecidable problem set, then you can avoid wasting time looking for an algorithm that solves every problem in the set. Often it is appropriate to look instead for an algorithm that solves just some of the problems in the set.

There are many famous undecidable problems, some quite surprising. An example is Hilbert's tenth problem. (It was the tenth in a list of problems constructed by mathematician David Hilbert in 1900 as a challenge to the mathematicians of the coming century.) The problem asks for an algorithm to determine whether a polynomial equation in more than one variable with integer coefficients has any integer solutions. This problem was shown to be undecidable in 1970 [113, 85], after great efforts had been made by mathematicians to come up with an algorithm to solve it.

The best-known undecidable problem (at least among computer scientists) is the *halting problem*: given a description of a Turing machine, and an input string, determine whether the machine eventually halts when given that string as input. More precisely, the problem is to specify a Turing machine M_H which, given a representation of an arbitrary Turing machine M and an input string s, will halt in a finite amount of time in state q_Y if M halts on input s, and will halt in a finite amount of time in state q_N if M does not halt on input s.

The crucial requirement here is that M_H always halts, whether or not M does. It is, in fact, fairly easy to construct a *Universal Turing machine* M_U that takes a description of machine M and a string s as input and reproduces the behavior of M on s: it produces the same answer M does if M halts and gets into an infinite loop if M gets into one. Likewise it is straightforward to construct a machine that for *some* inputs M and s will determine that M gets into an infinite loop. The impossible thing is to do it for *every* M and s.

We will present an informal argument showing why a machine M_H cannot exist. We will present all the important ideas but omit the bookkeeping details. (See Brainerd and Landweber [3] or Lewis and Papadimitriou [13] for a detailed proof.) As a first step, let's consider how M_U can be built. The main problem is to specify an arbitrary machine M in a way that M_U can understand. Suppose M has tape alphabet Γ. We don't want to make M_U so that it has to read any conceivable tape alphabet, so we number the symbols of Γ and replace each one with its number written in binary notation. The numbers will be separated by periods to make them distinct. For example, suppose M has an alphabet $\Gamma = \{a, b, c\}$, and the letters are numbered 0, 1, 2, in order. Then the tape sequence *acba* is represented in M_U by the tape sequence .0.10.1.0 .

The next problem is to represent the state transition function of M. We will number the states of M so that the special states q_0, q_Y, and q_N are encoded as states 0, 1, and 2, so that we need no extra information about them. A standard tabular format for description of the transition function is selected, and the description of the transition function is translated into a string of zeros, ones, and periods. The details of the translation are not important — what matters is that there is an algorithm that generates a description in this format for each Turing machine.

To finish building M_U, we design a machine that takes an input tape consisting of a machine M followed by an input string s. The machine M_U writes the initial state number and the initial string position on a work tape (or a part of the input tape past the input if it has no work tape). Then M_U looks up the current string position, reads that position from the input tape, looks up the current state, looks up the description of what M does when in the current state with the current input, and then updates the current string position and the current state number (based on the directions for what M does). It continues doing this until it comes to state 1 or 2 of M, when it halts.

Now we are ready to start the proof that M_H cannot exist. All strings of characters made up of zero, one, and period can be ordered (see Exercise 1.11–5). Each Turing machine description will be one of those strings. Some of the strings will not be in the correct format to be a Turing machine description, so let's extend our conventions and associate those strings with the Turing machine where state q_0 goes to state q_N for all tape symbols. Thus we have a Turing machine for each string, and it makes sense to call the machine M_i associated with the ith string the ith Turing machine. (Some machines have more than one number, but this is no problem.) Each string, of course, also corresponds to a possible Turing machine input string.

Machine M_U, the universal Turing machine, takes two strings as input: say, s_i and s_j. Then M_U uses s_i as a description of machine M_i, and simulates the actions of M_i on input s_j.

To show that M_H cannot exist, we first describe a problem that no Turing machine can solve. Finally we show that if M_H did exist, it could be used as a subroutine in a machine that could solve the problem that no Turing machine can solve. This is a contradiction if M_H exists, so M_H cannot exist.

What then is the unsolvable problem? Consider the set of strings $S = \{s_i | s_i$ is not accepted by $M_i\}$. Can there exist a Turing machine that accepts exactly the strings in S? If there does, it has a number, say, k. Is s_k in S? It is in S if and only if it is *not* accepted by M_k. But by definition, M_k *has to* accept all strings in S. This contradiction shows that no such M_k can exist.

To finish the proof, we show how, if M_H exists, we could construct a Turing machine to recognize S. Let us call this (nonexistent) machine M_S. It would work as follows. Let the input string be s_i. The machine M_S must determine whether s_i is in S. Machine M_S calls a subroutine that works like M_H (we will soon show that M_H does not exist, but for the purposes of this proof by contradiction we assume that it does exist) with (s_i, s_i) as input. If the M_H subroutine says that $M_i = s_i$ does not halt on input s_i, then M_S knows that M_i does not accept s_i. Therefore M_S accepts s_i. If M_H says that M_i does halt on s_i, then a subroutine that works like M_U is called. This subroutine runs until it halts. If it halts in the accepting state, M_S rejects s_i; if it halts in the rejecting state, M_S accepts s_i. In every case, M_S halts after correctly determining whether or not s_i is in S. The only part of M_S that is not straightforward to build is the M_H subroutine (and it can also be built if M_H exists); since M_S cannot exist, but it would be possible to build it if M_H existed, M_H cannot exist.

A basic part of this proof was the construction of the set $S = \{s_i | s_i$ is not accepted by $M_i\}$. Sets of the form $\{x_i | x_i$ is not compatible with $x_i\}$ can be thought of as the diagonal entries from sets of the form $S_j = \{x_i | x_i$ is not compatible with $y_j\}$. When x_i and y_i have the same form, such sets are often useful for *diagonalization proofs*, such as the one we had above. The proof that M_H does not exist depends on the fact that a machine can work on its own description. Such "self-referential" activity is common to many unsolvable problems.

Just as the typical proof of NP-completeness uses techniques of showing that one problem reduces to another, the typical undecidability proof also uses such reductions. It is, of course, necessary that some problem be proved undecidable by a direct method. Here we have indicated how to use diagonalization to directly show that the construction of M_S is an undecidable problem. Then we used a reduction to show that the halting problem is undecidable.

The halting problem has interesting practical implications. It says that it is impossible to build a program checker that will decide whether or not an arbitrary program can get into an infinite loop. This is unfortunate, since such a program checker might be very useful. But no one who knows about the halting problem will waste time trying to completely solve this problem. If they work on it at all, they will be satisfied to produce a program that has three outputs: (1) infinite loop, (2) no infinite loop, and (3) too hard to tell.

This practical consequence, of course, rests on our assumption that Turing machines are a general model of computation. This assumption, called *Church's thesis* (after logician Alonzo Church), is supported by powerful evidence. Basically, the evidence is that of the several general models of computation that have been proposed, many of which are superficially extremely different, all turn out to compute the same class of functions. It seems very likely then that this class (the class of *recursive* functions) is the class of functions for which algorithms exist. Nearly all computer scientists accept Church's thesis.

Undecidability is discussed in Brainerd and Landweber [3, Section 6.4], in Lewis and Papadimitriou [13, Chapter 6], and in Rogers [14, Chapter 2].

CHAPTER 11

Statistics

In this chapter we give a brief introduction to those parts of statistics that are most frequently used by algorithm analyzers.

Statistics is concerned with deducing the implications of measurements or data. There are several reasons for an algorithm analyzer to measure the performance of an algorithm. (1) Long analyses often contain errors. A few measurements will show that an error exists and that more checking is needed. (Measurements can also be used to check portions of an analysis and thereby help locate the error.) (2) Measurements can replace analysis in some cases. If you want to know how a simple algorithm performs under a small number of conditions, you should consider just measuring the performance. When you need to know the performance under many conditions, or when the method is harder to program than to analyze, measurements are not so useful. (3) Errors in approximations should usually be measured (see Section 4.8). (4) Measurements can suggest hypotheses that are then confirmed or rejected by analytical techniques.

One of the most common applications of statistics is in estimating an average using a few items from the set of items that is being averaged over. For example, suppose you want to know which of two sorting methods is quicker for sorting 100 numbers. There is an infinite set of possible lists of 100 numbers. For sorting methods based on comparisons, only the relative ordering of the numbers is important, but there are still 100! orderings of 100 distinct numbers. There is no way you can compare the performance of the two algorithms by generating all 100! orderings, running the programs, and averaging the results. Instead, if you want to use measurements to compare the two methods, the best you can do is to generate a random subset of the 100! cases, run the programs on the subset, and compare the two methods on the subset. From the results of such a comparison, you can never be *sure* which method is best. The problems in the random set might chance to be particularly difficult for the better method. With proper use of statistics, however, you can tell which method is most likely to be best and ensure that your choice is nearly always correct. Statistics is concerned with deducing from measurements such things as: (1) which method is probably best, (2) the chances that the wrong method was selected as best, and (3) an estimate of how good each method is.

Most elementary problems in statistics that are of interest to algorithm analyzers can be solved by making careful use of the information in the following sections. There is, however, a vast literature on statistics which you should consult whenever you run into a more difficult problem. A good general reference is Snedecor and Cochran [54]. The mathematical aspects of statistics are covered in Fisz [28].

11.1 DISTRIBUTION FUNCTIONS

Statistics is concerned with processes that have several possible outcomes (also called *events*). The set of possible outcomes is called the *sample space.* A *random variable* is a function from the sample space to the real numbers.

A trivial, but important, example is the space of results obtained by flipping a coin one time. The sample space (which we call S_1) contains two events, *heads* and *tails.* An example of a random variable (which we call X_1) on this space is the number of *heads.* Random variable X_1 can have the value 0 or 1. A more complicated example is the space (S_2) of ordered results obtained by flipping a coin ten times. This space contains 2^{10} elements: all length-ten sequences of *heads* and *tails.* One random variable (X_2) on this space is the number of *heads.* This random variable has an integer value in the range 0 to 10.

An example of a continuous sample space (S_3) is the set of results that can be obtained by throwing a dart at a dart board, where the events are the horizontal (x) and vertical (y) distance of the dart from the center of the board. One random variable for this space is the distance ($\sqrt{x^2+y^2}$) of the dart from the center of the board.

The *distribution function*, $F(x)$, of a random variable X is defined as

$$F(x) = \mathrm{Prob}(X \leq x). \tag{1}$$

The definition of distribution function ensures that $F(x)$ is a nondecreasing function of x, that $F(-\infty) = 0$, and that $F(+\infty) = 1$. Notice that the distribution function is associated with the random variable X. If we have two random variables, such as X and Y, we may wish to write $F_X(x)$ and $F_Y(x)$ to distinguish the two distribution functions.

To calculate the distribution function for a random variable, you need to know more than the elements of the sample space and the definition of the random variable. You must also understand the process that generates the events, so that you can assign probabilities to the events.

A true coin is one that lands *heads* up half the time. In sample space S_1 the random variable X_1 has the distribution function

$$F(x) = \begin{cases} 0 & \text{for } x < 0, \\ \frac{1}{2} & \text{for } 0 \leq x < 1, \\ 1 & \text{for } x \geq 1. \end{cases} \tag{2}$$

That is, the probability that X_1 is less than 0 is 0. The probability that X_1 is 0 or less is $\frac{1}{2}$. The probability that X_1 is less than 1 is $\frac{1}{2}$. The probability that X_1 is less than or equal to 1 is 1. The probability that X_1 is less than 2 (or any number greater than or equal to 1) is 1. For this example, the probability does not change except at $X_1 = 0$ and at $X_1 = 1$.

In sample space S_2 the random variable X_2 has the distribution function

$$F(x) = \sum_{i \leq x} \binom{10}{i} \left(\tfrac{1}{2}\right)^{10-i}. \tag{3}$$

That is, the probability that $X_2 < 0$ is 0; the probability that X_2 is less than 1 is $\left(\frac{1}{2}\right)^{10}$; the probability that it is less than 2 is $\left(\frac{1}{2}\right)^{10} + 10\left(\frac{1}{2}\right)^{9}$; etc.

For the space S_3 there is no way we can easily construct a realistic example of a distribution working from first principles. Considerable knowledge of the nature of darts and of the people who throw them is needed before you can construct the distribution of the distance of the dart from the center of the board.

EXERCISES

1. Give the distribution function for the sum of the numbers produced by two rolls of a six-sided die.

2. Give the distribution function for the number of heads obtained with n flips of a true coin.

3. Give the distribution function for the number of times Step 2 is done in the Random Hashing Algorithm (Algorithm 2.4).

4. Give the distribution function for the number of times Step 2 is done in the Non-repeating Random Hashing Algorithm (see Section 3.1.3).

11.2 HYPOTHESIS TESTING

Hypothesis testing gives a way to determine from measurements how likely it is that a hypothesis is true. We assume that the hypothesis has the form: Method A is better than Method B. Hypothesis testing requires a *null hypothesis*, which is intermediate between the hypothesis being tested and its negation, i.e., Method A and Method B are equally good. In some cases we are also interested in the negation of the original hypothesis: Method B is better than Method A. The negation of the original hypothesis is called the *alternative hypothesis*.

Here is the procedure for deciding whether the evidence supports the hypothesis, the alternative hypothesis, or neither.

1. Form a hypothesis, the corresponding null hypothesis, and the alternative hypothesis.

2. Make some measurements and calculate the ratio r of the number of tests that support the hypothesis to the total number of tests.

3. Calculate the probability that the null hypothesis will generate a ratio that is at least as large as r. In other words, first calculate the probability distribution $F(x) = \text{Prob}_{\text{Null}}(x \geq X)$ that the random result X of the null hypothesis will generate a ratio that is at least x. Second, evaluate $F(r)$. For example, let the hypothesis be that Method A is better than Method B, and the null hypothesis be that they are equally good. Suppose the two methods were compared n times and Method A was better m times. Then at this step you need to compute the probability $F(m/n)$ that the measurement will show Method A to be better m or more times out of n times under the assumption that both methods are equally good.

4. If the probability obtained in the previous step is small (less than some number α, see below), then accept the hypothesis. It is unlikely that the null hypothesis or the alternative hypothesis is true. If the probability is close to 1, then the data support the alternative hypothesis. (In this case m/n will be small.) If the probability is not small and not near to 1, then you cannot tell which method is best on the basis of your data. It may be that the two methods are equally good, or it may be that the two methods are so close to being equally good that you cannot tell which method is best with your current approach and your current amount of data.

The appropriate number (α) to compare the probability with depends on circumstances, but often 0.05 is used. Thus if the probability is below 0.05, you will accept the hypothesis; if it is above 0.95, you will accept the alternative hypothesis; and if it is between 0.05 and 0.95, you will say it is too close to call. The smaller you set your acceptance level, the less likely you are to accept a false hypotheses, i.e., to decide that Method A is better than Method B when in fact Method B is at least as good. On the other hand, the smaller you set your acceptance level, the more likely you are not to accept a true hypothesis.

With a hypothesis of the form "Method A is better than Method B", the probability that a statistical hypothesis test will correctly decide that Method A is better depends among other things on how much better Method A is than Method B. If the difference in the quality of the two methods is small compared to the variation in a single measurement, then it will be difficult (i.e., require a large amount of data) to determine which is better. If the difference in quality is large, then it will be easy to determine (with a high probability of success) which method is better. The *power* of a statistical test is the probability that Method A will be determined to be the better (under the assumption that Method A actually is better). The power depends on the particular statistical test that is used, the amount of data that is collected, and the difference in the quality of the two methods (compared to the variance of the measurements).

Often, instead of accepting or rejecting a hypothesis, an investigator will just report the probability that is found. This is fine, unless you plan to take some action based on the results. When the results of hypothesis testing are going to be used to decide which of two courses of action is going to be followed (i.e., whether to use Method A or Method B), then it is necessary to set the probability for accepting or rejecting the hypothesis at a level that will balance the ill effects of accepting a false hypothesis and of rejecting a true hypothesis.

The above procedure is known as *two-tailed* testing: either the hypothesis or its negation (or neither) may be accepted. In *one-tailed* testing, you either accept the hypothesis or reject it. The negation of the hypothesis is not of interest. You usually use one-tailed testing when you have reason to believe (before you begin your measurements) that one of the two methods is better. One-tailed testing should also be used when you are going to be forced to choose between two alternatives after a fixed number of measurements. A slight modification is needed to Step 4 of the above procedure in order to do one-tailed testing: if the probability obtained in Step 3 is small, then accept the hypothesis; otherwise, reject it. The *significance* of a hypothesis test is the probability that a true null hypothesis is rejected. For a one-tailed test this is the same as the probability that is computed at Step 3 of the procedure given above. For a two-tailed test, assuming that the null hypothesis is such that deviations in either direction are equally likely and assuming that the probability is below 0.5, the significance is double the probability obtained in Step 3.

To illustrate hypothesis testing, let's consider some data from Nau et al. [119]. The authors tested several different game-playing programs. They compared two methods, Method A and Method B, by having the two methods play randomly generated pairs of games. For each of the 1600 initial positions that they generated, they obtained two games: one in which Method A made the first move and one in which Method B made the first move. For most pairs of games, Method A won one of the two games and Method B won the other one. There were, however, 240 pairs where one method won both games of the pair. Of these 240 *critical* pairs, Method A won 140 pairs. This suggests that Method A is better, but perhaps it was just lucky. How likely is it that Method A was just lucky?

The appropriate null hypothesis is that Methods A and B are equally good (on the average) for those pairs of games where one method is able to win both games of the pair. Under the null hypothesis the probability that Method A can win m out of n games is

$$2^{-n} \sum_{i \geq m} \binom{n}{i} \tag{4}$$

(see Section 3.2.3). For $n = 240$ and $m = 140$, eq. (4) evaluates to 1.5×10^{-4}. Thus this one-tailed test says that it is very unlikely that the null hypothesis is true, i.e., we can be almost sure that the hypothesis is true. Notice that a small number (1.5×10^{-4}) corresponds to a highly significant result.

In this case, however, a one-tailed test is not appropriate. We had no strong reason to believe that Method A was better before we made the measurements. (We had hoped that Method A was better because one of us had invented it, but that is not a good enough reason.) For the two-tailed test, we want the probability under the null hypothesis that Method A would win either at least

m out of n games or no more than $n-m$ out of n games (assuming $m > n/2$; otherwise interchange m and $n-m$). Thus we need to compute

$$2^{-n}\left(\sum_{i\geq m}\binom{n}{i}+\sum_{i\leq n-m}\binom{n}{i}\right), \tag{5}$$

where $m > n/2$. (If $m < n/2$, then interchange $n-m$ and m on the summation limits, and if $m = n/2$, include the $i = n/2$ term in only one of the sums.) Thus our two-tailed significance is 3×10^{-4}. From the two-tailed test we still conclude that Method A is almost surely better than Method B.

Often the same data can be analyzed more than one way. Thus, using the same data from [119], we may combine all the games and conclude that Method A won 1620 out of 3200 games. Before reading on, you might wish to think about whether this is a better or worse way to analyze the data. We will come back to this question in a moment. With this hypothesis, the appropriate null hypothesis is that either method is equally likely to win a game. To compare Method A and Method B with a two-tailed test, we need to compute

$$2^{-3200}\left[\sum_{i\geq 1620}\binom{3200}{i}+\sum_{i\leq 1580}\binom{3200}{i}\right]\approx 0.91. \tag{6}$$

Thus, this way of analyzing the data gives almost no indication that one method is better than the other. Using this approach, the analysis indicates that either the null hypothesis is true, or that the difference in the two methods is too small to measure with a test that has only 3200 games.

Why did the second method of analyzing the data give such inconclusive results while the first method gave very reliable results? Some game positions are very delicate; the position will be won by the better method. Other game positions are difficult to lose; even a poor method will win if it goes first while even a good method will lose if it goes second. The first method of analysis divided the 3200 games into two groups (one group consisted of the games where a single player won both games of a pair and the other group consisted of the games where each player won one game of the pair) and then did the analysis on the group of games where the method of play was important. The second method of analysis lumped together all the cases. You can obtain significant results with less data by not lumping the different cases together. (It is not always easy, however, to find a suitable way to divide your data into cases.)

Some ways of doing data analysis lead to much more significant results than others. This may tempt you to collect some data and then to analyze it every way that you can think of. This is a *terrible* idea. A statistical analysis says how likely it is that you can explain the data collected on the basis of your hypothesis. If you test 100 different hypotheses, it is quite likely that one of them will be lucky and explain the data. This is very unlucky for you. If you try 100 different hypotheses, you are very likely to decide that one of them is true even if none of them actually are. When someone repeats your tests and tests only the hypothesis that you accepted, the person will most likely find that you have been fooled.

There is, however, a simple safe way to do data analysis that permits you to look at the data more than one way. First you collect some data, and try out all the hypotheses you wish on it. After you decide which hypothesis is most promising, go back and collect some new data. Test your hypothesis on the new data. If the new data supports the hypothesis, then you have good reason to believe that your hypothesis is true. If the new data does not support it, then you have been saved from being fooled.

Hypothesis testing is covered in Snedecor and Cochran [54, Chapter 5] and in Fisz [28, Chapter 16].

EXERCISES

1. Method A and Method B played 1600 pairs of games. For each pair, they started with the same position except that A went first one time and B went first the second time. For most pairs A and B each won one game, but in 70 cases A won both games of the pair and in 25 cases B won both games of the pair. We would like to know whether these results show that Method A is better than Method B, or if Method A was just lucky in this set of tests. Formulate the appropriate null hypothesis and compute the probability that these results would be obtained under your null hypothesis.

2. Suppose you have a coin that comes up heads two-thirds of the time. How many times will you need to flip it to be 95 percent sure that it comes up heads more often than it comes up tails? How many times will you need to flip it to be 95 percent sure that it is not a true coin? Why does it take more flips in the second case?

3. Suppose the coin in the previous problem comes up heads 51 percent of the time. How do the answers to the three questions change?

11.3 THE RIEMANN-STIELTJES INTEGRAL

The Stieltjes integral is a generalization of the traditional Riemann integral which unifies the notation for summation and integration. (The Riemann integral is the one taught in calculus courses.) The Stieltjes integral is commonly used in statistics, where it permits a single treatment of both discrete and continuous probability distributions. We will first remind you of the definition of the traditional Riemann integral.

A *partition* $p = [x_0, x_1, \ldots, x_n]$ of the interval $[a, b]$ is any sequence of numbers x_0, x_1, ..., x_n such that $x_0 = a$, $x_n = b$, and $x_0 < x_1 < \cdots < x_n$. The *norm* of the partition p, written $|p|$, is the maximum difference of consecutive x_i, that is

$$|p| = \max_{0 \leq i < n} \{x_{i+1} - x_i\}. \quad \textbf{(7)}$$

A *Riemann sum* of a function $f(x)$ is any sum of the form

$$R_p(f) = \sum_{0 \le i < n} f(z_i)(x_{i+1} - x_i), \tag{8}$$

where p is a partition and z_i is any point in the range $[x_i, x_{i+1}]$. The *Riemann integral* of $f(x)$ is given by

$$\int_a^b f(x)\,dx = \lim_{|p| \to 0} R_p(f), \tag{9}$$

provided the value of the limit is unique. (The value cannot, for example, depend on how the z_i are chosen.) Thus, the Riemann integral is the limit of any Riemann sum where the partition is made finer and finer. Most common functions, of course, have Riemann integrals (except near points where the function diverges to infinity).

The *Stieltjes sum* is

$$S_p(g, F) = \sum_{0 \le i < n} g(z_i)[F(x_{i+1}) - F(x_i)], \tag{10}$$

where p, z_i, and x_i are given above and F is a function. The *Stieltjes integral* is given by

$$\int_a^b g(x)\,dF(x) = \lim_{|p| \to 0} S_p(g, F), \tag{11}$$

provided the value of the limit is unique. By setting $g(x) = f(x)$ and $F(x) = x$, you can see that the Riemann integral is a special case of the Stieltjes integral. The Stieltjes integral is therefore usually called the Riemann-Stieltjes integral.

If the function $F(x)$ has a derivative $f(x)$ on the region $[a, b]$, then

$$\int_a^b g(x)\,dF(x) = \int_a^b g(x) f(x)\,dx, \tag{12}$$

where the second integral is a Riemann integral. The main interest in the Stieltjes integral arises, however, when $F(x)$ does not have a derivative everywhere in the region $[a, b]$.

For $F(x) = \lfloor x \rfloor$ we have

$$\int_0^n g(x)\,d\lfloor x \rfloor = \sum_{0 \le i < n} g(i). \tag{13}$$

Thus, the Riemann-Stieltjes integral is also a generalization of summation. Use of the Riemann-Stieltjes integral permits treatment of integrals and sums at the same time within the same theory.

A *step function* with a jump of size p at x_0 is a function of the form

$$f(x) = \begin{cases} c & \text{for } x < x_0, \\ c + p & \text{for } x \ge x_0, \end{cases} \tag{14}$$

where c is a constant. To understand the Stieltjes integral, it is useful to decompose $F(x)$ into the sum of a continuous function $C(x)$ and a function $S(x)$ that is the sum of step functions. Define $f(x) = dC(x)/dx$, and let the ith step function of $S(x)$ have a jump of size p_i at x_i. (For example, in Section 4.5.1 we considered the sawtooth function, which is equal to $x - \lfloor x \rfloor$. The function x is continuous and the function $\lfloor x \rfloor$ is the sum of an infinite number of step functions. The function $\lfloor x \rfloor$ has a jump of size 1 at each integer.) For any $F(x) = C(x) + S(x)$,

$$\int_a^b g(x)\,dF(x) = \int_a^b g(x) f(x)\,dx + \sum_{j \text{ such that } a \le x_j \le b} p_j g(x_j). \tag{15}$$

Eq. (15) says that the Stieltjes integral corresponds to a Riemann integral over the continuous part of $F(x)$ plus a sum over the jumps in $F(x)$. Eq. (15) is a generalization of both eq. (12) and eq. (13). Eq. (15) permits you to use your previous understanding of Riemann integrals and of summations to understand formulas that contain Riemann-Stieltjes integrals.

The Riemann-Stieltjes integral is covered in Apostol [17, Chapter 9].

11.4 USING DISTRIBUTION FUNCTIONS

Recall that the *distribution function*, $F(x)$, of a random variable X is defined as

$$F(x) = \text{Prob}(X < x), \tag{16}$$

which implies that $F(x)$ is a nondecreasing function of x, $F(-\infty) = 0$, and $F(+\infty) = 1$. Since $F(-\infty) = 0$ and $F(+\infty) = 1$, $\int_{-\infty}^{+\infty} dF(x) = 1$. The mean (also called the average or the expected value) of a distribution is given by

$$A = \int_{-\infty}^{+\infty} x\,dF(x). \tag{17}$$

The variance is given by

$$V = \int_{-\infty}^{+\infty} (x - A)^2\,dF(x), \tag{18}$$

$$= \int_{-\infty}^{+\infty} x^2\,dF(x) - A^2. \tag{19}$$

Eq. (19) is obtained from eq. (18) by using $(x - A)^2 = x^2 - 2Ax + A^2$ and integrating each term separately. [Compare this with the derivation of eq. (1.57).]

A distribution is a *discrete* distribution if $F(x)$ is constant except at a finite number of points (where it must then have jumps). Notice that for discrete distributions eqs. (17, 18) are equivalent to the corresponding definitions from Section 1.9 [eqs. (1.50, 1.51)].

The *probability density*, $f(x)$, of a distribution is given by

$$f(x) = \frac{dF(x)}{dx}. \tag{20}$$

The distribution function is the integral of the density function. Not all distribution functions have an associated finite probability density. In particular, for discrete distributions the function $F(x)$ consists of a series of jumps, and the density function does not exist. Use of the distribution function rather than the density function allows a unified treatment of discrete and continuous distributions.

Two random variables can have very similar distribution functions and yet have quite different density functions. For example, consider the density function $f(x) = 1$ for $0 \leq x \leq 1$ and the sequence of density functions $h_n(x) = 1 + \sin(2\pi nx)$ for $0 \leq x \leq 1$. The function $h_n(x)$ varies between 0 and 2 repeatedly as x varies between 0 and 1. The larger n is the faster $h_n(x)$ varies. The function $f(x)$ has the distribution function

$$F(x) = x \quad \text{for } 0 \leq x \leq 1. \tag{21}$$

The sequence $h_n(x)$ has the corresponding sequence of distribution functions

$$H_n(x) = x + \frac{\sin(2\pi nx)}{2\pi n} \quad \text{for } 0 \leq x \leq 1, \tag{22}$$

so in the limit as n goes to infinity the sequence of distribution functions $H_n(x)$ approaches the distribution function $F(x)$, while the sequence of density functions $h_n(x)$ does not approach the density function $f(x)$.

The *characteristic function*, $\phi(t)$, of a distribution function $F(x)$ is given by the following Riemann-Stieltjes integral:

$$\phi(t) = \int_{-\infty}^{+\infty} e^{itx}\, dF(x), \tag{23}$$

where i is the square root of -1. In other words, the characteristic function is the expected value of e^{itX}. Since $e^{itx} = (e^{it})^x$, the characteristic function is closely related to the generating function. In particular, for

$$F(x) = \sum_{0 \leq j < x} n_j \tag{24}$$

and $G(z) = \sum_{j \geq 0} n_j z^j$, we have

$$G(e^{it}) = \phi(t). \tag{25}$$

To see this, notice that when $F(x)$ is given by eq. (24), the value of $F(x)$ does not change except when x is an integer, and for integer x, $F(x)$ has a jump. Therefore, the Stieltjes integral reduces to the sum for $G(e^{it})$.

Since the characteristic function is in effect a generating function, it has many of the properties of a generating function. In particular, if X, Y, and Z are random variables with characteristic functions $\phi_x(t)$, $\phi_y(t)$, and $\phi_z(t)$, where $Z = X + Y$, then it follows from eq. (6.129) that

$$\phi_z(t) = \phi_x(t)\phi_y(t). \tag{26}$$

By writing the logarithm of the characteristic function as a power series, you obtain the *semi-invariants* of the distribution. Define κ_j by

$$\ln \phi(t) = \sum_{j \geq 0} \frac{\kappa_j}{j!} (it)^j. \tag{27}$$

Then κ_j is the jth semi-invariant. The two most important semi-invariants are the *mean* or *average*, κ_1, and the *variance*, κ_2.

From the characteristic function, you can recover the original density function:

$$f(x) = \frac{1}{2\pi} \int_{-\infty}^{+\infty} e^{-itx} \phi(t)\, dt. \tag{28}$$

It is also possible to recover the original distribution function even when the density function does not exist. (See Fisz [28, Section 4.5] for a proof of eq. (28) and a treatment of the more general case.)

If X is a random variable with mean μ_x and variance σ_x^2 and the random variable Z is given by $Z = aX + b$ where a and b are constants, then the mean and variance of Z are given by

$$\mu_z = a\mu_x + b, \tag{29}$$

$$\sigma_z^2 = a^2 \sigma_x^2. \tag{30}$$

[These results follow immediately from the definitions of mean and variance, eqs. (17, 18).]

EXERCISES

1. [See 92] Consider the distribution function

$$F(t) = \begin{cases} 0 & \text{for } t < 0, \\ 1 - e^{-\mu t} & \text{for } 0 \leq t < q, \\ 1 & \text{for } t \geq q. \end{cases} \tag{31}$$

Compute the mean and variance of the distribution.

2. Calculate the probability density for the distribution function

$$F(t) = \begin{cases} 0 & \text{for } t < 0, \\ 1 - e^{-\mu t} & \text{for } 0 \leq t. \end{cases}$$

3. Calculate the characteristic function for the binomial distribution

$$F(t) = \sum_{0 \leq i < t} \binom{n}{i} p^i (1-p)^{n-i}.$$

4. Calculate the semi-invariants of the binomial distribution.

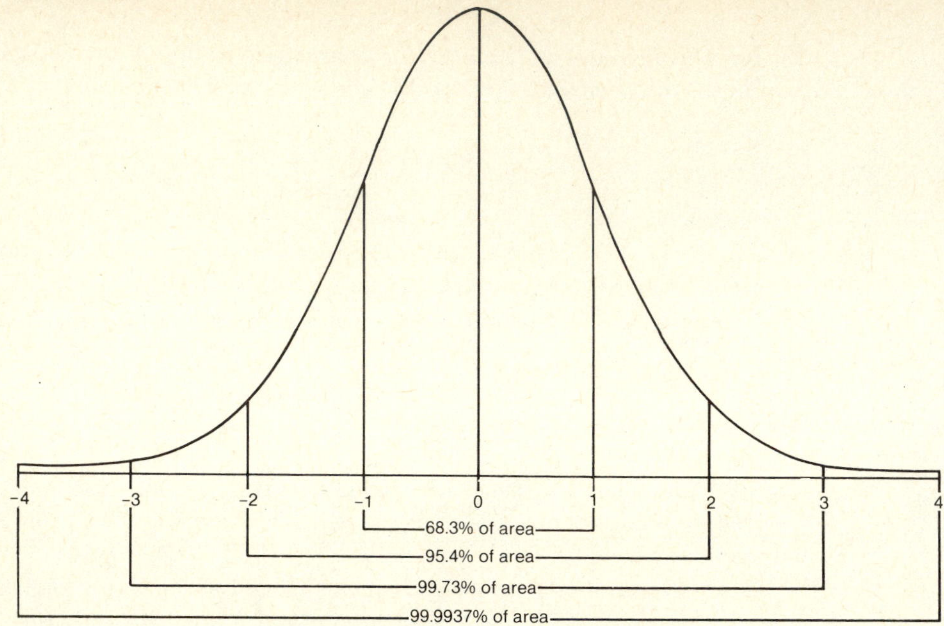

FIGURE 11.1 The density function for the normal distribution and the fraction of the area within different standard deviations.

11.5 THE NORMAL DISTRIBUTION

The *normal distribution* is the distribution for which all the semi-invariants except the mean and variance are zero. It is, therefore, in some sense the simplest nontrivial distribution. The normal distribution with mean μ and variance σ^2 has the distribution function

$$F(x) = \frac{1}{2\pi} \int_{-\infty}^{x} \exp\left(-\frac{(y-\mu)^2}{2\sigma^2}\right) dy. \tag{32}$$

This integral does not have any simple closed form, so extensive tables have been made for the case $\mu = 0$, $\sigma = 1$ (see [16, Table 26.1]). From such tables, it is easy to figure out the value of $F(x)$ for other values of μ and σ by using eqs. (29, 30). Since $d\int_a^x f(y) = f(x)\,dx$, the density function for the normal distribution is

$$f(x) = \frac{1}{2\pi} \exp\left(-\frac{(x-\mu)^2}{2\sigma^2}\right). \tag{33}$$

Figure 11.1 shows the distribution function for the normal distribution.

The characteristic function for the normal distribution is given by

$$\phi(t) = \frac{1}{2\pi} \int_{-\infty}^{+\infty} e^{itx} \exp\left(-\frac{(x-\mu)^2}{2\sigma^2}\right) dx \tag{34}$$

$$= \frac{1}{2\pi} \int_{-\infty}^{+\infty} \exp\left(-\frac{(x-\mu)^2}{2\sigma^2} + itx\right) dx. \tag{35}$$

To simplify this integral, notice that it is similar to

$$\int_{-\infty}^{+\infty} e^{-u^2}\, du = \sqrt{\pi}, \tag{36}$$

which is essentially the integral in eq. (3.34). This suggests using

$$\frac{(x-\mu)^2}{2\sigma^2} + itx = \frac{(x-\mu+it\sigma^2)^2}{2\sigma^2} + \frac{\sigma^2 t^2}{2} + i\mu t \tag{37}$$

to write eq. (37) as

$$\phi(t) = \frac{1}{2\pi}\int_{-\infty}^{+\infty} \exp\left(-\frac{(x-\mu+it\sigma^2)^2}{2\sigma^2} - \frac{\sigma^2 t^2}{2} + i\mu t\right) dx \tag{38}$$

$$= \frac{1}{2\pi} e^{-t^2/(2\sigma^2)-i\mu t} \int_{-\infty}^{+\infty} \exp\left(-\frac{(x-\mu+it\sigma^2)^2}{2\sigma^2}\right) dx \tag{39}$$

$$= e^{-\sigma^2 t^2/2 + i\mu t}. \tag{40}$$

For the special case of $\mu = 0$, $\sigma = 1$, the characteristic function is just $e^{-t^2/2}$.

If you have a random variable Z that is formed by adding the random variables X and Y (i.e., $Z = X + Y$), where X and Y are independent and have normal distributions, then Z also has a normal distribution. The easiest way to prove this is with characteristic functions. Let μ_w and σ_w^2 be the mean and variance of the random process W. Then the characteristic functions for X and Y are $e^{-t^2/(2\sigma_x^2)+i\mu_x t}$ and $e^{-t^2/(2\sigma_x^2)+i\mu_y t}$. By eq. (26) the characteristic function for Z is

$$e^{-t^2/(2\sigma_x^2)+i\mu_x t} e^{-t^2/(2\sigma_x^2)+i\mu_y t} = \exp\left[-\frac{t^2}{2}\left(\frac{1}{\sigma_x^2} + \frac{1}{\sigma_y^2}\right) + i(\mu_x + \mu_y)t\right] \tag{41}$$

which is equal to

$$e^{-t^2/2\sigma_z^2 + i\mu_x t} \tag{42}$$

where

$$\sigma_z^2 = \frac{1}{1/\sigma_x^2 + 1/\sigma_y^2}, \tag{43}$$

$$\mu_z = \mu_x + \mu_y. \tag{44}$$

EXERCISES

1. For a normal distribution with mean μ, what is the value of the distribution function at μ? [That is, what is the value of $F(\mu)$?] Hint: Use the symmetry of the integrand in eq. (32).

11.6 ESTIMATING MEAN AND VARIANCE

A common problem in statistics is to determine the characteristics of some complex process from measurements of the process. For example, you may wish to know the average running time of a program. Now in many cases you could use the analytical techniques discussed earlier in the book to determine the average running time, but the program may be too complex or you may not have any information on its inner workings. You would, therefore, like to run the program a few times on some random problems, measure the time for each problem (say $X_1, X_2, \ldots, X_n$), and from these measurements deduce the average time. Since statistical techniques *cannot* tell you for sure what the average time is, you would also like to know how accurate your answer is.

Now, probably you don't even know the distribution function for the running times of the program, but in this section we will assume that the process you are studying generates data that has a normal distribution. (If your data consists of running times, you can be sure that the process is not *exactly* described by a normal distribution, because any normal distribution with nonzero variance and finite mean sometimes generates negative numbers, and running times are never negative.) In the next section we will give a partial justification for why it is often reasonable to analyze your data in this way even when the process generating the data is not normal.

An *estimator* is any function of a series of random variables that attempts to determine the value of some parameter associated with the process that generates the random variables. There are two properties that we look for in estimators. The first is that the estimator be unbiased. An estimator is *unbiased* when the expected value of the estimator is equal to the value of the parameter that the estimator is attempting to estimate. The second desired property is that the estimator have a low variance. The *variance* of an estimator is the expected value of $(E - \mu)^2$ where E is the value from the estimator (E is a random variable since its value depends on the data that the estimator is using) and μ is the value of the parameter that we are estimating. Notice that in defining the standard deviation of an estimator, we use the actual value of the parameter, not the mean value returned by the estimator. When the estimator has a small standard deviation, then most values returned by the estimator are close to the true value.

The obvious way to estimate the mean of a process is to use the *sample mean*, i.e., the mean of the data collected. Let $x_1, x_2, \ldots, x_n$ be a sequence of independent measurements of a normal process. (Measurements are *independent* if the result of one measurement has no effect on the results of the other measurements.) Let μ be the unknown mean of the process and let σ be the unknown standard deviation of the process. Define

$$\bar{x} = \frac{1}{n} \sum_{1 \le j \le n} x_i. \tag{45}$$

Then from eqs. (29, 42) we know that $\bar{x}$ has a normal distribution with mean μ and standard deviation $\sigma/\sqrt{n}$. Since $\bar{x}$ obeys a distribution with mean μ, it is an unbiased estimator. Since the standard deviation of $\bar{x}$ varies like $1/\sqrt{n}$, the accuracy of $\bar{x}$ as an estimator increases as more measurements are made.

The accuracy of $\bar{x}$ increases slowly (like $\sqrt{n}$) with the number of measurements. For example, if we do 100 measurements and decide that we need 4 times as much accuracy, then we need to do 1600 measurements to obtained the desired accuracy. Thus statistical techniques often make it easy to obtain answers of moderate accuracy while not providing much help in obtaining answers of high accuracy.

Although we know that $\bar{x}$ comes from a normal distribution with standard deviation $\sigma/\sqrt{n}$, we do have one remaining problem in determining the accuracy of our estimate: we don't know the value of σ. We need a way to estimate σ from a series of measurements. Our success in using the sample mean to estimate the true mean suggests that we should try to use the sample variance to estimate the true variance. Define

$$v_x = \frac{1}{n} \sum_{1 \le i \le n} (x_i - \bar{x})^2. \tag{46}$$

A little thought will convince you that the average value of v_x *cannot* be equal to variance of the process. Consider the case $n = 1$. In this case $x_1 = \bar{x}$, so $v_x = 0$, regardless of the distribution of the process that generates the random variable x.

What we need to compute is the expected value of v_x, i.e.,

$$E(v_x) = \int_{-\infty}^{+\infty} \int_{-\infty}^{+\infty} \cdots \int_{-\infty}^{+\infty} v_x f(x_1) f(x_2) \cdots f(x_n)\, dx_1\, dx_2 \cdots dx_n, \tag{47}$$

where $f(x)$ is the density function for the normal distribution (see eq. (33)). After a long calculation this integral gives

$$E(v_x) = \frac{n-1}{n}\sigma^2. \tag{48}$$

Since eq. (47) implies that $\sqrt{nv_x/(n-1)}$ is an estimator for σ, and since the standard deviation of the sample mean is $\sigma/\sqrt{n}$, you can use

$$\sqrt{\frac{v_x}{n-1}} \tag{49}$$

as an estimate of the standard deviation of the sample mean.

These topics are discussed in much greater detail in most statistics books. For example, see Fisz [28, Chapter 9]. One thing you learn from these more detailed treatments is that it is best for n not to be too small when you use eq. (49). Two measurements give you very little idea about the true variance of a process. You should always make at least three measurements; make more than three if it is reasonable to do so.

11.7* THE CENTRAL LIMIT THEOREM

The last section explained how to estimate the mean and variance of a normal distribution, yet the running times of programs seldom obey a normal distribution. Now we will give an indication of why we are still interested in the normal distribution. To summarize the section, if you take the average of a series of independent measurements of a random process, then the distribution of the average approaches a normal distribution as the number of measurements increases. Thus, methods based on the normal distribution usually lead to reasonable answers.

The following theorem is useful for comparing distribution functions.

Theorem 11.1 (Lévy-Cramér) Let $\{X_n\}$ be a sequence of random variables, where X_n has distribution function $F_n(x)$ and characteristic function $\phi_n(t)$. If the sequence $\{\phi_n(t)\}$ is convergent at every point t $(-\infty < t < +\infty)$ to a function $\phi(t)$ continuous in some neighborhood $|t| < \tau$ of the origin, then $\{F_n(x)\}$ is convergent to the distribution $F(x)$ which corresponds to $\phi(t)$.

The proof is given in Fisz [28, Section 6.6]. This theorem is useful for showing that one sequence of distributions approaches some other distribution. When Theorem 11.1 is true, we say that the sequence of distributions $F_n(x)$ approaches the distribution $F(x)$.

One of the most common statistics is the sum of the results from repeated trials of an experiment. Some of the most important properties of this statistic follow from the central limit theorem:

Theorem 11.2 (Lindeberg-Lévy) Let

$$Y_n = X_1 + X_2 + \cdots + X_n, \tag{50}$$

where the X's are independent random variables with the same distribution and which have a mean μ and a nonzero standard deviation σ. Let

$$Z_n = \frac{Y_n - n\mu}{\sigma\sqrt{n}}. \tag{51}$$

Then the sequence $\{F_n(z)\}$ of distribution functions of the random variable Z_n approaches a normal distribution with mean zero and variance 1.

Proof. Let $\phi_x(t)$ be the characteristic function for the random variable $X_j - \mu$. (The characteristic function is the same for all j because the X's all have the same distribution.) Then $\phi_x(t)$ has the form

$$\phi_x(t) = 1 - \tfrac{1}{2}\sigma^2 t^2 + o(t^2), \tag{52}$$

because $X - \mu$ has mean zero and variance σ^2, and the higher-order terms in the power series have size $o(t^2)$. The random variable $(X - \mu)/(\sigma\sqrt{n})$ has the characteristic function $\phi_x(t/(\sigma\sqrt{n}))$. Finally since Z is the sum of $(X - \mu)/(\sigma\sqrt{n})$, by applying eq. (26) $n - 1$ times, we get

$$\phi_z(t) = \left[1 - \frac{t^2}{2n} + o\left(\frac{t^2}{n}\right)\right]^n = \exp\left(\tfrac{1}{2}t^2 + o(t^2)\right). \tag{53}$$

(This last result is from Exercise 4.2.2–7.) Note that in eq. (53) the little o term refers to the limit as n goes to infinity (for fixed t), so for any t the sequence of characteristic functions approachs $e^{-t^2/2}$, which is the characteristic function for the normal distribution with mean zero and variance 1.

From Theorem 11.2, with a change of variables, you can show that the random variable

$$U_n = \frac{X_1 + X_2 + \cdots + X_n}{n} \tag{54}$$

(where the X's are as before) approaches a normal distribution with mean μ and standard deviation $\sigma/\sqrt{n}$.

To make practical use of eq. (54), you need to know how large n should be for the distribution U_n to be reasonably close to normal. This question will be considered in the next section.

Greene and Knuth [6, Section 4.3.3] give a much more extensive treatment of the central limit theorem. This section was adapted from Fisz [28, Sections 6.6 and 6.8].

11.8* PRACTICAL CONSIDERATIONS

We will now cover some practical considerations that arise when analyzing data.

The central limit theorem gives a justification for using methods based on the normal distribution when analyzing data that comes from nonnormal processes. If enough data is collected, then the average of that data will be very much like data from a normal process. But how much data is enough? That depends on the process that generated the data. For example, we know from eqs. (42, 44) that if the original process is normal, then the average of any sample will be normal no matter how small the sample.

The processes that cause the most trouble are those that are very nonnormal. Two of the ways in which a distribution can depart from normal are to have very short tails or to have very long tails. Let's consider some simple but extreme distributions to illustrate what can happen. As an example of a distribution with a short tail (none in fact), consider the distribution with a rectangular density function:

$$f(x) = \begin{cases} 1/a & \text{for } 0 \le x \le a, \\ 0 & \text{for } x < 0 \text{ and } x > a. \end{cases} \tag{55}$$

The average is

$$A = \int_0^a \frac{x}{a}\, dx = \frac{1}{2}a, \tag{56}$$

and the variance is

$$V = \int_0^a \frac{(x - a/2)^2}{2a}\, dx = \frac{1}{12}a^2. \tag{57}$$

As an example of a very nasty distribution consider the distribution where $\text{Prob}(x = 1) = 1 - a$ and $\text{Prob}(x = 10^{10}) = a$, i.e.,

$$F(x) = \begin{cases} 0 & \text{for } x < 1, \\ 1 - a & \text{for } 1 \le x < 10^{10}, \\ 1 & \text{for } x \ge 10^{10}. \end{cases} \tag{58}$$

The mean of this distribution is $1-a+10^{10}a = 1+(10^{10}-1)a$ and the variance is (by eq. (1.57))

$$V = (1-a) + 10^{20}a - (1+(10^{10}-1)a)^2 \tag{59}$$
$$= (10^{20} - 2\cdot 10^{10} + 2)a - (10^{20} - 2\cdot 10^{10} + 1)a^2. \tag{60}$$

For $a = 10^{-5}$, the nasty distribution has a mean of just over 10^5 and a variance of just under 10^{15}.

Now consider what happens if we make 100 measurements on each distribution. On the rectangular distribution, we obtain 100 numbers between zero and a. Almost for sure their average is very close to $a/2$. From applying eq. (49) to our data, we will probably conclude that the average is within about $a/(20\sqrt{3})$ of the sample mean. On the nasty distribution we almost for sure obtain 100 numbers that are all 1. We (almost for sure) conclude that the average is 1 and the standard deviation is zero. The chances that we see the 10^{10} value one or more times is $1-(1-10^{-5})^{100} \approx 0.001$. The most likely estimates for both the mean and the standard deviation are way off from the true values, and there is only one chance in a thousand that we will notice this from the data. If we do notice it, then we will have at least one value that is 10^{10}, so we will obtain a sample mean that is greater than 10^8. Thus with the nasty distribution and 100 data measurements we usually conclude that the mean is way too low (1), but about one time out of a thousand we would conclude that the mean is way too high (about 10^8). By gathering more data we can make it more likely that we will obtain a reasonable answer.

Short tails are not a major problem, but long tails are. Most distributions, however, are not as bad as the nasty distribution. (People would do very little statistics if they were.) We are mainly concerned with doing statistics on running times, which are always positive. A useful test on such data is to compare the calculated standard deviation with the calculated mean. For positive data the sample standard deviation must be less than or equal to the sample mean. The easiest way to see this is to consider an arbitrary set of positive numbers and consider decreasing one number while increasing another. This will keep the sample mean the same if both numbers are changed at the same rate, but it will increase the sample variance if the larger number is the one being increased. Therefore to maximize the variance without changing the mean, you need to increase one number as much as possible, while decreasing the others. Since the numbers must be positive, the best you can do is to increase one number to $n\bar{x}$ (where $\bar{x}$ is the sample mean) and decrease the others to zero. This leads to a sample standard deviation that is equal to the sample mean.

If the sample standard deviation is almost equal to the sample mean, then you do not have enough data. Almost all of your variance is coming from your largest measurement. If you take more measurements, both your calculated mean and your calculated variance are likely to change greatly. If the sample standard deviation is much less than the sample mean, then probably everything is okay, but it could be that you just don't have enough data to show the trouble.

A second practical problem that arises in the analysis of data is determining whether all the data collected represents the process you are studying. For example, suppose you have someone roll a six-sided die, and they report back the following results:

$$5, 3, 4, 1, 1, 1, 5, 36, 1, 2, 2, 1.$$

You can be sure that the 36 is wrong. By proper treatment of erroneous data, you can improve the quality of your analysis. It is, however, rather complicated to separate out the effect of the erroneous data without affecting the correct data, so you should consult an advanced statistics book if you have this problem. Most computers have a very low error rate, so the problem of erroneous data is not as severe in the analysis of computer running times as it is in some other applications of statistics. When computers fail, you often get no answer rather than a wrong answer.

You should not forget, however, that computer programs often do have bugs. If you make measurements on a buggy program, you will obtain measurements for that program, which is not the same as obtaining correct measurements for the intended program.

Often an appropriate graph of the data from an experiment can lead to insights about the process that generates the data. See Tufte [57] for advice on representing data in graphical form.

EXERCISE

1. Purdom and Stigler [124] studied the buddy system of storage allocation. They developed formulas for E_0, the mean number of blocks of size 1 which are paired with empty blocks, and for W_1, the mean time between splits of blocks of size 2. They also ran the buddy system algorithm for 50,000 requests for blocks, with parameters set so that there were typically 100 nonempty blocks in the system. Their theoretical study predicted $E_0 = 5.394$ while their measurements gave $E_0 = 5.400$ with a standard deviation of 0.032. They predicted $W_1 = 0.2833$ and measured $W_1 = 0.1827$ with a standard deviation of 0.0011. How do these figures suggest that they have something to worry about? (Note: It was later determined that the formula for W_1 was incorrect.)

11.9 BOUNDING TAILS OF DISTRIBUTIONS

Often one needs to bound the tail of a distribution using simple properties of the distribution. For any nonnegative distribution with mean μ, the probability that $X \geq x$ obeys the *Markov bound*:

$$\text{Prob}(X \geq x) \leq \mu/x. \tag{61}$$

For most probability distributions the actual probability is much less than this limit, but Exercise 1 shows how to construct a distribution where the bound is tight.

Any distribution with mean μ and standard deviation σ obeys the *Chebyshev bound*:

$$\text{Prob}(|X - \mu| \geq x) \leq \frac{\sigma^2}{x^2}. \tag{62}$$

Again, for most probability distributions the actual probability is much less than the limit, but Exercise 2 shows how to construct a distribution where the bound is tight.

For distributions that are similar to the normal distribution, such as those obtained by summing independent random variables that come from the same distribution, the previous two bounds are very weak. The probability decreases at the rate x^{-1} in the first case and at the rate x^{-2} in the second case. The tail of the normal distribution decreases at an exponential rate.

For a random variable

$$Z = X_1 + X_2 + \cdots + X_n, \tag{63}$$

where the X_i are independent identically distributed random variables , a much better bound (the *Chernoff bound*) can be obtained if the characteristic function of X is known. Eq. (26) generalizes to give

$$\phi_Z(t) = \phi_X(t)^n \tag{64}$$

when Z is related to X by eq. (63). Let $F(z)$ be the distribution function for Z. The size of the tail of the distribution for Z is by definition

$$\text{Prob}(Z \geq z_0) = \int_{z_0}^{\infty} dF(z) = \int_{-\infty}^{\infty} S(z)\, dF(z), \tag{65}$$

where $S(z)$ is the step function that is equal to zero for $z < z_0$ and equal to 1 for $z \geq z_0$.

We can obtain an upper limit on the probability by replacing $S(z)$ by any function that is greater than or equal to $S(z)$ for each value z. To obtain the Chernoff bound, we replace $S(z)$ with $e^{a(z-z_0)}$. For any $a \geq 0$, this leads to the upper limit

$$\text{Prob}(Z \geq z_0) \leq \int_{-\infty}^{\infty} e^{a(z-z_0)}\, dF(z) = e^{-az_0} \int_{-\infty}^{\infty} e^{az}\, dF(z) \tag{66}$$

$$= e^{-az_0} \phi_Z(-ia) = e^{-az_0} \phi_X(-ia)^n. \tag{67}$$

Since the bound in eq. (67) works for any a, we can use derivatives to find the value of a that gives the smallest bound. First, write the bound as

$$\text{Prob}(Z \geq z_0) \leq e^{-az_0 + n \ln \phi_X(-ia)} \tag{68}$$

and set the derivative with respect to a of the exponent to zero to obtain the following equation for a:

$$z_0 = n \frac{d \ln \phi_X(-ia)}{da} = \frac{n}{\phi_X(-ia)} \frac{d\phi_X(-ia)}{da}. \tag{69}$$

Let's now consider what values are given by these various techniques on a sample problem, the probability that 100 flips of a coin will result in 70 or more heads. This is given by

$$\text{Prob}(X \geq 70) = \left(\frac{1}{2}\right)^{-100} \sum_{i \geq 70} \binom{100}{i} \approx 3.98 \times 10^{-5} \tag{70}$$

by direct calculation.

If we approximate the sum with a normal distribution, we have a distribution with mean 50 and standard deviation 5 (see Section 3.2.3). Obtaining 70 flips is 4 standard deviations from the mean. Checking against a table for the normal distribution [16, Section 26], we find that the probability of a deviation of 4 standard deviations for a normal distribution is

$$\text{Prob}(X \geq 70) \approx 3.17 \times 10^{-5} \tag{71}$$

so in this case the normal approximation has an error of about 20 percent.

Since our random variable is positive, we can get an upper limit on the size of the tail by using the Markov bound:

$$\text{Prob}(X \geq 70) \leq \frac{50}{70} \approx 0.714. \tag{72}$$

In this case the bound is not very good; it does not even give an indication that the value is small.

If we apply the Chebyshev inequality, we obtain

$$\text{Prob}(|X - 50| \geq 20) \leq \frac{25}{4900} \approx 5.10 \times 10^{-3} \tag{73}$$

which is a somewhat better bound. Since for this problem we know that the probability that X is greater than or equal to 70 is equal to the probability that x is less than or equal to 30 and since eq. (73) gives the probability for both events, we can (for this problem) reduce the limit to 2.55×10^{-3}.

To apply the Chernoff bound, we must first calculate the characteristic function for the random variable associated with flipping a single coin. The distribution function is given by eq. (2). The characteristic function is

$$\phi(t) = \int_{-\infty}^{\infty} e^{itx}\, dF(x) = \frac{1}{2} + \frac{1}{2}e^{it}. \tag{74}$$

Eq. (69) gives

$$70 = 100\frac{e^a}{1 + e^a} \tag{75}$$

which is equivalent to

$$e^a = \frac{7}{3} \qquad \text{or} \qquad a = \ln\frac{7}{3}. \tag{76}$$

Plugging this value into eq. (67) gives

$$\text{Prob}(X \geq 70) \leq e^{-70\ln(7/3)}(\tfrac{1}{2} + \tfrac{1}{2}e^{\ln(7/3)})^{100} = (\tfrac{7}{3})^{-70}(\tfrac{5}{3})^{100} = 2.70 \times 10^{-4} \tag{77}$$

This bound is a good bit better than the one given by the Chebyshev inequality, although it was also harder to obtain. If we had asked for the probability of obtaining 80 heads from 100 flips, the differences between the various bounds would be even more striking (see Exercise 3).

Let's now consider approximating the sum

$$\sum_{i \geq x} \binom{n}{i} p^i (1-p)^{n-i} \tag{78}$$

using various bounds. This sum gives the probability that flipping a biased coin (with probability p of landing heads) will result in at least x heads when the coin is flipped n times. The mean number of heads is np (see eq. (3.81)) and the variance of the number of heads is $p(1-p)n$ (see eq. (3.88)). The distribution function for one flip of the coin is

$$F(x) = \begin{cases} 0 & \text{for } x < 0 \\ 1-p & \text{for } 0 \leq x < 1 \\ 1 & \text{for } x \geq 1 \end{cases} \tag{79}$$

(compare with eq. (2)).

The Markov bound is

$$\sum_{i \geq x} \binom{n}{i} p^i (1-p)^{n-i} \leq \frac{np}{x}. \tag{80}$$

The Chebyshev bound is

$$\sum_{i \geq x} \binom{n}{i} p^i (1-p)^{n-i} \leq \frac{np(1-p)}{(x-np)^2}. \tag{81}$$

For small p and $x \gg np$, the Chebyshev bound is better than the Markov bound by about a factor of x. The Chebyshev bound is not so good for x near np.

To compute the Chernoff bound we need the integral

$$\int_{-\infty}^{\infty} e^{az} dF(z) = 1 - p + pe^a. \tag{82}$$

The bound is

$$\sum_{i \geq x} \binom{n}{i} p^i (1-p)^{n-i} \leq exp[-ax + n \ln(1 - p + pe^a)] \tag{83}$$

for any $a \geq 0$. The best bound is given by

$$a = \ln \frac{x(1-p)}{p(n-x)}, \tag{84}$$

which results in $a \geq 0$ provided $x \geq pn$. With this value of a, we get

$$\sum_{i \geq x} \binom{n}{i} p^i (1-p)^{n-i} \leq \exp\left[-x \ln \frac{x(1-p)}{p(n-x)} + n \ln\left(1 - p + \frac{x(1-p)}{n-x}\right)\right] \quad (85)$$

$$= \left(\frac{n(1-p)}{n-x}\right)^{n-x} \left(\frac{np}{x}\right)^x. \quad (86)$$

More information on the topics of this section is given in Kleinrock [39, Appendix II].

EXERCISES

1. Give a distribution where $\text{Prob}(X \geq x) = \mu/x$. Hint: Let $X = x$ with probability p and $X = 0$ with probability $1 - p$. Choose p appropriately.
2. Give a distribution where $\text{Prob}(|X - \mu| \geq x) = \sigma^2/x^2$. Hint: Let $X = x$ with probability p and $X = y$ with probability $1 - p$. Choose p and y appropriately.
3. Compute, approximate, and bound the probability of obtaining 80 heads out of 100 coin flips using all the methods given in this section (direct calculation, approximation with the normal distribution, Markov inequality, Chebyshev inequality, and Chernoff bound).

11.10* ESTIMATING BACKTRACK TREES

Backtracking (see Section 4.3.1) can solve some problems quite rapidly while taking almost forever on other similar problems. Often you would like to know how long the backtracking program will take before you run it. The following procedure gives a method of estimating the number of nodes in the backtrack tree generated by a backtracking program. Only minor modifications are needed to have it estimate the running time of the backtrack program. You should compare the following algorithm with the Backtracking Algorithm (Algorithm 4.1); they are quite similar.

Algorithm 11.1 Backtrack Estimation: Input: A set $\{x_1, \ldots, x_n\}$ of variables, where the values for the ith variable are $1, \ldots, v_i$, and a set of intermediate predicates P_i, where P_i is an intermediate predicate for the set $\{x_1, \ldots, x_i\}$. Local variable: The current level of the tree is i and $1/d_i$ is related to the fraction of level i of the tree that we are currently searching. (Level i consists of those nodes that are distance i from the root.) The variable m is the number of children of the current node that the algorithm will investigate, and r_i is used to help select the m children at random. Output: The estimated number of nodes in the backtrack tree t and also the array C, where c_i is an estimate of the number of nodes on level i of the backtrack tree.

Step **1.** [Initialize] Set $i \leftarrow 0$, $d_0 \leftarrow 1$, and $t \leftarrow 0$. For $1 \le i \le n$ set $c_j \leftarrow 0$.

Step **2.** [Solution?] If $i \neq n$, then go to Step 3. Otherwise, the current set of values is a solution. Go to Step 5.

Step **3.** [Search deeper] Set $i \leftarrow i + 1$. Set m to the number of values that will be investigated. (The algorithm will work for any value of m in the range $1 \le m \le v_i$, but it will give more accurate estimates for some values of m than others. How to best select m is discussed below.) Set $d_i \leftarrow d_{i-1}v_i/m$, $r_i \leftarrow m$.

Step **4.** [Starting value] Set $x_i \leftarrow v_i + 1$.

Step **5.** [Test value] If $x_i = 1$, then go to Step 8.

Step **6.** [Next value] Set $x_i \leftarrow x_i - 1$. With probability $1 - r_i/x_i$ reject x_i by going to Step 5. Otherwise set $r_i \leftarrow r_i - 1$, $t \leftarrow t + d_i$, and $c_i \leftarrow c_i + d_i$. (This step randomly selects exactly m of the nodes on level i to investigate. The changing of r_i and x_i interact in such a way that exactly m nodes will be selected and so that the selected nodes are chosen independently. The step also counts each node with multiplicity d_i. The factor d_i allows for the nodes that are skipped over.)

Step **7.** [Test] If $P_i(x_1, \ldots, x_i)$ is *true*, then go to Step 2. Otherwise go to Step 5.

Step **8.** [Backtrack] Set $i \leftarrow i - 1$. If $i = 0$, then stop. (The estimated number of nodes is t, and the estimated number of nodes on level i is c_i.) Otherwise go to Step 5.

Step 6 of the algorithm randomly selects for further investigation m of the possible v_i branches. See Vitter [137] for a complete discussion of algorithms for this problem. We used one of the simpler algorithms.

The operation of this algorithm is easiest to understand when m is always set to v_i in Step 3. In this case the algorithm operates just like the ordinary backtracking algorithm. The variable d_i is always 1, in Step 6 no nodes are rejected, and every single node of the backtrack tree is counted. Using this algorithm with $m = v_i$ is like running ordinary backtracking with a counter to count the number of nodes, except it is a little slower due to the extra overhead in Steps 3 and 6.

The next easiest case to understand is the $m = 1$ case. If $m = 1$, then the algorithm investigates one branch of the backtrack tree. As it goes down the branch, d_i is set to the product of the branching factors encountered. When estimating the tree size, each node is counted as though it were d_i nodes, thereby taking into account the nodes that are expected to be on the branches that were not investigated. Now sometimes the other branches will have more than d_i nodes, while other times they will have less than d_i nodes. When $m = 1$, Step 3

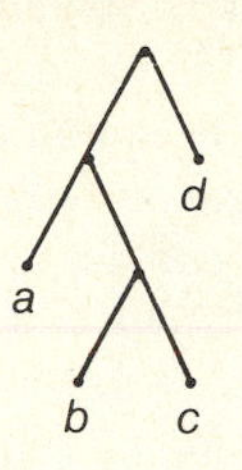

FIGURE 11.2 A small backtrack tree.

prepares to investigate one of v_i children. By multiplying the number of nodes found on the subtree for that one random child by v_i, we obtain an unbiased estimate of the number of nodes in the subtrees for all the children, because the expected number of times to investigate any child is $1/v_i$. In other words, we investigate each child one out of v_i times, so by multiplying the result for the child by v_i we obtain an unbiased estimate of the number of nodes in subtrees for the children. The factor d_i just has the product of the weights to allow for investigating a fraction of the nodes at each level.

When $m \neq 1$, the algorithm produces an unbiased estimate for much the same reason. Step 3 prepares to examine m of v_i children, so if the number of nodes in the subtrees for the examined children is multiplied by v_i/m, an unbiased estimate of the total number of nodes in the subtrees for all the children is obtained.

It is helpful to look at Figure 11.2 and notice how the algorithm estimates the size of the tree using $m = 1$. With $m = 1$, the algorithm randomly selects between the right branch and the left branch at each node of degree 2. Therefore, half the time it follows the path to node d, one-quarter of the time it follows the path to node a, one-eighth of the time it follows the path to node b, and one-eighth of the time it follows the path to node c. When the algorithm follows the path to node d, it estimates that the tree has one node on level 1, two nodes on level 2, and no nodes on deeper levels, for a total of three nodes. When the algorithm follows the path to node a, it estimates that the tree has one node on level 1, two nodes on level 2, four nodes on level 3, and no nodes on deeper levels, for a total of seven nodes. When the algorithm follows the path to node b, it estimates that the tree has one node on level 1, two nodes on level 2, four nodes on level 3, and eight nodes on level 4, and no nodes on deeper levels, for a total of 15 nodes. When the algorithm follows the path to node c, it obtains the same result as for node b. We can see that the algorithm is unbiased for this particular tree, because the algorithm estimates that the tree has 3 nodes one-half of the time, 7 nodes one-quarter of the time, and 15 nodes one-quarter of the time, for an average estimate of 7, which is the number of nodes in the tree.

Although the Backtrack Estimation Algorithm gives an unbiased estimate, the standard deviation for a single estimate can be very large, particularly if $m = 1$. The running time can be very large if $m > 1$ is used. One way to reduce the problems associated with the large standard deviation is to run the algorithm hundreds or thousands of times and average the results. This works well if the trees are not too large. By running the algorithm 100 times and averaging the results, we reduce the error in our estimate by a factor of 10.

Clever choice of m can improve the performance of the estimation algorithm. The best choice of m would often be a noninteger value, except that the algorithm does not work for noninteger m. We can, however, obtain almost the same effect by selecting m randomly in the appropriate way. For example, suppose we decide that $m = 1\frac{1}{3}$ is a good value. Then we just use $m = 1$ two-thirds of the time and $m = 2$ one-third of the time.

The algorithm returns values with a high standard deviation when $m = 1$ because the path in the tree being searched often dies out rather early, and so does not contain any information about the deeper parts of the tree. Even in Figure 11.2 this effect is beginning to show up. The algorithm does not get down to level 3 very often, so it does not provide a very accurate estimate of how many nodes are on level 3. From a single run we usually decide that level 3 has zero nodes, and we only obtain the right value on the average because one time out of four the algorithm decides that level 3 has eight nodes. The algorithm runs slowly for large values of m because the number of nodes examined on each level usually increases rapidly with the level number. One way to obtain an efficient estimation is to select m so that on the average the algorithm looks at about one node on each level.

One good way to use the algorithm is to do a few preliminary runs with $m = 1$ to estimate how many nodes are on each level. If the tree is tall, the preliminary run may provide accurate information only about the nodes near the root. From the preliminary run, you can estimate how to set m so that about one node per level is investigated. If on your next set of runs you sample each level of the tree about equally often, then you can go ahead and make a big run for your final estimate. If you are not sampling the levels evenly, then you can readjust m. The rule for selecting m at Step 4 can depend on the level number and whatever else is relevant. When the authors used this method to estimate the size of 600-level backtrack trees, several preliminary runs were needed to adjust the selection of m.

The original Backtrack Estimation Algorithm was developed by Knuth [106] (for the case $m = 1$). The algorithm for general m was developed by Purdom [122] when we found that the $m = 1$ algorithm was not efficient enough to give even a rough estimate of the size of some gigantic trees with over 10^{42} nodes.

11.10.1* Analysis

It is interesting to analyze the performance of the Backtrack Estimation Algorithm when it is used on random trees. A set of random trees is a set that contains trees where each tree has an associated probability. For our analysis we will use the following random sets of binary trees. Set T_1 will contain one tree, the tree consisting of a root node. This tree will, of course, have probability 1. The set T_i will contain the tree consisting of the root node. This tree will have

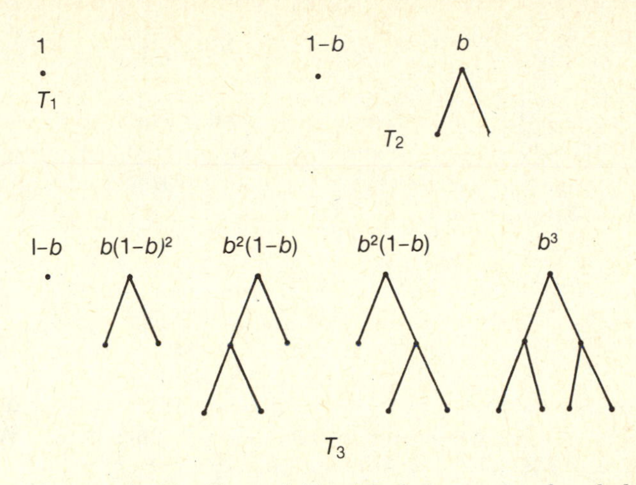

FIGURE 11.3 The trees in T_1, T_2, and T_3 with their associated probabilities. [From SIAM J. Comput. **7** (1978), p. 484.]

probability $1 - b$. In addition T_i will contain all the trees that can be formed by taking a root, a left subtree from T_{i-1}, and a right subtree from T_{i-1}. Each such tree will have probability $bp_{t_1}p_{t_2}$, where p_{t_1} is the probability (in T_{i-1}) of the left subtree and p_{t_2} is the probability of the right subtree. Figure 11.3 shows the trees in T_1, T_2, and T_3, along with their associated probabilities.

Before analyzing the algorithm, let's study the sets of trees. Let $p_{it}(n)$ be the probability that tree t is selected from set T_i and that tree t has n nodes. Now tree t will either have n nodes or not, so this probability will be the probability of tree t if t has n nodes and it will be zero if t does not have n nodes. One reason we are interested in $p_{it}(n)$ is that it obeys a nice recurrence relation, namely,

$$p_{it} = \begin{cases} \delta_{n1} & \text{if } i = 1 \text{ and } t \text{ is a root,} \\ (1-b)\delta_{n1} & \text{if } i > 1 \text{ and } t \text{ is a root,} \\ b\sum_m p_{i-1,t_1}(m)p_{i-1,t_2}(n-m-1) & \text{if } i > 1 \text{ and } t \text{ has left subtree } t_1 \text{ and right subtree } t_2. \end{cases} \tag{87}$$

We will use the notation (ϵ) to stand for the tree that just has a root and (t_1, t_2) to stand for a tree with left subtree t_1 and right subtree t_2. From eq. (87) we can obtain a recurrence equation for the number of nodes in T_i. Let n_i be expected number of nodes for a tree in T_i. We have

$$n_1 = \sum_{n,t} np_{1t}(n) = 1 \tag{88}$$

and

$$n_i = \sum_{n,t} np_{it}(n) \tag{89}$$

or

$$n_i = \sum_n n(1-b)\delta_{n1} + \sum_n \sum_{(t_1,t_2)} nb \sum_m p_{i-1,t_1}(m) p_{i-1,t_2}(n-m-1), \qquad (90)$$

where $i > 1$. The first sum in eq. (90) is equal to $(1-b)$.

To get the second sum into reasonable form, we need to change the order of summation and rearrange the sums so that each one of them is similar to eq. (88). We start with

$$\sum_n \sum_{(t_1,t_2)} nb \sum_m p_{i-1,t_1}(m) p_{i-1,t_2}(n-m-1)$$
$$= b \sum_m \sum_{t_1} \sum_n \sum_{t_2} n p_{i-1,t_1}(m) p_{i-1,t_2}(n-m-1). \qquad (91)$$

Now we need to change the variables so that the argument of p_{i-1,t_2} is a single variable while the argument of p_{i-1,t_1} is unchanged. We need $n' = n - m - 1$ $(n = n' + m + 1)$. With this change we get

$$\sum_n \sum_{(t_1,t_2)} nb \sum_m p_{i-1,t_1}(m) p_{i-1,t_2}(n-m-1)$$
$$= b \sum_m \sum_{t_1} \sum_n \sum_{t_2} (n+m+1) p_{i-1,t_1}(m) p_{i-1,t_2}(n) \qquad (92)$$
$$= b \sum_m \sum_{t_1} \sum_n \sum_{t_2} n p_{i-1,t_1}(m) p_{i-1,t_2}(n)$$
$$+ b \sum_m \sum_{t_1} \sum_n \sum_{t_2} m p_{i-1,t_1}(m) p_{i-1,t_2}(n)$$
$$+ b \sum_m \sum_{t_1} \sum_n \sum_{t_2} p_{i-1,t_1}(m) p_{i-1,t_2}(n). \qquad (93)$$

These sums can now be factored using eq. (3.115) to obtain

$$\sum_n \sum_{(t_1,t_2)} nb \sum_m p_{i-1,t_1}(m) p_{i-1,t_2}(n-m-1)$$
$$= b \Big(\sum_m \sum_{t_1} p_{i-1,t_1}(m) \Big) \Big(\sum_n \sum_{t_2} n p_{i-1,t_2}(n) \Big)$$
$$+ b \Big(\sum_m \sum_{t_1} m p_{i-1,t_1}(m) \Big) \Big(\sum_n \sum_{t_2} p_{i-1,t_2}(n) \Big)$$
$$+ b \Big(\sum_m \sum_{t_1} p_{i-1,t_1}(m) \Big) \Big(\sum_n \sum_{t_2} p_{i-1,t_2}(n) \Big). \qquad (94)$$

Now $\sum_n \sum_{t_2} p_{i-1,t_2}(n) = 1$ because it is the sum of the probabilities for all the trees in T_{i-1}. For the same reason, $\sum_m \sum_{t_1} p_{i-1,t_1}(m) = 1$. The sum $\sum_n \sum_{t_2} n p_{i-1,t_2}(n)$ is the average size of the trees in T_{i-1}, i.e., n_{i-1}. So is $\sum_m \sum_{t_1} m p_{i-1,t_1}(m)$. Combining these results with eq. (90), we obtain

$$n_i = 1 + 2bn_{i-1}. \qquad (95)$$

Thus we have a first order recurrence equation for n_i. The solution is

$$n_i = \frac{(2b)^i - 1}{2b - 1}. \tag{96}$$

So for b near 1 we have big bushy trees with high probability, while for b near zero we have little short trees with high probability. For b near $1/2$, long skinny trees are common.

The rest of the analysis is done using slight modifications of the technique used to compute the average tree size. Let's analyze the algorithm for the case where at each node m is set to 1 with probability p and m is set to 2 with probability $1-p$. For this way of setting m, the expected value of the estimate of tree size obeys the recurrence

$$E_{(\epsilon)} = 1, \tag{97}$$

$$E_{(t_1,t_2)} = 1 + \begin{cases} 2E_{t_1} & \text{with probability } (1-p)/2, \\ 2E_{t_2} & \text{with probability } (1-p)/2, \\ E_{t_1} + E_{t_2} & \text{with probability } p. \end{cases} \tag{98}$$

In other words, $(1-p)/2$ of the time the first backtrack tree is examined and 1 more than twice its estimated size is used as the estimate of the total size, $(1-p)/2$ of the time the second backtrack tree is examined and 1 more than twice its estimated size is used as the estimate of the total size, and p of the time both backtrack trees are examined and 1 plus the sum of the estimates for the subtrees is used as the estimate of the total size.

To judge the efficiency of the estimation algorithm, we need to calculate two items: the expected number of nodes examined by the algorithm (which is a proxy for the time spent arriving at the estimate) and the variance of the estimate. The average variance times the average number of nodes examined gives a measure of the inefficiency of the algorithm. For some purposes we might be more interested in estimating the average of the product, rather than the product of the estimates, but that appears to be much harder to do. (The product of two estimates is usually not exactly the same as the estimate of the product, but they are often close to the same.) The reason that we are interested in the variance, rather than the standard deviation or some other measure of the accuracy of the algorithm, is that by doubling the number of runs (thereby doubling the time required to obtain the estimate) we can reduce the expected variance in the estimate by a factor of 2. Thus the product of the variance and the number of nodes examined is the natural measure of the inefficiency of the estimation algorithm.

Let $p_{it}(l)$ be the probability that, when a random tree is selected from T_i, the tree selected is t and the algorithm examines l nodes. Then

$$p_{i,\epsilon}(l) = (1-b)\delta_{i0}, \tag{99}$$

because $(1-b)$ is the probability that tree ϵ, the tree with just a root, is selected, and if the tree ϵ is selected, then exactly one node is examined. Also

$$
\begin{aligned}
p_{i,(t_1,t_2)}(l) = \frac{b(1-p)}{2}\Bigg(&p_{i-1,t_1}(l-1)\sum_m p_{i-1,t_2}(m) \\
&+ p_{i-1,t_2}(l-1)\sum_m p_{i-1,t_1}(m)\Bigg) \\
&+ bp\sum_m p_{i-1,t_1}(m)p_{i-1,t_2}(l-m-1) \quad \text{for} \quad i>1. \qquad (100)
\end{aligned}
$$

The sums in the first two lines of eq. (100) correspond to the cases where the algorithm investigates only one child of the root: the first sum corresponds to investigating the first child, and the second sum to the second child. The probability of investigating the first child is $(1-p)/2$. If only the first child is investigated, the only way that the algorithm can examine a total of l nodes is to examine $l-1$ nodes in the subtree. The factor $bp_{i-1,t_1}(l-1)p_{i-1,t_2}(m)$ gives the probability of selecting the tree (t_1,t_2) and then examining $l-1$ nodes on the first subtree and m nodes on the second subtree, under the assumption that the algorithm will examine both subtrees. Now in the case corresponding to the first term of eq. (100) the algorithm does not examine the second subtree, and we do not care how many nodes it would find if it did. By summing over m, we count all the cases of interest. In other words, $bp_{i-1,t_1}(l-1)p_{i-1,t_2}(m)$ gives the probability of selecting the tree (t_1,t_2) and examining $l-1$ nodes on the first subtree. The explanation of the second sum is similar.

The sum on the last line of eq. (100) corresponds to the case where both children of the root are examined. The probability of investigating both children is p. In this case, we can obtain a total of l nodes by looking at any number of nodes in the first subtree [in eq. (100) m is used to represent this number] and then examining the appropriate number of nodes in the second subtree $(l-m-1)$. The factor $bpp_{i-1,t_1}(m)p_{i-1,t_2}(l-m-1)$ gives the probability of selecting a tree with left subtree t_1, right subtree t_2, and examining m nodes in the first subtree and $l-m-1$ nodes in the second subtree. By summing over m, we include all the cases where a total of l nodes are examined.

Now that we have a recurrence for $p_{it}(l)$ we can proceed in much the same way we did when we computed the average number of nodes in a random tree. Let l_i be the average number of nodes examined when estimating the size of a tree from T_i. Then l_i obeys the equation

$$
\begin{aligned}
l_i &= \sum_{lt} lp_{it}(l) \qquad (101) \\
&= \sum_l l(1-b)\delta_{l1} + \frac{1}{2}b(1-p)\sum_{l,t_1}\sum_{m,t_2} lp_{i-1,t_1}(l-1)p_{i-1,t_2}(m) \\
&\quad + \frac{1}{2}b(1-p)\sum_{l,t_1}\sum_{m,t_2} lp_{i-1,t_1}(m)p_{i-1,t_2}(l-1) \\
&\quad + bp\sum_{l,t_1}\sum_{m,t_2} lp_{i-1,t_1}(m)p_{i-1,t_2}(l-m-1). \qquad (102)
\end{aligned}
$$

The first sum is $1-b$. We now need to change variables so that the argument for each p_{it} is a simple variable. In the second and third sums we replace $l-1$ with l'. In the last sum we replace $l-m-1$ with l' $(l = l'+m+1)$. This gives

$$\begin{aligned} l_i = 1 - b &+ \frac{1}{2}b(1-p)\sum_{l,t_1}\sum_{m,t_2}(l+1)p_{i-1,t_1}(l)p_{i-1,t_2}(m) \\ &+ \frac{1}{2}b(1-p)\sum_{l,t_1}\sum_{m,t_2}(l+1)p_{i-1,t_1}(m)p_{i-1,t_2}(l) \\ &+ bp\sum_{l,t_1}\sum_{m,t_2}(l+m+1)p_{i-1,t_1}(m)p_{i-1,t_2}(l). \end{aligned} \tag{103}$$

Now we can break the sum containing $l+m+1$ into three sums, and factor all the sums to obtain

$$\begin{aligned} l_i = 1 - b &+ \frac{1}{2}b(1-p)\Big(\sum_{l,t_1}(l+1)p_{i-1,t_1}(l)\Big)\Big(\sum_{m,t_2}p_{i-1,t_2}(m)\Big) \\ &+ \frac{1}{2}b(1-p)\Big(\sum_{m,t_1}p_{i-1,t_1}(m)\Big)\Big(\sum_{l,t_2}(l+1)p_{i-1,t_2}(l)\Big) \\ &+ bp\sum_{m,t_1}p_{i-1,t_1}(m)\sum_{l,t_2}lp_{i-1,t_2}(l) \\ &+ bp\sum_{m,t_1}mp_{i-1,t_1}(m)\sum_{l,t_2}p_{i-1,t_2}(l) \\ &+ bp\sum_{m,t_1}p_{i-1,t_1}(m)\sum_{l,t_2}p_{i-1,t_2}(l). \end{aligned} \tag{104}$$

Each sum in eq. (104) is either 1 or l_{i-1}. Simplifying eq. (104) by making these replacements gives

$$l_i = 1 + b(1+p)l_{i-1}, \tag{105}$$

a first order linear recurrence equation for l_i. The boundary condition is $l_1 = 1$, and the solution is

$$l_i = \frac{[b(1+p)]^i - 1}{b(1+p) - 1}. \tag{106}$$

[For $b(1+p) = 1$, the solution is i.] Thus the algorithm does not look at very many nodes when $b(1+p) \le 1$. For $b(1+p) > 1$, the number of nodes examined increases exponentially with the tree height.

The parameter b indicates what type of trees are being measured. When b is near 1 the typical trees are very bushy; when b is near 0.5, the typical trees are long and skinny. Normally the user of the algorithm has no control over b; the backtrack tree is given and must be measured regardless of its shape. The user does, however, have control over p, which can be set to whatever value results in good performance. The user who wants to examine only a few nodes should set $p < 1 - 1/b$; the number of nodes examined is not very sensitive to how much p is less than $1 - 1/b$.

But the number of nodes examined is only part of the story. The user also wants an accurate estimate. So we need to analyze the accuracy of the algorithm as a function of p.

Let $p_{it}(e)$ be the probability that, when a random tree is selected from T_i, the tree selected is t, and the estimated size is e. It obeys the equation

$$p_{i,\epsilon}(e) = (1-b)\delta_{i0}, \tag{107}$$

$$\begin{aligned} p_{i,(t_1,t_2)}(e) = \frac{1}{2}b(1-p)\biggl(p_{i-1,t_1}\left(\frac{e-1}{2}\right)\sum_m p_{i-1,t_2}(m) \\ + p_{i-1,t_2}\left(\frac{e-1}{2}\right)\sum_m p_{i-1,t_1}(m)\biggr) \\ + bp\sum_m p_{i-1,t_1}(m)p_{i-1,t_2}(e-m-1) \end{aligned} \tag{108}$$

for $i > 1$. The first line of eq. (108) is for the case where one branch is followed; the total estimate is 1 more than twice the estimate for the subtree, so to obtain a total estimated size of e, the subtree must have estimated size $(e-1)/2$. The last term in eq. (108) is for the case where both subtrees are examined.

Eqs. (107, 108) lead to the following recurrence for e_i, the expected value of the estimate for a tree from T_i (see Exercise 1):

$$e_i = 1 + 2be_{i-1} \tag{109}$$

with boundary condition $e_1 = 1$. The solution is

$$e_i = \frac{(2b)^i - 1}{2b - 1}, \tag{110}$$

which is the average size of a tree in T_i. (If we did not obtain this size, we would know we had made an error somewhere, because the algorithm is supposed to give an unbiased estimate of the tree size.) Notice once again that the expected value of the estimate does not depend on p.

To compute the variance of the estimate, we first need to compute s_i, the expected value of the square of the estimate. It is given by

$$s_i = \sum_{et} e^2 p_{it}(e). \tag{111}$$

This sum can be converted to a recurrence with the same techniques we used before. The resulting recurrence equation is

$$s_i = 1 + 2b(2e_{i-1} + pe_{i-1}^2) + 2b(2-p)s_{i-1}. \tag{112}$$

with boundary condition $s_1 = 1$. This is a first order linear equation for s_i, because e_i is a known function (see eq. (110)). The solution is

$$\begin{aligned} s_i = \frac{1}{(2b-1)^3}\biggl[\frac{4b^4 - 2bp - 1}{4b - 2bp - 1} + \frac{2(3-p)(1-b)(2b-1)^2}{(2b-2p-1)(p-1)(2b+p-2)}[2b(2-p)]^i \\ + \frac{2(2b-p-1)}{p-1}(2b)^i + \frac{p}{2b+p-2}(4b)^i\biggr]. \end{aligned} \tag{113}$$

The variance is $s_i - e_i^2$. Notice that s_i (and therefore the variance) depends on p.

This analysis is from Purdom [122].

EXERCISES

1. Show eqs. (107, 108) lead to eq. (109).
2. Show that eq. (111) leads to eq. (112). The algebra is somewhat more complex for this problem, because after you do the needed changes of variables the e^2 factor is converted into a sum of factors, some of which are constant, some linear, and some quadratic. The constant factors sum to 1, the linear factors can be converted to e_{i-1} and the quadratic factors can be converted to s_{i-1}.

11.10.2* Interpretation

We are measuring the standard deviation of a process where a random tree is selected and its size is then estimated. There are two sources of variation in this process. Even if the estimation process were perfect, the variance would not be zero because T_i contains trees of various sizes. The fact that the estimation process is not perfect also contributes to the standard deviation.

We now consider how to disentangle that part of the variance that is due to the estimation process from that part of the variance that is due to the variation in the sizes of the trees in T_i. When running the estimation process on trees in T_i, there are only a finite number of possible outcomes. We are interested in the actual tree size and the estimated tree size, so we say that two runs that are done on trees of the same size and that produce the same answer are equivalent. Let z_k be the estimated size and x_k be the actual size associated with the outcome of the kth run. Define y_k to be the error of the kth outcome, i.e., $z_k = x_k + y_k$. We can specify an outcome with either the pair (x, z) or with the pair (x, y); we will use the latter. Let p_{xy} be the probability of outcome (x, y).

The probability p_{xy} must sum to 1, i.e.,

$$\sum_i p_{x_i,y_i} = 1. \tag{114}$$

The average value of x is

$$\bar{x} = \sum_i x_i p_{x_i,y_i}, \tag{115}$$

the average value of y is

$$\bar{y} = \sum_i y_i p_{x_i,y_i}, \tag{116}$$

and the average value of z is

$$\bar{z} = \sum_i z_i p_{x_i,y_i} = \sum_i (x_i + y_i) p_{x_i,y_i} = \sum_i x_i p_{x_i,y_i} + \sum_i y_i p_{x_i,y_i} \tag{117}$$

$$= \bar{x} + \bar{y}. \tag{118}$$

The variance of x is

$$v_x = \sum_i x_i^2 p_{x_i,y_i} - \bar{x}^2, \tag{119}$$

the variance of y is

$$v_y = \sum_i y_i^2 p_{x_i,y_i} - \bar{y}^2, \tag{120}$$

and the variance of z is

$$v_z = \sum_i z_i^2 p_{x_i,y_i} - \bar{z} = \sum_i (x_i + y_i)^2 p_{x_i,y_i} - \bar{z}^2 \tag{121}$$

$$= \sum_i x_i^2 p_{x_i,y_i} + \sum_i y_i^2 p_{x_i,y_i} + 2\sum_i x_i y_i p_{x_i,y_i} - \bar{x}^2 - \bar{y}^2 - 2\bar{x}\bar{y} \tag{122}$$

$$= v_x + v_y + 2\sum_i x_i y_i p_{x_i,y_i} - 2\bar{x}\bar{y}. \tag{123}$$

The results in this paragraph do not depend on the nature of the process. They apply to any process where there is a result that is the sum of a true value plus an error. (Actually, we would have to reformulate the equations in terms of Riemann-Stieltjes integrals to cover both discrete and continuous processes, but that is a technicality that we will not discuss further.)

For the estimation algorithm the terms $2\sum_i x_i y_i p_{x_i,y_i}$ and $2\bar{x}\bar{y}$ are zero. This is because the estimation algorithm produces an unbiased estimate of the tree size for any tree. Let

$$p_x = \sum_{i \text{ such that } x_i = x} p_{x_i,y_i}. \tag{124}$$

In other words p_x is the probability that a tree is of size x. Summing the statement of unbiasedness over all trees of size x gives

$$\sum_{i \text{ such that } x_i = x} x_i y_i p_{x_i,y_i} = 0. \tag{125}$$

Summing eq. (125) over x gives the result $\sum_i x_i y_i p_{x_i,y_i} = 0$. Since in eq. (125) x_i is a constant (equal to x), it can be factored out. Summing eq. (125) over i with the x_i factored out gives $x\bar{y} = 0$, and summing over x gives $\bar{x}\bar{y} = 0$. Thus,

$$v_z = v_x + v_y \tag{126}$$

for the Backtrack Estimation Algorithm.

We will call the variance in the estimation process as applied to the trees in T_i the *total variance*, the variance due to the Backtrack Estimation Algorithm the *internal variance*, and the variance due to the variation of the sizes of the trees in T_i the *external variance*. The total variance is $s_i - e_i^2$. You might think it is time for one more long calculation to determine the external variance. Well, you are wrong; it is time to think. When $p = 1$, the algorithm estimates the tree size by counting every node in the tree. There is no internal variance when

$p = 1$. The external variance is $s_i(p = 1) - e_i^2$. (Remember that e_i does not depend on p.) The internal variance is

$$v_i = s_i - s_i(p = 1) \tag{127}$$

$$= \frac{2(1-b)}{(2b-1)^3}\left[\frac{4b^4(1-p)}{4b-2bp-1} + \frac{(2b-1)^3}{(4b-2bp-1)(1-p)(2b+p-2)}[2b(2-p)]^i + (2b-1)\left(\frac{1+p}{1-p}+2i\right)(2b)^i + \frac{1-p}{2b+p-2}(4b)^i\right]. \tag{128}$$

To obtain an efficient estimation algorithm, we need to make $l_i v_i$ small. We will assume that b is in the range $\frac{1}{2} \le b \le 1$. (The sets of trees for small b are not very interesting.) For large i, if we want to keep the product small, then we must keep the exponential terms small. The exponential term in l_i is $[b(1+p)]^i$, which is small if $p \le 1 - b^{-1}$. The two most important exponential terms in v_i are $(4b^2)^i$ and $[2b(2-p)]^i$. Varying p does not affect the size of the first of these terms. The second term becomes smaller the larger p becomes. So from the viewpoint of the variance alone it is worthwhile to reduce p until $4b^2 = 2b(1-p)$, which means $p \ge 2 - 2b$.

Since $1 - b^{-1} \le 2 - 2b$ for $\frac{1}{2} \le b \le 1$, for large i it is definitely worthwhile to have p at least as large as $1 - b^{-1}$ since this reduces the variance without significantly increasing the number of nodes examined. A more detailed analysis of the product shows that in the limit as i goes to infinity, $l_i v_i$ has local minima at $p = 1 - b^{-1}$ and at $p = 2 - 2b$. For $\frac{1}{2} \le b < \frac{1}{2}\sqrt{2}$ the minimum at $p = 2 - 2b$ is smaller, while for $\frac{1}{2}\sqrt{2} < b \le 1$ the minimum at $p = 1 - b^{-1}$ is smaller.

The smallest minimum of all, however, occurs at $p = 1$. If you have enough time, you should estimate the tree size by counting all the nodes, because then you will have no error at all. The Backtrack Estimation Algorithm is useful when you do not have enough time to count all the nodes. Lack of time may also cause you to prefer the local minimum at $p = 1 - b^{-1}$ to the one at $p = 2 - 2b$ even when the latter one gives a better value for $l_i v_i$; l_i may be too large at $p = 2 - 2b$.

11.11* LINEAR LEAST SQUARES ESTIMATION

Least squares estimation is used when you have a function with some adjustable parameters and you have some data that the function is intended to explain. Each data item is the sum of what the function predicts plus an error term. The problem is to estimate the values of the parameters in the function. In particular, we wish to set the parameters so that, for the data available, the sum of the squares of the differences between the data and the prediction of the function is as small as possible. The hope is that this will give a good estimate of the values of the adjustable parameters.

Linear least squares applies when the answer computed by the function is a linear function of the parameters being estimated. (The function may contain other variables in a nonlinear manner; it is just the parameters that are being estimated that must occur linearly.) In other words, we have the equation

$$y(x_1, x_2, \ldots, x_k) = a_1 f_1(x_1, x_2, \ldots, x_k) + a_2 f_2(x_1, x_2, \ldots, x_k) + \cdots + a_n f_n(x_1, x_2, \ldots, x_k) + \epsilon, \quad (129)$$

where ϵ represents the unknown error. There are n adjustable parameters a_1, a_2, ..., a_n. We also have some data,

$$y_1(x_{11}, x_{12}, \ldots, x_{1k}), \quad (130)$$

$$y_2(x_{21}, x_{22}, \ldots, x_{2k}), \quad (131)$$

$$\vdots$$

$$y_m(x_{m1}, x_{m2}, \ldots, x_{mk}). \quad (132)$$

That is, we have m pieces of data giving the values of the x's and the resulting y. We allow for errors in the y's, but we assume that the x's are known exactly.

We wish to adjust the a's to minimize

$$\sum_{1 \le i \le m} [y_i - y(x_{i1}, x_{i2}, \ldots, x_{in})]^2 = \sum_{1 \le i \le m} [y_i - a_1 f_1(x_{i1}, x_{i2}, \ldots, x_{ik}) - a_2 f_2(x_{i1}, x_{i2}, \ldots, x_{ik}) - \cdots - a_n f_n(x_{i1}, x_{i2}, \ldots, x_{ik})]^2. \quad (133)$$

Eq. (133) is a quadratic function of the a's. For each a, there is a value that results in a minimum. (There is no maximum; the function approaches infinity whenever an a becomes very large or very small.) To find what values of the a's minimize eq. (133), we first need to take the partial derivative with respect to each a_j and set it equal to zero. After dividing out a factor of $-2a_j$, we get

$$\sum_{1 \le i \le m} [y_i - a_1 f_1(x_{i1}, x_{i2}, \ldots, x_{ik}) - a_2 f_2(x_{i1}, x_{i2}, \ldots, x_{ik}) - \cdots - a_n f_n(x_{i1}, x_{i2}, \ldots, x_{ik})] f_j(x_{i1}, x_{i2}, \ldots, x_{in}) = 0 \quad \mathbf{(134)}$$

for $1 \le j \le n$. This is a set of n equations in n unknowns (the a's), so they can be solved to obtain the value of the a's.

After you compute the a's you can obtain an estimate of how good the fit is by computing the sum of the squares of the errors, using eq. (133) with the computed values for the a's. If the number of data points is equal to the number of parameters, then your predicted values will be exactly equal to your measured values, but you will obtain no information about the error. When the number of data points is larger than the number of parameters, then you do obtain information about the error. An estimate for the variance of the error is the square root of the sum of the squares [eq. (133)] divided by $m - n$.

Gauss showed that if the equation being fit [the right side of eq. (129) without the ϵ term] has the correct functional form for the process that generates the data, if the error terms (the ϵ's) are statistically independent with a common variance, and the x's are known exactly, then the least squares method gives the smallest variance for the estimated values of the parameters [the a's in eq. (129)] of any method that estimates the parameters using a linear function of the data.

To illustrate the linear least squares method, let's see whether there is a linear relation between the size of a grammar and the number of states needed for an LR(1) parser for the grammar. Purdom [121] found experimentally that for a list of 84 LR(1) grammars the size of the grammar (measured by the total length of the productions of the grammar) was a good predictor of the size of the corresponding LR(1) parser. (Earlier Reynolds [128] showed that the worst-case size was an exponential function of the grammar size.)

Aho and Ullman [65, Chapter 6] have three LR(1) grammars and their parsers:

Grammar	Grammar Size	Parser Size
1	12	12
2	8	10
3	5	10

Let's see how well this data can be explained by assuming that the size of a typical parser is a linear function of the grammar size plus an error term. Let y_i be the size of the ith parser, and let x_i be the size of the ith grammar. We want to find a and b such that

$$\sum_{1 \le i \le 3} (y_i - a - bx_i)^2 \tag{135}$$

is as small as possible. Eq. (135) is eq. (133) with a_1 replaced by a and a_2 replaced by b. There is just one x, so no subscript is used. The function f_1 is the constant 1, and $f_2(x)$ is the function x.

Eq. (134) becomes

$$\sum_{1 \le m \le 3} [y_i - a - bx_i] = 0, \tag{136}$$

$$\sum_{1 \le m \le 3} [y_i - a - bx_i]x_i = 0. \tag{137}$$

[Eq. (136) is the $j = 1$ case of eq. (134), and eq. (137) is the $j = 2$ case.] Collecting the constants and the coefficients in eqs. (136, 137) gives

$$\sum_{1 \le m \le 3} y_i - a \sum_{1 \le m \le 3} 1 - b \sum_{1 \le m \le 3} x_i = 0, \tag{138}$$

$$\sum_{1 \le m \le 3} y_i x_i - a \sum_{1 \le m \le 3} x_i - b \sum_{1 \le m \le 3} x_i^2 = 0, \tag{139}$$

or

$$32 - 3a - 25b = 0, \tag{140}$$

$$274 - 25a - 233b = 0. \tag{141}$$

The solution of this set of equations is

$$a = \frac{303}{37} \approx 8.2, \qquad b = \frac{11}{37} \approx 0.30\,. \tag{142}$$

Using these values in eq. (135) gives a sum of the squares of the errors of 667/1369 ≈ 0.49 . Since $m - n = 1$ for this example, this is also the estimate of variance of the error.

These results suggest that there might be a linear relation between the grammar size and the parser size, although there are several defects in this small example that prevent us from coming to a very strong conclusion. First, the values of x cover a very small range. A function like $a + bx^2$ would also do a good job of explaining these three data points. This would not be very important if we had a strong theoretical justification for believing that the true function was linear, but in this case the function was pretty much pulled out of the air, so having data for a wide range of x is quite important. Second, we did not use very much data. This made for a nice example, but we would have had much more confidence in the result if there had been a lot more data.

Purdom [121] studied 84 grammars with sizes up to 323. He obtained $a = 0.02 \pm 0.45$ and $b = 0.5949 \pm 0.0048$, values quite different from what we got on our small sample. Either set of values of a and b will give close to the same results on the small data set. For this data set you can decrease a while increasing b without having a large effect on the quality of the fit. (This is again related to the fact that the x values in the data set used in the book cover such a small range.) This comparison of the results for 3 data points with the results for 84 data points shows the importance of having plenty of data when using least squares methods.

You should not use the results from least squares fits outside the range where the fitting took place unless you have good reason to believe that the functional form being used to fit the data is valid and unless you have studied the correlation in the least squares estimates of the parameters.

Least squares estimation is covered in Daniel and Wood [24], Montgomery and Peck [44], and Sedgewick [62, Chapter 6].

EXERCISES

1. Repeat the least squares error analysis in the text for the case where there is a fourth grammar of size 24 with a parser size of 24.
2. Consider the following data from Purdom and Stigler [124]:

ρ	E
20	2.441
40	3.433
100	5.394
200	7.600
400	10.720

Fit this data with the function $a_1\rho^{1/2} + a_2 + a_3\rho^{-1/2} + a_4\rho^{-3/2}$. How much do the coefficients change if a_4 or a_3 and a_4 are omitted? What if the data for $\rho = 400$ is omitted? What if the data for $\rho = 20$ is omitted?

11.12* MEASURING ASYMPTOTIC PERFORMANCE

Most of this book is concerned with finding an analytical formula that expresses the running time of an algorithm as a function of the size of its input. Ideally, we would like to be able to obtain such formulas for all interesting algorithms, but unfortunately present techniques are not strong enough to achieve this result. If we are not able to obtain a formula that describes the average running time of an algorithm analytically, we do not need to give up all hope; an empirical study can yield valuable information.

For such a study to be worthwhile, several points must be kept in mind. First, the inputs that are studied must be truly random. If the random number generator is faulty, the results of the study may say more about *its* properties than about those of the algorithm! One way to avoid spurious results of this type is to repeat the study using a second random number generator based on a different principle. (Knuth [10, Chapter 3] has a good discussion of various methods for generating pseudo-random numbers.)

Second, enough inputs must be used to give statistically meaningful results. Section 11.8 discusses how to determine how many inputs are enough.

Finally, it is important to extend the tests to large enough inputs to be sure that the asymptotic behavior of the algorithm is encountered. An algorithm with a running time of $e^{0.01n}$ can seem quite efficient when tested on inputs of size less than 100; if a study stopped with inputs of that size, it would lead to a very misleading idea of the algorithm's asymptotic behavior. Of course, it is not always clear when "asymptopia" has been reached; in general, a study should include inputs at least as large as the largest ones likely to be encountered in practice. In the next section we present an example of a statistical analysis of an algorithm.

11.12.1* Deletion in Binary Search Trees

Insertion in binary search trees was discussed in Section 7.1.1.2. Insertion is straightforward; there is only one reasonable way to do it. Deletion is more complicated. Let's define the *successor* of a node to be the smallest node in its right subtree. The *predecessor* of a node is the largest node in its left subtree. (Not all nodes have a successor or a predecessor.) A straightforward way to perform deletion is to replace the deleted node with its successor.

Algorithm 11.2 Rightdelete(*root*,*value*): Input: *root* points to the root of the search tree, and *value* is the data associated with the node that will be deleted. Result: A new search tree in which the uppermost node that has a *data* field equal to *value* has been deleted. The procedure can call itself or Delete.

Step **1.** [Test] If the relation between *value* and *data*(*root*) is

Step **2.** [Go left] *value* < *data*(*root*), then call Rightdelete(*llink*(*root*), *value*).

Step **3.** [Go right] *value* > *data*(*root*), then call Rightdelete(*rlink*(*root*), *value*).

Step **4.** [Node found] *value* = *data*(*root*), then call Delete(*root*).

Algorithm 11.3 Delete(*root*): Input: *root* points to the node to be deleted. Global variable: *avail* points to the available space list. Result: A new search tree in which the node pointed to by *root* has been deleted and returned to the available space list.

Step **1.** [Successor?] Set *temp* ← *root*. If *rlink*(*root*) is *empty*, then (there is no successor, so) set
data(*root*) ← *data*(*llink*(*root*)), *rlink*(*root*) ← *rlink*(*llink*(*root*)),
llink(*root*) ← *llink*(*llink*(*root*)), *llink*(*temp*) ← *avail*,
avail ← *temp*, and return.

Step **2.** [Right child] If *llink*(*rlink*(*root*)) is *empty*, then set
data(*llink*(*rlink*(*root*))) ← *data*(*llink*(*root*)),
llink(*llink*(*rlink*(*root*))) ← *llink*(*llink*(*root*)),
rlink(*llink*(*rlink*(*root*))) ← *rlink*(*llink*(*root*)),
data(*root*) ← *data*(*rlink*(*root*)), *llink*(*root*) ← *llink*(*rlink*(*root*)),
rlink(*root*) ← *rlink*(*rlink*(*root*)), *llink*(*temp*) ← *avail*,
avail ← *temp*, and return.

Step **3.** [Find successor] Set *succptr* ← *rlink*(*root*).

Step **4.** [Search down tree] While *llink*(*llink*(*succptr*)) is not *empty* set *succptr* ← *llink*(*succptr*).

Step **5.** [Successor found] Set *successor* ← *llink*(*succptr*), *llink*(*succptr*) ← *rlink*(*successor*), *llink*(*successor*) ← *llink*(*root*), *rlink*(*successor*) ← *rlink*(*root*), *root* ← *successor*, *llink*(*temp*) ← *avail*, *avail* ← *temp*, and return.

This Deletion Algorithm is *asymmetrical* in the sense that a deleted node is always replaced by its successor. A *symmetrical* Deletion Algorithm is one where the deleted node is alternately replaced by its successor and its predecessor.

If a binary tree is built using n distinct keys chosen at random from a uniform distribution, then, as we saw in Section 7.2.3, there are $C_n = \frac{1}{n+1}\binom{2n}{n}$ such trees, each with some probability of occurring; we will call this distribution D_n. Inserting a random node into such a tree gives a tree with $n+1$ nodes whose shape occurs with probability D_{n+1}; deleting a random node from a tree of size n gives a tree with $n-1$ nodes and distribution D_{n-1}. We will call trees with distribution D_n *random* binary trees.

Surprisingly, performing a sequence of random insertions and deletions on a random binary tree results in a tree whose shape is *not* random. When a random node is added to a random n-node tree, it is equally likely to be inserted in any of the $n-1$ spaces between existing nodes, before the smallest node, or after the largest node: $n+1$ possibilities in all. After a random deletion from an n-node tree, the distribution of *shapes* is D_{n-1}, but the probability of inserting a new node where the deleted node was is $2/(n+1)$, while the probability of inserting it in any of the other positions is $1/(n+1)$.

The time needed to look up an item in a binary search tree is proportional to the length of the path from the root to the node containing the item. The expected time to look up a node is the expected *average* path length. The *internal path length* (*IPL*) of a tree is the sum of the distances of the nodes from the root:

$$IPL = \sum_{\text{nodes in tree}} distance(root, node). \tag{143}$$

If we are considering trees with a fixed number of nodes, we may as well use *IPL* instead of IPL/n in comparing trees to see which has the best expected path length.

The expected *IPL* of a random n-node binary tree, I_n, was derived in Section 7.1.1.2; the value is approximately $1.386n\lg n - 2.846n$. To determine whether a distribution of trees is better or worse than random, we test whether the expected *IPL* of n-node trees in that distribution is less than or greater than I_n.

A study of the distribution of binary trees obtained after a sequence of random insertions and deletions was done by Eppinger [86]. In order to maintain the trees at a constant size, he used pairs of random insertions and deletions on random n-node trees. (If the nature of the operation, insertion or deletion, were also random, the size of the trees would vary, making comparisons more difficult). He proceeded by first building a random n-node tree using random insertions, after which i pairs of successive insertions and deletions were done. Then the *IPL* of the resulting tree was measured.

Earlier studies by Knott [105], which had been done for relatively small values of n and i, had indicated that the trees resulting from a sequence of insertions and deletions using the asymmetric deletion algorithm were *better* than those of random binary trees. Eppinger's data confirms this phenomenon for small values, but it also shows that for large i the trees become worse than random. Eppinger studied trees with 64, 128, 256, 512, 1024, and 2048 nodes. As n increased, he also increased i. The value of i ranged up to 10,000 for 64 node trees, and up to 9,000,00 for 2048-node trees. For each value of n, he went to large enough values of i so that the ratio of $\overline{IPL_{n,i}}$, the measured average *IPL* of n-node trees after i insertion-deletion pairs, to I_n, leveled off and remained constant for a considerable number of values of i. This was observed to occur after about n^2 pairs of insertions and deletions. The leveling off is a good indication that the region of asymptotic behavior was reached.

It is important to present data in a way that makes the main observations clear. Eppinger plotted the value of $\overline{IPL_{n,i}}/I_n$ as a function of i/n^2. (See

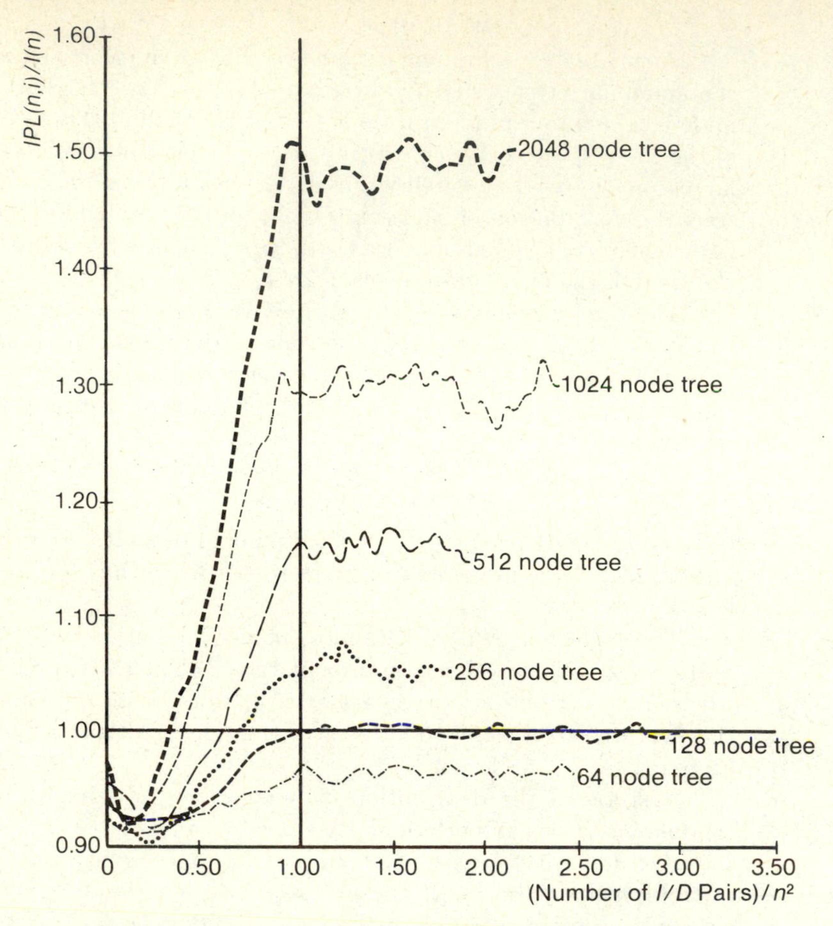

FIGURE 11.4 Comparison chart for asymmetric deletions. [From Jeffrey L. Eppinger, *An Empirical Study of Insertion and Deletion in Binary Search Trees*, Communications of the ACM **26** (1983) p. 666, copyright 1983, Association for Computing Machinery, Inc. Reprinted by permission.]

Figure 11.4.) Since the transition to asymptotic behavior occurs at $i = n^2$, this method of graphing normalizes the x axis so that one can directly compare the asymptotic values of $\overline{IPL_{n,i}}/I_n$ for all values of n that were studied.

It is instructive to examine the data on this graph for small values of n and i. The ratio for 64-node trees leveled off at a value well below 1; a study that had measured only trees of that size or smaller would have concluded, falsely, that trees resulting from random pairs of insertions and deletions tend to be better than random trees. Likewise, the ratio tends to decrease for small i up to about $n^2/4$; this was the region measured in earlier studies. The moral is that a worthwhile experimental study needs to measure enough large cases to be reasonably confident that the asymptotic behavior has been encountered.

Eppinger also studied symmetric deletions. The corresponding graph is shown in Figure 11.5. The data demonstrates that using symmetric deletions

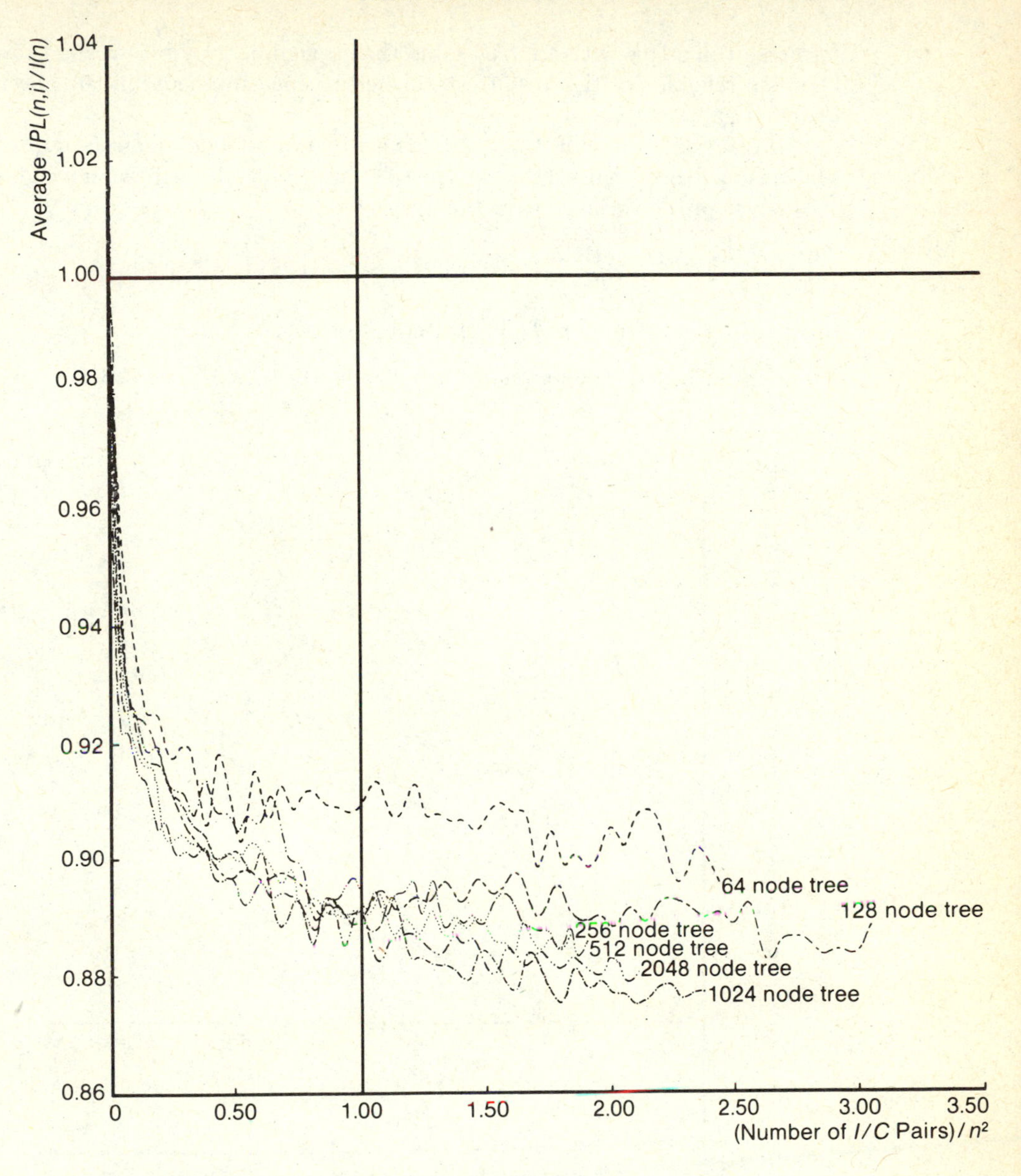

FIGURE 11.5 Comparison chart for symmetric deletions. [From Jeffrey L. Eppinger, *An Empirical Study of Insertion and Deletion in Binary Search Trees*, Communications of the ACM **26** (1983) p. 668, copyright 1983, Association for Computing Machinery, Inc. Reprinted by permission.]

does result in a tree with an *IPL* value that is significantly better than random. This is a reliable positive result established by measurements in the absence of a theoretical analysis.

Plotting the value of $\overline{IPL_{n,i}}/I_n$ as a function of $\lg n$ suggests a quadratic relationship. (See Figure 11.6.) A least-squares multiple regression weighted by the inverse of the variance gave the approximation

$$\lim_{i\to\infty} \frac{\overline{IPL_{n,i}}}{I_n} \approx 0.0202(\lg n)^2 - 0.241\lg n + 1.69. \tag{144}$$

Substituting our value for I_n in this equation gives

$$\lim_{i\to\infty} \overline{IPL_{n,i}} \approx 0.0280n(\lg n)^3 - 0.92n(\lg n)^2 + 3.03\lg n - 4.81n. \tag{145}$$

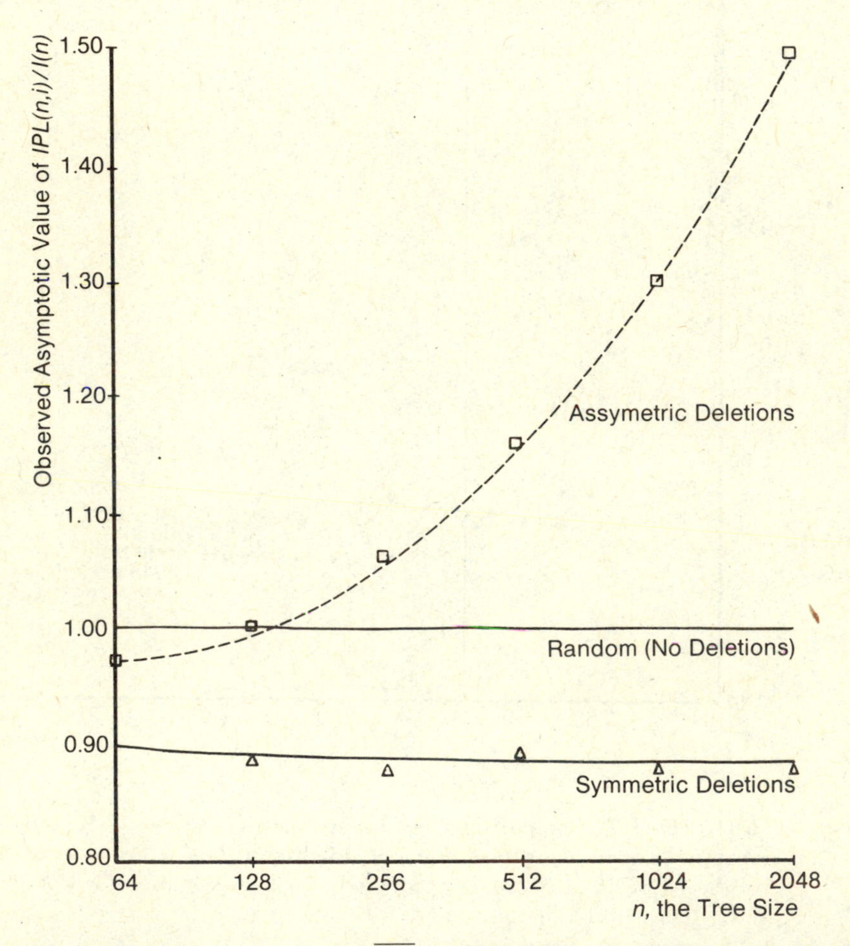

FIGURE 11.6 Asymptotic values of $\overline{IPL_{n,i}}$. [From Jeffrey L. Eppinger, *An Empirical Study of Insertion and Deletion in Binary Search Trees*, Communications of the ACM **26** (1983) p. 668, copyright 1983, Association for Computing Machinery, Inc. Reprinted by permission.]

Appendix

Here we give a brief discussion of a few items of importance to algorithm analysis which are not covered in the main text.

A.1 LINEAR PROGRAMMING

Linear programming solves the problem of minimizing a linear function subject to linear inequalities, i.e.,

$$\min_{x_1, x_2, \ldots, x_n} \{c_1 x_1 + c_2 x_2 + \cdots + c_n x_n\} \tag{1}$$

subject to

$$\begin{aligned} a_{11}x_1 &+ a_{12}x_2 + \cdots + a_{1n}x_n \leq b_1 \\ a_{21}x_1 &+ a_{22}x_2 + \cdots + a_{2n}x_n \leq b_2 \\ &\vdots \\ a_{m1}x_m &+ a_{m2}x_2 + \cdots + a_{mn}x_n \leq b_m. \end{aligned} \tag{2}$$

This form also covers maximization (reverse the sign of the c's) and inequalities with greater-than relations (reverse the signs of the a_{ij} and b_i for some i to have the same effect as changing the less than to a greater than in row i).

Many important problems can be expressed as linear programming problems. Sophisticated software packages are available for solving linear programming problems. See Gale [31] for a treatment of linear programming that pays careful attention to both the mathematics and the intuition for linear programming. See Hadley [34] or Simonnard [52] for a treatment of the mathematics.

If some or all of the variables of a linear programming problem are required to be integer, then the problem may become much harder to solve. Integer programming is concerned with such problems. See Simonnard [52], Hu [35], or Salkin [51]. If either the conditions or the function being optimized are changed to be nonlinear, then the problem becomes one of nonlinear programming. See McCormick [42].

A.2* CONTOUR INTEGRATION

Contour integration gives an algebraically simple way to evaluate many difficult sums and integrals. This section has a very brief introduction to complex variable theory and contour integration. Apostol [17] is a good source of additional information. This section contains a brief explanation of some of the key facts from Apostol and examples showing how to apply contour integration to summation and integration problems.

The most general function of a complex variable $z = x + iy$ (where x and y are real and i is the square root of -1) is $f(z) = u(x,y) + iv(x,y)$, where $u(x,y)$ and $v(x,y)$ are real-valued functions. Functions of this form, however, are too general to have many interesting properties. Therefore, a *function of a complex variable* $f(z)$ is defined to be a function that has a derivative on some set S. That is, the limit

$$f' = \lim_{z \to z_0} \frac{f(z) - f(z_0)}{z - z_0} \qquad \textbf{(3)}$$

exists on some set S, and the value of the limit does not depend on which way z approaches z_0. Such a function is also called *analytic.* An analytic function is continuous on the set where it is analytic.

Complex variable theory treats infinity as a single point. For a function $f(z)$ to be continuous at infinity, $\lim_{z \to \infty} f(z)$ must exist and have a value that is independent of how z goes to infinity. For example, z is not continuous at infinity, because you get a large value ($+\infty$) when z approaches infinity through positive values, and a small value ($-\infty$) when z approaches infinity through negative values. On the other hand, z^{-1} is continuous at infinity; it goes to zero no matter how z approaches infinity. It is often useful to think of the complex plane as though it is a large sphere obtained by joining together all the points that have infinite absolute value.

Most common functions are analytic except at a few points. For example, z^n is analytic except at infinity, e^z is analytic except at infinity, and z^{-n} is analytic except at zero. A constant function is analytic everywhere. (No other functions are analytic everywhere.) If f and g are two analytic functions, then $f+g$, $f-g$, and fg are analytic functions. The function f/g is analytic except where $g = 0$. The function $\log z$ is analytic except when z is a real number that is less than or equal to zero.

If a function is analytic, then there is a close relation between its real and imaginary parts.

Theorem A.1 (Cauchy-Riemann) For $f(z) = u(x,y) + iv(x,y)$, $z = x + iy$, where x, y, $u(x,y)$, and $v(x,y)$ are real and $f(z)$ is analytic,

$$\frac{\partial u(x,y)}{\partial x} = \frac{\partial v(x,y)}{\partial y} \quad \text{and} \quad \frac{\partial v(x,y)}{\partial x} = -\frac{\partial u(x,y)}{\partial y}, \qquad \textbf{(4)}$$

where $\partial u(x,y)/\partial x$ denotes the *partial derivative* of u with respect to x.

Integrals over regions are related to integrals around curves. In all that follows concerning integrals around closed curves, we assume that all curves are reasonably smooth. We omit the technical statement of this condition (see Apostol [17, Section 8.9]).

Theorem A.2 (Green) If $P(x,y)$ and $Q(x,y)$ are continuous functions with bounded partial derivatives, then

$$\int\int_R \left(\frac{\partial Q(x,y)}{\partial x} - \frac{\partial P(x,y)}{\partial y} \right) dx\, dy = \oint_{\Gamma(R)} [P(x,y)\, dx + Q(x,y)\, dy] \qquad \textbf{(5)}$$

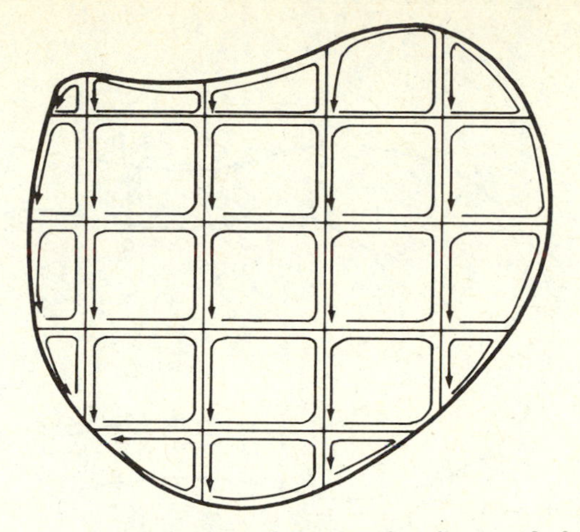

FIGURE A.1 Dividing a curve into small regions for the proof of Green's theorem.

where R is a region of two-dimensional space and $\Gamma(R)$ is the curve that bounds the region. When integrating along a closed curve, the value of the integral depends on which way you go around the curve. The integral along a curve is called a *contour integral.* If you reverse the direction, you reverse the sign of the answer. We assume that you proceed in a counterclockwise direction.

The proof of Green's theorem is based on first dividing the region into many small rectangles and showing that the theorem holds on small rectangles (see Figure A.1.) Then the results for all the rectangles are added. Plugging into eq. (5) and adding the left side gives the integral over the region, while adding the right sides results in cancellation of the contributions from adjacent rectangles, so that only the contribution from the boundary is left. (For adjacent rectangles, the integral for one rectangle follows the boundary in one direction while for the adjacent region it follows it in the opposite direction. The contributions from the two regions sum to zero.)

Theorem A.3 If Γ is a unit circle around the point z_0, then

$$\oint_\Gamma \frac{dz}{z - z_0} = 2\pi i. \tag{6}$$

Proof. We can represent z as $z_0 + e^{2\pi i\theta}$. Letting z go around a unit circle around z_0 corresponds to θ going from zero to 1, so

$$\oint_\Gamma \frac{dz}{z - z_0} = \int_0^1 e^{-2\pi i\theta}\, d(e^{2\pi i\theta}) = \int_0^1 d\theta = 2\pi i. \tag{7}$$

Integrating a function around a closed curve gives a result of zero if the function is analytic on and inside the region bounded by the curve.

Theorem A.4 (Cauchy) If Γ is a closed curve and $f(z)$ is analytic on the region bounded by Γ (i.e., analytic on the boundary and everywhere inside the boundary), then

$$\oint_\Gamma f(z)\, dz = 0. \tag{8}$$

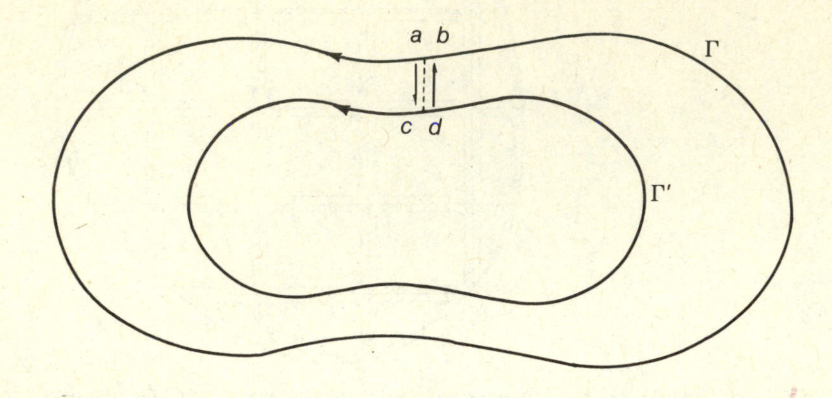

FIGURE A.2 Curve Γ goes from point a to point b. (Points a and b are the same point, but we use two names to distinguish the start of Γ from the end of Γ.) Curve Γ' goes from c to d.

Cauchy's theorem is straightforward to prove from Green's theorem using $f(z) = P(x,y) + iQ(x,y)$ and $z = x + iy$.

An important consequence of Cauchy's theorem is that for *any* function, the integral around closed curve Γ is equal to the integral around closed curve Γ' provided the two curves enclose the same set of singularities. Here we give a sketch of the proof. Consider the curve of Figure A.2 that follows Γ from a to b, then goes from b to d, then follows Γ' in the backwards direction from d to c and finally returns from c to a along the same path that was followed from b to d (but the path is followed in the reverse direction this time.) The integral of a function $f(z)$ around this path is equal to the integral of the function along Γ plus the integral along the path from b to d plus the integral along Γ' plus the integral along the path from c to a. But the integral along the path from b to d is just the negative of the integral along the path from c to a, because the two paths are the same except for their direction. Also the integral along Γ' in the backward direction is the negative of the integral in the positive direction. So the integral along the path is equal to the integral along Γ plus the integral along Γ' (going in the forward direction). By Cauchy's theorem, the integral is also zero if the function is analytic everywhere between the two curves (including the boundaries). The value of contour integrals is determined by the nature of the function at the places inside the boundary where the function is not analytic.

For analytic functions, there is a close connection between the value of the function in one region and its value in other nearby regions, as indicated by the following theorem.

Theorem A.5 (Cauchy's integral formula) If Γ is a closed curve and $f(z)$ is analytic on the region bounded by Γ (i.e., analytic on the boundary and everywhere inside the boundary), then

$$f(z_0) = \frac{1}{2\pi i} \oint_\Gamma \frac{f(z)}{z - z_0}\, dz. \tag{9}$$

Notice that $f(z)/(z - z_0)$ is not analytic at z_0 unless $f(z_0) = 0$. To prove Cauchy's integral formula, you need to integrate

$$g(z) = \begin{cases} \dfrac{f(z) - f(z_0)}{z - z_0} - f'(z_0) & \text{if } z \neq z_0 \\ 0 & \text{if } z \neq z_0 \end{cases} \tag{10}$$

The definition of $g(z)$ ensures that $g(z)$ is continuous at $z = z_0$, so Cauchy's theorem says that the integral of $g(z)$ around Γ is zero. The integral of $f(z)/(z - z_0)$ is the left side of eq. (9). Since $f'(z_0)$ is a constant, Cauchy's integral formula says its integral around Γ is zero. The remaining part of $g(z)$ is $f(z_0)/(z - z_0)$. By theorem 9.5, this integral is $2\pi i f(z_0)$.

Cauchy's integral theorem can be extended to give

$$f^{(n)}(z_0) = \frac{n!}{2\pi i} \oint_\Gamma \frac{f(z)}{(z - z_0)^{n+1}}\, dz \tag{11}$$

where $f^{(n)}(z_0)$ is the nth derivative at z_0 and $f(z)$ is analytic inside Γ (and on the boundary).

If $f(z)$ is analytic on an annulus around the point z_0 (the region between two circles centered on z_0), then $f(z)$ has a *Laurent expansion*:

$$f(z) = \sum_{n \geq 0} a_n (z - z_0)^n + \sum_{n \geq 1} a_{-n} (z - z_0)^{-n}, \tag{12}$$

where the constant a_n is given by

$$a_n = \frac{1}{2\pi i} \oint_\Gamma \frac{f(z)}{(z - z_0)^{n+1}}\, dz. \tag{13}$$

When $f(z)$ is analytic over a disk around z_0, then we have a power series (i.e., $a_n = 0$ for $n < 0$).

If, in the Laurent expansion, $a_{-n} \neq 0$ for some $n > 0$, then $f(z)$ is *singular* somewhere inside the inner circle of the annulus.

If the Laurent expansion is for an annulus having only one singularity inside the inner circle, then the expansion can be used to indicate the type of the singularity. For such an expansion, if $a_{-n} \neq 0$ but $a_{-m} = 0$ for $m > n$, then point z_0 is called a *pole* of order n. For example, z^{-1} has a pole of order 1 at $z = 0$. A singular point that is not a pole is called an *essential singularity*.

The coefficient a_{-1} is called the *residue*, i.e.,

$$a_{-1} = \operatorname*{Res}_{z=z_0} f(z). \tag{14}$$

Theorem A.6 (Cauchy residue theorem)

$$\oint_\Gamma f(z)\,dz = 2\pi i \sum_k \operatorname*{Res}_{z=z_k} f(z), \tag{15}$$

where the sum over k is over all the singularities inside the curve Γ.

In other words, the integral around a closed contour is equal to $2\pi i$ times the sum of the residues at the singularities inside of Γ. The integral must go in the counterclockwise direction (otherwise the negative of the above answer will be found), no singularities can occur on Γ, and Γ must be a reasonably smooth curve.

For poles of order 1, the residue can be computed with the limit

$$\operatorname*{Res}_{z=z_0} f(z) = \lim_{z\to z_0} zf(z) \qquad \text{provided } z_0 \text{ is a pole of order 1.} \tag{16}$$

Cauchy's residue theorem shows that the coefficient of z^{-1} in a Laurent expansion is the sum of the residues inside of the inner circle of the annulus. A Laurent expansion converges only on an annulus (considering a disk and the region outside of a disk as two degenerate cases of an annulus). If you have a Laurent expansion on an annulus A, and you start increasing the radius of the outer circle, the Laurent expansion will continue to converge on the annulus until the radius becomes so large that there is a singularity in the annulus (and then the expansion will no longer converge everywhere in the annulus). Likewise, you can decrease the radius of the inner circle until a further decrease would result in a singularity in the annulus.

Although eq. (15) is hard to derive (we gave only a sketch of the derivation), it is easy to use. Let's consider a few examples. Consider evaluating the integral

$$\int_{-\infty}^{+\infty} \frac{dx}{(x^2+a^2)}. \tag{17}$$

To convert this into a contour integral, we will integrate along the real axis, from $-\infty$ to $+\infty$, and then follow a large circle from $+\infty$ to $-\infty$ going counterclockwise (see Figure A.3). To be more precise, the path will go from $-R$ to $+R$ along the real axis, and then follow a circle from R to iR to $-R$; and we will take the limit as R goes to infinity. In this case for large x our function becomes small fast enough that the integral around the circle is zero. That is, for the circle of radius R the absolute value of $1/(x^2+a^2)$ is about R^{-2} and the length of the integration path is πR, so the value of the integral around the circle is no more than R^{-1}, which goes to zero as R goes to infinity.

The function $1/(x^2+a^2)$ can be factored to give

$$\frac{1}{x^2+a^2} = \frac{1}{(x-ia)(x+ia)}, \tag{18}$$

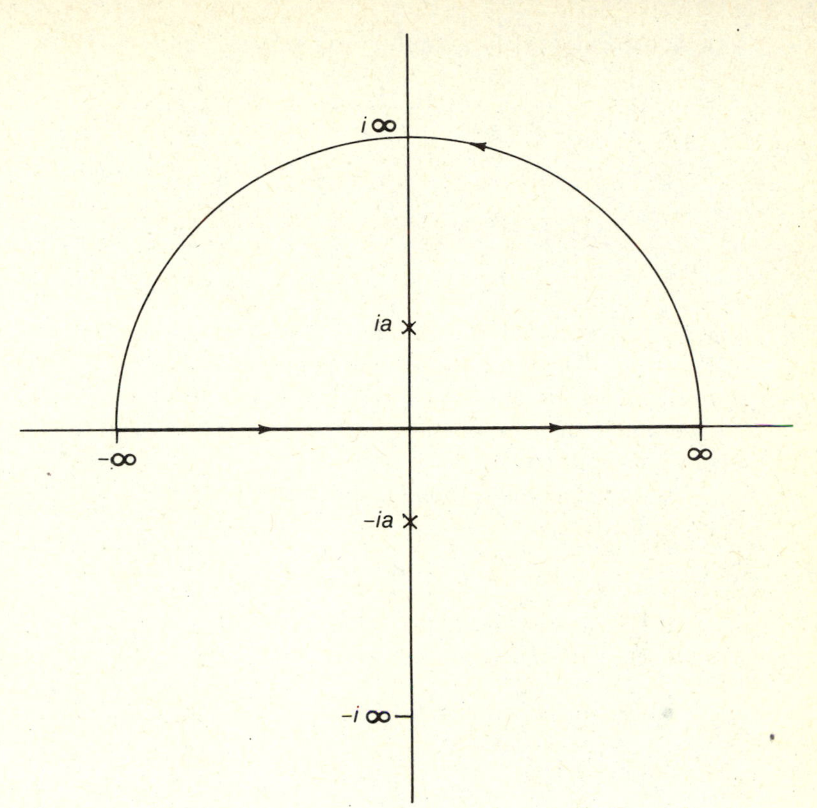

FIGURE A.3 The path for integrating $1/(x^2 + a^2)$.

so the function has singularities at $x = ia$ and $x = -ia$. The singularity at $x = ia$ is inside our contour. The other singularity is outside of it. The residue at $x = ia$ can be calculated as follows:

$$\operatorname*{Res}_{x=ia} \frac{1}{x^2 + a^2} = \lim_{x \to ia} \frac{x - ia}{x^2 + a^2} = \lim_{x \to ia} \frac{1}{x + ia} = \frac{1}{2ia}. \tag{19}$$

Therefore, the Cauchy residue theorem gives

$$\int_{-\infty}^{+\infty} \frac{dx}{(x^2 + a^2)} = \frac{\pi}{a}. \tag{20}$$

For a second example, let's consider evaluating the sum

$$\sum_{1 \le j \le n} \Bigg(x_j^m \Big/ \prod_{\substack{1 \le k \le n \\ k \ne j}} (x_j - x_k) \Bigg). \tag{21}$$

This sum is the sum of the residues of the function

$$f(x) = \frac{x^m}{(x - x_1)(x - x_2) \ldots (x - x_n)}, \tag{22}$$

so let's expand $f(x)$ in a Laurent expansion. Using

$$\frac{1}{x - x_j} = \frac{1/x}{1 - x_j/x} = x^{-1}(1 + x_j + x_j^2 + \cdots), \tag{23}$$

we get

$$f(x) = x^{m-n}\left(\frac{1}{1 - x_1/x}\right)\left(\frac{1}{1 - x_2/x}\right)\cdots\left(\frac{1}{1 - x_n/x}\right) \tag{24}$$

$$= x^{m-n} + (x_1 + x_2 + \cdots + x_n)x^{m-n+1} + \cdots, \tag{25}$$

which converges when the absolute value of x is greater than the absolute value of all the x_i. The residue is the coefficient of x^{-1}. For $m < n - 1$, the answer is zero; the series has a zero z^{-1} term. For $m = n - 1$ the answer is 1. For $m = n$ the answer is the sum of the x_j. The answer for larger m can also be figured out by this method.

A source for this section is Apostol [17].

EXERCISES

1. What is the value of eq. (21) when $m = n + 1$?

2. What is the value of eq. (21) when $m = n + 2$?

References

ANALYSIS OF ALGORITHMS

This section of the references lists those other books that you may wish to refer to for additional information about analysis of algorithms.

1. Alfred V. Aho, John E. Hopcroft, and Jeffrey D. Ullman, *The Design and Analysis of Computer Algorithms*, Addison-Wesley, Reading, Mass. (1976). This is one of the basic advanced books on the analysis of algorithms. It concentrates on big O analyses.
2. Sara Baase, *Computer Algorithms*, Addison-Wesley, Reading, Mass. (1978). This book is an elementary introduction to the analysis of algorithms.
3. Walter S. Brainerd and Lawrence H. Landweber, *Theory of Computation*, John Wiley & Sons, New York (1974). This book has a good introduction to undecidability.
4. Michael R. Garey and David S. Johnson, *Computers and Intractability*, W. H. Freeman and Co., San Francisco (1979). This is the basic book on NP-completeness. It contains both elementary and advanced material.
5. G. H. Gonnet, *Handbook of Algorithms and Data Structures*, Addison-Wesley, Reading, Mass. (1984). This book gives the statement of many basic algorithms and the results of analyses of them. References are given for each algorithm.
6. Daniel H. Greene and Donald E. Knuth, *Mathematics for the Analysis of Algorithms*, Birkhäuser, Boston (1981). This short book is one of the best sources on advanced techniques for analysis of algorithms.
7. Ellis Horowitz and Sartaj Sahni, *Fundamentals of Computer Algorithms*, Computer Science Press, Potomac Md. (1978).
8. T. C. Hu, *Combinatorial Algorithms*, Addison-Wesley, Reading, Mass. (1982).
9. Donald E. Knuth, *The Art of Computer Programming, Vol. 1, Fundamental Algorithm*, Addison-Wesley, Reading, Mass. second edition (1973). The first chapter of this book contains all the elementary mathematics and some of the advanced mathematics used in the analysis of algorithms. This is one of the basic references in the field. The second chapter has a good introduction to the data structures used in many sophisticated algorithms.
10. Donald E. Knuth, *The Art of Computer Programming, Vol. 2, Seminumerical Algorithms*, Addison-Wesley, Reading, Mass., second edition (1981).

This book has a complete discussion of random number generation and of the algorithms used for computer arithmetic.

11. Donald E. Knuth, *The Art of Computer Programming, Vol. 3, Searching and Sorting*, Addison-Wesley, Reading, Mass. (1973). This book gives the analyses of the performance of most searching and sorting algorithms. A person studying analysis of algorithms who wants to learn just one type of algorithm for practicing mathematical techniques would do well to consider searching and sorting.
12. Eugene Lawler, *Combinatorial Optimization*, Holt, Rinehart and Winston, New York (1976).
13. Harry R. Lewis and Christos H. Papadimitriou, *Elements of the Theory of Computation*, Prentice-Hall, Englewood-Cliffs, N.J. (1981). This book has a good introduction to undecidability.
14. Hartley Rogers, Jr., *Theory of Recursive Functions and Effective Computability*, McGraw-Hill Book Company, New York(1967).
15. Robert Endre Tarjan, *Data Structures and Network Algorithms*, SIAM, Philadelphia (1983).

MATHEMATICS

A person who analyzes algorithms oftens needs information from other branches of mathematics. Here we list several books that the authors have found to be particularly helpful. These books are not in all cases the most up-to-date ones on their subject. In some cases the authors used them when they were in college. The authors have, however, found them to be very useful.

16. Milton Abramowitz and Irene A. Stegun, *Handbook of Mathematical Functions*, Dover, New York (1965). This is a very useful collection of the fundamental formulas of mathematics and a set of tables of the values of various functions.
17. Tom M. Apostol, *Mathematical Analysis*, Addison-Wesley, Reading, Mass. (1957). This book covers real and complex variable theory.
18. Tom M. Apostol, *Calculus*, Blaisdell, New York (1962).
19. Emil Artin, *The Gamma Function*, Holt, Rinehart and Winston, 1964. Translated by Michael Butler.
20. R. Bellman and S. E. Dreyfus, *Applied Dynamic Programming*, Princeton University Press, Princeton, N. J. (1962).
21. Carl M. Bender and Steven A. Orszag, *Advanced Mathematical Methods for Scientists and Engineers*, McGraw-Hill Book Company, New York, 1978. This book covers the solution for differential and difference equations. Both exact and asymptotic methods are considered.
22. Garrett Birkhoff and Saunders Mac Lane, *A Survey of Modern Algebra*, revised edition, Macmillan, New York (1962).

23. Daniel I. A. Cohen, *Basic Techniques of Combinatorial Theory*, John Wiley & Sons, New York (1978). This book shows how to derive combinatorial formulas in many different ways, leading to a deeper understanding of the formulas. It covers all of standard combinatorics.
24. Cuthbert Daniel and Fred S. Wood, *Fitting Equations to Data*, John Wiley & Sons, New York (1980).
25. N. G. de Bruijn, *Asymptotic Methods in Analysis*, North-Holland Publishing Co., Amsterdam (1961). This book has a thorough discussion of asymptotic methods for various mathematical problems including the summation of series.
26. Douglas F. Elliott and K. Ramamohan Rao, *Fast Transforms: Algorithms, Analyses, Applications*, Academic Press, New York (1982).
27. William Feller, *An Introduction to Probability Theory and Its Applications*, John Wiley & Sons, New York, third edition (revised printing) (1968).
28. Marek Fisz, *Probability Theory and Mathematical Statistics*, John Wiley & Sons, New York, third edition (1963). This is a concise high-level introduction to probability and statistics.
29. Lester R. Ford, *Differential Equations*, McGraw-Hill Book Company, New York, second edition (1955). This has almost everything the authors have ever needed on solving differential equations.
30. Tomlinson Fort, *Finite Differences*, Oxford University Press, New York (1948).
31. David Gale, *The Theory of Linear Economic Models*, McGraw-Hill Book Company, New York (1960).
32. Jack L. Goldberg and Arthur J. Schwartz, *Systems of Ordinary Differential Equations: An Introduction*, Harper and Row, New York (1972).
33. Ian P. Goulden and David M. Jackson, *Combinatorial Enumeration*, John Wiley & Sons, New York (1983). This book contains a terse high-level treatment of formal power series, generating functions, and combinatorics.
34. G. Hadley, *Linear Programming*, Addison-Wesley, Reading, Mass. (1962).
35. T. C. Hu, *Integer Programming and Network Flow*, Addison-Wesley, Reading, Mass. (1970).
36. Glenn James (ed.), *Mathematics Dictionary*, Van Nostrand Reinhold, New York, fourth edition (1968). This is a good source of mathematical definitions and short discussions.
37. Richard E. Johnson and Fred L. Kiokemeister, *Calculus with Analytic Geometry*, Allyn and Bacon, Boston (1960).
38. L. B. W. Jolley, *Summation of Series*, Dover, New York, second revised edition (1961).
39. Leonard Kleinrock, *Queueing Systems, Vol. 1*, John Wiley & Sons, New York (1975).
40. Konrad Knopp, *Theory and Application of Infinite Series*, Blackie, Glasgow, second edition (1951).
41. C. L. Liu, *Introduction to Combinatorial Mathematics*, McGraw-Hill Book Company, New York (1968).

42. Garth P. McCormick, *Nonlinear Programming: Theory, Algorithms, and Applications*, John Wiley & Sons, New York (1983).
43. Louis Melville Milne-Thomson, *The Calculus of Finite Differences*, Macmillan, New York (1933).
44. Douglas C. Montgomery and Elizabeth A. Peck, *Introduction to Linear Regression Analysis*, John Wiley & Sons, New York (1982).
45. Ben Noble, *Applied Linear Algebra*, Prentice-Hall, Englewood Cliffs, N. J. (1969).
46. F. W. J. Olver, *Asymptotics and Special Functions*, Academic Press, New York (1974).
47. Carl E. Pearson, ed., *Handbook of Applied Mathematics*, Van Nostrand Reinhold, New York (1983).
48. John Riordan, *An Introduction to Combinatorial Analysis*, John Wiley & Sons, New York (1958).
49. John Riordan, *Combinatorial Identities*, John Wiley & Sons, New York (1968).
50. Thomas L. Saaty, *Elements of Queueing Theory*, McGraw-Hill Book Company, New York (1961).
51. Harvey M. Salkin, *Integer Programming*, Addison-Wesley, Reading, Mass. (1975).
52. Michel Simonnard, *Linear Programming*, Prentice-Hall, Englewood Cliffs, N. J. (1966).
53. Lucy Joan Slater, *Generalized Hypergeometric Functions*, Cambridge University Press, London (1966). This is the best source on the generalized hypergeometric function.
54. George W. Snedecor and William G. Cochran, *Statistical Methods*, The Iowa State University Press, Ames, Iowa (1980).
55. B. M. Steward, *Theory of Numbers*, Macmillan, New York (1962).
56. Gilbert Strang, *Linear Algebra and Its Applications*, Academic Press, New York (1976).
57. E. R. Tufte, *The Visual Display of Quantitative Information*, Graphics Press, Cheshire, Conn. (1983).
58. Mitchell Wand, *Induction, Recursion, and Programming*, North Holland, New York (1980).

ALGORITHMS

59. Alfred V. Aho, John E. Hopcroft, and Jeffrey D. Ullman, *Data Structures and Algorithms*, Addison-Wesley, Reading, Mass. (1983).
60. Ellis Horowitz and Sartaj Sahni, *Fundamentals of Data Structures*, Computer Science Press, Potomac, Md. (1976).
61. Edward M. Reingold and Wilfred J. Hansen, *Data Structures*, Little, Brown and Company, Boston (1983).
62. Robert Sedgewick, *Algorithms*, Addison-Wesley, Reading, Mass. (1983).

63. Thomas A. Standish, *Data Structure Techniques*, Addison-Welsey, Reading, Mass. (1980).

OTHER REFERENCES

This section contains the references to more specialized works, usually papers on some aspect of analysis of algorithms. The following is a partial list of the abbreviations that are used in journal names and organizations: ACM (Association for Computing Machinery), Comm. ACM (Communications of the ACM), IEEE (Institute of Electrical and Electronic Engineers), J. ACM (Journal of the ACM), Math. Comp. (Mathematics of Computation), Proc. Natl. Acad. Sci. USA (Proceedings of the National Academy of Sciences), SIAM (Society for Industrial and Applied Math), SIAM J. Comput. (SIAM Journal on Computing), SIGACT (ACM Special Interest Group for Automata and Computability Theory),

64. Alfred V. Aho and N. J. A. Sloane, *Some Doubly Exponential Sequences*, Fibonacci Quarterly **11** (1973), pp. 429–437.

65. Alfred V. Aho and Jeffrey D. Ullman, *Principles of Compiler Design*, Addison-Wesley, Reading, Mass. (1977).

66. George E. Andrews, *Applications of the Basic Hypergeometric Functions*, SIAM Review **16** (1974), pp. 441–484.

67. Mikhail J. Atallah and S. Rao Kosaraju, *An Adversary-based Lower Bound for Sorting*, Information Processing Letters **13** (1981), pp. 55–57.

68. Edward A. Bender, *Asymptotic Methods in Enumeration*, SIAM Review **16** (1974), pp. 485–513.

69. Jon L. Bentley, *Programming Pearls*, Comm. ACM **26** (1983), pp. 623–628. Programming Pearls is a monthly column giving ways to use simple analysis to develop efficient algorithms.

70. Jon L. Bentley and Donna J. Brown, *A General Class of Resource Tradeoffs*, 21st Ann. Symp. on Foundations of Computer Science, IEEE Computer Society (1980), pp. 217–228.

71. Jon Louis Bentley, Dorothea Haken, and James B. Saxe, *A General Method for Solving Divide-and-Conquer Recurrences*, SIGACT News, (Fall 1980), pp. 36–44.

72. Jon Louis Bentley, David S. Johnson, Tom Leighton, and Catherine Cole Mc Geoch, *An Experimental Study of Bin Packing*, Technical Report (1984).

73. Manuel Blum, Robert W. Floyd, Vaughan Pratt, Ronald L. Rivest, and Robert E. Tarjan, *Time Bounds for Selection*, Journal of Computer and System Sciences **7** (1973), pp. 448–461.

74. A. D. Booth and A. J. T. Colin, *On the Efficiency of a New Method of Dictionary Construction*, Information and Control **3** (1960), pp. 327–334.

75. I. Borosh and A. S. Fraenkel, *Exact Solutions to Linear Equations with Rational Coefficients by Congruence Techniques*, Math. Comp. **20** (1966), pp. 107–112.

76. K. Brinck and N. Y. Foo, *Analysis of Algorithms in Threaded Trees*, Computer Journal **24** (1981), pp. 148–155.

77. K. Brinck, *Analysis of Algorithms in Lexicographic Trees*, submitted to Computer Journal.

78. Andrei Z. Broder, *A General Expression for Abelian Identities* in *Combinatorics on Words* (Larry J. Cummings, ed.), Academic Press, New York (1983).

79. Cynthia A. Brown and Paul Walton Purdom Jr., *An Average Time Analysis of Backtracking*, SIAM J. Comput. **10** (1981), pp. 583–593.

81. J. Cocke and J. I. Schwartz, *Programming Languages and Their Compilers*, Courant Institute of Mathematical Sciences, New York University, New York (1970).

80. Norman H. Cohen, *Eliminating Redundant Recursive Calls*, ACM Trans. Programming Languages and Systems **5** (1983), pp. 265–299.

82. S. A. Cook, *The Complexity of Theorem Proving Procedures*, Proc. 3rd Ann. ACM Symp. on Theory of Computing, Assoc. for Comput. Mach., New York (1971), pp. 151–158.

83. Stephen Cook and Cynthia Dwork, *Bounds on the Time for Parallel RAM's to Compute Simple Functions*, Proc. of the 14th Symp. on Theory of Computing, Assoc. for Comput. Mach., New York (1982), pp. 231–233.

84. D. Coppersmith and S. Winograd, *On the Asymptotic Complexity of Matrix Multiplication*, SIAM J. Comput. **11** (1982), pp. 472–492.

85. Martin Davis, *Hilbert's Tenth Problem Is Unsolvable*, American Mathematical Monthly **80** (1973) pp. 233–269.

86. Jeffrey L. Eppinger, *An Empirical Study of Insertion and Deletion in Binary Search Trees*, Comm. ACM **26** (1983), pp. 663–669.

87. R. W. Floyd, *Algorithm 97: Shortest Path*, Comm. ACM **5** (1962), p. 345.

88. Allen Goldberg, Paul Purdom, and Cynthia Brown, *Average Time Analyses of Simplified Davis-Putnam Procedures*, Information Processing Letters **15** (1982), pp. 72–75. Corrections in **16** (1983), p. 213.

89. I. J. Good, *The Interaction Algorithm and Practical Fourier Series*, J. R. Statist. Soc. Sect. B **20** (1958), pp. 361–372; **22** (1960), pp. 372–375.

90. R. William Gosper, Jr., *Decision Procedure for Indefinite Hypergeometric Summation*, Proc. Natl. Acad. Sci. USA **75** (1978), pp. 40–42.

91. Susan L. Graham, Michael A. Harrison, and Walter L. Ruzzo, *An Improved Context-Free Recognizer*, ACM Trans. Programming Languages and Systems **2** (1980), pp. 415–462.

92. Harry C. Heacox and Paul W. Purdom Jr., *Analysis of Two Time-Sharing Queueing Models*, J. ACM **19** (1972), pp. 70–91.

93. Thomas N. Hibbard, *Some Combinatorial Properties of Trees with Applications to Searching and Sorting*, J. ACM **9** (1962), pp. 13–28.

94. C. A. R. Hoare, *Quicksort*, The Computer Journal, **5** (1962), pp. 10–15.

95. J. E. Hopcroft and J. K. Wong, *Linear Time Algorithm for Isomorphism of Planar Graphs*, Proc. 6th Ann. ACM Symp. on Theory of Computing, Assoc. for Comput. Mach., New York (1974), pp. 172–184.

96. T. C. Hu and M. T. Shing, *Computations of Matrix Chain Products, Part I*, SIAM J. on Comput. **11** (1982), pp. 362–373; *Part II*, **13** (1984), pp. 228–251.

97. Gérard Huet, *Confluent Reductions: Abstract Properties and Applications to Term Rewriting Systems*, J. ACM **27** (1980), pp. 797–821.

98. F. K. Hwang and Shen Lin, *A Simple Algorithm for Merging Two Disjoint Linearly-Ordered Sets*, SIAM J. Comput. **1** (1972), pp. 31–39.

99. D. S. Johnson, A. Demers, J. D. Ullman, M. R. Garey, and R. L. Graham, *Worst-Case Performance Bounds for One-Dimensional Packing Algorithms*, SIAM J. Comput. **3** (1974), pp. 299–325.

100. A. Karatsuba and Yu. Ofman, *Multiplication of Multidigit Numbers on Automata*, Sov. Phys., Dokl. **7** (1963), pp. 595–596. Originally in Dokl. Akad. Nauk SSSR **145** (1962), pp. 293–294.

101. R. M. Karp, *Reducibility among Combinatorial Problems*, in R. E. Miller and J. W. Thatcher (eds.), *Complexity of Computer Computations*, Plenum Press, New York (1972), pp. 85–103.

102. Richard M. Karp and Judea Pearl, *Searching for an Optimal Path in a Tree with Random Costs*, Artificial Intelligence **21** (1983), pp. 99–116. [Also published as *Search and Heuristics* (J. Pearl, ed.), North-Holland, New York (1983), same page numbers.]

103. Michael Karr, *Summation in Finite Terms*, J. ACM **28** (1981), pp. 305–350. This has a general theory for simplifying indefinite sums (sums with general limits) and linear first order difference equations. The theory not only provides a closed-form sum when possible, but also tells when no closed-form answer exists.

104. T. Kasami, *An Efficient Recognition and Syntax Analysis Algorithm for Context Free Languages*, Sci. Rep. AF CRL-65-758 Air Force Cambridge Research Laboratory, Bedford, Mass. (1965).

105. G. D. Knott, *Deletion in Binary Storage Trees*, Ph.D. thesis, Stanford University (1975), STAN-CS-75-491.

106. D. E. Knuth, *Estimating the Efficiency of Backtracking Programs*, Math. Comp. **29** (1975), pp. 121–136.

107. D. E. Knuth, *Big Omicron and Big Omega and Big Theta*, SIGACT News **8** No. 2 (1976), pp. 18–23. This is the first printed proposal for the definition of the big theta notation.

108. D. E. Knuth, J. H. Morris Jr., and V. R. Pratt, *Fast Pattern Match in Strings*, Stanford University CS-74-440 (1976).

109. I. Lah, *Eine neue Art von Zahren, ihre Eigenschaften and Anwendung in der Mathematischen Statiskik*, Mitteilungsbl. Math. Statist. **7** (1955), pp. 203–212.

110. George S. Lueker, *Some Techniques for Solving Recurrences*, Computing Surveys **12** (1980), pp. 419–436.

111. George S. Lueker, *An Average-Case Analysis of Bin Packing with Uniformly Distributed Item Sizes*, University of California at Irvine Computer and Information Sciences Technical Report No. 181 (1982).

112. William C. Lynch, *More Combinatorial Properties of Certain Trees*, Computer Journal **7** (1965), pp. 299-302.

113. Y. Matiyasevich, *Diophantine Representation of Rec. Enumerable Predicates*, Proc. of the Scand. Logic Symp., J. E. Fenstad, ed. North Holland, New York (1978).

114. Michael T. McClellan, *The Exact Solutions of Systems of Linear Equations*, J. ACM **20** (1973), pp. 563–588.

115. John R. Metzer and Bruce H. Barnes, *Decision Table Languages and Systems*, Academic Press, New York (1977).

116. Barry M. Minton, *Generalized Hypergeometric Function of Unit Argument*, Journal of Mathematical Physic **11** (1970), pp. 1375–1376.

117. Louis Monier, *Combinatorial Solutions of Multidimensional Divide-and-Conquer Recurrences*, Journal of Algorithms **1** (1980), pp. 60–74.

118. Robert Morris, *Scatter Storage Techniques*, Comm. ACM **11** (1968), pp. 38–44.

119. Dana S. Nau, Paul Purdom, and Chun Hung Tzeng, *Experiments on Alternatives to Minimax*, University of Maryland Tech. Report No. 1333, 1983.

120. W. W. Peterson, *Addressing for Random-Access Storage*, IBM Journal of Research and Development, **17** (1957), pp. 130–146.

121. Paul Purdom, *The Size of LALR(1) Parsers*, BIT **14** (1974), pp. 326–337.

122. Paul W. Purdom, *Tree Size by Partial Backtracking*, SIAM J. Comput. **7** (1978), pp. 481–491.

123. P. W. Purdom and J. H. Williams, *Cycle Length in a Random Function*, Transactions of the American Mathematics Society **133** (1968), pp. 547–551.

124. Paul W. Purdom Jr. and Stephen M. Stigler, *Statistical Properties of the Buddy System*, J. ACM **17** (1970), pp. 683–697.

125. Paul W. Purdom Jr. and Edward F. Moore, *Immediate Predominators in a Directed Graph*, Comm. ACM **15** (1972), pp. 777–778.

126. Paul Walton Purdom Jr., Cynthia A. Brown, and Edward L. Robertson, *Backtracking with Multi-Level Dynamic Search Rearrangement*, Acta Informatica **15** (1981), pp. 99–113.

127. Edward M. Reingold and Robert E. Tarjan, *On a Greedy Heuristic for Complete Matching*, SIAM J. Comput. **10** (1981), pp. 676–681.

128. John Reynolds, sited in J. C. Early, *An Efficient Context-Free Parsing Algorithm*, thesis, Carnegie-Melon U. (1968), pp. 128–129.

129. A. Schönhage, M. Patterson, and N. Pippenger, *Finding the Median*, Journal of Computer and System Sciences **13** (1976), pp. 184–199.

130. Daniel Dominic Sleator and Robert Endre Tarjan, *Amortized Efficiency of Update Rules*, Proc. 16th Ann. ACM Symp. on Theory of Computing, Assoc. for Comput. Mach., New York (1984), pp. 488-492.

131. J. Spiess, *Untersuchungen des Zeitgewinns durch neue Algorithmen zur Matrix-Multiplikation*, Computing **17** (1976), pp. 23–36.

132. Francis R. Stevenson and Donald E. Knuth, *Optimal Measurement Points for Program Frequency Counts*, BIT **13** (1973), pp. 313–322.

133. Volker Strassen, *Gaussian Elimination Is Not Optimal*, Numerische Mathematik, **13** (1969), pp. 354–356.
134. Robert Endre Tarjan, *Efficiency of a Good But Not Linear Set Union Algorithm*, J. ACM **22** (1975), pp. 215–225.
135. Robert E. Tarjan and Jan van Leeuween, *Worst-case Analysis of Set Union Algorithms*, J. ACM **31** (1984), pp. 245–281.
136. Jeffery Scott Vitter, *Analysis of the Search Performance of Coalesced Hashing*, J. ACM **30** (1983), pp. 231–258.
137. Jeffery Scott Vitter, *Faster Methods for Random Sampling*, Comm. ACM, **27** (1984) pp. 703–718.
138. J. W. J. Williams, *Algorithm 232: Heapsort*, Comm. ACM, **7** (1964), p. 701.
139. P. F. Windley, *Trees, Forest and Rearranging*, Computer Journal, **3** (1960), pp. 84–88.
140. S. Winograd, *On the Multiplication of* 2×2 *Matrices*, IBM Research Report RC267 (1970).
141. D. H. Younger, *Recognition of Context-free Languages in Time* n^3, Information and Control **10** (1967), pp. 189–208.

Math Index

The purpose of this index is to help you find formulas in the text and to help you remember the formulas that you already know how to use. Many details concerning the meaning of the formulas and the conditions under which they apply have been omitted to save space. You should therefore always consider referring back to the text when looking up a formula.

NOTATION

Look in the section on definitions for additional examples of notation.

Absolute value: $|x|$.
Ancestor (have common): $x \uparrow y$.
Bernoulli number: B_n.
Bernoulli polynomial: $B_m(x)$.
Bilateral function: ${}_2H_2\begin{bmatrix} a, b; 1 \\ c, d \end{bmatrix}$.
Binomial coefficient: $\binom{n}{k}$.
Catalan number: C_n.
Characteristic function: $\phi(t)$.
Complex conjugate: z^*.
Complex number: $x + iy$.
Contour integral: $\oint_{\Gamma(R)} f(x)\,dx$.
Convolution: $A * B$.
Correlation coefficient: ρ.
Cyclic convolution: $A \circledast B$.
Decreasing power: $n^{\underline{k}}$.
Delta function: δ_{ij}.
Derivative: $f'(x)$, $\dfrac{df(x)}{dx}$, $f^{(i)}(x)$.
Descendant (have common): $x \downarrow y$.
Determinant: $\det|A|$.
Equivalent: $x \equiv y$, $P \Longleftrightarrow Q$.
Euler's constant: γ.
Exponent: x^y, $e^x = \exp(x)$, $(f(x))^n$.
Factorial: $n!$.
Fibonacci number: F_n.
Gamma function: $\Gamma(x)$.
Gaussian binomial coefficient: $\binom{n}{k}_q$.
Harmonic number: H_n.
Hypergeometric function:
$_mF_n\begin{bmatrix} a_1, \ldots, a_m; z \\ b_1, \ldots, b_n \end{bmatrix}$.
$_mF_n[a_1, \ldots, a_m; b_1, \ldots, b_n; z]$.
Implies: $P \Longrightarrow Q$.
Increasing power: $n^{\overline{k}}$.
Limit: $\lim_{x \to b} f(x)$.
Logarithm, base b: $\log_b x$.
base 2: $\lg x$.
base e: $\ln x$.
Magnitude: $|x|$.
Matrix element: a_{ij}, A_{ij}, $[A]_{ij}$.
Mod: $x \bmod y$.
Multinomial coefficient:
$\binom{n}{i_1, i_2, \ldots, i_m}$.
Number of elements in a set: $|S|$.
Partial derivative: $\dfrac{\partial f(x)}{\partial x}$.
Product: $\prod_{1 \le i \le n} a_i = a_1 a_2 \ldots a_n$.
Reflexive transitive closure: $x \xrightarrow{*} y$.
Relation: $R(x,y)$, xRy, $x \underset{R}{\rightarrow} y$, $x \rightarrow y$.

Repeated function application: $g^{[k]}(n)$.

Residue: $\mathrm{Res}_{z=z_0} f(z)$.

Root of unity ω.

Sawtooth function: $\{x\}$.

Standard deviation: $a \pm b$.

Stirling number of the first kind: $\left[{n \atop k}\right]$.

Stirling number of the second kind: $\left\{{n \atop k}\right\}$.

Summation: $\sum_{1\le i\le n} a_i = a_1 + a_2 + \cdots + a_n$.

Symmetric reflexive transitive closure: $x \overset{*}{\leftrightarrow} y$.

Transitive closure: $x \overset{+}{\to} y$.

DEFINITIONS

Also see the sections on conventions and asymptotics.

Average:

$$A = \sum_i p_i T_i. \tag{1.48}$$

$$A = \sum_{0\le i<n} q_i. \tag{2.79}$$

$$A = \int_{-\infty}^{+\infty} x\, dF(x). \tag{11.17}$$

Bernoulli polynomial:

$$B_m(x) = \sum_i \binom{m}{i} B_i x^{m-i}. \tag{4.159}$$

Bilateral function:

$${}_mH_n\left[{a_1, \ldots, a_m; z \atop b_1, \ldots, b_n}\right] = \sum_i \frac{a_1^{\overline{i}} \ldots a_m^{\overline{i}} z^i}{b_1^{\overline{i}} \ldots b_n^{\overline{i}}}. \tag{3.208}$$

Binomial coefficient:

$$\binom{n}{k} = \frac{n!}{k!(n-k)!}. \tag{3.40}$$

$$\binom{r}{k} = \prod_{1\le i\le k} \left(\frac{r+1-i}{i}\right) \tag{3.41}$$

Casoratian:

$$W_n = \det \begin{vmatrix} B_1(n) & B_2(n) & \cdots & B_k(n) \\ B_1(n+1) & B_2(n+1) & \cdots & B_k(n+1) \\ \vdots & \vdots & & \vdots \\ B_1(n+k-1) & B_2(n+k-1) & \cdots & B_k(n+k-1) \end{vmatrix} \tag{6.33}$$

Characteristic function:

$$\phi(t) = \int_{-\infty}^{+\infty} e^{itx}\, dF(x). \tag{11.23}$$

Coin distribution function:

$$F(x) = \begin{cases} 0 & \text{for } x < 0, \\ 1-p & \text{for } 0 \le x < 1, \\ 1 & \text{for } x \ge 1. \end{cases} \tag{11.79}$$

Continuous:

$$\lim_{x \to b} f(x) = f(b). \tag{1.83}$$

Convolution:

$$c_n = \sum_{0 \le i \le n} a_i b_{n-i}. \tag{6.128}$$

Cyclic:

$$c_k = \sum_{0 \le j < n} a_j b_{(k-j) \bmod n} \qquad \text{for } 0 \le k < n. \tag{9.32}$$

Correlation coefficient:

$$\rho = \frac{C_{12}}{\sigma_1 \sigma_2}. \tag{1.64}$$

Covariance:

$$C_{12} = \sum_i p_i (T_{\text{average1}} - T_{i1})(T_{\text{average2}} - T_{i2}). \tag{1.62}$$

Decreasing power:

$$n^{\underline{k}} = n(n-1)\cdots(n-k+1). \tag{2.54}$$

Delta function:

$$\delta_{ij} = \begin{cases} 0 & \text{if } i \ne j, \\ 1 & \text{if } i = j. \end{cases} \tag{3.62}$$

Density (probability):

$$f(x) = \frac{dF(x)}{dx}. \tag{11.20}$$

Derivative:

$$f' = \lim_{z \to z_0} \frac{f(z) - f(z_0)}{z - z_0} \tag{A.3}$$

Distribution function:

$$F(x) = \text{Prob}(X \le x). \tag{11.1}$$

Exponent:

$$n^k = \prod_{1 \le i \le k} n. \tag{3.3}$$

Factorial:

$$n! = \prod_{1 \le i \le n} i. \tag{3.6}$$

Fibonacci number:

$$F_n = F_{n-1} + F_{n-2} \qquad \text{for } n \ge 2, \tag{5.123}$$
$$F_0 = 0, \qquad F_1 = 1.$$

Fourier transform:

$$y_j = \sum_{0 \le k < n} x_k \omega^{jk} \qquad \text{for } 0 \le j < n. \tag{9.30}$$

Inverse Fourier transform:

$$x_j = \frac{1}{n} \sum_{0 \le k < n} y_k \omega^{-jk} \qquad \text{for } 0 \le j < n. \tag{9.31}$$

Gamma function:

$$\Gamma(x) = \lim_{m\to\infty} \frac{m^x m!}{x(x+1)(x+2)\cdots(x+m)}. \tag{3.22}$$

$$= \int_0^\infty e^{-t}t^{x-1}\,dt. \tag{3.32}$$

Gaussian binomial coefficient:

$$\binom{n}{k}_q = \frac{(q^n-1)}{(q-1)}\cdot\frac{(q^{n-1}-1)}{(q^2-1)}\cdots\frac{(q^{n-k-1}-1)}{(q^k-1)} \qquad \text{(Ex. 8.2.2.1–3)}$$

Harmonic number:

$$H_n = \sum_{1\le i\le n} \frac{1}{i}. \tag{4.169}$$

Hypergeometric function:

$${}_mF_n\left[\begin{matrix} a_1,\cdots,a_m; z \\ b_1,\cdots,b_n \end{matrix}\right] = \sum_{i\ge 0} \frac{a_1^{\bar{i}}\cdots a_m^{\bar{i}} z^i}{b_1^{\bar{i}}\cdots b_n^{\bar{i}}\, i!}. \tag{3.203}$$

Increasing power:

$$n^{\bar{k}} = n(n+1)\cdots(n+k-1). \tag{2.55}$$

Integer:

$$\sum_{0\le i\le n} d_i b^i. \tag{1.7}$$

Integral: Riemann sum:

$$R_p(f) = \sum_{0\le i<n} f(z_i)(x_{i+1}-x_i). \tag{11.8}$$

Riemann integral:

$$\int_a^b f(x)\,dx = \lim_{|p|\to 0} R_p(f). \tag{11.9}$$

Stieltjes sum:

$$S_p(g,F) = \sum_{0\le i<n} g(z_i)[F(x_{i+1}) - F(x_i)]. \tag{11.10}$$

Stieltjes integral:

$$\int_a^b g(x)\,dF(x) = \lim_{|p|\to 0} S_p(g,F). \tag{11.11}$$

Inverse (modulo m)

$$xy \equiv 1 \pmod{m'}. \tag{1.17}$$

Inversions, number of:

$$\sum_i \sum_{j>i} I_{ij} \quad \text{where} \quad I_{ij} = \begin{cases} 0 & \text{if } k_i \le k_j, \\ 1 & \text{if } k_i > k_j. \end{cases} \tag{9.6}$$

Laurent expansion:

$$f(z) = \sum_{n\ge 0} a_n(z-z_0)^n + \sum_{n\ge 1} a_{-n}(z-z_0)^{-n}, \tag{A.12}$$

$$a_n = \frac{1}{2\pi i}\oint_\Gamma \frac{f(z)}{(z-z_0)^{n+1}}\,dz.$$

Limit:

$$\lim_{n\to\infty} f(n) = a \iff |f(n) - a| < \epsilon \text{ for } n \geq n(\epsilon), \tag{1.76}$$

$$\lim_{x\to b} f(x) = a \iff |f(n) - a| < \epsilon \text{ for } |x - b| < \delta(\epsilon) \tag{1.82}$$

Linear programming:

$$\min_{x_1, x_2, \ldots, x_n} \{c_1x_1 + c_2x_2 + \cdots + c_nx_n\} \tag{A.1}$$

subject to

$$\begin{array}{llll} a_{11}x_1 & + a_{12}x_2 & + \cdots + a_{1n}x_n & \leq b_1, \\ a_{21}x_1 & + a_{22}x_2 & + \cdots + a_{2n}x_n & \leq b_2, \\ \vdots & \vdots & \vdots & \vdots \\ a_{m1}x_m & + a_{m2}x_2 & + \cdots + a_{mn}x_n & \leq b_m. \end{array}$$

Matrix product:

$$c_{ij} = \sum_{1\leq k\leq p} a_{ik}b_{kj}. \tag{1.35}$$

Mod operation:

$$x \bmod y = x - y \left\lfloor \frac{x}{y} \right\rfloor. \tag{1.9}$$

Modular representation:

$$\begin{aligned} n_1 &= n \bmod m_1, \\ n_2 &= n \bmod m_2, \\ &\vdots \\ n_k &= n \bmod m_k. \end{aligned} \tag{9.54}$$

Norm of a partition:

$$|p| = \max_{0\leq i<n} \{x_{i+1} - x_i\}. \tag{11.7}$$

Normal distribution:

$$F(x) = \frac{1}{2\pi} \int_{-\infty}^{x} \exp\left(\frac{-(y-\mu)^2}{2\sigma}\right) dy, \tag{11.32}$$

$$f(x) = \frac{1}{2\pi} \exp\left(-\frac{(x-\mu)^2}{2\sigma}\right), \tag{11.33}$$

$$\phi(t) = e^{-t^2/(2\sigma) - i\mu t}. \tag{11.40}$$

Number of combinations:

$$\binom{n}{k} = \frac{n!}{k!(n-k)!}. \tag{3.40}$$

Number of partitions into nonempty subsets:

$$\left\{ n \atop i \right\}. \qquad \text{(See Section 3.7)}$$

Number of permutations:

$$\prod_{0\leq i<k} (n - i) = \frac{n!}{(n-k)!}. \tag{3.14}$$

Operators:

$$\mathbf{E}\{a_n\} = \{a_1,\ a_2,\ a_3,\ \ldots\}, \tag{2.33}$$

$$\mathbf{I}\{a_n\} = \{a_0,\ a_1,\ a_2,\ \ldots\}, \tag{2.34}$$

$$\mathbf{P}_k\{a_n\} = \{a_k,\ a_k,\ a_k,\ \ldots\}, \tag{2.37}$$

$$\mathbf{\Delta}\{a_n\} = \{a_1 - a_0,\ a_2 - a_1,\ a_3 - a_2,\ \ldots\}, \tag{2.35}$$

$$\mathbf{\Sigma}\{a_n\} = \left\{ \sum_{0 \le i \le n} a_i \right\}. \tag{2.32}$$

Partial fraction decomposition:

$$\frac{R(i)}{D(i)} = \sum_{1 \le j \le n} \frac{T_j(i)}{P_j^{a_j}(i)}. \tag{2.95}$$

Polynomial:

$$p(x) = \sum_{0 \le i \le n} a_i x^i. \tag{1.3}$$

Power series:

$$f(x) = \sum_{i \ge 0} a_i x^i. \tag{4.42}$$

Sample mean:

$$\bar{x} = \frac{1}{n} \sum_{1 \le j \le n} x_i. \tag{11.45}$$

Sawtooth function:

$$\{x\} = x \bmod 1 = x - \lfloor x \rfloor. \tag{4.149}$$

Semi-invariant:

$$\ln \phi(t) = \sum_{j \ge 0} \frac{\kappa_j}{j!} (it)^j. \tag{11.27}$$

Standard deviation:

$$\sigma = \sqrt{V}. \tag{1.52}$$

Step function:

$$f(x) = \begin{cases} c & \text{for } x < x_0. \\ c + p & \text{for } x \ge x_0. \end{cases} \tag{11.14}$$

Stirling number of the first kind:

$$n! \binom{x}{n} = \sum_i (-1)^{n-i} \left[{n \atop i} \right] x^i. \tag{3.176}$$

Stirling number of the second kind:

$$x^n = \sum_i \left\{ {n \atop i} \right\} \binom{x}{i} i!. \tag{3.187}$$

Time (total):

$$T = \sum_{1 \le j \le k} n_j t_j. \tag{1.1}$$

Time (worst case):

$$T_{\text{worst-case}} = \max_{1 \le j \le i} T_j. \tag{1.45}$$

Variance:

$$V = \sum_i p_i (T_{\text{average}} - T_i)^2, \tag{1.51}$$

$$V = \int_{-\infty}^{+\infty} (x - A)^2 \, dF(x), \tag{11.18}$$

$$V = \int_{-\infty}^{+\infty} x^2 \, dF(x) - A^2. \tag{11.19}$$

CONVENTIONS

Factorials:

$$0! = 1. \tag{3.10}$$

$$n! = \infty \quad \text{for} \quad n < 0. \tag{3.12}$$

Increasing powers:

$$x^{-\bar{i}} \quad \text{means} \quad \frac{(-1)^i}{(1-a)^{\bar{i}}}. \tag{3.209}$$

Infinite Sums:

$$\sum_{R(i)} x_i \quad \text{means} \quad \Big(\lim_{n\to\infty} \sum_{\substack{R(i)\\ 0\le i\le n}} x_i\Big) + \Big(\lim_{n\to\infty} \sum_{\substack{R(i)\\ 0>i\ge -n}} x_i\Big). \tag{2.27}$$

Products:

$$\prod_{m\le i<n} a_i \quad \text{means} \quad \prod_{n\le i<m} a_i^{-1} \tag{3.4}$$

[Occasionally $\prod_{R(i)} a_i$ means 1 when $R(i)$ is never *true*.]

Scope of summations:

$$\sum_{1\le i\le n} x_i + y = \Big(\sum_{1\le i\le n} x_i\Big) + y \tag{2.3}$$

$$\sum_{1\le i\le n} x_i y = \sum_{1\le i\le n} (x_i y) \tag{2.4}$$

Summations:

$$\sum_{m\le i<n} a_i \quad \text{means} \quad -\sum_{n\le i<m} a_i \tag{2.30}$$

[Occasionally $\sum_{R(i)} a_i$ means zero when $R(i)$ is never *true*.]

IDENTITIES

Binomial coefficients, summations, and asymptotics are listed separately.

Analytic function:

$$\frac{\partial u(x,y)}{\partial x} = \frac{\partial v(x,y)}{\partial y} \quad \text{and} \quad \frac{\partial v(x,y)}{\partial x} = -\frac{\partial u(x,y)}{\partial y}, \tag{A.4}$$

where $f(z) = u(x,y) + iv(x,y)$.

Average:

$$A = A_1 + A_2. \tag{1.50}$$

Bilateral function:

$${}_2H_2\begin{bmatrix} a, b; 1 \\ c, d \end{bmatrix} = \frac{\Gamma(c)\Gamma(d)\Gamma(1-a)\Gamma(1-b)\Gamma(c+d-a-b-1)}{\Gamma(c-a)\Gamma(d-a)\Gamma(c-b)\Gamma(d-b)}. \tag{3.212}$$

Catalan number:

$$C_n = \frac{1}{n+1}\binom{2n}{n}. \tag{7.166}$$

Cauchy's integral formula:

$$f(z_0) = \frac{1}{2\pi i}\oint_\Gamma \frac{f(z)}{z - z_0}\,dz. \tag{A.9}$$

Cauchy's residue theorem:

$$\oint_\Gamma f(z)\,dz = 2\pi i \sum_k \operatorname*{Res}_{z=z_k} f(z). \tag{A.15}$$

Cauchy's theorem:

$$\oint_\Gamma f(z)\,dz = 0. \tag{A.8}$$

Central limit theorem:

$$U_n = \frac{X_1 + X_2 + \cdots + X_n}{n} \tag{11.54}$$

approaches a normal distribution with average μ and standard deviation $\sigma/\sqrt{n}$.

Contour integral:

$$\oint_\Gamma \frac{dz}{z - z_0} = 2\pi i. \tag{A.6}$$

Decreasing power:

$$n^{\underline{k}} = \frac{n!}{(n-k)!}. \tag{3.8}$$

Density:

$$f(x) = \frac{1}{2\pi}\int_{-\infty}^{+\infty} e^{-itx}\phi(t)\,dt. \tag{11.28}$$

Distribution functions:

$$\text{For } Z = X + Y \qquad \phi_z(t) = \phi_x(t)\phi_y(t), \tag{11.26}$$

$$\text{For } Z = aX + b \qquad \mu_z = a\mu_x + b \tag{11.29}$$

$$\sigma_z^2 = a^2\sigma_x^2, \tag{11.30}$$

For $Z = X + Y$ where X and Y are independent and normal,

$$\sigma_z = \frac{1}{1/\sigma_x + 1/\sigma_y}, \tag{11.43}$$

$$\mu_z = \mu_x + \mu_y. \tag{11.44}$$

Euler ϕ function:

$$\phi(n) = n - \sum_{1\le i\le k} (-1)^i \sum_{1\le j_1<j_2<\cdots<j_i\le k} \frac{n}{p_{j1}p_{j2}\cdots p_{ji}}. \tag{3.245}$$

Exponents:

$$x^{y+z} = x^y x^z \quad \text{and} \quad (x^y)^z = x^{yz}, \tag{1.20}$$

$$e^{x+iy} = e^x(\cos y + i\sin y). \tag{3.157}$$

Factorial:

$$n! = n(n-1)!, \tag{3.11}$$

$$(ai + ad + c)! = (ad + c)!(a^a)^i \left(d + \frac{c+1}{a}\right)^{\bar{i}} \left(d + \frac{c+2}{a}\right)^{\bar{i}} \cdots \left(d + 1 + \frac{c}{a}\right)^{\bar{i}}, \tag{3.215}$$

$$(b - ai)! = \frac{(-1)^{ai} b!}{(a^a)^i \left(-\frac{b}{a}\right)^{\bar{i}} \left(-\frac{b-1}{a}\right)^{\bar{i}} \cdots \left(-\frac{b-a+1}{a}\right)^{\bar{i}}}. \tag{3.218}$$

Fibonacci number:

$$F_n = \frac{1}{\sqrt{5}}(\phi^n - \hat{\phi}^n), \tag{5.140}$$

$$F_n = F_m F_{n-m+1} + F_{m-1} F_{n-m}, \tag{5.143}$$

$$\phi^n + \hat{\phi}^n = F_n + 2F_{n-1}. \tag{5.150}$$

Fourier transform:

$$\mathbf{F}(A \circledast B) = (\mathbf{F}A) \cdot (\mathbf{F}B). \tag{9.36}$$

Gamma function:

$$n! = \Gamma(n+1), \tag{3.21}$$

$$\Gamma(x+1) = x\Gamma(x), \tag{3.25}$$

$$\Gamma\left(\tfrac{1}{2}\right) = \sqrt{\pi}. \tag{3.38}$$

Gauss's generalization of the binomial theorem:

$$(1+x)(1+qx)\ldots(1+q^{n-1}) = \sum_{0 \le i \le n} \binom{n}{i}_q q^{i(i-1)/2} x^i. \tag{Ex. 8.2.2.1–7}$$

Green's theorem:

$$\int\int_R \left(\frac{\partial Q(x,y)}{\partial x} - \frac{\partial P(x,y)}{\partial y}\right) dx\,dy = \oint_{\Gamma(R)} [P(x,y)\,dx + Q(x,y)\,dy]. \tag{A.5}$$

Hypergeometric:

$${}_2F_1\begin{bmatrix} a, b; 1 \\ c \end{bmatrix} = \frac{\Gamma(c)\Gamma(c-a-b)}{\Gamma(c-a)\Gamma(c-b)}. \tag{3.207}$$

Inclusion and Exclusion:

$$N = \sum_{i \ge 0} (-1)^i \sum_{1 \le j_1 < j_2 < \cdots < j_i} N(A_{j_1} A_{j_2} \ldots A_{j_i}). \tag{3.243}$$

Independent properties:

$$N(A_1 A_2 \ldots A_i) = N(A_1) N(A_2) \cdots N(A_i). \tag{3.244}$$

Increasing power:

$$n^{\bar{k}} = \frac{(n+k-1)!}{(n-1)!}, \tag{3.9}$$

$$\frac{\Gamma(-a+b)}{\Gamma(-a)} = (-1)^b \frac{\Gamma(a+1)}{\Gamma(a-b+1)}. \tag{3.213}$$

Integrals:

$$\int_1^x \frac{du}{u} = \log_e x, \tag{1.25}$$

$$\int_0^n x^k\,dx = \frac{n^{k+1}}{k+1}, \tag{2.57}$$

$$\int_a^b g(x)\,dF(x) = \int_a^b g(x)f(x)\,dx, \tag{11.12}$$

$$\int_0^n g(x)\,d\lfloor x\rfloor = \sum_{0\le i<n} g(i), \tag{11.13}$$

$$\int_a^b g(x)\,dF(x) = \int_a^b g(x)f(x)\,dx + \sum_{j \text{ such that } a\le x_j\le b} p_j g(x_j). \tag{11.15}$$

Least Squares (linear):
minimize

$$\sum_{1\le i\le m} [y_i - a_1 f_1(x_{i1}, x_{i2}, \ldots, x_{ik}) + a_2 f_2(x_{i1}, x_{i2}, \ldots, x_{ik}) + \cdots + a_n f_n(x_{i1}, x_{i2}, \ldots, x_{ik})]^2. \tag{11.133}$$

L'Hôpital's rule:

$$\lim_{x\to b}\frac{f(x)}{g(x)} = \lim_{x\to b}\frac{f'(x)}{g'(x)}. \tag{1.84}$$

Logarithms:

$$\log_b(xy) = \log_b x + \log_b y, \tag{1.22}$$

$$\log_b(x^y) = y\log_b x, \tag{1.23}$$

$$\log_b x = \frac{\log_a x}{\log_a b}. \tag{1.26}$$

Modulo:

$$a + x \equiv b + y \pmod m, \tag{1.13}$$

$$a - x \equiv b - y \pmod m, \tag{1.14}$$

$$ax \equiv by \pmod m, \tag{1.15}$$

$$a \equiv b \pmod m \iff an \equiv bn \pmod{mn}, \tag{1.16}$$

For r relatively prime to s, $\quad a \equiv b \pmod{rs} \iff$
$(a \equiv b \pmod r)$ and $a \equiv b \pmod s$. (Theorem 1.4)

Multinomial:

$$\binom{i_1+i_2+\cdots+i_m}{i_1, i_2, \ldots, i_m} = \binom{i_1+i_2+\cdots+i_m}{i_1}\binom{i_2+\cdots+i_m}{i_2, \ldots, i_m}, \tag{3.172}$$

$$\binom{i_1+i_2+\cdots+i_m}{i_1, i_2, \ldots, i_m} = \binom{i_1+i_2+\cdots+i_m}{i_1}\binom{i_2+\cdots+i_m}{i_2}\cdots\binom{i_{m-1}+i_m}{i_m}, \tag{3.173}$$

$$\binom{i_1+i_2+\cdots+i_m}{i_1, i_2, \ldots, i_m} = \frac{(i_1+i_2+\cdots+i_m)!}{i_1!i_2!\cdots i_m!}, \tag{3.174}$$

$$(x_1+x_2+\cdots+x_m)^n = \sum_{i_1, i_2, \ldots, i_m}\binom{n}{i_1, i_2, \ldots, i_m} x_1^{i_1}x_2^{i_2}\cdots x_m^{i_m}. \tag{3.175}$$

Operator:

$$\mathbf{\Delta\Sigma} = \mathbf{E}, \tag{2.36}$$

$$\mathbf{\Sigma\Delta} = \mathbf{E} - \mathbf{P}_0. \tag{2.38}$$

Polynomials:

$$p(x) = p_{\mathrm{even}}(x^2) + x p_{\mathrm{odd}}(x^2), \tag{9.43}$$

and

$$p(-x) = p_{\mathrm{even}}(x^2) - x p_{\mathrm{odd}}(x^2). \tag{9.44}$$

Probability:

$$\mathrm{Prob}(A \cup B) = \mathrm{Prob}(A) + \mathrm{Prob}(B) - \mathrm{Prob}(A \cap B), \tag{1.42}$$

$$\mathrm{Prob}(A \cap B) = \mathrm{Prob}(B) \cdot \mathrm{Prob}(A|B) = \mathrm{Prob}(A) \cdot \mathrm{Prob}(B|A). \tag{1.44}$$

Product:

$$\prod_{R(i)} a_i = \exp\left(\sum_{R(i)} \ln a_i\right), \tag{3.5}$$

$$\frac{n!}{(n-k)!} = \prod_{n-k+1 \le i \le n} i, \tag{3.7}$$

$$\sum_{0 \le j < n} a^j = \prod_{0 \le j < k} \left(1 + a^{2^j}\right) \qquad \text{provided } n = 2^k. \tag{9.51}$$

Residue:

$$\operatorname*{Res}_{z=z_0} f(z) = \lim_{z \to z_0} z f(z) \qquad \text{provided } z_0 \text{ is a pole of order 1.} \tag{A.16}$$

Root of unity:

$$\omega = e^{2\pi i/n} = \cos(2\pi/n) + i \sin(2\pi/n). \tag{9.48}$$

$$\sum_{0 \le j < n} \omega^{jn} \equiv n \pmod{m}. \tag{9.49}$$

Stirling number:

$$\begin{bmatrix} n \\ n \end{bmatrix} = 1, \quad \begin{bmatrix} n \\ n-1 \end{bmatrix} = \binom{n}{2}, \tag{3.178}$$

$$\begin{bmatrix} n \\ 1 \end{bmatrix} = (n-1)! \quad (\text{for } n > 0), \quad \begin{bmatrix} n \\ 0 \end{bmatrix} = \delta_{n0}, \tag{3.179}$$

$$\begin{bmatrix} n \\ n-k \end{bmatrix} = \sum_{1 \le i_1 < i_2 < \cdots < i_k < n} i_1 i_2 \cdots i_k, \tag{3.180}$$

$$\begin{bmatrix} n \\ i \end{bmatrix} = (n-1)\begin{bmatrix} n-1 \\ i \end{bmatrix} + \begin{bmatrix} n-1 \\ i-1 \end{bmatrix} \quad \text{for } n > 0, \tag{3.185}$$

$$\begin{Bmatrix} n \\ i \end{Bmatrix} = i\begin{Bmatrix} n-1 \\ i \end{Bmatrix} + \begin{Bmatrix} n-1 \\ i-1 \end{Bmatrix}, \tag{3.189}$$

$$\begin{Bmatrix} n \\ n-k \end{Bmatrix} = \sum_{1 \le i_1 \le i_2 \le \cdots \le i_k < n} i_1 i_2 \cdots i_k, \tag{3.186}$$

$$\sum_i \begin{bmatrix} n \\ i \end{bmatrix} \begin{Bmatrix} i \\ m \end{Bmatrix} (-1)^i = (-1)^n \delta_{mn}, \tag{3.190}$$

$$\sum_i \begin{Bmatrix} n \\ i \end{Bmatrix} \begin{bmatrix} i \\ m \end{bmatrix} (-1)^i = (-1)^n \delta_{mn}, \tag{3.191}$$

$$\sum_i \begin{bmatrix} n \\ i \end{bmatrix} \binom{i}{m} = \begin{bmatrix} n+1 \\ m+1 \end{bmatrix}, \tag{3.192}$$

$$\sum_i \left\{ {n \atop i} \right\} \binom{i}{m} = \left\{ {n+1 \atop m+1} \right\}. \tag{3.193}$$

Variance:

$$V = \sum_i p_i T_i^2 - T_{\text{average}}^2, \tag{1.57}$$

$$V = V_1 + V_2 + 2C_{12}. \tag{1.63}$$

Variance (measured):

$$E(v_x) = \frac{n-1}{n}\sigma^2. \tag{11.48}$$

RELATIONS AND ORDERINGS

Characteristic function:

$$f_R(a, b) \iff [a, b] \in R. \qquad \text{(Section 1.11)}$$

Confluent:

$$x \uparrow y \implies x \downarrow y. \qquad \text{(Section 1.12.2)}$$

Dictionary order:

$$\begin{aligned} a_1 a_2 a_3 \ldots a_i \prec b_1 b_2 b_3 \ldots b_j \quad & \text{if, for some } k \geq 0, \\ & a_1 = b_1,\ a_2 = b_2,\ \ldots,\ a_{k-1} = b_{k-1},\ a_k < b_k \text{ or} \\ & a_1 = b_1,\ a_2 = b_2,\ \ldots,\ a_i = b_i,\ \text{and } i < j. \end{aligned} \tag{1.87}$$

Lexicographic order:

$$(m_1, n_1) \prec (m_2, n_2) \quad \begin{cases} \text{if} & m_1 < m_2, \\ \text{if} & m_1 = m_2 \text{ and } n_1 < n_2. \end{cases} \tag{1.86}$$

Locally confluent:

$$a \xrightarrow{+} x \text{ and } a \xrightarrow{+} y \implies x \downarrow y. \qquad \text{(Section 1.12.2)}$$

Reflexive:

$$x \to x. \qquad \text{(Section 1.11)}$$

Relation:

$$R \subseteq A \times B. \qquad \text{(Section 1.11)}$$

Relation-complete:

$$x \xrightarrow{*} y \text{ and } P(y) \implies P(x). \qquad \text{(Section 1.12.2)}$$

Symmetric:

$$x \to y \iff y \to x. \qquad \text{(Section 1.11)}$$

Transitive:

$$(x \to y) \wedge (y \to z) \implies (x \to z). \qquad \text{(Section 1.11)}$$

BINOMIALS

Also see Binomial Sums.

Adding formula:

$$\binom{r}{i} = \binom{r-1}{i} + \binom{r-1}{i-1}, \quad \text{integer } i. \tag{3.65}$$

Moving factors:

$$i\binom{r}{i} = r\binom{r-1}{i-1}, \quad \text{integer } i. \tag{3.55}$$

Negating top index:

$$\binom{-r}{k} = (-1)^k \binom{r+k-1}{k}, \tag{3.69}$$

$$\binom{n}{m} = (-1)^{n-m} \binom{-m-1}{n-m}, \tag{3.71}$$

$$\binom{-n}{i} = (-1)^{n+i+1} \binom{-i-1}{n-1}. \tag{3.72}$$

Product:

$$\binom{r}{m}\binom{m}{k} = \binom{r}{k}\binom{r-k}{m-k}. \tag{3.73}$$

Special values:

$$\binom{r}{k} = 0 \quad \text{for integer } k < 0, \tag{3.42}$$

$$\binom{n}{k} = 0 \quad \text{for integer } n, k, \quad k > n, \tag{3.43}$$

$$\binom{r}{0} = 1, \quad \binom{r}{1} = r, \quad \binom{r}{2} = \frac{r(r-1)}{2}. \tag{3.44}$$

Symmetry:

$$\binom{n}{i} = \binom{n}{n-i}, \quad \text{integer } n \geq 0, \quad \text{integer } i. \tag{3.53}$$

SUMMATIONS

Particular Results

Constant:

$$\sum_{1 \leq i \leq n} m = nm. \tag{1.37}$$

Harmonic:

$$\sum_{1 \leq i \leq n} H_i = (n+1)H_n - n. \tag{4.177}$$

Linear:

$$\sum_{1\le i\le n} i = \frac{n(n+1)}{2}. \tag{1.41}$$

Quadratic:

$$\sum_{0\le i\le n} i^2 = \frac{2n^3+3n^2+n}{6}. \tag{2.18}$$

Decreasing powers:

$$\sum_{0\le i<n} i^{\underline{k}} = \frac{n^{\underline{k+1}}}{k+1}. \tag{2.56}$$

Exponential:

$$\sum_{0\le i\le n} x^i = \frac{x^{n+1}-1}{x-1}. \tag{2.61}$$

Fibonacci:

$$\sum_{0\le k\le n} F_k F_{n-k} = \frac{n-1}{5}F_n + \frac{2n}{5}F_{n-1}. \tag{5.157}$$

Linear exponential:

$$\sum_{0\le i<n} ix^i = \frac{(n-1)x^{n+1}-nx^n+x}{(x-1)^2}. \tag{2.75}$$

Logarithm:

$$\sum_{1\le r<N} \lfloor \lg r \rfloor == N\lfloor \lg N \rfloor - 2^{\lfloor \lg N\rfloor+1} + 2. \tag{2.83}$$

Reciprocal:

$$\sum_{1\le i\le n} \frac{1}{i(i+1)} = 1 - \frac{1}{n+1}, \tag{2.91}$$

$$\sum_{i\ge 0} i^{-2k} = \frac{(-1)^{k+1}B_{2k}(2\pi)^{2p}}{2(2p)!}. \tag{Ex. 4.54–4}$$

General Laws

Combining ranges:

$$\sum_{m\le i<n} a_i = \sum_{m\le i<j} a_i + \sum_{j\le i<n} a_i \tag{2.29}$$

Derivative:

$$\frac{dS(x)}{dx} = \sum_{1\le i\le n} \frac{da_i(x)}{dx}. \tag{A.3}$$

Distributive law:

$$\left(\sum_{R(i)} a_i\right)\left(\sum_{S(j)} b_j\right) = \sum_{R(i)}\sum_{S(j)} a_i b_j. \tag{3.115}$$

Euler summation formula:

$$\sum_{1\le i<n} f(i) = \int_1^n f(x)\,dx - \tfrac{1}{2}(f(n)-f(1)) + \int_1^n B_1(\{x\})f'(x)\,dx, \qquad (4.153)$$

$$\sum_{1\le i<n} f(i) = \int_1^n f(x)\,dx + \sum_{1\le i\le m} \frac{B_i}{i!}\Big(f^{(i-1)}(n) - f^{(i-1)}(1)\Big) + R_m, \qquad (4.164)$$

$$R_m = \frac{(-1)^{m+1}}{m!}\int_1^n B_m(\{x\})f^{(m)}(x)\,dx, \qquad (4.165)$$

$$R_{2i} = \theta\frac{B_{2i+2}}{(2i+2)!}(f^{(2i+1)}(n) - f^{(2i+1)}(1)) \quad \text{where } 0<\theta<1, \qquad (4.167)$$

$$|R_{2i}| \le \left|\frac{B_{2i}}{(2i)!}\Big(f^{(2i-1)}(n) - f^{(2i-1)}(1)\Big)\right|. \qquad (4.168)$$

Integral:

$$\int S(x)\,dx = \sum_{1\le i\le n}\int a_i(x)\,dx. \qquad (2.118)$$

Interchanging order:

$$\sum_{R(i)}\sum_{S(j)} a_{ij} = \sum_{S(j)}\sum_{R(i)} a_{ij}, \qquad (3.118)$$

$$\sum_{R(i)}\sum_{S(i,j)} a_{ij} = \sum_{S'(j)}\sum_{R'(i,j)} a_{ij}, \qquad (3.119)$$

$$\sum_{R(i,j)} a_{ij} = \sum_{R'(i',j')} a_{i'j'}. \qquad (3.120)$$

See (3.127) for linear changes of indices.

Periodic terms:

$$\sum_{j\text{ even}} a_j = \frac{1}{2}\sum_j a_j + \frac{1}{2}\sum_j (-1)^j a_j, \qquad (3.150)$$

$$\sum_{j\text{ odd}} a_j = \frac{1}{2}\sum_j a_j - \frac{1}{2}\sum_j (-1)^j a_j, \qquad (3.151)$$

$$\sum_{j \bmod 4=0} a_j = \frac{1}{4}\sum_j (1)^j a_j + \frac{1}{4}\sum_j (i)^j a_j + \frac{1}{4}\sum_j (-1)^j a_j + \frac{1}{4}\sum_j (-i)^j a_j, \qquad (3.155)$$

$$\sum_{j \bmod n=0} a_j = \frac{1}{n}\sum_{0\le k<n}\sum_j (\omega^k)^j a_j, \qquad (3.158)$$

$$\sum_j \binom{n}{mj+k} = \frac{1}{m}\sum_{0\le j<m}\left(2\cos\frac{\pi j}{m}\right)^n \cos\frac{\pi j(n-2k)}{m}. \qquad (\text{Ex. } 3.4.2\text{–}2)$$

Summation by parts:

$$\sum_{m\le i<n} (a_{i+1}-a_i)b_i = a_n b_n - a_m b_m - \sum_{m\le i<n} a_{i+1}(b_{i+1}-b_i). \qquad (2.72)$$

BINOMIAL SUMS

Binomial theorem:

$$(x+y)^n = \sum_i \binom{n}{i} x^i y^{n-i}, \tag{3.46}$$

$$(x+y)^r = \sum_{i \geq 0} \binom{r}{i} x^i y^{r-i}, \tag{3.47}$$

$$\sum_i \binom{n+i}{i} x^i = (1-x)^{-n-1}. \tag{3.70}$$

Both indices:

$$\sum_{0 \leq i \leq n} \binom{r+i}{i} = \binom{r+n+1}{n}. \tag{3.67}$$

Top index:

$$\sum_{0 \leq i < n} \binom{i}{k} = \binom{n}{k+1}. \tag{3.50}$$

Products:

$$\sum_i \binom{r}{i} \binom{s}{n-i} = \binom{r+s}{n}, \tag{3.166}$$

$$\sum_i \binom{r}{i} \binom{s}{n+i} = \binom{r+s}{r+n}, \tag{3.167}$$

$$\sum_{-(s-n) \leq i \leq r-m} \binom{r-i}{m} \binom{s+i}{n} = \binom{r+s+1}{m+n+1}, \tag{3.168}$$

$$\sum_i \binom{r}{i} \binom{s+i}{n} (-1)^i = (-1)^r \binom{s}{n-r}, \tag{3.169}$$

$$\sum_{0 \leq i \leq r} \binom{r-i}{m} \binom{s}{i-t} (-1)^i = (-1)^t \binom{r-t-s}{r-t-m}. \tag{3.170}$$

BOUNDS

Particular Results

Bernoulli polynomials:

$$\left| \frac{B_m(\{x\})}{m!} \right| < \left| \frac{4}{(2\pi)^m} \right|. \tag{4.166}$$

Binomial tail:

(Markov)

$$\sum_{i \geq x} \binom{n}{i} p^i (1-p)^{n-i} \leq \frac{np}{x}. \tag{11.80}$$

(Chebyshev)

$$\sum_{i \geq x} \binom{n}{i} p^i (1-p)^{n-i} \leq \frac{np(1-p)}{(x-np)^2}. \tag{11.81}$$

(Chernoff)

$$\sum_{i \geq x} \binom{n}{i} p^i (1-p)^{n-i} \leq \left(\frac{n(1-p)}{n-x}\right)^{n-x} \left(\frac{np}{x}\right)^x, \tag{11.86}$$

Chebyshev inequality:

$$\text{Prob}(|X-\mu| \geq x) \leq \frac{\sigma^2}{x^2}. \tag{11.62}$$

Chernoff bound:

$$\text{Prob}(Z \geq z_0) \leq e^{-az_0 + n \ln \phi_X(-ia)} \tag{11.68}$$

$$\text{where } z_0 = \frac{n}{\phi_X(-ia)} \frac{d\phi_X(-ia)}{da}. \tag{11.69}$$

Exponential:

$$e^x \geq 1+x \quad \text{and} \quad e^x \leq \frac{1}{1-x} \quad \text{for} \quad x \geq 0. \tag{4.15}$$

Logarithm:

$$\frac{x}{1+x} \leq \ln(1+x) \leq x. \tag{4.17}$$

Markov inequality:

$$\text{Prob}(X \geq x) \leq \frac{\mu}{x}. \tag{11.61}$$

General Laws

$$f_L(x) + g_L(x) \leq f(x) + g(x) \leq f_U(x) + g_U(x) \quad \text{for } x_0 \leq x \leq x_1, \tag{4.3}$$

$$cf_L(x) \leq cf(x) \leq cf_U(x) \quad \text{for } x_0 \leq x \leq x_1 \text{ where } c \geq 0, \tag{4.4}$$

$$cf_U(x) \leq cf(x) \leq cf_L(x) \quad \text{for } x_0 \leq x \leq x_1 \text{ where } c \leq 0, \tag{4.5}$$

$$f_L(x)g_L(x) \leq f(x)g(x) \leq f_U(x)g_U(x) \quad \text{for } x_0 \leq x \leq x_1, \tag{4.6}$$

$$f(g_L(x)) \leq f(g(x)) \leq f(g_U(x)) \quad \text{for } x_0 \leq x \leq x_1, \tag{4.7}$$

where $f(y)$ is an increasing function in the range

$$\min_{x_0 \leq x \leq x_1} \{g_L(x)\} \leq y \leq \max_{x_0 \leq x \leq x_1} \{g_U(x)\}.$$

Alternating sums:

$$\sum_{0 \leq i \leq k} (-1)^i a_i > \sum_{0 \leq i \leq n} (-1)^i a_i > \sum_{0 \leq i \leq k+1} (-1)^i a_i$$

$$\text{where } a_i > a_{i+1},\ k \text{ is even, and } 0 < k < n-2. \tag{4.203}$$

Decreasing functions:

$$\int_a^b f(x)\,dx \leq \sum_{a\leq i<b} f(i) \leq \int_{a+1}^{b+1} f(x)\,dx. \tag{4.146}$$

Increasing functions:

$$\int_{a-1}^{b-1} f(x)\,dx \leq \sum_{a\leq i<b} f(i) \leq \int_a^b f(x)\,dx. \tag{4.145}$$

ASYMPTOTICS

Catalan number:

$$C_n = 4^n/(n\sqrt{\pi n}) + O(4^n n^{-5/2}). \tag{7.167}$$

Definitions:

$$f(x) = O(g(x)) \Longrightarrow |f(x)| \leq Cg(x), \tag{4.22}$$

$$f(x) = \Theta(g(x)) \Longrightarrow Cg(x) \leq f(x) \leq C'g(x), \tag{4.23}$$

$$f(x) = \Omega(g(x)) \Longrightarrow f(x) \geq Cg(x), \tag{4.24}$$

$$f(x) = o(g(x)) \Longrightarrow \lim_{x\to\infty} \frac{f(x)}{g(x)} = 0. \tag{4.25}$$

Harmonic numbers:

$$H_n = \ln n + \gamma + \frac{1}{2n} - \frac{1}{12n^2} + \frac{1}{120n^4} - \frac{1}{252n^6} + O\left(\frac{1}{n^8}\right). \tag{4.175}$$

Identities:

$$cf(x) = \Theta(f(x)), \tag{4.33}$$

$$g(x)f(x) = O(f(x)) \quad \text{provided } |g(x)| \leq c \quad \text{for } x \geq x_0, \tag{4.34}$$

$$g(x)f(x) = \Omega(f(x)) \quad \text{provided } g(x) \geq c \quad \text{for } x \geq x_0, \tag{4.35}$$

$$\Theta(\Theta(f(x))) = \Theta(f(x)), \tag{4.36}$$

$$\Theta(f(x))\Theta(g(x)) = \Theta(f(x)g(x)), \tag{4.37}$$

$$\Theta(f(x))\Theta(g(x)) = f(x)\Theta(g(x)), \tag{4.38}$$

$$\Theta(f(x)) \pm \Theta(g(x)) = \Theta(f(x)), \tag{4.39}$$

$$\Theta(f(x)) + \Theta(f(x)) = \Theta(f(x)), \tag{4.40}$$

$$O(f(x)) \pm O(f(x)) = O(f(x)), \tag{4.41}$$

Power series:

$$f(x) = \sum_{0\leq i\leq m} a_i x^i + \Theta(x^{m+1}) \quad \text{for } |x| \leq r', \tag{4.45}$$

$$e^x = \sum_{0\leq i\leq m} \frac{x^i}{i!} + \Theta(x^{m+1}) \quad \text{for } |x| \leq r, \tag{4.47}$$

$$\ln(1+x) = -\sum_{1\leq i\leq m} (-1)^i \frac{x^i}{i} + (-1)^m \Theta(x^{m+1}) \quad \text{for } |x| \leq r < 1, \tag{4.48}$$

$$(1+x)^r = \sum_{0\leq i<m} \binom{r}{i} x^i + \Theta(x^m), \tag{4.49}$$

$$\frac{1}{(1-x)^r} = \sum_{0\le i<m} \binom{i+r-1}{i} x^i + \Theta(x^m), \qquad (4.50)$$

$$(e^x - 1)^m = \sum_{i\ge 0} \left\{ {m+i \atop m} \right\} \frac{m!}{(m+i)!} x^{m+i}. \qquad \text{(Ex. 4.2.1–5)}$$

Stirling's approximation:

$$n! = \sqrt{2\pi n}\left(\frac{n}{e}\right)^n \left(1 + \frac{1}{12n} + O\left(\frac{1}{n^2}\right)\right). \qquad (4.201)$$

Taylor's theorem:

$$f(x) = \sum_{0\le i\le n} \frac{f^{(i)}(x_0)}{i!}(x-x_0)^i + \frac{f^{(n+1)}(c)}{(n+1)!}(x-x_0)^{n+1}. \qquad (4.10)$$

RECURRENCES

Ackermann:

$$A(i,x) = A(i-1, A(i, x-1)) \quad \text{for } i \ge 1, \quad x \ge 2. \qquad (9.29)$$

Constant coefficients:

$$\sum_{0\le i\le k} a_i T_{n-i} = b_i. \qquad (5.179)$$

Characteristic equation:

$$\sum_{0\le i\le k} a_{k-i} z^i = 0. \qquad (64)$$

Solution:

$$T_k = \sum_{1\le p\le j} \sum_{0\le q\le \beta_p - 1} f_{pq} k^q \lambda_p^k. \qquad (5.208)$$

Divide and conquer:

$$T_n = aT_{n/c} + n^x. \qquad (5.81)$$

Solution:

$$T_n = O(n^{\log_c a}) \qquad \text{when } a > c^x, \qquad (5.82)$$

$$T_n = O(n^x \log n) \qquad \text{when } a = c^x, \qquad (5.83)$$

$$T_n = O(n^x) \qquad \text{when } a < c^x. \qquad (5.84)$$

Uneven parts:

$$T_n = \sum_{1\le i\le k} a_i T_{c_i n} + bn^x. \qquad (5.240)$$

Euler:

$$c_N n^{\overline{N}} \Delta^N F_n + c_{N-1} n^{\overline{N-1}} \Delta^{N-1} F_n + \cdots + c_1 n \Delta F_n + c_0 F_n = 0. \qquad \text{(Ex. 6.2.2–4)}$$

Extended first order:

$$f(n, T_n, T_{g(n)}) = 0, \qquad (5.6)$$

Solution:

$$T_n = T_{f^{[k]}(n)} \prod_{0\le i<k} a_{f^{[i]}(n)} + \sum_{0\le i<k} b_{f^{[i]}(n)} \prod_{0\le j<i} a_{f^{[j]}(n)}. \qquad (5.38)$$

Extended k^{th} order:

$$f(n, T_n, T_{g(n)}, T_{g(g(n))}, \ldots, T_{g^{[k]}(n)}) = 0. \tag{5.8}$$

k^{th} order:

$$f(n, T_n, T_{n-1}, \ldots, T_{n-k}) = 0. \tag{5.9}$$

Extended linear first-order:

$$T_n = a_n T_{f(n)} + b_n. \tag{5.32}$$

First-order:

$$f(n, T_n, T_{n-1}) = 0. \tag{5.7}$$

Full-history:

$$f(n, T_n, T_{n-1}, \ldots, T_0) = 0, \tag{5.10}$$

$$T_n = \sum_{0 \le i < n} T_i \qquad \text{for } n \ge 1, \tag{7.1}$$

$$C_n = n + 1 + \frac{2}{n} \sum_{0 \le i < n} C_i. \tag{7.7}$$

Hermite polynomials:

$$H_{n+1}(z) = zH_n(z) - nH_{n-1}(z). \qquad \text{(Ex. 7.1.3-1)}$$

Integer part:

$$T_n = 1 + \frac{\lceil n/2 \rceil}{n} T_{\lceil n/2 \rceil} + \frac{\lfloor n/2 \rfloor}{n} T_{\lfloor n/2 \rfloor}, \tag{7.181}$$

$$T_n = n + T_{\lceil n/2 \rceil + 1} + T_{\lceil n/2 \rceil} + T_{\lfloor n/2 \rfloor}, \tag{7.185}$$

$$T_i = 1 + \frac{\lfloor i/2 \rfloor}{i} T_{\lfloor i/2 \rfloor} + \frac{\lceil i/2 \rceil}{i} T_{\lceil i/2 \rceil}. \tag{7.193}$$

Linear first-order:

$$T_n = a_n T_{n-1} + b_n. \tag{5.23}$$

Solution:

$$T_n = T_0 \prod_{1 \le i \le n} a_i + \sum_{1 \le i \le n} b_i \prod_{i < j \le n} a_j. \tag{5.29}$$

Linear k^{th}-order:

$$\sum_{0 \le i \le k} a_i(n) T_{n-i} = b(n), \tag{5.11}$$

$$\sum_{0 \le i \le k} a_i'(n) T_{n+i} = b'(n). \tag{5.12}$$

(to transform between forms, use)

$$a_i'(n) = a_{k-i}(n+k), \tag{5.13}$$

$$b'(n) = b(n+k). \tag{5.14}$$

Min and max:

$$C_n = 1 + \min_{0 \le i \le n} \left\{ \frac{i}{n} C_i + \frac{n-i}{n} C_{n-i} \right\} \quad \text{for } n > 1. \tag{7.190}$$

Multidimensional:

Three-term linear (see Section 8.2.1):

$$x(n,i) H_{n,i} = y(n,i) H_{n-a,i-b} + z(n,i) H_{n-c,i-d}, \tag{8.132}$$

$$F_{ni} = F_{n-1,i} + F_{n-1,i-1}, \tag{8.170}$$

$$F_{ni} = (n-1)F_{n-1,i} + F_{n-1,i-1}, \tag{8.171}$$

$$F_{ni} = iF_{n-1,i} + F_{n-1,i-1}, \tag{8.172}$$

$$\left\langle {n \atop k} \right\rangle = k\left\langle {n-1 \atop k} \right\rangle + (n-k+1)\left\langle {n-1 \atop k-1} \right\rangle, \tag{8.177}$$

$$L_{n,k} = (n+k-1)L_{n-1,k} + L_{n-1,k-1}, \tag{8.178}$$

$$F_{n,i} = F_{n-1,i} + (n+i-1)F_{n-1,i-1}, \tag{8.180}$$

$$\binom{n}{k}_q = \binom{n-1}{k}_q + q^{n-k}\binom{n-1}{k-1}_q, \tag{Ex. 8.2.2.1–5}$$

$$\binom{n}{k}_q = q^k\binom{n-1}{k}_q + \binom{n-1}{k-1}_q, \tag{Ex. 8.2.2.1–6}$$

$$x(n,i)F_{n,i} = y(n,i)F_{n-1,i} + z(n,i)F_{n-1,i-1}, \tag{8.173}$$

$$F(x_1,\ldots,x_n) = \begin{cases} \text{if } p(x_1,\ldots,x_n) \text{ is } \textit{true} \text{ then } a(x_1,\ldots,x_n) \\ \text{otherwise } b(x_1,\ldots,x_n,F(C_1(x_1,\ldots,x_n)),\ldots,F(C_m(x_1,\ldots,x_n))). \end{cases} \tag{8.190}$$

Nonlinear:

$$A(t,v) = at + 2\sum_i \binom{t}{i} p^i(1-p)^{t-i}A(t-i,v-1), \tag{7.83}$$

$$T_{n+1} = a_nT_n + f(n,T_n), \tag{7.139}$$

$$C_n = \sum_{0\le i\le n-1} C_iC_{n-i-1} \qquad \text{for } n \ge 1, \tag{7.156}$$

$$\sum_{0\le i\le n} c_{in}g(T_i) = h_n, \tag{7.115}$$

$$T_nT_{n+2} = a_nT_{n+1}^2, \tag{7.125}$$

$$f_{n+1}f_n + af_{n+1} + bf_n + c = 0, \tag{Ex. 7.2.1–1}$$

$$T_n = T_0T_1\ldots T_{n-1} + r, \tag{Ex. 7.2.2.0–3}$$

$$p_{it} = \begin{cases} \delta_{n1} & \text{if } i=1 \text{ and } t \text{ is a root,} \\ (1-b)\delta_{n1} & \text{if } i>1 \text{ and } t \text{ is a root,} \\ b\sum_m p_{i-1,t_1}(m)p_{i-1,t_2}(n-m-1) & \text{if } i>1 \text{ and } t \text{ has left subtree } t_1 \text{ and right subtree } t_2. \end{cases} \tag{11.87}$$

Reduction of order with a solution: see Section 6.2.2.

Reduction of order by factoring: see Section 6.2.3.

Superexponential:

$$A_{i+1} = 2^{A_i}. \tag{9.24}$$

Systems of equations:

$$S_1(t+1) = a_{11}(t)S_1(t) + a_{12}(t)S_2(t) + \cdots + a_{1n}(t)S_n(t), \tag{8.37}$$
$$S_2(t+1) = a_{21}(t)S_1(t) + a_{22}(t)S_2(t) + \cdots + a_{2n}(t)S_n(t),$$
$$\vdots$$
$$S_n(t+1) = a_{n1}(t)S_1(t) + a_{n2}(t)S_2(t) + \cdots + a_{nn}(t)S_n(t).$$

Characteristic function:

$$\det|A - \lambda I| = 0. \tag{8.58}$$

Variation of parameters: see Section 6.2.5.

GENERATING FUNCTIONS

Catalan numbers:

$$G(z) = \frac{1}{2z}\left(1 - \sqrt{1-4z}\right). \tag{7.161}$$

Characteristic function:

$$G(e^{it}) = \phi(t). \tag{11.25}$$

Derivative:

$$z\frac{dG(z)}{dz} = \sum_{i\geq 1} ia_i z^i = \sum_{i\geq 0} ia_i z^i. \tag{6.3}$$

Exponential:

$$G(z) = \sum_{n\geq 0} \frac{T_n z^n}{n!}. \tag{7.76}$$

Product:

$$\text{for} \quad C(z) = A(z)B(z), \qquad c_n = \sum_{0\leq i\leq n} \binom{n}{i} a_i b_{n-i}. \tag{7.82}$$

Fibonacci numbers:

$$G(z) = \frac{z}{1 - z - z^2}. \tag{5.133}$$

Ordinary:

$$G(z) = \sum_{i\geq 0} a_i z^i. \tag{5.103}$$

Product:

$$\text{for} \quad C(z) = A(z)B(z), \qquad c_n = \sum_{0\leq i\leq n} a_i b_{n-i}. \tag{6.128}$$

Probability:

$$G(z) = \sum_{i\geq 0} p_i z^i, \tag{6.112}$$

$$G'(1) = A, \tag{6.115}$$

$$V = G''(1) + G'(1) - G'(1)^2. \tag{6.118}$$

Sums of scores:

$$A_C = A_A + A_B, \tag{6.130}$$

$$V_C = V_A + V_B. \tag{6.131}$$

Composite games:

$$A_C = A_A A_B, \tag{6.134}$$

$$V_C = V_A A_B^2 + V_B A_A. \tag{6.135}$$

DIFFERENTIAL EQUATIONS

First order linear:

$$A(x)\frac{dy}{dx} + B(x) + C(x) = 0, \tag{6.27}$$

$$\frac{dy}{dx} + P(x)y = Q(x). \tag{6.28}$$

Solution:

$$y = Ce^{-\int P(x)\,dx} + e^{-\int P(x)\,dx} \int e^{\int P(x)\,dx} Q(x)\,dx. \tag{6.29}$$

n^{th} order linear:

$$y^{(n)} + p_{n-1}(x)y^{(n-1)} + \cdots + p_0(x)y = f(x). \tag{6.30}$$

Algorithm Index

Index